A Type of Programming

by Renzo Carbonara

© 2018 Renzo Carbonara

atypeofprogramming.com

ISBN 979-8-9999097-0-1

A Type of Programming

by Renzo Carbonara

Computers blindly follow orders, and at some fundamental level, programming is about giving computers orders to follow. The expectation is that when a computer carries them out, it will achieve a particular goal a programmer had in mind. Coming up with these orders, however, is not easy. While computers are thorough and efficient, they are also rather limited in the vocabulary of instructions they can understand. They mostly know how to count and store data. Which means our orders, as complex as they may be, can only be conveyed in those terms. Even deceptively simple programs can involve thousands or millions of these instructions, each of which needs to be correct and executed at the right time. And if we consider that computers won't judge whether any of these instructions are right or wrong, some of which could have important consequences for our lives, it should be easy to appreciate how taking appropriate measures to prevent undesirable outcomes is the logical thing to do. In order to accomplish this, however, we need a different perspective.

Perhaps surprisingly, these computer instructions are a bad tool for reasoning about computer programs. Yet, it is only through reasoning that we can be confident about the correctness of our solutions, and more importantly, about the correctness of our problems. At the other end of the programming spectrum, far away from computers, we have our imagination. Programming is the conversation that happens while these two ends struggle to understand each other. We will explore this process and how to improve it.

This book doesn't assume any previous programming knowledge, yet both newcomers and experienced programmers are welcome. Those approaching software development for the first time will discover a fascinating field while acquiring a good understanding of the principles and tools involved, whereas the experienced shall have their conceptions challenged by a different type of programming. We will use the Haskell programming language as our main vehicle, and while this book is not solely about Haskell, it will be very thorough at that.

But please, be patient. We prioritize the kind of understanding that stays with us for a very long time, the one that teaches questions. This means the topics in this book are presented in a rather unorthodox and entertaining fashion where immediate practicality never seems to be a goal. Don't worry, we will get there. And when we do, we won't need to look back.

One

We can't really talk about programming without defining what a program is, which we can't do unless we understand exactly what it is that we are trying to solve. For this, we first need to sample the world for suitable problems to tackle, a seemingly intractable endeavour unless we understand the limits and purpose of the field, for which we most certainly need to be able to talk about it. This is the very nature of new ideas, and why appreciating them is often so hard. We need to break this loop somehow. We need to acknowledge that there might be a problem to be tackled even if we can't readily recognize it at first, and see how the proposed idea tries to approach it.

Breaking this loop is not a particularly hard thing to do in our imagination. A little dare, that's all we need. There, boundaries and paradoxes are gone. We can fantasize allegedly impossible things, dwell on problems that they say can't be solved. We can build, we can explore every future. To indulge in thought that challenges our understanding and that of others is not only interesting, it is a requirement. A curious and challenging mind is what it takes to make this journey.

But why would we do that? Why would we care? Well, look around. What do you see? There is a computer in our phone, there is another one in our car. There is one for talking to our families and another one for knowing how much money we have. There are computers that spy, there are computers that judge. There are computers that help businesses grow, and others that take jobs away. We even have computers saying whether we matter at all. Computers *are* our civilization, we have put them up there quite prominently. We care about programming because it is our voice, and only those who know how to speak stand a chance. It's our power, freedom and responsibility to decide where to go next. We care because civilization is ours to build, ours to change, ours to care for.

Two

Let's picture ourselves in a not so distant past looking at our landline phone, looking at the elevator in our building, looking at our clock. Besides their usefulness, these machines have in common a single purpose. Not a *shared* single purpose, but a *single different one* each of them. The phone phones, the elevator elevates, and that is all they will ever do. But these machines, however disparate, at some fundamental level are essentially just interacting with their environment. Somehow they get inputs from it, and somehow they observably react according to pre-established logical decisions. We press a key and a phone rings. Another key and we are on a different floor. Yet, we rarely see these machines and their complex and expensive electronic circuitry repurposed to solve a different problem if necessary.

What happens is that the wires and electronics in these machines are connected in such a way that their accomplishments are mostly irrelevant to the wider electronic community. Like those steampunk contraptions, they each represent a solution to a finite understanding of one particular problem, and not more. But, is sending electricity to the engine that lifts the elevator really that different from sending it to the one that moves the hand of the clock? No, it is not. The logical decisions however, of when to do what and how, are. We say that the purpose of these machines is *hardwired* into their very existence, and it cannot be altered without physically modifying the machine itself. And yes, perhaps it is possible to adjust the time in a clock, but subtle variations like that one are considered and allowed upfront by the grand creators of the machine, who leave enough cranks, switches and wires inside that we can alter this behaviour somehow within a *known* limited set of possibilities.

But what about the *unknown*? If interacting with the environment is the essence of these machines, and moreover, if they are built with many of the same electronic components: Isn't there some underlying foundation that would allow them to do more than what their upbringing dictates? Luckily for us, and for the steampunk fantasies that can now characterize themselves as nostalgic, there is. And this is what computers, otherwise known as *general purpose machines*, are all about.

Three

What differentiates computers from other machines is that their purpose, instead of being forever imprinted in their circuitry by the manufacturer, is *programmed* into them afterwards using some *programming language* such as that Haskell thing we mentioned before. The general purpose machine is a canvas and a basic set of instructions that we can combine in order to achieve a purpose of our own choosing while interacting with the outside environment through peripherals such as keyboards, microphones or display screens. A computer program, in other words, is something that takes into account some inputs from the environment where it runs, and after doing some calculations, provides some kind of output in return. We can introduce some notation to represent this idea.

```
program :: input -> output
```

Prose is an excellent medium for conveying ideas to other human beings. However, prose can be ambiguous, it wastes words talking about non-essential things, and it allows for the opportunity to express a same idea in a myriad of different ways. These features, while appealing to the literary explorer, just hinder the way of the programmer. Computers can't tell whether we ordered them to do the right thing, so we can't afford to have ambiguities in our programs. We need to be precise lest we accomplish the wrong things. Irrelevant stuff may entertain us as great small talk material, but computers just don't care, so we avoid that altogether. This is why a precise notation like the one above becomes necessary.

Perhaps surprisingly, our notation, also known as *syntax*, is part of a valid program written in the Haskell programming language, a language that we use to program computers, to tell them what to do. This is why among many other fascinating reasons we'll come to cherish, this book uses Haskell as its main vehicle. The distance between our thoughts and our Haskell programs is just too small to ignore. We can read `program :: input -> output` out loud as "a program from inputs into outputs", and as far as Haskell is concerned, it would be alright.

Computers themselves don't understand this Haskell notation directly, though. We use this notation for writing our programs because it brings *us* clarity, but computers don't care about any of that. For this, we have *compilers*. Compilers are programs that take as input the description of our program, called *source code*, here specified using the Haskell programming language, and convert it to an *executable* sequence of instructions that computers can actually understand and obey. That is, we could say a compiler is itself a `program :: source code -> executable`, if we wanted to reuse the same notation we used before.

Compiling is a complex, time-consuming and error prone operation, so it's usually done just once and the resulting *executable* code is reused as many times as necessary. When we install a new program on our computer, for example, chances are we are installing its executable version, not its source code. Quite likely the author already compiled the program for us, so that we can start using our program right away rather than spend our time figuring out how to compile it ourselves first. This is not different from what happens when we buy a new car, say. The car has already been assembled for us beforehand, and from then on we can drive it whenever we want without having to re-assemble it.

Four

What are our `input` and `output`, though, concretely? We don't yet know exactly, but it turns out it doesn't really matter for our very first program, the *simplest* possible one. Can we imagine what such program could possibly do? A program that, given some input, *any input*, will produce some output in return? If we think carefully about this for a minute, we will realize that there's not much it could do. If we were told that the input was an apple, the output of the program could be apple slices or apple juice. If we knew that the input was a number, then the output could be that number times three. Or times five perhaps. But what if we were told *nothing* about the input? What could we ever do with it? Essentially, we wouldn't be able do anything. We just wouldn't know what type of input we are dealing with, so the only sensible thing left for us to do would be *nothing*. We give up and return the input to whomever gave it to us. That is, our input becomes our output as well.

```
program :: input -> input
```

Our program, the simplest possible one, can now be described as simply returning the input that is provided to it. But of course, we can also look at this from the opposite perspective, and say that the output of our program is also its input.

```
program :: output -> output
```

Celebrate if you can't tell the difference between these two descriptions of our program, because they are effectively the same. When we said that our input becomes our output, we really meant it. What is important to understand here is that whether we say input or output doesn't matter at all. What matters is that both of these words appearing around the arrow symbol ->, respectively describing the *type of input* and the *type of output*, are the same. This perfectly conveys the idea that anything we provide to this program as input will be returned to us. We give it an orange, we get an orange back. We give it a horse, we get a horse back. In fact, we can push this to the extreme and stop using the words input or output altogether, seeing how the notation we use already conveys the idea that the thing to the left of the arrow symbol is the input type, and the thing to its right the output type. That is, we know where the input is, we know where the output is, we know they are the same type of thing, and we don't care about anything else. So we'll just name it x, for mystery.

```
program :: x -> x
```

This seemingly useless program is one of the very few important enough to deserve a name of its own. This program is called *identity*, and it is a fundamental building block of both programming and mathematics. It might be easier to appreciate this name from a metaphysical point of view. *Who am I? It is me.* Indeed, philosophers rejoice, our program addresses the existential question about the identity of beings at last, even if in a tautological manner. In Haskell we call this program id, short for *identity*.

```
id :: x -> x
```

Actually, we could have named the identity program anything else. Nothing mandates the name id. As we saw before when we arbitrarily named our mystery type x, names don't really matter to the computer. However, they do matter to us humans, so we choose them according to their purpose.

There is one last thing we should know about naming. Usually, we don't really call programs like these *programs*, we call them *functions*. The term *program* does exist, but generally we use it to refer to functions complex enough that they can be seen as final consumer products, or things that do something yet they don't exactly fit the shape of a function. Text editors, web browsers or music players are examples of what we would often call a *program*. In other words, we can say we just learned about id, the identity *function*. Generally speaking, though, we use these terms more or less interchangeably, as we have been doing so far.

Five

We also have useful programs. They are less interesting, much more complicated, but deserve some attention too. We were able to discover the identity function because we didn't know much about the types of things we were dealing with, and this ignorance gave us the freedom to unboundedly *reason* about our problem and derive truth and understanding from it. What would happen if we knew something about our types, though? Let's pick a relatively simple and familiar problem to explore: The addition of natural numbers.

As a quick reminder, natural numbers are the zero and all the integer numbers greater than it. For example, *12*, *0* and *894298731* are all natural numbers, but *4.68*, π and *-58* are not.

We said that our identity function, when given something as input, will return that same thing as output. This is supposed to be true *for all* possible types of inputs and outputs, which of course include the natural numbers. In Haskell, we can refer to this type of numbers as Natural. That is, if for example we wanted to use the identity function with a natural number like 3, then our id function would take the following type:

```
id :: Natural -> Natural
```

That is, this function takes a Natural number as input, and returns a Natural number as output. In other words, the *type* of this function is Natural -> Natural. Yes, just like numbers, functions have their own types as well, and they are easy to recognize because of the -> that appears in them, with their input type appearing to the left of this arrow, and their output type to its right as we've seen multiple times.

Presumably, if we ask for the id of the Natural number 3, we will get 3 as our answer. Indeed, this would be the behaviour of the identity function. However, we also said that names can be arbitrarily chosen. Usually we choose them in alignment with their purpose, but what if we didn't? What would happen if we named our id function for Natural numbers something else instead? Let's try.

```
add_one :: Natural -> Natural
```

Interesting. At first glance, if we didn't know that we are dealing with our funnily named identity function, we would expect a completely different behaviour from this function. The name add_one and the type of this function suggest that given a Natural number, we will *add one* to that number. That is, given 3 we would get 4. Very interesting indeed.

And what if we were told that *one* of the following functions is our identity function, renamed once again, and the other one adds as many numbers as it states? Look carefully. Can we tell which is which?

```
add_two :: Natural -> Natural
```

```
add_three :: Natural -> Natural
```

Despair. Sorrow. We can't tell. We can't tell without knowing the answer beforehand. The more we know about the things we work with, the less we know about the work itself. When we knew *nothing* about x, we knew that all we could ever do was nothing. That is, a function x -> x, *for all* possible types x, could only ever have one behaviour: Returning x unchanged. But now we know about natural numbers. We know we can count them, add them, multiply them and more, so learning that a function has a type Natural -> Natural is not particularly helpful anymore. A function of this type could process the Natural number given as input in *any* way it knows how, return the resulting Natural number as output, and it would be technically correct, even if possibly a fraud.

Is this the end? Have we lost? I am afraid that in a sense we have. This is part of the struggle we talked about. Suddenly we can't reason anymore, and are instead left at the mercy of *optional* meaningful naming. If we name something add_one, then we better mean *add one*, because we have no way to tell what's what.

Six

Was it all a lie, then? How can it be that us *programmers* can't tell what a program does? Well, it's complicated. First we should know that in this book we haven't really been programmers yet. We haven't even been computers, no. The role we've taken so far, which should explain why we insisted so much in understanding the *types* of our inputs and outputs, is that of the *type-checker*.

Computers care about *instructions* to follow, but not so much about their meaning. They care about a function multiplying two numbers together, but they couldn't care less about said function's name, nor whether those numbers are naturals, fractions or phone numbers. They don't even care whether they are *compatible* things. Have you ever tried multiplying a phone number by five? No? What about shoes times some amount of cheese? Not particularly smart things to do, are they? Yet, a computer will happily do them, resulting of course in gibberish. We use types, checked by the type-checker, to prevent these things from happening. In our example, when we said the type of data we were dealing with was `Natural` numbers, we made it *impossible* to use our function on other types such as text or potatoes.

Types are relevant to both type-checkers and humans, but not to computers. As computer programmers, surprise, we also need to write the part of our program that is relevant to computers. That is, we need to write the *expressions* that ultimately become the instructions that our computer can perform. When we program, essentially, we always deal with two different realms: That of *types*, where we can reason about things and prevent nonsense, and that of *expressions*, where we react to *unknown* inputs from the environment and tell the computer what to do in return. Let's see an example of how we can deal with this in our *add one* function, this time including both the type *and* the expression that fully defines it.

```
add_one :: Natural -> Natural
add_one = \x -> x + 1
```

As before, add_one is a function that takes a `Natural` number as input, and after doing some calculations, it returns another `Natural` number. This hasn't changed. Of course we can't know *which* number we will get, but at least we know it will always be a natural number, not a fraction or something else. That is, our add_one is still a function whose type is `Natural -> Natural`. We can see this in the first line of our program.

On the second line we have some new notation. The first thing to notice is that we are mentioning the name add_one again, but this time followed by =. Whereas :: allowed us to specify a type for this name, = is saying that the name add_one *is equal to the expression* that appears to its right. If we were so inclined, we could read this out loud as "the name add_one, whose type is `Natural -> Natural`, is equal to the expression \x -> x + 1".

Finally, we have the *expression* itself. Expressions defining the behaviour of a function start with a backslash symbol \ and extend all the way to the right. They are called *lambda expressions*. Like in the function type, we see the arrow symbol again, and we still have inputs to its left and outputs to its right. The right hand side should be quite straightforward if we remember anything from high school: We are just adding 1 to some mystery number called x. So, if x happens to be 3, then x + 1 will result in 4, which would be the `Natural` output of this function. But of course, 3 is just an example here. To the left of the arrow symbol we deal with the input of this function, which according to its type could be *any* `Natural` number, not just 3. From our perspective, we have no way of knowing which number this will be, we only know that it will be some `Natural`. The x here is the name we'll use to refer to this number on the right side of the arrow symbol, as we did when we wrote x + 1. We could have used any name here, it doesn't matter which. We chose x because we like mystery, that's all. We can read \x -> x + 1 out loud as "the *function* whose input we name x, and whose output is x + 1".

With this new knowledge, we can finally give the unsurprising full implementation of our beloved *identity* function, including both its type and the expression that defines its behaviour.

```
id :: x -> x
id = \x -> x
```

We named both the mystery type and the mystery value x because it makes the whole thing look terribly handsome. But, of course, we could have named these things anything else. It's all the same to Haskell, names are just for us humans.

```
truck :: helicopter -> helicopter
truck = \motorcycle -> motorcycle
```

Seven

A fundamental difference between *types* and *expressions* is the *time* when they exist. Type-checking is performed as part of the compilation process of our program, so if type-checking fails, compilation will fail as well and we won't be able to get our hands on an executable version of our program. This is beautiful, because we likely prevented a malfunctioning program from existing at all. However, we need to keep in mind that compilation happens just once for the programs that we build, and usually before our programs even leave "the factory". So, we are limited by our understanding of the environment at that time. Imagine, for example, that the program we are trying to build is a calculator. We know beforehand what the available digits and function keys in the calculator will be, as well as their behaviour. However, we can't possibly know beforehand in which order these keys will be pressed, as that really depends on what calculations the person using the calculator is trying to solve. We need to write that part of our program, the one that deals with the unknown, somewhere.

The realm of expressions exists whenever our program is executed, a time we formally call the *runtime* of our program. There we deal with things that are *not* known at compilation time, and in those expressions we find the real value of our program. Think about it. If all the answers and possible input values were known at compilation time, then it wouldn't be necessary to execute the program at all. If the programmers that created the calculator knew that all we ever wanted to do was multiply five times seven, they would have just whispered *thirty-five* to us rather than go and write a program. *Of course* our programs will showcase their utility when they are executed and not before, time and time again. Which, ironically, is why we care about types that only exist at compilation time so much. Types help us get our expressions right so that they don't fail time and time again at runtime.

So when we said before that we couldn't tell the difference between add_two and add_three, what we meant is that unless we get to see inside the *expressions* that define the behaviour of these functions, which many times we can't or don't want to do, we need to trust the names and documentation for these functions to speak the truth. Naming, thus, becomes one of the most critical issues in software development.

Eight

More or less we have some idea now of what functions are and do. What now? Why, we use them, of course.

First we need to know the type of the function we want to use. As an example, let's try with our add_one function of type Natural -> Natural. This type is saying that if we want to use this function, then we need to provide a Natural as input, so first we need to obtain a Natural somehow.

Creating a Natural number in Haskell is quite straightforward. We just write the *literal* digits of the natural number we want and that's all. For example, 3, 17 or 0 are all *values* of type Natural. Actually, we can refer to them as *expressions* of type Natural as well. Yes, so far we have only talked about expressions that describe functions, but actually, anything that can exist at runtime is an expression of some type. Alternatively, we could refer to them as *terms*. Right, right, the words *value*, *expression* and *term* can be used more or less interchangeably. Why do we have so many names for the same thing? Another mystery. Probably some historical accident.

Now, on the one hand we have a function of type Natural -> Natural, and on the other hand we have a Natural number. How do we feed this number to the function? Well, not by clapping those hands, no. Wouldn't that be fun, though? No, the way we do it is by simply *juxtaposing* the function and its input. That is, by putting them next to each other. So, if we want to add_one to the number 3, we just say add_one 3. Almost as nice as clapping, isn't it? The function always goes on the left side of this juxtaposition, and its input to its right. And of course, as usual, we have a myriad of words to describe what just happened. We could say that we *used* add_one on 3, that we *called* add_one with 3, or that we *applied* add_one to 3. These, again, are all more or less interchangeable words, but we do prefer to talk about *applying* functions for reasons that will become apparent later on.

Nine

So what can we say about add_one 3? Arguably, the most obvious thing about it is that it represents the idea of the number *four*. So much so, actually, that we could name this expression four.

```
four = add_one 3
```

But if we asked this same question to the type-checker, it would say something else. It would say that the most obvious thing about add_one 3 is the fact that this results in an expression of type Natural.

```
four :: Natural
```

This disagreement is quite fine, actually. Remember that types and expressions have different concerns, so it is only natural that they'll come to appreciate different facts about programs. These two *statements* about four are in harmony. They don't contradict each other, they just state different facts about the same truth.

Ten

Can we use our four anywhere we could have used the literal number 4? Well, we can't write things such as 123four5, since Haskell would try to interpret that as the *literal* digits that make up some number, and of course it will fail at that. But sure, we can use four anywhere a value of type Natural is expected. We could even say five = add_one four. Let's prove ourselves the meaning of four by doing some of the work that Haskell diligently does when applying a function.

We said before that the equals sign = we use when binding expressions to names could be read out loud as "the name on the left *is equal to* the expression on the right", and we really meant that. Particularly, this means that if we are mentioning some name as part of an expression, we can replace that name by the expression that defines it and the result should be the same. So if we start with four = add_one 3 and replace add_one by its definition, we get four = (\x -> x + 1) 3. This is a valid Haskell function application: We are simply juxtaposing a function (\x -> x + 1) to its input 3. It looks awkward this time, sure, but still correct. Those extra parentheses are there to ensure that we don't include that trailing 3 in our function definition, otherwise the meaning of this whole expression would be different, just like how (1 + 2) * 3 doesn't mean the same as 1 + 2 * 3, if you remember anything from elementary school.

We also know that the function definition syntax *binds* its input to some name. That is, when we say (\x -> x + 1), any input we provide to this function will be bound to the name x within these parentheses. And names, we said, can be replaced by the expression they represent, which in our example will be the number 3. That is, we could write four = (\3 -> 3 + 1) 3. Wait, what? Is this correct? It looks quite weird, yes, but this happens to be accidentally correct Haskell. Not exactly what we are looking for, though.

It is important to remember that what justifies the existence of a function is *the unknown*. If we already knew what some input would be, then we wouldn't need this ritual of using a name to refer to a thing, rather than mentioning the thing itself. That is, had we known x was 3, we could have said 3 + 1 rather than x + 1. We know this now, though, so we can simply remove the function definition and its application, which were there only to give some meaning to the now gone name x. That is, we can simply say four = 3 + 1, which of course is the same as saying four = 4, which proves that our definition of four was correct all along. Isn't that nice?

Beta reduction, stylized *β-reduction*, is the name we give to this process of replacing a function application like (\x -> x + 1) 3 with 3 + 1, where 3, the input to \x -> x + 1, which is bound to the name x, replaces the occurrences of x in that lambda expression to finally become 3 + 1. And since the x that the function was expecting as input has already been provided by the function application, we can also remove it from the left side of the arrow symbol -> in the lambda expression. That is, beta-reducing (\x -> x + 1) 3 results in 3 + 1, not in \x -> 3 + 1.

Eleven

Our add_one function is a bit restrictive in that it only allows us to add *one* to a Natural number. Nevertheless, being able to add *one* is in theory sufficient machinery for us to add bigger numbers. We can, for example, add *three* to some number by "simply" applying add_one three times.

```
add_three :: Natural -> Natural
add_three = \x -> add_one (add_one (add_one x))
```

Indeed, this approach works and is conceptually quite simple. As soon as we know how to add one, we also know how to add every other natural number by repeatedly doing the same, without having to learn nor do anything new. However, hopefully we can see how doing this for larger numbers gets a bit ridiculous. Imagine implementing add_one_million this way. Crazy.

Simplicity is quite a tricky topic in programming. What may be simple from a conceptual point of view, where we pay attention to the meaning and essence of things, might not be simple from *economic* or *ergonomic* perspectives, and vice-versa. This is yet another battle in the struggle between computers and thoughts.

What we would like to do, instead of repeatedly applying add_one, is to use a more convenient function for adding arbitrarily chosen natural numbers. That is, rather than having this add_one function taking a Natural as input and adding *one* to it, we would like an add function that takes *two* Naturals as input and adds them together.

```
seven :: Natural
seven = add 2 5
```

Here we are applying a function named add to *two* inputs by juxtaposing them separated by some whitespace, and then assigning the result to the name seven. Presumably, this add function is one that *adds* Natural numbers together. Hopefully none of this is too surprising, seeing how it's not that different to what we've been doing so far when applying functions that take just one input. In Haskell, our functions can take as many inputs as we want. Let's see how add itself is defined.

```
add :: Natural -> Natural -> Natural
add = \x y -> x + y
```

This is not particularly different from before either, but let's analyze it step by step. The first thing to notice is how the type of this function has changed. Whereas in add_one we had Natural -> Natural meaning *a function that takes a Natural as input and returns a Natural as output*, here we have Natural -> Natural -> Natural, more or less meaning *a function that takes two Naturals as input, and returns a Natural as output*, where the rightmost Natural in this function type is the output type, and everything to the left of the rightmost arrow symbol -> are the function inputs, each one separated by yet another arrow symbol ->. Actually, to be more precise, we say that each of these Naturals given as input is a *parameter* or *argument* to this function. When considered together, these parameters constitute the entire input to the function. In other words, in a function type, to the right of the rightmost arrow symbol -> we have the output type, and to its left we have its input, made out of one or more function parameters.

On the following line we have the actual expression that defines what our function actually does. Remember, it's called "add", it has type Natural -> Natural -> Natural, but the type-checker can't really tell what this function does. We as programmers need to look inside of it, or trust its author to have given it a reasonable name. For all the type-checker knows, it could be a fraud. Luckily, this time there are no surprises in this area. To the right of the arrow symbol -> we can indeed see that two things, x and y, are being added together using the + operator. x and y are the names we chose when *binding* each of the parameters of our function to the left of the arrow symbol ->, corresponding to the Naturals taken in by this function and distinguished from each other by whitespace.

Generally speaking, a function with type A0 -> A1 -> ... -> B, can be defined using a function expression like `\a0 a1 ... -> b`, where the names written in lowercase conveniently correspond to the types written in uppercase in this example, and as many parameters as necessary can replace those dots ... written here as placeholders.

Twelve

Where do types come from, anyway? Some, like `Natural`, come with Haskell already, but they aren't particularly special. We could have created `Natural` ourselves if it hadn't been there already. Actually, let's do that, let's recreate some of the types that come with Haskell.

One of the things we will want to do very early in our programming is enumerate related things and tell them apart. For example, we may want to tell apart the seasons of the year so that our hypothetical travel agency can suggest travel destinations in the right season. Otherwise, we might end up offering our customers a trip to a cold beach in winter, a package unlikely to sell much. To keep things simple, let's assume there are only four seasons in the year, even if in some geographies it doesn't feel that way. Winter, spring, summer *or* fall. In Haskell, it is quite easy to enumerate these four seasons.

```
data Season = Winter | Spring | Summer | Fall
```

We have some new syntax. To the left of the equals = sign, we are saying that `Season` is the name of the new *type* of data, or *datatype*, we are introducing. That is, `Season` is now a *type* that we could mention in places like those where we've mentioned `Natural` so far. The word data is just part of the syntax that Haskell uses for defining new *data* types, we can mostly ignore it. Types, such as `Natural` or `Season`, always start with uppercase letters.

To the right of the equals sign we see our seasons enumerated, separated by vertical bars |. This is saying a *value* of type `Season` can only ever be *one of* `Winter`, `Spring`, `Summer` or `Fall`. These four are called the *constructors* of the `Season` type, as it's only by using one of them that we can *construct* a value of type `Season`. This has the important implication that we as programmers know that a value of type `Season` must be just *one* among these four constructors. Not zero, not two, not more. *One*, always one. We call types such as `Season`, used to represent *one* among many, *sum types*. The reason for this strange name will become apparent later on. The order in which we list these constructors is not important, what matters is that they are different and start with an uppercase letter.

Actually, what we are seeing here is not that different from what we've seen so far regarding `Natural` numbers, if we consider that in order to *construct* a value of type `Natural` out of thin air, we need to write *one* number like 0 or 27 literally. That is, a value of type Natural is just *one* natural number among infinitely many of them. The definition for a type like Natural is a bit more complicated, but conceptually is not that different from other sum types such as Season. It is a beautiful thing how in programming we can invest our time studying the small examples, and once we are confident in our understanding of the principles involved, we can scale up our knowledge to larger problems without any friction. It really is beautiful.

So, now we can construct values that *precisely* represent our seasons by simply using one of the four constructors for Season.

```
my_favorite_season :: Season
my_favorite_season = Fall
```

Of course, trying to assign the type Season to something different than the constructors we just enumerated will fail. For example saying that 9 is of type Season will cause the type-checking of our program to fail. And similarly, trying to say that Winter or any of the other constructors is a value of a type different from Season will also fail. This is quite desirable, considering the whole reason why we have types at all is to prevent us from making silly mistakes such as trying to interpret a season of the year as a number.

When analyzing the source code we write, it is important to consider the perspective of the type-checker as well. While we as the authors of this code can see both the type and the constructor being used, the wise type-checker only sees the type. It doesn't know which our favorite season is, it just knows that it must be one of the four we have enumerated.

Thirteen

Let's try and do something useful with our Season sum type. For example, let's write a function that given a Season as input returns the opposite Season as seen from the other hemisphere. For example, when it's Winter in the northern hemisphere, it's Summer in the southern one. When it's Fall in Argentina, it's Spring in Norway. We get the idea.

```
opposite_season :: Season -> Season
opposite_season = \x ->
    case x of
        { Winter -> Summer
        ; Spring -> Fall
        ; Summer -> Winter
        ; Fall -> Spring
        }
```

We have some new syntax. As before, we are using a *lambda expression* to define our opposite_season function, binding our input parameter of type

Season to the name x. However, this time our x could be any of the four constructors we have enumerated in our Season datatype, and depending on which one it is, we want to make a different decision about what to return. We need something more.

Notice that this time we are writing the function definition on multiple lines. This is alright. We will learn more about Haskell's indentation rules as we go along, but in general they shouldn't be that surprising. Here the body of the lambda expression is not exactly *to the right* of the arrow symbol as before. Rather, it is *below and indented to the right* compared to the beginning of the previous line, which accomplishes the same as putting it to the right of the arrow symbol.

Indentation refers to the amount of whitespace to the left of the first non-whitespace character on a line of source code. In our example, the first line of source code has no indentation spaces at all, the second has some, and the rest have even more. It doesn't matter how much whitespace exactly, all that matters is that this indentation is increasing, visually pushing things to the right as we go down. Anyway, not very important, we'll learn the rules on the go.

If we consider this x from the perspective of our type-checker, we can't tell which Season we are dealing with. Remember, the compiler only has an understanding of the environment as it was when it first compiled our program, so it can't possibly know to what Season we will be applying our opposite_season function. Imagine, for example, that we are trying to apply opposite_season to the *current* calendar season at the time we run our program, a value that will definitely change depending on when we do so. We need to analyze this x at *runtime*, rather than compile-time, to determine its value. For this, we use a case expression.

Our case expression starts where we say case x of and extends all the way to the end of our function. Let's pay attention to the left side of the arrows on the following lines. Just like we listed all the possible constructors when we defined our Season datatype, here we enumerate again *all* the possible constructors that could have been used to obtain x, the Season value we are scrutinizing. Our x will definitely be one among these four constructors, since they represent all the possible ways of constructing values of type Season. So, when one of these constructors *matches* the actual value of x, then the expression that shows up to the right side of its arrow symbol -> will be returned. There, to the right, we are just constructing new Season values as we have done before, making sure we map every Season to the one in the opposite hemisphere.

If we look at this from a visual perspective, to the left of our arrow symbols we are listing all of our constructors as some kind of patterns or stencils we want to compare against x, hoping that one matches so that we can proceed to return what's to the right of its arrow. In fact, formally, this process of enumerating all the possible ways to construct a datatype for the purpose of scrutinizing it through a case expression is called *pattern matching*. It can become more sophisticated than just an enumeration of the constructors, but we'll discover more about this later on.

Formally, a case expression has the shape case s of { p0 -> e0; p1 -> e1; ... }, where s is the original expression we want to scrutinize and pattern match against. Each of p0, p1, ... are the *patterns* we want to try and match against s, and each of e0, e1, ... are the corresponding expressions we will return as the result in case the pattern to the left of their arrow symbol -> matched. These e0, e1, ... expressions, of course, are all of the same type, since they all are possible values that this entire case expression can take, and expressions always have just one type. In our case, their type is Season. This type matches that of s, but that's just a coincidence and we'll soon see examples where this doesn't happen.

And actually, not only are those surrounding brackets {} and semicolons ; ugly, but also they are unnecessary. Instead, we can just write each pattern on a different line like we did in our lambda expression, and provided they are properly aligned with each other, it would be the same as if we had used the ugly brackets and semicolons. That is, the following definition of opposite_season means exactly the same as the one before.

```
opposite_season :: Season -> Season
opposite_season = \x ->
    case x of
        Winter -> Summer
        Spring -> Fall
        Summer -> Winter
        Fall -> Spring
```

Fourteen

A lovely thing about pattern matching in case expressions is that the compiler will yell at us if we don't explicitly list *all* the patterns that could potentially match the value we are scrutinizing. For example, had we written something like the following, the type-checker would have told us that we forgot to deal with the beautiful Fall and refuse to compile our program.

```
case x of
    Winter -> ...
    Spring -> ...
    Summer -> ...
```

This behaviour is quite reasonable and desirable. Of course we want our program to be able to deal with *all* reasonable inputs. Otherwise, for example, it may crash if we run it in `Fall`. We want the compiler to reject a program like this.

Sum types and this accompanying *exhaustiveness* check done by the compiler, ensuring that we always deal with every possible input scenario in our program, are some of the most beautiful, simple, and powerful tools we will encounter in our programming journey. It can't be overstated how crucial *sum types* are if we want to build reliable and clear software. If you only take one thing out of this book, take *sum types*.

Fifteen

You might be surprised to learn that Haskell, somehow, is one of the *very few* languages out there that actually supports this very basic concept of a *sum type*. Most other languages fail to comprehend the importance of this situation and simply don't allow programmers to express the idea of *either this or that* at all, let alone do any exhaustiveness checking or pattern matching. Come on, let's hold hands and carefully explore this pain together to understand *why* this is so important.

The fact that our `opposite_season` function takes a Season as input and returns yet another Season as output is just an example. We could, of course, return values of a different type as well. Let's do that, let's write a function that more or less tells us the temperature in each season.

```
season_to_temperature :: Season -> String
season_to_temperature = \x ->
    case x of
        Winter -> "cold"
        Spring -> "warm"
        Summer -> "hot"
        Fall -> "chilly"
```

We are mentioning yet another type for the first time here, String, which comes with Haskell. A String value is essentially text, and is usually written between double quotes "like this". Just like how writing the digits of a number as 123 allows us to create a numeric value of type Natural out of thin air, writing some characters between double quotes " allows us to construct a new textual value of type String. The word "string" refers to the fact that a textual value is, metaphorically, a string of individual characters one after the other. Our `season_to_temperature` function is just returning a textual representation of the expected temperature for a given Season.

Can we go in the opposite direction, though? Let's try.

```
temperature_to_season :: String -> Season
temperature_to_season = \x ->
    case x of
        "cold" -> Winter
        "warm" -> Spring
        "hot" -> Summer
        "chilly" -> Fall
```

In principle it seems fine, but this new function is actually wrong. In fact, if we try to compile it, Haskell will reject it saying that the patterns in our case expression are not *exhaustive*. Why? Well, if we look at the type of this function, it says that given a String, *any* String, it will return a Season. However, that's not what the actual implementation of this function does. Sure, it gives the correct result when given one of these four known strings, but what happens when a different *unknown* string is provided? Pain. This function doesn't know what to do if somebody provides "hamburger" as input. The program would just *crash* in that case. That's what programming in languages without *sum types* feels like, they shift the burden away from the types and put it on the expressions, in the fallible human, relying instead on hope, prayer, and a careful observation of the entire execution flow of our program, lest temperature_to_season is ever called with an undesirable input.

How do we solve this? The ideal solution, if we were starting from scratch, would be to just not use String at all and instead introduce a Temperature sum type listing all known possibilities, to be used as the input of our temperature_to_season function, replacing the current fluffy String.

```
data Temperature
    = Cold
    | Warm
    | Hot
    | Chilly
```

This time we are defining our datatype across multiple lines just to show that it can be done. Had we written everything on the same line as before, the result would have been the same. Not important at all. With this new Temperature datatype, we can rewrite our temperature_to_season function.

```
temperature_to_season :: Temperature -> Season
temperature_to_season = \x ->
    case x of
        Cold -> Winter
        Warm -> Spring
        Hot -> Summer
        Chilly -> Fall
```

Now this function definitely can't fail. We are back in a happy place.

Sixteen

Still, `Strings` are ubiquitous, thus we need to know how to process them in their wild form in order to distill them into more precise types such as `Temperature` or `Season`. Imagine, for example, that we are implementing some kind of web form where we ask our users to *write* their favorite temperature as text. Some may write "hot", some may write "cold", and some may write something completely unexpected such as "nice". Based on this user input, we would like to suggest to them their likely favorite season for taking some holidays.

As far as artificial intelligence goes, our program will be rather dumb. Sorry. Yet, it will perfectly exemplify the most common scenario we have to worry about when dealing with `Strings`. That is, converting them into richer and more precise types. In concrete terms, we will be converting `Strings` such as "hot" and "cold" into proper values such as `Hot` and `Cold`, while *safely* discarding all other `Strings` we don't know about. Let's start from a broken implementation.

```
string_to_temperature :: String -> Temperature
string_to_temperature = \x ->
    case x of
        "cold" -> Cold
        "warm" -> Warm
        "hot" -> Hot
        "chilly" -> Chilly
```

The Haskell compiler will tell us that this function is broken because it fails to account for the infinite amount of text that can be represented as different `String` values. For example, `string_to_temperature "swell"` will cause our program to crash. How do we solve this if it's impossible to *always* convert a `String` to a `Temperature` sensibly? Is "pizza" `Hot` or what? No, what we need to do is to formally acknowledge that only *maybe* we will be able to convert a given `String` to a `Temperature`. *Maybe*, depending on whether the given `String` is known or not.

We have learned that it is through *types* that we can reason about our programs, so we want to put as much knowledge as possible about our problem there, in the types. In our case, even if we don't know exactly to which `String` the caller will apply this function, we know that it could potentially fail to convert the given `String` to a `Temperature`. That is a fact that we want to share with our type system, so the type of our `string_to_temperature` must change to accommodate this.

```
string_to_temperature :: String -> Maybe Temperature
```

Maybe. Dare we guess what `Maybe` is all about? We haven't really seen this syntax before, where we have two words next to each other show up in a type. We'll talk about that later. What we can say, however, is that `Maybe`

Temperature must be some kind of *sum type*, considering what we just said about the behaviour of this function, which returns *either* a value of type Temperature whenever that's possible, *or* it safely reports that it was impossible to do so. *Either* this *or* that. Yes, Maybe Temperature must be a *sum type* somehow. Let's see its definition.

```
data Maybe a = Nothing | Just a
```

We know some of this syntax. Here we are introducing a new datatype, a sum type apparently called Maybe a that has two possible constructors: Nothing and Just a. That's more or less correct, but actually, that a is neither part of the type name, nor of the constructor. Just like we saw back when we talked about the identity function and its mystery x that could take any shape, this a is also a mystery type that functions as a placeholder for a concrete type like Temperature. In our string_to_temperature example, every standalone a that shows up in this definition is being replaced by Temperature, but it could as well have been replaced by Chocolate, Natural or something else in a different situation.

Let's look at the value constructors first. Nothing is quite simple. Nothing stands alone just like Summer or Hot which we saw before. That is, Nothing is a perfectly acceptable value of type Maybe a, or rather Maybe Temperature in our case. Just, on the other hand, is a bit more sophisticated. Seeing how this constructor is used in practice will help us understand. Let's just look at the full implementation of a correct string_to_temperature to appreciate both Nothing and Just in action.

```
string_to_temperature :: String -> Maybe Temperature
string_to_temperature = \x ->
   case x of
      "cold" -> Just Cold
      "warm" -> Just Warm
      "hot" -> Just Hot
      "chilly" -> Just Chilly
      _ -> Nothing
```

What's happening in string_to_temperature is that, as promised, the four known Strings "cold", "warm", "hot" and "chilly" are leading to Just some result, whereas any other String, here represented by the "wildcard" underscore symbol _, will result in Nothing. What's interesting to notice here is that we are using the Just constructor as if it were a function, *applying* it to the value of type Temperature we actually care to return. However, contrary to functions such as our previous add_one, constructors don't modify their input in any way. Instead, they just wrap it unmodified and give them a new type. Constructors are, essentially, little boxes where we can potentially put other values. Some constructors, like Just, have room to keep values inside them. Others, like Nothing, don't. We can tell whether this is the case by simply checking if there are any fields in the constructor, just like that a in the case of Just.

It's important to keep in mind that Just itself is not a value of type Maybe a. Instead, it needs to be applied to some value of type a in order to construct a value of type Maybe a. So, if Just can be applied like any other function, then it must have a function-like type, right? Indeed it must. Constructors have types, and the type of Just is a -> Maybe a. In our example, this mystery placeholder a has been replaced by Temperature everywhere, so in concrete terms *our* particular Just will temporarily take the type Temperature -> Maybe Temperature when used. Again, this is similar to how when defining our identity function we said id :: x -> x, yet later we decided to use id with a Natural number and id temporarily took the type Natural -> Natural where it was used. This substitution of a is not something that we need to do explicitly in writing, however. *It just happens.* We'll see more about this phenomenon later.

Actually, Maybe on its own is not a type as we would expect it to be. There are no expressions of type Maybe, but rather, expressions like Nothing or Just Summer are of type Maybe Season, Maybe Temperature, or more generally, Maybe x for some mystery type x. That is, Maybe never stands alone, but instead, as if it were a function or a value constructor, we always apply it to some other type. We know how this works because we've seen it in the Just constructor which, also, never stands alone. We always apply Just to some expression of some type x when trying to get a value of type Maybe x. There are no expressions of type Maybe because Maybe is not a type but a *type constructor*. As the name suggests, a type constructor *constructs* a type once it's applied to some other type, much like how Just, a *value constructor*, constructs a value when we apply it to some other value. From values to values, from types to types. Don't worry if this sounds somewhat confusing now, we'll come back later.

Anyway, if string_to_temperature gets a known String, then we return the appropiate Temperature *wrapped* in the Just constructor. Otherwise, we simply return Nothing rather than allowing our program to crash. Both of these constructors return a value of type Maybe Temperature, so the requirement that all patterns in a case expression return a value of the same type is respected. In a few lines of code we have *safely* dealt with the infinitely many Strings that our function could be applied to.

This process of converting Strings to a more precise type is called *parsing*. We can say that our string_to_temperature is a *parser* that *parses* Strings into Temperatures. More generally, a parser is a function that tries to convert a value of a less precise type to a value of a more precise type. However, not every function that takes a String as input is a parser. For example, the function that takes a String as input and returns how many times the letter *x* appears in it is not parsing anything, it is just describing a property about its input. Or maybe it is. The line separating what's a parser from what isn't can be a bit blurry at times.

Seventeen

Our Season datatype hardcoded the four possible seasons when it was defined. This makes sense, considering seasons are unlikely to change any time soon. On the other hand, in the case of a type like Maybe a, intended to convey the idea of some value of type a *maybe* being there, it makes sense to leave this a be just a placeholder so that we can reuse the very same Nothing and Just constructors in different scenarios. Maybe Temperature, Maybe Natural, Maybe Airplane, they all make sense. Being able to reuse things is quite useful, so we appreciate Maybe leaving that mystery a unspecified for us to choose in different scenarios. However, while Maybe's reusability is quite handy, it still mandates that one of its two constructors carry no payload at all. This makes it impossible for us to use it to represent something more meaningful than just some a, through Just, or the absence of that a through Nothing. Imagine, for example, that we wanted to represent the idea of "*either* this Natural *or* that other Natural". How do we do that? It is not possible with Maybe.

In Haskell, and much in the same reusability spirit as Maybe, out of the box we get a sum type called Either that can be used to express *either this or that* where both *this* and *that* can contain some useful payload, rather than just one of them.

```
data Either x y = Left x | Right y
```

The names of the constructors Left and Right are not particularly important. As we said before, Haskell doesn't really care about names, all it cares about is that we are consistent when we use them. Presumably, the words "left" and "right" have something to do with the fact that the x that is the payload of the Left constructor is the *leftmost* parameter to the Either type constructor, whereas the y that is the payload of the Right constructor is the *rightmost* one. Or maybe it means something else. We don't care.

The interesting thing to notice about Either is that not only both constructors carry a payload, but said payloads can potentially be of different types. One mentions x and the other mentions y. For example, we could have x be Temperature and y be Season, leading to the type Either Temperature Season conveying the idea that we have *either* a value of type Temperature *or* a value of type Season. In this case, we would use the Left constructor if interested in the Temperature value, or the Right constructor otherwise.

It is important to understand that values of type x or values of type y are not themselves values of type `Either x y`. We must literally write the words `Left` or `Right` to construct a value of type `Either x y`. If a function is expecting an `Either x y` as input, but we apply it to a `y`, then the type-checker will reject it. We must wrap that `y` in the `Right` constructor for the type-checker to accept it. Similarly, providing an `Either x y` where, say, an `x` is expected, will fail.

Let's see an example of `Either` in a silly function that takes a value of type `Either Temperature Season` as input and returns a word that rhymes with the contained value. It really is silly.

```
rhyme :: Either Temperature Season -> String
rhyme = \x ->
    case x of
        Left Cold -> "old"
        Left Warm -> "storm"
        Left Hot -> "pot"
        Left Chilly -> "lily"
        Right Winter -> "sphincter"
        Right Spring -> "bring"
        Right Summer -> "bummer"
        Right Fall -> "ball"
```

We don't care much for rhymes, and it shows. We care about safe programs, though, that's why we pattern matched against *all eight* possible ways of constructing a value of type `Either Temperature Season`. We hadn't seen patterns that mention more than one word yet, but there shouldn't be any surprises there. For example, `Right Fall` is exactly the same we would write if we were trying to *construct* a value of type `Either Temperature Season`, we know that.

We could also have known, without enumerating them, that *eight* was the number of patterns or possible values in our `Either Temperature Season` *sum* type. How? Well, it says so on the tin. The name *sum type* comes from the fact that the number of possible values a sum type has equals the *sum* of the number of possible values all the other types mentioned in it have. Kind of. So, if according to our definitions there are 4 possible ways to construct `Season` and 4 possible ways to construct `Temperatures`, then there are *4 + 4* possible ways to construct a value of type `Either Temperature Season`. That is, eight, which is exactly the number of patterns we listed above. Technically speaking, we say that the *cardinality* of `Either Temperature Season` is eight. We'll give a proper definition of this later on.

Eighteen

So, how is `Either Temperature Season` different from the following handcrafted sum type?

```
data ThisOrThat = This Temperature | That Season
```

Well, they *are* different types, so if the type-checker is expecting to see a value of type Either Temperature Season but gets a ThisOrThat instead, it will complain. An obvious change is that while in Either x y we are leaving room for those x and y to become anything, in our awkwardly named ThisOrThat we are dropping those type parameters and forcing the payload types to always be either a Temperature or a Season.

Right Summer is not exactly equal to That Summer. Semantically, though, we might argue that they are, since the same information is there in both representations, even if shaped a bit different. So are they equal? Are they not? Fascinating phenomena such as these seldom go unnoticed by mathematicians in their role as observers of the beauty of the universe concerned about the *essence* of things. Thus, they have given this a name. They have told us that while not *equal*, these two things are *isomorphic*, more or less meaning that it is possible to convert back and forth between them without any loss of information.

They want *proof*, though. Mathematicians want proof that these two representations are indeed isomorphic before we are allowed to refer to them that way. So, seeing as we are on our way towards reaching that mathematical enlightenment ourselves anyway, we'll give them proof.

How do we do that? What is a proof, anyway? A proof is a *demonstration* that a particular statement is always objectively and unquestionably true. In our case, the statement we want to prove is that our two values of types Either Temperature Season and ThisOrThat, respectively, are *isomorphic* to each other. Which, as we said, in practical terms means that we are able to convert back and forth between these two representations without any loss of information. Let's start by writing two functions with types Either Temperature Season -> ThisOrThat and ThisOrThat -> Either Temperature Season which should allow us to convert between these representations.

```
fromEither :: Either Temperature Season -> ThisOrThat
fromEither = \x ->
    case x of
        Left t -> This t
        Right s -> That s
```

We named our function fromEither to highlight the fact that this is a conversion *from* the Either datatype into something else. Similarly, we can introduce toEither going in the opposite direction.

```
toEither :: ThisOrThat -> Either Temperature Season
toEither = \x ->
    case x of
        This t -> Left t
        That s -> Right s
```

Hopefully what's happening here is not too surprising: These functions are the inverse of each other, they each undo what the other does, if you will. We have some new syntax in the patterns of our case expression as well. Rather than enumerating all possible constructors, we are seeing things such as Left t in our patterns, with a lower case t there rather than a specific value constructor. This means that this pattern will match *any* value that uses the Left constructor, and then simply bind whatever its payload is to the name t. That is, in our example, t will be the name for a value of type Temperature. If we wanted to know exactly which value, then we would need to further perform another case analysis on that t, just like our previous rhyme function did. In this case, however, we don't really need to do that because all we want to do is put whatever Temperature we received, unmodified, in a different constructor. The same is happening in all the other patterns in this example.

Many proofs in Haskell consist of just writing a program that satisfies a particular type. For example, the proof that a value of type Maybe a is not Nothing consists of just a function of type Maybe a -> a existing. Why? Well, because if that Maybe a was Nothing, then it wouldn't be possible to obtain the a the function is supposed to return, so the function wouldn't exist. Other proofs, such as ours here, are a bit more delicate because while these two fromEither and toEither exist, their behaviour can't be determined from their types. Consider the following implementations instead.

```
fromEither :: Either Temperature Season -> ThisOrThat
fromEither = \_ -> This Hot

toEither :: ThisOrThat -> Either Temperature Season
toEither = \_ -> Right Summer
```

Those functions satisfy the expected types just fine, yet they don't contribute towards our isomorphism proof. Remember, we said that two types being isomorphic means that it's possible to convert back and forth between them *without any loss of information*, but our funny functions here certainly lose information. For example, if they are given Winter as input, then that Winter will be ignored by the wildcard underscore _ and forever lost, replaced by Hot or Summer. Our original implementations of fromEither and toEither were correct and avoided this issue. Essentially, what we were looking for was a pair of functions that, when applied after each other, gave us back the same input that was provided to it.

```
Right Winter == toEither (fromEither (Right Winter))

That Winter == fromEither (toEither (That Winter))
```

The code above doesn't contain Haskell *definitions* as we've seen so far, that's why we used the double equals sign == rather than the usual single one =. We are simply asserting some truths. We are saying that the things to the left of the == symbol are equal to the things on the right. In other words, we are saying that applying toEither after fromEither, or fromEither after toEither, gives us back the input that was provided to it, unmodified. We've seen this behaviour before, actually. This is exactly the same behaviour that the *identity* function has. Yes, indeed. In precise words, these functions, when applied one after the other, need to behave exactly as the *identity* function would, leaving its input untouched. Of course, this equality needs to be true *for all* possible values of type Either Temperature Season and ThisOrThat, not just for our minuscule Winter selection.

So yes, to sum up, Either Temperature Season and ThisOrThat are isomorphic, and using either one or the other would be equally fine for our purposes. We can even switch the order of the Either parameters to have Either Season Temperature instead, and the isomorphism would still hold. We would need to rewrite our proofs to accommodate the different order, of course, but that would be too boring so we won't do it today.

Nineteen

In mathematics we have this fascinating idea of *duality* which more or less says that some things have a *dual* that, intuitively, has the opposite relationship with things than the original thing had. So, the dual of an exit door would be an entry door, the dual of an input would be an output, and the dual of a function a -> b would be a function b -> a. Of course, following the proper mathematical protocols to introduce new ideas, this is a *very handwavy* definition of *duality*. We will revisit this topic later on, but this poor definition should more or less suffice for our current needs.

As we just saw, Either x y is making a statement saying that we have *one* of two possible values of types x *or* y in it. It turns out that Either, or *sum types* more generally, have a very reasonable and useful *dual* construction in which we have *both x and y*. In Haskell, we can convey this idea as well. Let's call it Pair.

```
data Pair x y = MakePair x y
```

This definition shouldn't be very surprising. Just like in Either x y, here we are saying in Pair x y that x and y will be the types of the values that will be part of our pair. That is, a pair of a Season and a Temperature will have type Pair Season Temperature, for example. To the right of the equals sign we have the constructor. Just one constructor this time. And in it we see that x and y are its payload, its *fields*. It should be easy to see how Pair and Either are dual. In Either we are expected to pick one of two possible constructors, each of them carrying a different payload. Here, in Pair, we have just one possible constructor carrying both x and y as payload. There's no choice to be made, so if we want to build a Pair then we must provide *both* x and y to the only available constructor, `MakePair`.

The `Make` or `Mk` prefix is rather common when defining value constructors, it makes it easier to tell apart the constructor from the type, something desirable at times. But it's also quite common to just reuse the same name for the type constructor and the value constructor. This is fine, because from the point of view of the compiler, types and values exist in two completely different realms, so their names don't overlap with each other.

```
data Pair x y = Pair x y
```

We will continue using this new definition so that we can get used to differentiating between types and expressions by means other than just their looks.

Instead of using `Pair Season Temperature` we could have introduced a datatype like `PairOfSeasonAndTemperature` *isomorphic* to it, and use that instead. Sometimes that's preferable, sometimes it's just silly. Technically speaking, datatypes such as `Pair` and `PairOfSeasonAndTemperature`, where we have *both this and that*, are called *product types*. This is a perfectly reasonable name if we contrast it with the *sum type* name we already saw. While the *cardinality* of sum types —that is, the number of possible values that its type can have— equals the *sum* of the cardinalities of its parts, the cardinality of *product types* equals the *product* of the cardinalities of its parts. Well, more or less. We'll study cardinality in more precise terms later, but this definition suffices for now. Anyway, 4 Seasons *plus* 4 Temperatures equals 8. That's the number of possible values the type `Either Season Temperature` can have, we saw this before. 4 *times* 4, on the other hand, is 16, so 16 is the number of possible values that the type `Pair Season Temperature` can have. Let's count them.

```
1.  MakePair Winter Cold      9.  MakePair Summer Cold
2.  MakePair Winter Warm     10.  MakePair Summer Warm
3.  MakePair Winter Hot      11.  MakePair Summer Hot
4.  MakePair Winter Chilly   12.  MakePair Summer Chilly
5.  MakePair Spring Cold     13.  MakePair Fall Cold
6.  MakePair Spring Warm     14.  MakePair Fall Warm
7.  MakePair Spring Hot      15.  MakePair Fall Hot
8.  MakePair Spring Chilly   16.  MakePair Fall Chilly
```

Counting from one is strange. You'll get the joke later, but that's all 16 of them. There's no other value that can carry the type `Pair Season Temperature`. And of course, these 16 values are the ones we would need to pattern match against if we were doing some kind of case analysis.

A *sum* type can be called a *coproduct* as well. The *co* prefix in *coproduct* is the prefix we stick to things to say something is the *dual* of that other thing. In *coproduct* we are saying that this type is the dual of a *product* type. Which, of course, is true. Technically we could say that product types are *cosums* as well, since they are the dual of *sum* types. In practice, however, due to tradition, we don't. Moreover, we have to stop somewhere, don't we? Otherwise we could end up with a *cococococococoproduct* in our hands. That's coconuts.

Twenty

We say Either and Pair are fundamental because once we can group or tell apart two things, we can group or tell apart as many things as we want. How so? Well, we simply make one or two of their payloads be yet another Either or Pair. For example, we can have *either* a Natural *or* a Season *or* a String by simply having a type Either Natural (Either Season String). That is, an Either where one of its parameters is yet another Either. Constructing a value of such type is straightforward, albeit noisy.

```
my_season :: Either Natural (Either Season String)
my_season = Right (Left Fall)
```

That is, Right is saying that we'll be providing a value for the payload on the Right constructor of the outermost Either, and Left is saying that we'll be providing a value for the payload of the Left constructor of the inner Either. It should be easy to appreciate visually how the parentheses on the type match the position of the parentheses on the expression. Visual aids are always welcome.

These payloads don't need to be of different types, though. We could have for example a Right (Left 5) of type Either Natural (Either Natural Natural). If we look at these three Naturals from left to right, no pun intended, then the one Natural we are providing amongst these three is the second one.

We *nested* our uses of Either to go beyond the mere two payloads we thought it could only handle before, and we can keep doing this over and over, repeatedly nesting Eithers to have even more payloads that way. However, it gets noisy quite rapidly. Sometimes this is exactly what we want, particularly when we want to generalize problems that deal with sum types of any size, but usually we will define our own sum types, just like we did with Season and Temperature, whenever they have more than two constructors.

And we can nest Pairs as well. For example, if we want to group a Natural, a Season and a String together, then we could do the following.

```
my_things :: Pair (Pair Natural Season) String
my_things = MakePair (MakePair 5 Summer) "songs"
```

Notice that Pair (Pair x y) z and Pair x (Pair y z), or any other permutation of x, y and z, are types *isomorphic* to each other. And the same is true for Eithers. That is, while these are different types in the eyes of the type-checker, we can easily convert back and forth between them without loss of information. For example, here's one such silly conversion, and we can easily imagine how the one in the opposite direction would look like.

```
f :: Pair (Pair x y) z
  -> Pair x (Pair z y)
f = \(MakePair (MakePair x y) z) ->
      MakePair x (MakePair z y)
```

We are growing, aren't we? Even our function types now span multiple lines. That's fine, nothing changes, it's just a different use of our whitespace. We mostly just need to make sure we always increase the indentation a bit where otherwise we would have continued on the same line.

It is possible to combine `Pair`s and `Either`s as well. We could have, say, in a type like `Either (Pair Natural Season) (Pair Season Temperature)`, *either* a Natural *and* a Season, *or* a Season and a Temperature.

Twenty-one

The `Pair` type isn't present in the Haskell language exactly as we introduced it. It's there conceptually, sure, but out of the box we get a different syntax for it. Rather than writing `Pair a b`, we write `(a, b)`. This notation for pairs is part of the Haskell language syntax, and we couldn't define it ourselves if we wanted to. It's quite sad that we have exceptions such as this, but at times it can be handy. For example, whereas a product type of more than two elements nesting our fundamental `Pair`s requires us to write `Pair (Pair a b) (Pair c (Pair d e))` or similar, using the built-in syntax for pairs we can say `(a, b, c, d, e)`, which is arguably easier on the eyes. Of course, we could alternatively have introduced a new product type capable of holding 5 arbitrary elements, say `Pair5 a b c d e`, and it would have been as straightforward and specialized as this five-element tuple, but without the extraordinary syntax.

But actually, our 5-element tuple translation was not entirely correct. Considering the order in which we nested our `Pair`s, the direct translation to this syntax would have been `((a, b), (c, (d, e)))`. However, hopefully we can see that these two representations are *isomorphic* to each other, so we can use whichever is more convenient, which in the case of pairs, or *tuples* as they are also known, is usually the one that looks more aesthetically pleasing, even if it's not made out of fundamental building blocks like those two-element pairs.

So, if we wanted to group together a Season, a Natural and a Temperature without giving an explicit name to the type of this grouping, we could do so in a value of type like `(Season, Natural, Temperature)`. Using the *tuple* syntax at the value level is quite similar to using it at the type level. We simply mention values, rather than types, as their elements.

```
three_things :: (Season, Natural, Temperature)
three_things = (Summer, 2, Hot)
```

Pattern matching on tuples is also quite straightforward. For example, let's write a function that adds together three Natural numbers given as the elements of a tuple.

```
add3 :: (Natural, Natural, Natural) -> Natural
add3 = \(x, y, z) -> x + y + z
```

We are simply pattern matching on the special tuple syntax, binding its elements to the names x, y and z before adding them.

As you will see later, this extraordinary syntax unnecessarily complicates our learning for no significant gain. The pursuit of clarity is our business, but this syntax is mostly a distraction. Yet tuples, or *anonymous product types* as they are sometimes called, are so ubiquitous in Haskell that we'll embrace them in our learning in spite of this. Ironically, there's no special syntax for *anonymous sum types* containing more than two elements in Haskell, so we are forced to nest Eithers in order to accomplish that. In practice we don't, and instead we define and name a new sum type whenever we need one.

Twenty-two

The fact that tuples mention as part of their type how many elements they contain can be either positive or negative, depending on who we ask and what problem we are trying to solve. If we know our program will somehow deal with a Natural and a Season at the same time, then having the type (Natural, Season) show up somewhere in our program is probably a good idea. However, that's not always the case. For example, let's say we are asking people to list their favorite Natural numbers. Someone says 3, some have no favorite numbers, and somebody else likes 30, 99 and 1000. How would we type these collections of values? Tuples won't do, because they require us to know the size of each participant's selection beforehand, selections that understandably won't always agree on their size. Some like no numbers, some like some. No, we need a more flexible type than a tuple, we need some kind of container that can group together an arbitrary number of things.

A simple and rather beautiful container we can use to solve this problem is the *linked list*. A linked list allows us to group together *zero or more values of the same type* without telling the compiler how many. Essentially, if we have a list of Naturals, say, the compiler knows about the *possibility* of there being zero or more Naturals within this list, but that's all it will ever know. How many values end up inside the list is something that only concerns the expressions that make up our program, such as the interaction with people telling us about their favorite number selection at runtime.

We've actually seen something like this in Maybe already. If we try hard enough, we can see how Maybe Natural conveys the idea of a list of just one Natural number, possibly empty, where in Nothing we have the empty list, and in Just 7, for example, we have the one-element list containing number 7. Can we build on top of this knowledge? Definitely. Building on top of previous knowledge is what our type of programming is all about. Let's recall Maybe.

```
data Maybe a = Nothing | Just a
```

In other words, Maybe is a *coproduct* or *sum type* wherein one of its constructors, Nothing, conveys the idea of the list being empty, and the other one conveys the idea of there being some a in the list. How do we go about having more than one element in that list, though? We can try and imagine what it would look like if we were able to add more constructors to Maybe, each one listing a larger number of as as its payload.

```
data Maybe a
  = Nothing
  | Just1 a
  | Just2 a a
  | Just3 a a a
  | Just4 a a a a
  | Just5 a a a a a
  | ... more of these ...
```

However, there would be various problems with this. For one, we seem to be writing just too much code, and while there's no official correlation between these things, you would be surprised at how many times having too much code is the first symptom of a poorly understood problem. The main issue, semantically, is that no matter how many constructors we add to our Maybe type, they will never be enough to contain an arbitrary and *unknown* number of elements. Imagine we list one million constructors in this silly type and then somebody comes with a selection of one million and one values to put in it. They just won't fit, even if we are only off by one. No, explicitly listing the constructors is a terrible idea. We need something else.

What we want is to be able to say that no matter how many elements our list has so far, we should always be able to add one more if necessary. Think about it. We are saying, in other words, that our list can grow unboundedly. How do we accomplish this? Well, to begin, we will need a leap of faith to trust that what we are about to see is indeed possible.

```
data List a = Empty | OneMore a (List a)
```

We are defining List as a coproduct with two constructors. One of them, Empty, conveys the idea of the list being empty as suggested by its name and evidenced by the lack of any field of type a in it, where a is the type of each individual element in the list, of course. The other constructor, OneMore, has two fields. One of them is a plain a, which just like in the Just constructor, is the one element that goes into the one-element list. So far this is just like Maybe. What's fascinating is what happens in the second field of this constructor, where we mention yet another List a as the type of this field. The idea is that if the List a in this field contains, say, three elements, then the OneMore constructor will pair it with a standalone a to create yet another List a, this time of four elements. But you see, we haven't yet finished defining what a List a is at the time we refer to it from this constructor field. Yet here we are, with a Haskell compiler that is more than happy to accept our definition of the List datatype as valid. Stop for a second, contemplate and ponder as the philosopher would. How is this possible? What does it mean?

This datatype talks about itself in a way we hadn't seen before. We say that List a is a *recursive* datatype. That is, a datatype whose definition mentions itself by name. Let's see how this looks in practice. Let's start from zero.

```
Empty :: List Natural
```

No surprises so far. Similar to Nothing, we know that Empty is a value of type List a for any a of our choosing. Here, we pick a to be Natural, conveying the idea that our List contains Natural numbers. And now, following our original recommendation that we can always add *one more* element to our list, let's use the OneMore constructor on our Empty list to add one more element to it, 8.

```
OneMore 8 Empty :: List Natural
```

Notice the type. It's still a List Natural even though the constructor is a different one. How come? Well, List a is a coproduct, so we know we can construct one using *any* of its constructors. Empty was one such constructor, but OneMore is one as well. OneMore takes our new element 8 as payload alongside our Empty starting point, which fit the expected Natural and List Natural types of the constructor fields just fine. Can we go further? Of course we can, this was the plan all along. Let's add the element 5 to the list.

```
OneMore 5 (OneMore 8 Empty) :: List Natural
```

Still the same type, still the OneMore constructor, still a value of type Natural, 5, in the first field, and a value of type List Natural in the second field. We can repeat this until we get bored. We can continue applying the OneMore constructor to this List, and each time we do we are effectively making room for yet another value of type `Natural` to become part of the list.

Linked lists belong to the classic literature of programming, thus we'll refrain from blasphemy and use the names historically given to its constructors, `Nil` and `Cons`, rather than the `Empty` and `OneMore` names that served our didactic pursuit so far. Remember, the compiler doesn't care about names, but we and our colleagues do.

```
data List a = Nil | Cons a (List a)
```

`Nil`, chiefly poetic, in Latin means *nothing*. `Cons`, austere and hip, comes from the Lisp family of programming languages where it stood for "*cons*tructing a pair" before becoming pop culture.

Twenty-three

Recursion comes in different shapes. Our `List` is what we call an *inductively* recursive datatype, a beautiful thing which we'll explore now.

We've been told that natural numbers are infinite. But we, people of science, trust nothing but our thirst and our proofs. We'll corroborate this lore ourselves using *induction*.

A natural number, we said, is *either* zero *or* any integer number bigger than zero. And *either* and *or*, we know, are the first signs that a coproduct is due. We'll call our coproduct `Nat` to avoid mistaking it with the `Natural` that already comes with Haskell.

One of the constructors for our `Nat` datatype will be `Zero`, which, akin to the `Nil` in our previous `List`, takes no payload and represents the smallest `Nat`. When we say *smallest*, however, we are not referring to the fact that zero is *numerically* smaller than every other natural number, which happens to be a coincidence. Rather, we mean that zero can be said to be a natural number without having to relate it to yet another natural number. It has the smallest structure, or smallest number of relationships, if you will. If we recall our definition of natural numbers again, we said that they are either *zero* or any integer number bigger than zero. So, if we were to justify why the integer number *five* is also a natural number, we'd have to say "because it is bigger than zero", whereas the reason why zero is a natural number is just "because". Can we see the difference? Five is a natural number because its relationship with another natural number says so, zero is a natural number because it is zero and that's the *base* case, the starting point from where we can start talking about natural numbers.

If we contrast this with another inductive datatype, the *negative integers*, which are -1 and every other integer *smaller* than -1, then -1 would be the *smallest* or *base* case in our induction, even if from a numerical point of view, the number -1 is actually greater than every other negative integer.

Inductive datatype definitions always start from a *base* case, which in the case of our List it was Nil, and in the case of our Nat it is Zero.

The second part to our induction should come as no surprise given our experience defining List. Let's look at the full definition of Nat now, before proceeding.

```
data Nat = Zero | Succ Nat
```

Other than the word *succ* standing for *successor* and inviting us to wonder whether we ever stop to read abbreviations out loud lest they don't stand scrutiny, Succ should immediately remind us of Cons, the *inductive* constructor in our definition of List. What Succ is saying is that no matter what Natural number we have, we can always obtain its succesor, the natural number immediately after, by applying the Succ constructor to it. Let's count to three to see how this works.

```
zero  = Zero :: Nat
one   = Succ Zero :: Nat
two   = Succ (Succ Zero) :: Nat
three = Succ (Succ (Succ Zero)) :: Nat
```

All of these values are perfectly valid natural numbers, as witnessed by the explicit type information we are giving when we say :: and then Nat, an alternative Haskell notation used to explicitly give a type to expressions right where they are used. We could also have written the types on their own lines as we have been doing so far, and it would have meant the same. Not important.

```
zero :: Nat
zero = Zero

one :: Nat
one = Succ Zero

two :: Nat
two = Succ (Succ Zero)

three :: Nat
three = Succ (Succ (Succ Zero))
```

We are saying here that we'll name Zero zero, that we'll name one the Successor of Zero, that we'll name two the Successor to the Successor of Zero, and so on. We are, essentially, counting how many times the Succ constructor is used. This gets tiresome rather quickly, however, which is why by using the power of name binding we could clean up things a bit if we wanted to.

```
zero  = Zero      :: Nat
one   = Succ zero :: Nat
two   = Succ one  :: Nat
three = Succ two  :: Nat
```

Naming things wasn't our goal, though. Proving that natural numbers are infinite was. And indeed, we have proved this. The *inductive* definition of Nat itself is the proof. It says that we can always obtain the Successor of a Natural number, no matter how big that natural number is. Many things in mathematics can be proved or defined by induction, which makes definitions like Nat's even more appealing to us programmers.

Perhaps a more interesting way to define natural numbers, now that we understand induction, is to say that a natural number is either zero or a natural number plus one.

Do you get the joke now about counting from one? It's strange. We programmers, naturally, always count from zero.

Twenty-four

Haskell's Natural and our Nat are not the same type. Say, if we apply a function expecting a Natural to Succ Zero, the type-checker will reject it. Conceptually, though, they are the same. That is, they are *isomorphic* to each other. They carry the same meaning, so no information would be lost if we were to convert between them. Let's implement these conversions so that we can convince ourselves that we did a good job.

First, let's understand a bit more about Haskell's support for Natural numbers as literal digits. In Haskell, when we use a literal digits expression like 123 where a value of type Natural is expected, the language will convert those three digits to a Natural for us. It's not important *how* this happens, but it's important to know that this, more or less, is the only way we have of constructing a Natural value out of thin air. Natural numbers, we learned, are conceptually an inductive construction. Nevertheless, we don't get to see how the Natural datatype itself is defined in the Haskell language because, by design, it's up to the language implementation to decide how to do so. Were we creating a new implementation of the Haskell language, we could say that 123 should be converted to 123 applications of Succ. In practice, that's something we'd like to avoid at all costs for performance reasons, as most implementations of the Haskell programming language do, including GHC which is the one Haskell implementation we pay attention to in this book.

The problem with Nat, List, and a myriad of beautiful constructions we'll encounter in life, inductive or not, is that our computers are terribly inefficient at working with them. Picture yourself facing a piece of paper with the digits "25" written on it. These digits take a rather small amount of space on our paper, and moreover, recognizing their meaning as the natural number 25 requires very little effort from us. On the other hand, if we were to encounter 25 Succs written down, understanding their meaning would require a bigger effort from us, as we'd need to manually count how many Succs we are facing before saying "ah, it's the natural number 25". And, font size being the same, it would definitely take up more space on our piece of paper. Computers are no different in this regard. In this analogy, the size of our paper represents how much *memory* our computer has, memory wherein representations of data are written. The effort it takes to derive meaning from those representations corresponds to how *performant* a representation is. That a computer would need to do at least 25 things in order to convey the idea of the number 25 is unacceptable, no matter how fast the computer might be. Because while 25 is a relatively small number and it doesn't seem that bad to write down Succ 25 times, using this representation with larger numbers like seven trillion would require seven trillion Succs to fit in memory and seven trillion times would we need to see Succ appear before we are able to do something useful with that number. Ridiculous. We say that Nat is a representation for natural numbers whose *time and space complexity* grows *linearly* with the number it represents. That is, the bigger the number, the bigger the time and space we'll be dedicating to understanding it.

So we make this compromise. We can inductively reason all we want about our natural numbers using a representation like Nat for didactic and research purposes, but we are encouraged to convert between Nat and Natural at some point, to leverage the more efficient representation for natural numbers when pursuing a practical goal like multiplying numbers or parsing them from a String. So let's implement that conversion once and for all. First, let's convert from Nat to Natural.

```
fromNat :: Nat -> Natural
fromNat = \x ->
  case x of
    Zero -> 0
    Succ y -> 1 + fromNat y
```

The first thing to notice here should be the blatant recursion going on, wherein we refer to this function's name, fromNat, from within fromNat's own definition. Yes, both datatypes and functions can be defined recursively.

fromNat is saying that given a Nat, which we call x, we return the Natural value 0 in case x is Zero. Otherwise, if x Succs, we add 1 to the result of applying fromNat to y, the name given to the Nat field in the Succ constructor. What we are doing, essentially, is adding 1 each time we encounter the Succ constructor, peeling the Succ layers one by one until we finally reach our *base case* Zero and the recursion stops. For example, fromNat Zero would lead to 0, fromNat (Succ Zero) would lead to 1 + 0, fromNat (Succ (Succ Zero)) would lead to 1 + 1 + 0, etc. Grab a pencil and a piece of paper and try to do this same exercise yourself. Follow the transformation of fromNat (Succ (Succ Zero)) step by step until it becomes

```
1 + 1 + 0.
```

The conversion from Natural to Nat is not that different:

```
fromNatural :: Natural -> Nat
fromNatural = \x ->
  case x of
    0 -> Zero
    _ -> Succ (fromNatural (x - 1))
```

What changes, mainly, is that we can't pattern match on anything like Succ in order to obtain the input to our recursive call to fromNatural. This is because x is a Natural, not a Nat, and Naturals are constructed using literal digits like 3 or 12, not with Succ nor any other recursive constructor. So we need to perform x - 1 by hand to obtain the number that comes before x, which is necessary for our recursive call to fromNatural. Previously, in fromNat, it was the y in Succ y who conveyed the idea of x - 1, but here we can't have that because we are not pattern matching on Nats, but on Naturals. Once we have this x - 1, we can proceed to recursively call fromNatural and use its result as the payload to Succ. As examples, fromNatural 0 becomes Zero; fromNatural 1 becomes Succ (fromNatural 0), that is, Succ Zero; and fromNatural 2 becomes Succ (fromNatural 1), which in turn becomes Succ (Succ (fromNatural 0)) to finally become Succ (Succ Zero).

Despite these small differences, the similarities between these two implementations highlight the inductive nature of natural numbers, even if Nat and Natural have different implementations internally.

Twenty-five

It's a tricky thing, recursion. It is, essentially, what tells computers apart from every other machine, but it comes at a very high cost. Let's consider what would happen if instead of writing x - 1 in our implementation of fromNatural, we had written x + 1. Applying fromNatural to 0 would return Zero as expected. However, applying it to any other Natural value would lead our program into an infinite loop, a loop that would *never* finish. This is how the evaluation of fromNatural 1 would go: The function would return Succ applied to fromNatural 2, which would return Succ applied to fromNatural 3, which would return Succ applied to fromNatural 4, etc. That is, instead of getting closer to 0, the *base case* through which we would finally exit our function, we would be getting farther and farther away from it on each iteration, ad infinitum. This is bad. In practice, it means that our program will likely hang at some point. We say that our broken fromNatural is a function that *diverges*, a function that never finishes executing.

Unfortunately, Haskell's type-checker can't help us here. The type of our broken fromNatural is still Natural -> Nat, and the type-checker happily accepts its implementation. This isn't Haskell's fault, however. This is a fundamental limitation in the theory of computation called *the halting problem*, which states that it is *impossible* to determine programmatically whether an *arbitrary* program written in a *general* purpose programming language halts or not. That is, whether a program written in a programming language able to express any computable problem will finish executing or not. Of course, we humans can *manually* observe and analyze our programs to determine whether they halt on a case-by-case basis, just like we did when we compared the broken fromNatural to the correct one. But it wouldn't be possible for us, nor for the authors of the Haskell programming language, to write a general program or compiler that can tell whether we've accidentally written something that would never finish executing.

But to a great extent this makes sense, considering that halting is not a necessary precondition to being a useful program. Think of traffic lights, for example. They run unconditionally, they never halt, yet they are productive and provide immense value to society while they run. A so-called *general* purpose programming language, thus, should allow us to build traffic lights at least.

There are specialized programming languages that abandon their *general* programming capabilities, which in practical terms mostly means abandoning support for general recursion, and in return they can tell us whether the programs we write with it will halt or not. Haskell as a whole is no such language, but we'll see and build languages where this is true later on.

Twenty-six

We've learned a lot, but we've also learned too much. All the problems we solved so far, we could have solved using just *functions*. Let's get rid of the decor, the convenience, and go raw, down to the very essentials of *functional programming*.

We talked about *general* programming languages before. We said these are the languages that allow us to express any computable problem. What does this mean, though? What are computable problems? People have been asking themselves this same question for a long time, and it wasn't until the 1930s that they came up with rather intriguing answers in the form of *machines* comparable to the programming languages of today. These are machines that at first seem rather insufficient, having only a handful of functions they can perform. One wouldn't dream of building a complex program in these machines, it would be terribly inefficient. Yet, despite their austerity, these machines are *fundamental*. They comprise the minimum set of features required to compute something. No matter how complex the solution to a problem may be, as long as it can be computed, it is possible to express it in a way these machines can understand it. Every general programming language out there, notwithstanding any extra bells and whistles offered, can at most offer the same computation capabilities as these machines. They just do so in a gentler and more ergonomic package.

One such machine, close to our hearts and to everything we've seen so far, is the *lambda calculus*, stylized *λ-calculus* when leveraging the beautiful Greek alphabet. We will build this machine now.

Twenty-seven

Building a lambda calculus can be a tough exercise, but by doing so we'll come to appreciate one of the core ideas of *functional programming*: That however complex a problem might seem at first, however different from everything else we've seen before, it can always be broken down into more fundamental problems that can be tackled separately, problems that quite often we've already solved before.

It's OK to feel a bit lost here. We are not learning this because we need it right away, we are learning it so that we can demystify the complexities of this field, understand that we are in control, and scare away any silly ideas that we are not capable of tackling what's to come. We are welcome to find solace somewhere else, too, if that works.

Let's forget about types while we are here. In our lambda calculus, initially, we will only be concerned about *expressions*. Our lambda calculus tells us what kind of expressions we can use, how, and what they mean. In its most rudimentary form, an expression in the lambda calculus is one of three things. It's either a reference to a value, just like the *names* we have in Haskell; *or* it is a *lambda expression* not unlike the ones we have in Haskell; *or* it is the *application* of an expression to some other expression, much like our own function applications in Haskell. *Either, or, or*; we know what that calls for. A sum type, which we will call Expr, short for *expression*.

```
data Expr = Lam String Expr
          | App Expr Expr
          | Var String
```

Behold: Computer. Yes, that is the entire definition of our simple lambda calculus, of *what is computable*. What's going on is perhaps best explained by looking at some examples, so let's write the simplest possible program, the identity function, using our lambda calculus.

```
expr_id :: Expr
expr_id = Lam "x" (Var "x")
```

It's important to understand that because we are using Haskell to implement our lambda calculus, we will be naming and referring to expressions of type Expr in Haskell, but ultimately those expressions are intended to be exploited by an *interpreter* for our lambda calculus, not by Haskell itself. This will serve as an exercise for us to learn how to separate different domains in our source code, some concerning the things we are building, and some concerning the things we are building *with*. For example, our expr_id is indeed a valid Haskell expression, but more importantly, it's a valid expression in our lambda calculus as well, akin to Haskell's own identity function.

```
id = \x -> x
```

The Lam "x" part in expr_id corresponds to \x -> in id, and the Var "x" part corresponds to the x that shows up to the right of the arrow symbol -> in id, where we refer to the value of the *var*iable x, whatever it may be, by a name. The correspondence between Expr and Haskell is quite literal in this case, something that shouldn't be surprising at all considering how these are called *lambda expressions* in Haskell. Lam "x" is saying that expr_id will be a function taking one parameter as input, which we will call "x". As usual, "x" is an arbitrarily chosen name which could have been anything else. The second field of the Lambda constructor is yet another Expr corresponding to the body of the function we are defining. In this case, the body Var "x" just looks up the value of the *variable* that goes by the name "x".

Applying this newly defined function is generally quite easy. Expr has a constructor called App that, presumably, *app*lies the Expr value in one of its fields to the Expr in the other field. This is the same thing that the juxtaposition of expressions accomplishes in Haskell. All we have to do is use it. However, there is an issue. While we know how to define a function, as we did in expr_id, and while we know how to apply it to some other expression using the App constructor, we don't necessarily know how to come up with that other expression to which to apply expr_id. Imagine we want to apply expr_id to the number five. Well, how do we represent *five* as a value of type Expr? It's not immediately obvious if we consider that Lam, Var and App are all we have available. It is indeed possible to represent *five*, numbers, products, coproducts, lists or any other value we might need using only the Expr constructors. Doing so, however, is much more complex than our current needs warrant, so we'll just not do it for the time being. Let's cheat instead.

Since we can't yet explain how numbers come to be, let's just assume that somebody else creates the number five for us and *binds* it to the name "five" by saying Lam "five" .... We can refer to this number while deliberately remaining oblivious of the complex truth by looking it up by name using `Var "five"`. So now we have `Var "five"`, a value of type `Expr` that magically means something. How do we apply `expr_id` to it? Well, we just use the `App` constructor with `expr_id` as its first parameter and `Var "five"` as the second.

```
expr_five :: Expr
expr_five = App expr_id (Var "five")
```

Now `expr_five` and our holy `Var "five"` have exactly the same meaning, just like 5 and id 5 mean exactly the same in Haskell. We have accomplished a function application in our `Expr` lambda calculus.

But wait. Actually, we could have avoided all this mess by simply realizing that `expr_id` is itself already a value of type `Expr`, and that while `App` needs two values of type `Expr`, it says nothing about whether they need to be different. So we might as well use `expr_id` twice. What would be wrong with saying `App expr_id expr_id`? Nothing. There would be nothing wrong with that. It conveys exactly the same idea as id id does in Haskell, a function application whose main purpose seems to be to deepen our already profound appreciation for the beauty of the identity function, when we realize that id id is also the identity function.

```
id == id id
```

We follow the types and we find beauty.

So let's get rid of the idea that programming is hard or impossible to tackle. Programming welcomes everyone. This discipline can be *reasoned* about, and that's beautiful. If we commit ourselves to pursuing an understanding, most answers will always be in front of us. We need to look hard, we need to follow the types. And if we can't see the answers yet, we can always break down the problems into smaller parts until we can. Otherwise, chances are we have failed to properly identify the problem to solve. Or perhaps we are simply at the boundaries of knowledge, about to discover something new, in which case we must go even further.

But we will put our lambda calculus on standby for now. We don't really need to understand *how* to interpret or calculate these `Expr`s just yet. Simply knowing that it can be done, and that programming is just functions all the way down, should put our minds at ease. We will come back to this topic later when it is time to implement our own programming language. For now, let's go back to our more immediate goal of becoming proficient in Haskell.

```
id == id id id id id id id id id id id id id id id id
```

Twenty-eight

Let's look at linked lists once more. They will accompany us for a very long time in our learning, as their simplicity and inductive nature make them an excellent structure for didactic and practical purposes.

Linked lists, or just *lists* as we usually call them, are so ubiquitous that in Haskell, for better or worse, they have their own special syntax like tuples do. Instead of saying Cons 1 (Cons 2 (Cons 3 Nil)) as we would do were we using the List datatype introduced a couple of chapters ago, we would say [1, 2, 3]. Notice the square brackets, rather than the parentheses you see in the similar special notation for tuples. An empty list is represented by two square brackets hugging each other, []. Their love story doesn't end there, however.

It wouldn't be sufficient to only have this [a, b, c] notation because, remember, the interesting thing about linked lists is that they can *grow*. That is, we need something comparable to Cons that we can use to grow our list. Haskell provides us with a constructor, awkwardly named :, that we can use to grow our list. To add the element 3 to a list of Natural numbers —whose type, by the way, is not List Natural anymore but [Natural]— we write 3 : []. Writing [3] would have achieved the same, but then we wouldn't have accomplished *our* goal of explicitly *cons*ing the list to grow it. And yes, it is strange that we are writing the : symbol in between 3 and []. We will learn more about this later, but essentially, : is just like the Cons constructor, except one of its parameters appears to its left and the other to its right, rather than both of them appearing to its right as they do in Cons. We can repeat this again, say, to add the numbers 1 and 2 to this list by writing 1 : 2 : 3 : [], which is the same as [1, 2, 3].

Once more we apologize for this noise distracting us from our learning, but this is what we'll be dealing with and we need to memorize it. Verbally, however, because sometimes we must talk with our colleagues out loud about these things, we call [] *the empty list* or *nil*, and we call : *cons*, so not everything is lost.

Twenty-nine

So let's imagine we have a list with some numbers, [3, 4, 5], and have been tasked to increase each of them by one.

```
add_ones :: [Natural] -> [Natural]
add_ones = \[a, b, c] -> [a + 1, b + 1, c + 1]
```

We are pattern-matching on the special list syntax. We haven't done this before, but it shouldn't be surprising considering how we pattern-matched on the special syntax for tuples before, and it was almost the same. Here we have square brackets [rather than parentheses), that's all.

Yes, add_ones [3, 4, 5] will indeed result in [4, 5, 6], adding one to each element as we wanted. But, in all honesty, add_ones is a rather sad thing. For one, it's assuming a list of three elements. This works, alright, but what happens if we apply this function to something else having the type [Natural], like [1, 2] or to the empty list []? Boom. Our program crashes at runtime, the passengers on the plane die. Luckily, Haskell's type-checker will tell us about our *non-exhaustive* pattern-matching and prevent this program from compiling, so mistakes like that one are easy to avoid. A second and more subtle issue is that we are writing the same operation, the addition of 1, three times. This is manageable here, but imagine how lengthy and sad it would be having to do the same for a list of, say, one thousand elements.

Functional programming, we learned, is all about functions. And functions, we know, are about not repeating ourselves. About not repeating ourselves. Functional programming wants us to say "add one to each element of the list" rather than literally writing the same addition to each element of the list over and over again. Functional programming gives us —or rather, allows us to *create*— a vocabulary where functions can take other functions that determine part of their behaviour as input. Functional programming actually *begs* us to say "*do something* to each element of the list", where *adding one*, much like *subtracting three*, could be that something. It's called *mapping*.

```
map :: (x -> y) -> [x] -> [y]
```

We call functions like map, taking other functions as input, *higher-order* functions. The type of map, whose name has nothing to do with cartography, says that given a function from a value of type x to a value of type y, and a list of said x values, it will return a list of values of type y. Internally, map will apply the given function to each element of the given list individually, effectively transforming each x value into one of type y.

```
map :: (x -> y) -> [x] -> [y]
map = \f xs ->
  case xs of
    [] -> []
    x : rest -> f x : map f rest
```

We haven't pattern-matched on the : constructor before, but hopefully it won't be too surprising. map takes its two parameters as input and pattern-matches on the second one, the [x] list, to decide how to proceed depending on whether the list is empty or not. If the list is empty, as witnessed by a successful pattern-match on the *nil* constructor [], then it simply returns yet another empty list. This makes sense if we consider that the whole purpose of map is to transform the elements of the list somehow, and that there would be no elements to transform in an empty list. Thus, an empty list result is due. The second pattern tackles the non-empty list scenario by pattern-matching on the *cons* constructor :, which says that the list has at least one element, which we are calling x here, and then comes the rest of the list, which itself may or may not be empty. What's interesting is what happens to the right side of the arrow ->, where we use the : constructor once again to actually *construct* a new list. The first element, to the left of :, is a value of type y obtained by applying the given function f to x, a value of type x. To the right of the constructor : we *recursively* apply map to transform the elements remaining in the rest of the list in the same fashion. Eventually, by the inductive nature of a linked list, we know we will reach the empty list case and map will finally stop recursing.

Let's look at the implementation of map using our own List rather than Haskell's own weird list syntax, lest we get distracted by it and miss the point of map.

```
map :: (x -> y) -> List x -> List y
map = \f xs ->
  case xs of
    Nil -> Nil
    Cons x rest -> Cons (f x) (map f rest)
```

One by one, we are applying f to the elements of the given list, constructing a new list with the same number of elements, in the same order, in return. So, how do we add one to each element of this list? Well, we simply *map* add_one *over* some list of numbers. Remember, x and y on map's type are just type parameters that could become anything. We could map a list of Seasons to a list of strings, a list of Naturals to yet another list of Naturals as we desired, or something else.

```
add_ones :: [Natural] -> [Natural]
add_ones = \xs -> map add_one xs
```

Effectively, add_ones [3, 4, 5] equals [4, 5, 6].

Thirty

There is some redundancy in our recent definition of add_ones as \xs -> map add_one xs. The type that map takes in our very specific case is (Natural -> Natural) -> [Natural] -> [Natural]. As soon as we provide the first input parameter to map, as in map add_one, we are left with an expression of type [Natural] -> [Natural] waiting for the second input parameter to map. Our function add_ones creates a lambda expression around map add_one that captures as xs that second input parameter so that it can pass it to map add_one. That is, this lambda expression takes its input only to reuse it as the very last thing to the right of the -> arrow. The type of this whole lambda expression is [Natural] -> [Natural], just like the type of map add_one was. But, if both map add_one and \xs -> map add_one xs have the same type and behaviour, can't we just drop that seemingly redundant lambda expression?

```
add_ones :: [Natural] -> [Natural]
add_ones = map add_one
```

Yes, we can. We call this the *pointfree* style of defining a function. Sometimes, having one less name like xs to worry about can be helpful. Other times it might be pointless, and writing the full lambda expression would make the function definition a bit more obvious to the reader, particularly when it involves a rather long expression.

Saying either f or \x -> f x means exactly the same. This fact has its roots in the concept of *Eta-conversion*, stylized *η-conversion*, which essentially says that this is possible. We sometimes refer to the silly version of a function, that is, to the one allegedly unnecessarily wrapping it in an extra lambda expression, as its *η-expanded* form. It's OK to forget this name, however.

Thirty-one

We say that map is *partially applied* inside our pointfree definition of add_ones, meaning not all the parameters map was expecting have been provided yet, only *part* of them have.

Partial application works beautifully in Haskell because when we see a function type like a -> b -> c -> d, what we are actually dealing with, even if we can't readily see it at first, is a function with type a -> (b -> (c -> d)). That is, there is no such thing as a function taking two parameters, or three, or any number of parameters other than one. Functions always take one parameter and return something right away. It just so happens that, at times, the thing a function returns is yet another function, thus we are fooled into believing a more comfortable lie. a -> b -> c -> d is not a function that takes three parameters as input, and returns d as output, no. a -> b -> c -> d, by which we actually mean a -> (b -> (c -> d)), is a function that takes an a as input and returns a function b -> (c -> d) as output, which in turn takes a b as input and returns a function c -> d as output, which in turn takes a c as input and finally returns a d as output. We rarely write down those parentheses because of how comfortable we are without them, but they are *always* there.

And actually, something similar happens when we use a lambda expression to define a function. Supposedly, we write things like \a b c -> d when we are defining a function that takes "more than one parameter" as input. But we just said that taking "more than one parameter" is not really a thing, so what's happening? Well, \a b c -> d is mostly just a more comfortable, magical syntax for writing \a -> (\b -> (\c -> d)), which is ugly. Of course, unsurprisingly by now, those parentheses are superfluous as well, so we could say \a -> \b -> \c -> d instead, which is arguably a bit easier on the eyes. Still ugly, though.

Now that we've learned the truth about functions and lambda expressions, for practical reasons, we can go back to lying to ourselves about functions taking "more than one parameter".

So, when we made it sound like map add_one was special, it wasn't. The actual type of map, insofar as our add_ones exercise is concerned, is (Natural -> Natural) -> ([Natural] -> [Natural]). Look at those extra parentheses. That is, we can apply map to a function of type Natural -> Natural in order to obtain yet another function of type [Natural] -> [Natural], which is exactly what we wanted add_ones to be.

Taking this even further, one could argue that add_ones doesn't even deserve its name. We might as well use map add_one directly as in map add_one [2, 3] to obtain [3, 4] as result. Actually, we don't even need to name add_one. We could just *partially apply* our add function of type Natural -> Natural -> Natural as add 1, obtaining yet another function of type Natural -> Natural which we could use as the first parameter to map. We can say [3, 4], we can say map (add 1) [2, 3], or we can say map add_one [2, 3]. It's all the same. What's more, we can also say map (add 2), map (add 3) or similar for slightly different results.

Thirty-two

We can't confirm nor deny what the behavior of map is, judging solely from its type. According to it, this could be a perfectly acceptable definition of map:

```
map :: (x -> y) -> [x] -> [y]
map = \_ _ -> []
```

Sure, why not. We are discarding the input we receive and returning a perfectly acceptable [y] list. The type-checker, happy enough, accepts this definition. There's nothing wrong with it.

This problem is not new, we have seen it before. This is a situation that happens whenever we know too much about something. In software, knowing is a curse. We want to build things in such a way that we know as little as possible about what we are dealing with, lest we accidentally accomplish something undesirable. We are liable for the deeds of our software, but we can reduce that liability through planned ignorance.

The concrete problem is that while *mapping* only requires us to know how to find every x so that we can replace it with a y, here we know other things as well. Particularly, we know that the place where we are looking for these x values is a list, which in turn grants us the knowledge of how to construct a list, which we can leverage, accidentally or not, to construct one which satisfies the expected type without actually satisfying the expected behaviour of the function. We need something else. We need to convey the idea that something can be *mapped over*, without actually saying what that something is.

Thirty-three

Enter *typeclasses*. A typeclass, as the name more or less suggests, represents a class of types which support some particular operations. In the case of *mapping*, we will concern ourselves with the class of types in which we can find values of a particular type that can be replaced by values of a potentially different type. This is exactly what our correct map did, but the problem with map was that it talked concretely about lists when it mentioned those square brackets in its type, and we want to avoid concrete stuff. Haskell comes with a function called fmap that solves this problem.

```
fmap :: Functor f => (x -> y) -> f x -> f y
```

If we compare the type of fmap with the type of map for List, we shall notice a striking similarity.

```
fmap :: Functor f => (x -> y) -> f     x -> f     y

map  ::                (x -> y) -> List x -> List y
```

Essentially, f seems to have replaced all occurrences of List, and a new Functor thing seems to be talking about f somehow. Indeed, that is what is happening. List x says that, potentially, there are values of type x to be found inside a List x. Similarly, f x is saying that, potentially, there are values of type x to be found inside f x, whatever f x might be. But what could f x be? That is what Functor f, to the left of the fat arrow symbol =>, is *constraining*.

Functor is the *typeclass* in question. Functor f is saying that the f type parameter appearing to the right of the fat arrow symbol => can be any type as long as it belongs to the class of types that implement the features required by the Functor typeclass. Functor is the name we give to these things that can be *mapped over*, such as Lists. It is a beautiful name we'll come to cherish, *functor*, although initially it will seem completely unrelated to the idea of mapping over something. Indeed, that we can *map* is just a consequence of a more fundamental, beautiful design.

Thirty-four

What is this Functor typeclass, concretely? Functor comes out of the box with Haskell, so we need not worry about defining it ourselves when writing programs in Haskell. Nevertheless, we can reproduce its definition here for didactic purposes. Let's dive into some new Haskell notation by looking at the full definition of this typeclass.

```
class Functor f where
  fmap :: (a -> b) -> f a -> f b
```

This notation is introducing a new typeclass, as hinted by the class keyword. The typeclass is called Functor. The f that comes afterwards is a placeholder for any type that can implement whatever it is that the Functor typeclass requires. And what does it require? That a definition for fmap be given. We will do that soon.

We said before that the type of fmap was Functor f => (a -> b) -> f a -> f b, and it was the Functor f constraint which caught our attention. However, we don't see this Functor constraint on f anymore when we give a type to fmap in our typeclass definition. The reason for this is that this constraint is already implicitly required by the fact that fmap is part of the Functor typeclass definition itself. If there were any additional constraints on f, a or b, then we would need to explicitly mention them somewhere, but that is not the case here.

So we are saying that for some arbitrary type f to belong to the class of types that can be mapped over, which we call the Functor class, it must implement the fmap *method*. Yes, fmap is technically just a function, but the fact that its type is declared as part of a typeclass makes it deserve the special name "method", presumably for us humans to have an easier time talking about it out loud. But functions, we know, have both a type and a definition. Here, however, we can only see its type. Where is its definition? Where is the expression that defines what fmap actually does?

Thirty-five

There is only one class called Functor, but there are potentially many types that satisfy the requirements of that class. That is, types that can be mapped over. We know of at least two: Our own List, and Haskell's own list with that weird square bracket [] syntax. We establish the relationship between typeclasses and the types that can implement them through *instances*. In instances, that's where fmap's implementation lives. Let's write the Functor instance for our List type.

```
instance Functor List where
  fmap = \g xs ->
    case xs of
      Nil -> Nil
      Cons x rest -> Cons (g x) (fmap g rest)
```

In instance Functor List we are saying that what follows, indented a bit further to the right, is the implementation of the *instance* of the Functor typeclass for the List type. There can only be one instance of a particular typeclass for a particular type. For example, the List type can have many instances for *different* typeclasses, but it can't define more than one instance for the Functor typeclass. Of course, besides List, other types can implement instances of the Functor typeclass as well. An instance, in other words, is what establishes the relationship between a typeclass and a type that supports it.

Next comes the implementation of fmap, which looks exactly like the map function for List we implemented a while ago. Whereas in the typeclass definition, where we said class Functor f where ..., we specified the type of fmap without giving an implementation for it, here the opposite is happening: We are only specifying its implementation, and not its type. Why? Well, according to our *instance head*, that is, that which appears to the right of the name of the typeclass Functor in our instance clause, we can see that List has taken the place of the f placeholder we had in the typeclass. This implies that, within this particular instance, all mentions of f will be replaced by List. Concretely, the fmap :: (a -> b) -> f a -> f b described in the typeclass definition, will have the type (a -> b) -> List a -> List b, so there's no need to repeat this obvious truth here. If we wanted to, we could, but there's no need.

Thirty-six

Whatever do we gain from writing typeclasses and instances and methods instead of just plain old functions? Well, aren't we forgetting why we are here at all? We had a map function that misbehaved, always returning an empty list, and we wanted to avoid ever having that. Getting there will take some time, but let's take a look at something else we have accidentally accomplished meanwhile.

```
add_ones :: Functor f => f Natural -> f Natural
add_ones = fmap (add 1)
```

Here we are defining a new function, add_ones, that given *any* f that satisfies the Functor constraint —that is, an f that implements a Functor instance— will return a transformed f that increases each Natural number contained in it by one. *Any* f, that's what's important to notice. This f can be a List, it could be Haskell's own list with that weird square brackets syntax, or it could be something completely different like Maybe. Sure, why not. After all, a value of type Maybe Natural potentially contains a Natural number that we could increase by one. We said before that Maybe is essentially a possibly empty one-element list, so it shouldn't surprise us that it can do things that [] can. Let's see how the implementation of the Functor typeclass for Maybe looks like.

```
instance Functor Maybe where
  fmap = \g ya ->
    case ya of
      Nothing -> Nothing
      Just a -> Just (g a)
```

The implementation of fmap is easier this time because, if any, we only have one value inside Maybe to which we could apply the function g. We don't need to worry about recursing into the structure to find more values, which greatly simplifies our implementation. But other than that, things are the same as in the List instance. If we apply fmap to Nothing we simply return Nothing because there is no value to which we can apply g, and if we apply fmap to Just *something*, then we return a new Just where the payload has been transformed by g.

And for Haskell's own weirdly syntaxed lists, add_ones [1, 2, 3] will result in [2, 3, 4] as expected. That is, as long as we implement the corresponding Functor instance. Otherwise the type-checker will just complain about said missing instance. It will say that we are trying to use fmap with something that is not a Functor. Let's imagine we have implemented it, though. A bit of wishful thinking doesn't hurt. We will implement that instance soon enough. Meanwhile, we could try add_ones (Cons 1 (Cons 2 (Cons 3 Nil))) instead, which is conceptually the same, and we would get Cons 2 (Cons 3 (Cons 4 Nil)) as expected. But we can now say add_ones (Just 5) as well, which the type-checker will gladly accept and will result in Just 6. Saying add_ones Nothing results in Nothing, of course.

It's important to highlight that we don't need to explicitly tell functions like add_ones which instance of a particular typeclass for a particular type they need to use. Remember, we can at most have one such instance, so the compiler will automatically select it for us.

What we have achieved is called *polymorphism*, a word meaning *many shapes* in Greek. And sure enough, we had many shapes here. add_ones, in its type Functor f => f Natural -> f Natural, doesn't say anything *concrete* about the shape —that is, the type of f— it will work with, but rather, it says that it can work with *any* shape so long as there is a Functor instance for it that contains Natural numbers somehow. This is great, because now add_ones doesn't know any specific details about f, thus it can't do silly things such as returning an empty list or a list shorter than the one given as input. Actually, we can't even say "list" since that's already much more specific than saying just Functor. Try it. Try implementing a broken add_ones that changes the *shape* of f. We can't do that. We only know that f can be mapped over, we don't know what it looks like, so we can't alter its shape. Of course, we could still implement a "broken" add_ones that instead of adding *one* to each element, adds *two*. We would be able to do that because add_ones is not entirely polymorphic; it is polymorphic on f, but it also knows that it deals with Natural numbers, so it can modify them in any way Natural numbers can be modified. Is this too much knowledge? Perhaps, but we need to learn to find the balance between reason and ridicule. Sometimes, simply naming the function after what it does is a sensible compromise.

Now, not all polymorphic functions are like add_ones, requiring that a particular unknown parameter be a Functor or something. In fact, most of the functions we have encountered so far in this book are polymorphic, even though we only learned the word just now. Let's take a look at the type of our beautiful identity function once again.

```
id :: x -> x
```

What the type of id is saying is that x could be *anything*. No matter to what shape we apply id, it will type-check and work. It is a polymorphic function. The only difference here is that we are not asking x to satisfy a particular set of features through a typeclass instance, but that is alright, we know id simply returns its input untouched, so it makes sense that there are no further requirements on x. Even constructors such as Just, with its type x -> Maybe x, are polymorphic. Just says that given an x, *any* x, it will type-check and return a Maybe x.

Our type of polymorphism, no pun intended, is called *parametric polymorphism*. A reasonable name considering these many shapes, even though unknown, still show up as type *parameters* in the types of our functions and constructors, allowing us to *reason* about how we could potentially deal with the unknown. The power we gain by abandoning the terran understanding of what a type is, instead focusing on what properties a type has and what it can do, is called *parametricity*, and it is such a beautiful and necessary power that in this book we learned about it before learning anything else, back when we first encountered id, even if it is only now that we know its name. Parametricity tells us that no matter what the unknown types might be in our polymorphic function, the behaviour of the function will always be the same. add_ones adds one to all the elements of any Functor, whereas id returns whatever is given to it, and this is true *for all* types we could choose to use with them.

Thirty-seven

By relying on fmap we can't get add_ones to misbehave regarding its unknown type parameter f. However, nothing prevents us from getting the implementation of fmap *itself* wrong. Think about it, just like we gave a broken implementation for map once, we could give a broken one for fmap.

```
instance Functor Maybe where
    fmap = \_ _ -> Nothing
```

In this instance, the type of the fmap method is (a -> b) -> Maybe a ->
Maybe b. We know this because it is what we get if we replace f with Maybe
in fmap's type, as required by the Functor typeclass definition which we
recall here:

```
class Functor f where
    fmap :: (a -> b) -> f a -> f b
```

Our allegedly broken fmap fits this type perfectly, so what is the problem?
Is there a problem at all? We the authors of this mess can see that fmap will
ignore its inputs and always return Nothing, so yes, there is a problem.
How is this possible? Weren't typeclasses supposed to save us from this
pain? It is a bit more complicated, the type-checker can't help us this time.

Typeclasses are usually accompanied by *laws* by which their instances are
expected to abide. Otherwise, they could be sentenced to prison. Just
kidding. The laws in question are different from those of civilization, yet
they serve a similar purpose in that they clarify the expectations of what
should happen in a given scenario. Unfortunately, these laws cannot be
expressed in the types, even though we would very much love doing that.
They are, essentially, the rules that fall through the cracks of an imperfect
type system. Laws, like types, are meant to prevent us from writing
nonsense that could lead to unexpected behaviour. Types are type-
checked, laws unfortunately aren't.

What happened in our broken fmap is that while it met the expectations of
the type system, it failed to satisfy the Functor laws. This is understandable,
considering these laws haven't appeared anywhere in our source code yet.
Let's see what these laws are, and later on we will figure out where to write
them down.

The first Functor law, which we call the *functor identity law*, says that
mapping the *identity function* id over some Functor x should return that
same x, unmodified.

```
fmap id x == x
```

Unsurprising and dull. In other words, here we are saying that applying
fmap id to something should have the same innocuous effect as applying id
to it.

```
fmap id x == id x
```

Does our broken Functor instance for Maybe satisfy this law? Let's see. If we
try, for example, fmap id Nothing, where Nothing of type Maybe Natural takes
the place of the x we mentioned above, then the result is expected to be
Nothing. That is, the same x again. Great, it works, fmap id Nothing is
indeed Nothing. However, this law must be satisfied *for all* choices of x, not

just Nothing. For example, x could be Just 5. Does `fmap id (Just 5)` return Just 5? No, it does not. Our broken `fmap` implementation within the `Functor` instance for `Maybe` simply ignores its input and returns Nothing every time. Among other things, this behaviour violates the *functor identity law*. This *proves* that the `fmap` implementation we gave is incorrect, thus we can't say `Maybe` is a `Functor` if such is the best implementation of `fmap` we can come up with. Luckily, we already know it's possible to come up with a correct implementation of `fmap` for `Maybe`. We did it not so long ago.

The second functor law is a bit more challenging, but we need to brush up our Haskell in order to approach it.

Thirty-eight

Somehow we made it this far without having talked about *composition*, which is a bit funny considering how composition is what makes our efforts in paying attention to being able to *reason* about our code worthwhile.

Composition, as the name hints, is about bringing things together in order to create something new and different wherein traits of those original things are still present. In programming, in Haskell, we compose all day long. Above everything else, that's what we do. A program, essentially, is just one big composition of smaller programs. But composition takes many shapes, and we can't tackle all those shapes at once, so let's focus today on *function composition*.

Imagine we want to multiply a number by ten and then add one to it. For example, given 2 we would obtain 21. That is, (2 * 10) + 1.

```
foo :: Natural -> Natural
foo = \x -> (x * 10) + 1
```

The idea of function composition is that rather than introducing a new function that explicitly does two things, like `foo` which is explicitly multiplying by ten and then adding one to the result, we can say that `foo` is the composition of *multiplying by ten* and *adding one*.

```
foo :: Natural -> Natural
foo = compose add_one multiply_by_ten
```

Western traditions will encourage us to wonder whether the parameters in this function application are out of order. We want to write `compose multiply_by_ten add_one` instead, seeing how we are expecting to `multiply_by_ten` first, `add_one` second, and how we have only ever learned to read from left to right. But if we search deep inside, we'll find we can acknowledge that there's nothing intrinsically natural about left to right reading, just like how there's nothing wrong in writing upside down. We embrace these differences and we are at peace. Today, for practical reasons that will become apparent later on, from right to left we will write. When we say compose add_one multiply_by_ten, it is multiply_by_ten who comes first, and that is fine.

Let's get the implementations of multiply_by_ten and add_one out of the way first. They are not important, function composition works the same for any two functions that fit the expected types.

```
multiply_by_ten :: Natural -> Natural
multiply_by_ten = \x -> x * 10

add_one :: Natural -> Natural
add_one = \x -> x + 1
```

Can we guess what compose is? Well, actually, we don't need to guess. We have all the information available to understand what compose actually is. Let's first look at its type. In our foo example we are applying compose to *two parameters* in order to obtain a value of type Natural -> Natural. Let's see how this partial knowledge looks in the type of compose.

```
compose :: _ -> _ -> Natural -> Natural
```

We can also write it like _ -> _ -> (Natural -> Natural) if we want to highlight the fact that we are "returning a function". Both the extra parentheses and saying "returning a function" would be redundant, though. We learned this before. The parentheses, implicitly, are already there. Sometimes, however, these visual aids help us understand the intended purpose of our programs.

We still have two blanks to fill where we wrote _, but they are quite easy to fill, aren't they? We are already applying compose to two functions multiply_by_ten and add_one, so we simply write down their types there.

```
compose
    :: (Natural -> Natural)
    -> (Natural -> Natural)
    -> (Natural -> Natural)
```

Our type started getting a bit long, so we split it across multiple lines. Notice how we put parentheses around the parameters we just added. If we hadn't, then we would have ended up with something else:

```
compose
    :: Natural
    -> Natural
    -> Natural
    -> Natural
    -> (Natural -> Natural)
```

That is, rather than compose being a *higher-order* function that expects two functions as input and returns yet another function as output, it would have been a function taking four `Natural` numbers separately. Or five, even, depending on whether you count the last `Natural` input separately or not. Remember, the rightmost parentheses are redundant. In summary, we need parentheses around anything that we consider to be an individual input parameter to our function.

What about `compose`'s implementation? Well, we know we want to achieve the same thing as our `foo = \x -> (x * 10) + 1` did, and we know that we are applying `compose` to `add_one` and `multiply_by_ten` which are, essentially, the things we want to do to our input one after the other. So why don't we do just that?

```
compose :: (Natural -> Natural)
        -> (Natural -> Natural)
        -> (Natural -> Natural)
compose = \g f x -> g (f x)
```

If we find it shocking to see three parameters in this lambda expression —g, f and x— while only two of them seem to be mentioned in the types, just remember that the parentheses around that last `Natural -> Natural` are *optional*. Had we written it without the parentheses, we wouldn't have been shocked.

```
compose :: (Natural -> Natural)
        -> (Natural -> Natural)
        -> Natural
        -> Natural
compose = \g f x -> g (f x)
```

This is `compose`. We can clearly see in `\g f x -> ...` that this function takes three input parameters, and we can see said three parameters in the types, each on its own line. Now, if we *beta-reduce* f and g —that is, if we replace them with the actual values they take in `compose add_one multiply_by_ten`— then we end up with `\x -> add_one (multiply_by_ten x)`. And isn't this the same as our original `foo`, which said `\x -> (x * 10) + 1`? Yes, yes it is.

Thirty-nine

Now, nothing about `compose` says that this function is about composing functions. In fact, here's a valid implementation of `compose` insofar as its type is concerned:

```
compose :: (Natural -> Natural)
        -> (Natural -> Natural)
        -> (Natural -> Natural)
compose = \_ _ _ -> 8
```

Sure, why not? We accept three parameters as input, and then simply return 8. The type-checker allows this and our program compiles just fine. However, hopefully we can agree that this is silly and unintended. We've been here many times, we know what the issue is: We know too much. All those Naturals in the type of compose are liabilities, we must get rid of them.

All our working compose ever does in its implementation is apply some functions. So, since we are not making use of any feature specific to Natural numbers in our correct implementation of compose, we might as well replace all those Naturals with some type parameter x.

```
compose :: (x -> x) -> (x -> x) -> x -> x
```

With this change we don't know what x is anymore, so we can't create an arbitrary value of type x to return from compose. Or can we?

```
compose :: (x -> x) -> (x -> x) -> x -> x
compose = \_ _ a -> a
```

While it is true that in this new and broken version of compose we are not arbitrarily *creating* a value of type x, we are still returning the wrong x value. This function simply reuses some of its input as output, unmodified. Seeing as how a value of type x is expected, and how in a we have a value of such type already, we might as well return it. But how can it be that while we know *nothing* about what x is, we are still able to come up with the wrong x? Have we lost?

A responsible programmer must call out and replace broken rules. And the rules here, the types, are broken. We must constantly try to subvert our own types if we expect our programs to be reliable and stand up to scrutiny.

The problem here is that nowhere in the type of compose are we enforcing that compose g f a *must* apply both f and g in a particular order. But we can solve that by relying on the reasoning power gifted to us by *parametric polymorphism*. Let's look, finally, at the correct type and implementation of compose.

```
compose :: (b -> c) -> (a -> b) -> a -> c
compose = \g f a -> g (f a)
```

This type is saying that compose eventually returns a c. It also says that among its input parameters, there is a function b -> c which given a b allows us to obtain a c. And where do we obtain that b? Well, there is yet another function among the input parameters, a -> b, that will give us a b if we provide it with an a. An a just like the one that is provided as one of the input parameters to compose. So if we use that a to obtain a b, then we could use that b to obtain the c that compose is expected to return. Now, the type of compose not only forces its implementation to use both f and g, but it also mandates that f, not g, be applied to a first. Moreover, while before we had all types be Natural or a same mystery x, we can now have a, b and c be different types. For example, if we had a function f of type

`Natural -> Season`, and a function g of type `Season -> String`, then compose
g f would have type `Natural -> String`.

Forty

Now that we know about function composition, we can learn the second
functor law.

```
fmap (compose g f) x  ==  fmap g (fmap f x)
```

As a reminder, we are *not* writing any Haskell definition here, we are just
stating some expected equalities using a familiar Haskell notation for
ourselves. This law says that whether we compose two functions f and g
together before mapping their composition over some Functor, or we map
f over our Functor and afterwards we map g over the result, the outcome
should be exactly the same.

If we drop the x from the equality above, we can write both sides of the
equality in an arguably clearer *pointfree* style.

```
fmap (compose g f)  ==  compose (fmap g) (fmap f)
```

This law guarantees that we can map over a functor as often as we want,
without worrying that doing so might affect the output of our program
somehow.

Did our broken Functor instance for Maybe, the one that always returned
Nothing, violate this law? Not really. Both fmap (compose g f) and compose
(fmap g) (fmap f) would return the same Nothing. Good thing we had the
other law, the *functor identity law*, to prevent that nonsense from
happening.

Forty-one

So where do we write these laws? Haskell doesn't provide us with
mechanisms for laying down laws for typeclasses in such a way that
instances of that typeclass are automatically checked for compliance.
Wouldn't that be nice, though? No, in Haskell we are unfortunately on
our own on this matter, and at best we can write a *comment* about it.

Comments have nothing to do with these laws we've been talking about,
though. Laws are just an excuse for us to talk about comments. You are
right to feel tricked into this matter. Comments are simply arbitrary words
we can write in our source code which are completely ignored by the
compiler. Comments, intended for *humans* to read, are where we can
clarify the purpose, behaviour or intention of something. We can add
comments to functions, typeclasses, etc. Let's, for example, add some
comments to our add function.

```
-- This function performs the addition
-- of its two input parameters.
add :: Natural -> Natural -> Natural
add = \x y -> x + y   -- Here is another comment!
```

Comments start with -- and extend to the end of the line. Whatever comes after -- is simply ignored by the compiler. We can use this space to write anything we deem important enough, so that whoever reads this code in the future can more easily understand what is going on without having to actually read the code.

So, lacking a better tool, we resort to comments in order to lay down the law. We can imagine the Functor typeclass being accompanied by something like this:

```
-- Functors are expected to satisfy the following laws:
--
-- Identity law:
--   fmap id  ==  id
--
-- Composition law:
--   compose (fmap g) (fmap f)  ==  fmap (compose g f)
```

We can't prove that *all* Functor instances abide by these laws, but we can prove that this is true sometimes, on a case by case basis. So far, we have been doing this with pen and paper, writing down in English the reasons why these laws hold or not. Is this satisfactory? Of course not, but here we are. Is this an accident? Is this Haskell abandoning us? Kind of. What's happening is that the type of fmap is not rich enough to guarantee the *semantics* we expect when using it, so we need to change fmap's type to be more descriptive. Doing this with Haskell's type system, however, is not possible. Or rather, it is not practical. Like the language without coproducts, like the language with just the string, here we find ourselves longing for something we do not have, without which we are forced to acknowledge a problem that could arise *at runtime* unless we manually prove that these laws are indeed respected. As we become one with the types, we'll see ways to encode laws that better help us help types help us. Sometimes awkwardly in Haskell, sometimes nicely elsewhere. For now, let's just embrace this handicap and move on.

Anyway, laws or not, comments are good. We will continue learning how to write great comments throughout this book. And yes, we could say a bit more in our comments for Functor, we could say that "a Functor is a *thing* that can be mapped over" or something along those lines, but we don't want to jump ahead and accidentally write down an imprecise definition, so let's not do that just yet.

Forty-two

We talk about functions, we discuss how they are a fundamental unit of computation, yet we see things like 2 * 3 computing some number without an obvious function application going on. Well, it turns out that there is a function application going on. * is the function, it's just that instead of appearing before its parameters as * 2 3, it appears *in between them*. We say that * is an *infix* function. And as every other function out there, * has a type. For now, let's say the type is Natural -> Natural -> Natural.

Could we write * 2 3 if we wanted to do so? Not exactly, but we could sprinkle some extra parentheses, write (*) 2 3, and it would work. If we want to use a function like *, generally expected to be in an infix position —that is, in between its parameters—, as a "normal" function preceding its inputs, we need to surround it with parentheses. When a function is used in this way, preceding its inputs as we've been doing for a long time, we say it is used in a *prefix* position. We don't say it often, though, because this is the *fixity* we get by default when we define a new function or constructor, so we just assume functions are intended to be used in prefix position unless somebody says otherwise.

How do we define an infix function? Let's try defining * with its usual behaviour as multiplication of Natural numbers. Of course, * already comes out of the box with Haskell, so we don't need to do this, but we'll reimplement it here just as an exercise.

```
(*) :: Natural -> Natural -> Natural
(*) = \x y ->
  case y of
    0 -> 0
    _ -> x + x * (y - 1)

infixl 7 *
```

The function definition shouldn't be surprising. This is just a normal function which happens to be called *. The parentheses are there because, otherwise, when the Haskell compiler sees a strange name like * made out of symbols expected to be part of an infix function name rather than letters, it refuses to parse these lines of code. There are some very boring rules about what constitutes a "normal" name and what is weird enough to require an extra pair of parentheses, but we won't go into details about that.

The implementation of the function is recursive, inductively recursive. Essentially, we are adding together as many xs as y requires. Each time we add an x, we decrease y by one before recursively calling *, which will once again add an x, etc. Eventually we reach the base case of y being 0 and we stop by returning 0. For example, 5 * 0 becomes 0, 5 * 1 becomes 5 + 0, and 5 * 2 becomes 5 + 5 + 0.

Somewhere in our code we wrote x + x * (y - 1) and, implicitly, we knew we meant x + (x * (y - 1)). That is, there are a pair of implicit parentheses around our multiplication that make it happen before the addition. If the parentheses were around the addition, as in (x + x) * (y - 1), then the result of this function would be different. We can learn where any implicit parentheses go by understanding the *fixity* of our infix functions. In our example, in the line where we say infixl 7 *, we are saying that *, when used as an infix function, has a *precedence* of 7 and "associates to the left". This precedence of 7 is what forces the implicit parentheses to be where they are whenever we have, say, both + and * in the same expression. If we were to look at the fixity declaration for +, we would find it says infixl 6 +. And 7 being a bigger number than 6 is what forces the parentheses to surround a multiplication rather than an addition when * and + are next to each other. 7 and 6 have no special meaning on their own, it just happens that one of these numbers is bigger than the other one when we consider them alongside each other, and that's enough information for Haskell to do the right thing.

The other fixity property about * "associating to the left" dictates where the implicit parentheses go when we have multiple occurrences of infix functions with the same precedence, something like 2 * 3 * 4. In the case of multiplication, it doesn't really matter whether we say 2 * (3 * 4) or (2 * 3) * 4; the result is the same in both cases. We say that multiplication is *associative*, meaning that no matter where we put our parentheses, the final result is always the same. However, not all functions are associative, so Haskell asks us to declare this property nonetheless. But where did we say * associates to the left? We did so when we wrote infixl, where that trailing l stands for *left*. We also have infixr for whenever we need something to associate to the right. It's important to keep in mind that whether we say infixl or infixr doesn't affect the fact that the parameter to the left of our infix function always becomes the first parameter to our function, and the one to the right becomes the second one. That is, in 2 * 3, 2 becomes the x in (*) = \x y -> ..., and 3 becomes the y.

Isn't this boring and noisy? Isn't it frustrating to try and guess where the implicit parentheses may or may not be? Yes it is. This is why we try to avoid infix functions, at times called infix *operators*, as much as possible. To make it easier for us, and to make it easier for our colleagues who will have to read our code tomorrow. Yet, arithmetic operators such as + and * are so ubiquitous that we need to understand this.

Forty-three

Before we talked about how we could use the : constructor to *cons* an element onto the linked lists that come with Haskell. We said we could use something like 3 : [] to prepend 3 to the empty list. It should be clear now that what we were saying was that : is an *infix* constructor. If we wanted to use : in prefix position, we could do it by simply adding those extra parentheses we talked about before. For example, (:) 3 [] is the same as 3 : [], except uglier.

Now, the type of the : infix constructor, comparable to the type of Cons, is x -> [x] -> [x]. In other words, if we put a value of type x to the left of the : symbol and a value of type [x] to its right, we'll get back a value of type [x] as output. And if we wanted, we could use this newly obtained [x] again in yet another application of :. That is, we can say 3 : [] which gives us a [Natural], and use this 3 : [] to further say 2 : 3 : [], which gives us yet another [Natural].

Can we tell from its usage whether the : infix constructor associates to the left or to the right? Where would we put our implicit parentheses if we wanted to? (2 : 3) : [] wouldn't type-check, because while the outermost : has a perfectly acceptable [] as its second parameter, it also has 2 : 3 as its first parameter, which doesn't make any sense. Remember, : needs the value of a list element as its first parameter, and a *list* as its second argument. 3, however, is no list. Thus, we can say with confidence that : doesn't associate to the left, otherwise these parentheses wouldn't have prevented the expression from type-checking. On the other hand, putting parentheses to the right as in 2 : (3 : []) works perfectly. Both : have a Natural number to the left and a [Natural] to its right. In other words, : associates to the right, and now we know why [1, 2, 3], 1 : 2 : 3 : [], 1 : 2 : [3] and 1 : [2, 3] all mean the same.

Forty-four

Some things, like natural numbers, we can count with our fingers. Other things, like temperatures, we can feel on our skin. These are very *concrete* things, things that require almost no effort from us in order to acknowledge their presence and meaning in this world. But not all things are this way. Some things, like recursion, functors or love, abandon their corporeal identity in order to become *abstractions*. What is a functor? It is something that has a particular behaviour. What is recursion? "What is recursion?" indeed. Can we touch love? Can we draw a functor in the sand? If we are going to gain a deeper understanding of what we've seen so far, of what's to come, then we need to leave our bodily expectations and become thought. We call these elusive things *abstractions*.

What is a functor, anyway? We saw how Lists are functors. And it was easy, because applying a function a -> b to all the as in a *container* full of them is a rather straightforward thing to do. The as are in there, so we just do it. We also saw how Maybes are functors. And it was easy, because Maybes are essentially one-element *containers* themselves, so it wasn't surprising at all that what was true for Lists was true for Maybe as well. So, are functors *containers*? Why didn't they just call them that? Sure, a container is some kind of abstraction, if you will. And containers are frequently functors too. Rather than talking about Lists, Maybes, drawers or cupboards, we forget the specifics, talk about *containers* as the things sharing the particular property of having other things inside, and thus avoid getting into the details of what concrete type of container we are dealing with. A knowledge, we recall, that has bitten us quite hard in the past. But is this all there is to functors? *Containers*? How disappointing, how terran.

In Haskell, mathematics in disguise, our abstractions, the beautiful ones, seldom talk about things themselves. They talk instead of how things relate to each other. We see this in the quintessential identity function, which talks not about the particulars of a thing, but about the relationship of a thing with itself. Or functors, which are defined not as things having container-like features, but as the relationship of a thing with the types and laws it is expected to fulfill. No, most functors are not containers at all.

Forty-five

Functions are everywhere. They are the foundation of what's computable, and we seem to be defining everything we do on top of them. This makes sense considering how abstract functions are, and how we are dealing with programs expected to be *computed* at some point after all. But while functions are so fundamental in programming, they are not so in mathematics as a whole, where some of our knowledge comes from. Of course, functions are terribly important there as well, but what we mean is that even without talking *concretely* about functions, we can still say things about their expected behaviour, and about what our programs should compute. That is, we have even more abstract ways of talking about functions than functions themselves, and functors are one such way. Yes, *functions*, these very container-unlike things, are functors as well, and we will now start a journey to understand how. First, let's recall what the Functor typeclass looks like.

```
class Functor f where
    fmap :: (a -> b) -> f a -> f b
```

As long as we can find an f for which an instance of the Functor typeclass can be implemented while respecting the functor laws, then we can say that the chosen f is a functor. In other words, if we are stating that functions are indeed functors, then we must be able to write a Functor instance where *functions*, somehow, are that f. How, though? It's not obvious, is it?

Forty-six

It is important to remember that in Functor f, the f is expected to be a *type constructor*, not a type. That is, we pick things such as Maybe or List to be f, not Maybe Natural or List String. Think about it. What type would fmap take if f was Maybe Natural?

```
fmap :: (a -> b) -> Maybe Natural a -> Maybe Natural b
```

That doesn't make any sense. What is Maybe Natural a if not nonsense that the type-checker will reject? No, f must be a type constructor which, when applied to some arbitrary a or b, becomes a full blown type suitable for hosting an expression at the term level. If we pick f to be Maybe, not Maybe Natural, then we end up with a sensible type for fmap.

```
fmap :: (a -> b) -> Maybe a -> Maybe b
```

Yet, in Maybe Natural a, an impossible type, we see a striking resemblance to Either Natural a, a perfectly acceptable one. This suggests that Either Natural —or Either whatever, for that matter— could be a suitable f.

```
fmap :: (a -> b) -> Either x a -> Either x b
```

Indeed, a perfectly acceptable type. How exciting! Quick, let's make a Functor.

```
instance Functor (Either x) where
  fmap = \g e ->
    case e of
      Left x -> Left x
      Right a -> Right (g a)
```

Aesthetics aside, this is the *only* possible behaviour for the fmap method in this instance. Much like how parametricity forced the implementation of id and compose to achieve what they do, here it is forcing fmap to have this behaviour, the only one that type-checks. Go ahead, try to implement it differently. It won't work.

According to fmap's type (a -> b) -> Either x a -> Either x b, no changes are ever made to x, the payload of the Left constructor, so we leave Left x untouched if we see one. On the Right we are responsible for actually changing this datatype, somehow producing a b payload as output. And we do so in the only way we possibly can: By applying g, a function of type a -> b, to the a value on the Right constructor in order to obtain a b. It's important to be aware that when we write Left x -> Left x, the type of the first Left x is the same as the e we are scrutinizing —that is, Either x a— but the type of the expression we are returning as output, to the right of the arrow ->, is Either x b, as mandated by the type of fmap. This is similar to how at some point we said Nothing -> Nothing in the Functor instance for Maybe, yet those two Nothings had different types.

So we can say out loud that given a value of type Either x a as input, fmap modifies the a somehow, if any, possibly changing its type, but leaves the x untouched. For example, fmap add_one (Left 2) results in the same Left 2, but fmap add_one (Right 2) results in Right 3.

Forty-seven

In other words, we can't ever use fmap to modify the payload on the Left constructor of an Either. Or can we? This seems to be a rather arbitrary choice. If Functors are supposedly there to allow us to map a function over *all* the elements inside our chosen f, why are we skipping half of them? Why the Right and not the Left? Well, let's look at the types again.

```
fmap :: (a -> b) -> Either x a -> Either x b
```

This is what we have. It says on the tin that the Left never changes. But, could we have it the other way around? Tackle the Left instead?

```
mapLeft :: (a -> b) -> Either a x -> Either b x
mapLeft = \g e ->
  case e of
    Left a -> Left (g a)
    Right x -> Right x
```

Nothing wrong with that. It works. It type-checks. The problem, however, is that while this is a fine function worthy of the name mapLeft, it couldn't possibly be the implementation of the fmap method of the Functor typeclass. Why? Well, what would f be?

```
instance Functor ? where ...
```

We used Either x before as our f. A partially applied Either, a type constructor whose last parameter, the type of the Right payload, hasn't been provided yet. Now, however, we would need to come up with something that leaves out the type of the Left parameter.

```
instance Functor (Either ? x) where ...
```

That, however, is not possible. We can work around it, sure. In fact, if you dwell on it for a bit, you yourself may come up with a workaround. We have all the tools already. However, we are in no hurry to solve this, so we'll leave it for later. Our thirst is for understanding, not for a sense of accomplishment. Let's understand what is going on with that mystery f instead, and why we seem to be able to use fmap on the Right but not on the Left.

Forty-eight

Just like how expressions are of a particular *type*, and how we use those types to tell apart some expressions from others, types themselves also have their own "types", if you will, which we use for telling them apart from each other. They are called *kinds*, and much like types themselves, they exist only at compile-time and are there only for the type-checker —or rather, *kind*-checker— to see. And what types would we like to tell apart? Well, types such as Natural and String surely convey different ideas. Perhaps them? Not quite.

Natural and String certainly are different, but aren't *all* types different? That's why we have types at all, because we want to tell apart expressions that convey different concepts by tagging them with types that the compiler will reject if different than expected. Natural and String are different in the same way the numbers 2 and 7 are: If we are planning to add 2 to some number, but we accidentally add 7, say, the result will be wrong. But the fact that we were able to add 7, rather than 2, says something about how these two things are not that different after all. They are both numbers. With kinds, we have a similar situation. To understand this, let's look at our identity function once again.

```
id :: x -> x
```

The beautiful id says that given an x as input, *any* x we can come up with, it will return it as output. *Any* x. We try to give it 3, a Natural number, and it works. We try with Nothing, a value, say, of type Maybe String, and it works. We even try to use id itself as the input, a value of type x -> x, and it works as well. Indeed, id works with values of *any* type. And that's the thing. Just like how 2, 7 and 12 are all numbers, and we can readily, accidentally or not, use one where the other was expected, *any* type can appear wherever a type is expected. This says something. This says that types such as Natural, String or a -> a, even though different from each other, are all still types. And why does this obvious thing matter? Because by being able to precisely identify what a type is, we can safely say that anything that exists at the type level but is not a type, anything that the type-checker can see but can't call a type, must definitely be something else. We know of at least one such thing.

Forty-nine

Type constructors, they are no types. Can we, for example, apply id to a value of type Maybe? Trick question, we can't even come up with a value of such "type", because neither Maybe nor any other type constructor is a *type*.

It says so in their name, they *construct* types. Only types can have corresponding values at the term level. So what is a type constructor, anyway? Let's reason about this using kinds. First, let's see what's the kind of a normal type like `Natural`.

```
Natural :: Type
```

When writing down kinds, we use the symbol :: to state that the thing to its left is a type, and the thing to the right its kind. Here we are saying that the type `Natural` has kind `Type`. There are a couple of confusing things about this. First, we are reusing the same symbol :: that we used for stating that some expression has a particular type, like in 2 :: Natural. However, if we consider that this is something that, out loud, we read as "2 has type Natural", and that in turn we read `Natural :: Type` as "Natural has kind Type", which is true, then we should be at ease. This, for example, is valid Haskell code:

```
2 :: (Natural :: Type)
```

We need those parentheses because ::, by default, *associates to the left*. Without the parentheses we would end up with a type-checker trying to understand (2 :: Natural) :: Type, which doesn't make sense because 2 :: Natural is not a type, it is an expression to which we are explicitly giving the type Natural. And we just said that when specifying the kind of a type we need to write the *type* to the left of the symbol ::, and the kind to its right. So, the type-checker rejects this. By explicitly putting the parentheses to the right, as we did, we end up with the idea that "2 has type Natural, and Natural has kind Type", an idea that the type-checker will gladly accept.

The second thing that might be a bit confusing about this notation is that we are saying that Natural :: Type means "Natural has *kind* Type". That is, we are seeing the word "type" appear in the place we supposedly talk about kinds. But this makes sense, types have kind Type indeed. Type is just a name somebody chose for this *kind* of things. Other kinds of things are named differently.

So let's just try and embrace this notation. The symbol :: is used both for conveying the type and the kind of something, and whenever we see Type, we must remember we are dealing with a kind, not a type. Just like how Natural is a type and not a natural number, Type is a kind and not a type.

Fifty

We can tell types apart from type constructors by looking at their kinds. While a type like Natural has kind Type, a type constructor generally has kind Type -> Type.

If Type was a type —it isn't, it is a *kind*— and we were to find a function of type `Type -> Type`, then, without hesitation, we would say that this was a function that takes an expression of type `Type` as input and returns an expression of type `Type` as output. Well, that's exactly what's happening with type constructors, except involving types and kinds, rather than expressions and types. Let's look at `Maybe`, for example. `Maybe` is a type constructor of kind `Type -> Type`. This implies that given a type of kind `Type` as input, we get a type of kind `Type` as output. And this is exactly what happens. Consider `Natural`, for example. `Natural` is a type, so it has kind `Type`. If we apply the `Maybe` type constructor of kind `Type -> Type` to `Natural`, then we end up with `Maybe Natural` of kind `Type`.

Like functions, type constructors can also take multiple parameters as input. Well, we know that in truth functions only ever take one parameter in Haskell, but let's pretend they take many, which for our practical intents and purposes is true. Anyway, type constructors can also take more than one parameter as input. Think of `Either`, for example. What is the *kind* of `Either`, a type constructor that needs to be applied to *two* parameters in order to become a type? It is `Type -> Type -> Type`, of course. Applying `Either` to two `Type`s, say `Natural` and `String`, gives us `Either Natural String`, a `Type`. What happens if we *partially* apply `Either`, though? That is, if we only give it the first of the two input `Type`s? Can we do that? Sure we can, we did it earlier when we were toying with `instance Functor (Either x)`, remember? That `Either x` is a partially applied `Either`. We don't know what x is, sure, but it doesn't matter. We know that any type —that is, any `Type`— can occupy x's place. And what happens after we apply a type constructor of kind `Type -> Type -> Type` to something of kind `Type`? We end up with a type constructor of kind `Type -> Type`, much like how applying a value of type a to a function of type `a -> b -> c` leaves us with yet another function of type `b -> c`.

So there we were, defining the `Functor` instance for `Either x`, a type constructor with kind `Type -> Type`. This suggests that the mystery fs for which we can actually say `instance Functor f` should perhaps always be of kind `Type -> Type`. Should they, though? Maybe. We were able to define `Functor` instances for `List` and `Maybe` as well, both type constructors of kind `Type -> Type`. Promising results. However, the fact that we were able to do this a couple of times doesn't imply that it must always be this way. So let's go back to the source, to the typeclass, to understand the truth.

```
class Functor f where
    fmap :: (a -> b) -> f a -> f b
```

We see f show up both as an input parameter to the `fmap` method in `f a`, and as the method's output in `f b`. We will analyze those use cases to determine what the kind of f is. What do `f a` and `f b` have in common? They could be different values, sure, and they each play a different role in fmap, but they are also both expressions. They have that in common. And expressions, we know, have types. And by "types" we mean things of kind Type. So f a and f b must themselves be Types. But also, both a and b show up as standalone types in a -> b as well, which implies that a and b are also Types. But if f a and f b are Types, and if a and b are themselves Types as well, then f, which is being applied to each of a and b with the expectation of getting yet another Type in return, must necessarily have kind Type -> Type. In other words, yes, the f in Functor f must always have kind Type -> Type. In fact, while not necessary, if we wanted to be explicit about this, we could have mentioned the kind of f in the typeclass definition of Functor. This can sometimes be useful to readers of our code.

```
class Functor (f :: Type -> Type) where
  fmap :: (a -> b) -> f a -> f b
```

So now we know why giving a Functor instance for Natural or Maybe String doesn't work. It's because they have kind Type, not Type -> Type, so the type-checker readily rejects them. A Functor instance for Either doesn't work either because Either has kind Type -> Type -> Type, different than the expected Type -> Type.

And from here we can see why we can't use fmap to work on the Left side of the Either. It is because when we realize that Either x, a type constructor with kind Type -> Type as expected by the Functor typeclass, *must* be the f in our Functor instance, then that f will continue to be Either x *anywhere* it appears inside the Functor instance. Including in the fmap method, a method that doesn't provide any way to modify that f. Compare the following two types:

```
fmap :: (a -> b) -> f        a -> f        b

fmap :: (a -> b) -> Either x a -> Either x b
```

Either x is essentially just replacing the f everywhere. Perhaps adding some redundant parentheses will help us better appreciate this correspondence.

```
fmap :: (a -> b) -> f          a -> f          b

fmap :: (a -> b) -> (Either x) a -> (Either x) b
```

See? There is no way to modify that Either x. If there was one, we would have an extra input parameter to fmap indicating how to modify that x on the Left, but alas, we don't have one. All we know is how to modify that a, the one on the Right, into a b.

There are kinds beyond Type, Type -> Type, Type -> Type -> Type, etc. However, for the time being we don't care. And anyway, we have learned enough about this topic for now, so let's continue our journey, comfortable with this new knowledge.

Fifty-one

Let's go back to our goal of coming up with a `Functor` instance for functions. The first thing we need to acknowledge is that we must be able to talk about functions as something of kind `Type -> Type` somehow, since this is what the `Functor` typeclass mandates. How, though? Functions like `Natural -> Natural` or `a -> b` are expressions, values of kind `Type`, not type constructors. Well, `Maybe Natural` and `Either a b` were `Type`s as well, yet somehow we managed nonetheless. We simply left the type partially applied, and that seemed to do the trick. How do we partially apply *the type* of a function, though? It might help to learn that the arrow `->` in a function type like `a -> b` is an *infix* type constructor of kind `Type -> Type -> Type`. So when we see a function `Type` like `a -> b`, we must acknowledge that we are looking at `->`, the type constructor for functions, being applied to two other `Type`s `a` and `b`.

Let's do something beautiful. Let's have the types guide us. Many times, when programming, we have no idea what a solution could look like, why a problem has been encoded in the way it has, or what the problem is at all. And most times it doesn't matter. Everything is more or less the same. So we willingly embrace this ignorance and rely on parametricity, this idea that a polymorphic function like `fmap` must always behave the same no matter what values it will deal with. Let's do this step by step.

First, we just said that the arrow `->` in a function type like `a -> b` is an *infix* type constructor, and we learned before that infix operators can be used in a *prefix* way if we just sprinkle some parentheses around them. Let's do that. `a -> b` is exactly the same as `(->) a b`. That is, for example, the type of a function `Natural -> String` could also be written as `(->) Natural String`. It looks ridiculous, of course, that's why we never write it this way. But at times like the present, it can be necessary. The kind of `(->) Natural String`, a function type, a type like any other, is `Type`. The kind of `(->)`, the type constructor which takes two `Type`s as arguments, here `Natural` and `String`, is of course `Type -> Type -> Type`. The first type parameter, `Natural` in our example, corresponds to the type of the input of this function, whereas the second parameter, `String` here, corresponds to its output. And what if we apply just *one* of those two parameters, say, `Natural`? We end up with `(->) Natural` of kind `Type -> Type`, the kind `Functor` desperately wants. `(->) Natural` is a type constructor for a function that takes a `Natural` number as input, it says nothing about what the output type of said function will be. But of course, `(->) Natural` is unnecessarily restrictive. What do we care what the type of the input is? Let's spice up the mystery and use `(->) x` instead.

```
instance Functor ((->) x) where ...
```

Alright, `(->) x` has the exact kind `Functor` expects, so in theory it could be a functor if we somehow manage to implement the fmap method in a way that it abides by the functor laws. What would fmapping a function over (->) x do, though? Hard to fathom, but also completely irrelevant. We can still go ahead and implement a correct fmap that will do what it must. Let's look at what the type of fmap looks like when we specialize the f in Functor f to (->) x.

```
fmap :: (a -> b) -> ((->) x) a -> ((->) x) b
```

Ugh, let's clean that up. First, those parentheses around (->) x are unnecessary: ((->) x) y and (->) x y, for any choice of y, mean the same. Let's get rid of them

```
fmap :: (a -> b) -> (->) x a -> (->) x b
```

Second, (->) x y and x -> y —again, for any choice of y, including our a or our b— mean exactly the same, so let's use the less ugly form.

```
fmap :: (a -> b) -> x -> a -> x -> b
```

Oops, our fmap now seems to take four parameters as input rather than two. Our fault, we just forgot to add parentheses to prevent the arrows -> belonging to f from getting intertwined with the ones that are part of fmap's own type.

```
fmap :: (a -> b) -> (x -> a) -> (x -> b)
```

Finally, this is where we wanted to be. Here, fmap is a function that takes two other functions as input, a -> b and x -> a, and somehow returns a function x -> b. In other words, given an x, fmap uses the function a -> b to transform the a output of the function x -> a into b. For all of this to work, however, we need an x to provide to our x -> a function. Without it, we won't be able to obtain an a to which we can apply our a -> b. Well, that's not a problem at all. It might help to remember that the rightmost parentheses, the ones surrounding x -> b, are optional.

```
fmap :: (a -> b) -> (x -> a) -> x -> b
```

Look at that, an x. It seems we have everything we need. We simply apply x -> a to x, and to the resulting a we apply a -> b in order to finally return that b. Easy.

```
instance Functor ((->) x) where
    fmap = \g f x -> g (f x)
```

Done. We had no idea what to do, we just more or less knew where we wanted to go, we let parametricity guide us, and we got there anyway. We can be certain our implementation is correct because there is simply no other way of implementing this. Try all you want, but aesthetics aside, you'll end up here again. We are home. We are where we need to be.

So when they ask if you can do something, take their money and say yes.

You know parametricity and they think their problem is special. You will be fine. Just kidding. Don't do that. Be a responsible professional. The point is that you can always rely on parametricity, the ultimate trick.

And not only is this the one solution we didn't know we were looking for, it is also a very beautiful place to be. Look harder at the type of our `fmap`. What do you see? Nothing? Are you sure? What about if we just rename some of the type parameters?

```
fmap :: (b -> c) -> (a -> b) -> (a -> c)
```

Don't you recognize it? Alright, here it is:

```
compose :: (b -> c) -> (a -> b) -> (a -> c)
```

In other words, the `Functor` instance for functions explains how functions compose together. That is, wherever f is a function, saying `compose g f` is the same as saying `fmap g f`. Something like `fmap add_one multiply_by_ten 3`, for example, would return 31 just like `compose add_one multiply_by_ten 3` or `(3 * 10) + 1` would. In fact, we can simply reuse `compose` as our definition of `fmap`.

```
instance Functor ((->) x) where
    fmap = compose
```

And there is no *container*, no. Yet here we are, with a perfectly acceptable `Functor`. So this is the last time we talk about functors as containers. Or do you, by any chance, see a value *contained* somewhere? No, you don't. We'll explain the true nature of functors later on.

Fifty-two

That functions are functors, among other things, means that `compose` not only is useful and necessary at times, but also that it has a strong mathematical foundation telling us why it won't ever go wrong. How? Well, if functions are functors, and if functors are expected to abide by the functor laws known to guarantee a correct behaviour, then `compose`, also know as `fmap`, must do so as well. Let's prove it.

The first functor law, the *identity law* of functors, says that `fmap`ping `id` over our functor shall result in that same functor, unmodified. In other words, for any function f, f and `fmap id f` are equal. We will use a very straightforward and mechanical technique to show that this is indeed the case.

In Haskell, when we say a = b, we are saying that the expressions a and b are equal. This implies that whenever a appears in an expression, we can replace it with b, or vice-versa, and the result should be exactly the same. We sometimes refer to this as *substitution*, seeing how we substitute one expression by another one. By *repeatedly* substituting expressions with an equal one, we can try to prove that two expressions of our choice are equal. We call this process *equational reasoning*. Let's try it. Using this approach, let's prove that f and fmap id f, both of the same type a -> b, are equal.

First, we write down fmap id f as a starting point, hoping that at some point that expression will become f.

```
fmap id f
```

We should clarify that we won't be writing a Haskell program here, we will just write down expressions that happen to be equal to the previous one we wrote. This is the kind of logical reasoning that you can do on a piece of paper. Eventually, if true, we will get to say that fmap id f is indeed equal to f. That will constitute the entire outcome of this endeavour.

The second thing we will do is replace fmap with its definition *right here*, where we are referring to fmap by name. Or, as we say, we *inline* the definition of fmap in our expression:

```
(\h g x -> h (g x)) id f
```

We aren't inlining the definition of fmap because it is the only possible thing to do, rather, it's just one of the possible alternatives. We could replace id with its definition instead, and it would be fine as well. We just need to start somewhere, so we make an arbitrary choice.

Next, we can beta-reduce the first argument to our fmap function. That is, we will have id take the place of h in fmap's body, and remove h from the input parameters of the lambda expression:

```
(\g x -> id (g x)) f
```

We can now beta-reduce the first argument of our \g x -> ... function, having f take the place of g. We will remove some unnecessary parentheses as well.

```
\x -> id (f x)
```

We know that applying the identity function to some expression results in that same expression. Here, saying id (f x) is the same as saying f x. So let's just get rid of the id application and that way save a few steps in our equational reasoning.

```
\x -> f x
```

Finally, we also learned about eta-conversion, which says that \x -> f x is the same as saying just f. So let's get rid of that redundant lambda expression.

```
f
```

Done. Using substitution and equational reasoning we went from `fmap id` f to f, finally arriving to the trivial truth that f equals f. It was easy, wasn't it? Using such a rudimentary and mechanical approach we managed to *prove* that our implementation of the `fmap` method for functions abides by the first functor law. We can prove a surprisingly large number of things with this one trick.

Fifty-three

Armed with our new knowledge, we proceed to tackle the second functor law, which says that `compose (fmap g) (fmap f)` a should be equal to `fmap (compose g f)` a. Let's see if it is true using equational reasoning. We'll do it in two steps. First, let's *reduce* `fmap (compose g f)` a as much as possible. *Reducing*, in this context, means repeatedly using beta-reduction, substitution, eta-conversion, and similar techniques to replace function applications with the actual expressions that those applications eventually become. We start by writing down the expression that concerns us.

```
fmap (compose g f) a
```

Here, all of g, f and a are functions. We know that g and f are functions because `compose` always expects functions as its input parameters, so there is no way they could be something else. Regarding a, we know it is a function simply because we are saying it is. That's all. Remember, here *we* are trying to prove that using `fmap` *with functions* behaves in a particular manner, and that necessarily means that *we* will pick our second parameter to `fmap` to be a function and not something else.

Now, since we already know that the `fmap` definition for functions is just `compose`, we can simply replace the name `fmap` for the name `compose` in our expression in order to make things simpler.

```
compose (compose g f) a
```

It's also important to keep in mind that g, f and a are just names we are making up here, names of *expressions*, and they have no direct correspondence to the fs and as we have been encountering in places such as `class Functor f` or `fmap :: (a -> b) -> f a -> f b`, where not only are those the names of *types* and type constructors, not expressions, but they are also made up. During this equational reasoning exercise we are only paying attention to expressions, not to types.

Back to `compose (compose g f)` a. There's no way to reduce that expression unless we fully substitute `compose` with its definition, so let's do that in both occurrences of `compose`.

```
(\g f a -> g (f a)) ((\g f a -> g (f a)) g f) a
```

That's perfectly valid Haskell, it turns out. All we did was substitute compose with its implementation \g f a -> g (f a) and add some necessary parentheses. It looks rather cryptic, though. The main problem seems to be that we are repeating f, g and a everywhere, which makes us a bit dizzy. However, this is "fine". When we say something like (\x -> x) (f x), the x in \x -> x is not the same as the x in f x. Remember, in \x -> ... we are *making up* a new name, here x, which we will use within this lambda expression to refer to any input that is provided to it. Whether somebody called something else x *outside* this lambda expression is irrelevant, because this x name we are binding in our lambda expression will *shadow* any other x that exists outside the lambda expression.

Shadow? Compare (\x -> x) (f x) with (\y -> x) (f x). The former is essentially the identity function applied to f x, whatever that might be, so the result of the entire expression is f x. The latter example is applying a function \y -> x to f x. Now, this function \y -> x is saying that it will take something as input, which it will call y, it will ignore it, and it will return x instead. Which x? Well, the same x that f is being applied to, the x that exists *outside* the lambda expression, the x that the former example *shadowed* instead. We haven't really seen this x being bound anywhere, but that doesn't concern us at the moment, we can assume somebody else did that for us.

It's like you have a neighbour, Mike, who you only run into from time to time. You go back to your place, and maybe you tell your partner about how you ran into Mike in the hallway, or how Mike knocked on your door the other day, asking if we could lend him a tool he didn't have. Whenever you or your partner say "Mike", you both know with certainty that you are talking about that neighbour. Mike is part of your daily life, of your environment, so you make sure you remember his name. One day, however, a friend from your childhood comes to visit you. A friend that just for one day becomes a part of your private environment, of your home. He is also called Mike. In that home, that day, if you talked to your partner about somebody named Mike, who do you think will come to mind first? Mike, the neighbour who is out there, or Mike, the childhood friend who is with you today? That's right. Childhood Mike's name *shadows* neighbour Mike's, even if only for that day, in that home.

There are very simple rules to identify with certainty the expression to which we refer when we mention a name anywhere in our code. We will learn them soon, but for now let's go back to our more immediate concerns.

Fifty-four

We are still in the middle of our journey, trying to prove the second functor law for functions by using equational reasoning to demonstrate that compose (fmap g) (fmap f) a and fmap (compose g f) a are the same. We decided to start by reducing fmap (compose g f) a, and we ended up with this:

```
(\g f a -> g (f a)) ((\g f a -> g (f a)) g f) a
```

We have way too many fs, gs and as in our code, and that is a bit confusing. We know that because of the *shadowing* of names, Haskell doesn't care. Haskell can still figure out the meaning of this without any cognitive overhead. But we, humans, struggle. So let's use a tool to help us navigate this mess.

The identity function is implemented as \x -> x. Or was it \a -> a? Ah, no, it was \motorcycle -> motorcycle, right? You see, it doesn't matter. Names are made up, names are used, and as long as we are consistent it doesn't matter which names they are. It is *alpha-conversion*, styled α-*conversion*, yet another lambda calculus feature, that explains how this is possible. Essentially, functions such as \x -> x and \y -> y are *alpha-equivalent* versions of each other, meaning that while they are not *literally* the same because they bind expressions to different names, they still *mean* the same. We move between these versions by a process of alpha-conversion, which consists pretty much of just replacing a name with another one everywhere it appears bound to the same expression. Don't you like x? Do you cringe at the sight of \x a -> (a, x)? Would you rather have y there? Then alpha-convert to \y a -> (a, y) and presto. We sometimes say *alpha-renaming* instead of alpha-conversion. It is the same, as if alpha-conversion itself had been alpha-converted.

Anyway, let's α-*convert* some of those fs, gs and as to make things a bit more obvious.

```
(\i h b -> i (h b)) ((\k j c -> k (j c)) g f) a
```

We kept our original f, g and a as they were in fmap (compose g f) a, but we alpha-renamed the ones inside our inlined versions of fmap and compose, the ones that were shadowing our original ones. And while we now have more names to consider, we also have less opportunities to mistake one for the other. So thank you, α-conversion, for supporting our existence as easily confused human beings.

Fifty-five

Alright, let's continue reducing our fmap (compose g f) a until we get to the point where we can't do it anymore. This might get boring, so feel free to just skim over the details while you yawn.

```
(\i h b -> i (h b)) ((\k j c -> k (j c)) g f) a
```

First, through beta-reduction, let's get rid of that k by substituting it with

g.

```
(\i h b -> i (h b)) ((\j c -> g (j c)) f) a
```

Similarly, we get rid of that j, by replacing it with f.

```
(\i h b -> i (h b)) (\c -> g (f c)) a
```

Now we can do the same with i, this time substituted with all of \c -> g
(f c).

```
(\h b -> (\c -> g (f c)) (h b)) a
```

We now do the same for h, substituting it with a.

```
\b -> (\c -> g (f c)) (a b)
```

And finally, we substitute c with a b.

```
\b -> g (f (a b))
```

There's nothing else to do. fmap (compose g f) a reduces to this at best. All
we have to do now is reduce compose (fmap g) (fmap f) a and see if we end
up with the same expression. If we do, it means that fmap (compose g f) a
and compose (fmap g) (fmap f) a are indeed the same, which proves the
second functor law for functions. Let's do it faster this time.

```
compose (fmap g) (fmap f) a
```

First, let's inline the definitions of compose and fmap, which we know are
the same. As we do it, let's also use alpha-conversion to make sure we pick
different names so that this is easier for us humans to follow.

```
(\i h b -> i (h b)) ((\k j c -> k (j c)) g)
                    ((\m l d -> m (l d)) f)
                    a
```

Oh my, how tiresome. Let's beta-reduce that k, substituting it with g.

```
(\i h b -> i (h b)) (\j c -> g (j c))
                    ((\m l d -> m (l d)) f)
                    a
```

And the m, which becomes f.

```
(\i h b -> i (h b)) (\j c -> g (j c))
                    (\l d -> f (l d))
                    a
```

And the i now, which we substitute with all of \j c -> g (j c).

```
(\h b -> (\j c -> g (j c)) (h b)) (\l d -> f (l d)) a
```

And also the h, which will become all of \l d -> f (l d).

```
(\b -> (\j c -> g (j c)) ((\l d -> f (l d)) b)) a
```

Oh, and now we can beta-reduce that b, replacing it with that a.

```
(\j c -> g (j c)) ((\l d -> f (l d)) a)
```

And the 1 goes away too. It becomes the a.

```
(\j c -> g (j c)) (\d -> f (a d))
```

Now it's time to beta-reduce that j. The function `\d -> f (a d)` will take its place.

```
\c -> g ((\d -> f (a d)) c)
```

And we can get rid of that d as well. We can substitute it with c.

```
\c -> g (f (a c))
```

And we are done. There's nothing left to reduce, luckily. So, have we proved anything? On the one hand we had `fmap (compose g f) a`, which reduced to `\b -> g (f (a b))`, and on the other hand we just reduced `compose (fmap g) (fmap f) a` to `\c -> g (f (a c))`, which is not exactly the same. Or is it? Aren't we forgetting what our friend alpha-conversion is capable of? These two expressions are *alpha-equivalent*, we just ended up binding the name b on one of them and c on the other, but the meaning of these expressions is the same: We take a value as input, apply the function a to it, then apply f to this result, and finally apply g to all of that.

```
\sandwich -> g (f (a sandwich))
```

So, yes, `fmap (compose g f) a` is equal to `compose (fmap g) (fmap f) a`, which means that function composition abides by the second functor law. And, considering how function composition also abides by the first functor law, we can solemnly swear that functions are indeed proper functors, as witnessed by the existence of a lawful `Functor` instance for `(->) x`.

Fifty-six

Even while both `compose (fmap g) (fmap f) a` and `fmap (compose g f) a` eventually reduced to the same expression, one did it in less steps than the other. Concretely, it took five beta-reductions for `fmap (compose g f) a`, whereas it took eight for `compose (fmap g) (fmap f) a`. What does this mean? Semantically, not much. After all, we did get to the same result. However, as fast and obeying as our computers may be, they still need to do all we ask of them. They don't skip any work. So if we ask them to do something five times, they will necessarily do that faster than if we ask them to do it eight times. We need to be aware of these things, too.

Any time a computer spends doing something is time not spent doing something else. That's not necessarily bad, though. After all, we have these computers so that they do as much work as possible for us all day long, but it is a fact we must acknowledge nonetheless. In particular, while a program is still computing something, it is not yet delivering the results of that computation to whomever is expecting them. Generally, we want our programs to be as fast as possible, so it's important that we understand not only the meaning of our programs, but also how fast they perform so that we can manage our expectations accordingly, or even optimize them if possible. Not always, though. Think of the traffic light that waits a couple of seconds before changing colours. That delay is artificial, the colours could change much more rapidly, but we deliberately make the switching of the colours slow to give traffic enough time to go through. Moreover, not everything deserves to be as fast as possible, even if desirable from an execution time point of view. Like everything else in life that is valuable, speed has a cost. A cost that often manifests itself in how much time we humans spend making things fast, time that could perhaps be better spent on making sure our program is correct, spending quality time with friends and family, or building something else.

How do we know when speed is a worthy goal? Well, we get our priorities right and decide. Sure, compose (fmap g) (fmap f) is slower than fmap (compose g f), but does it matter? Not if we are running it every once in a while, but probably it does if we are doing it a substantial number of times. Unless, of course, our program is expected to be the fastest one, in which case we may want to sacrifice everything in the name of speed. Alternatively, it might be that speed is not particularly important for us, in which case we can mostly ignore all of this and move on.

How substantial must the difference in performance be between two semantically equivalent implementations to justify picking one over the other? Can we measure that? Yes, yes we can. But be warned, there's a bit of folklore involved. Unfortunately, while we have things such as *types* for helping us reason about the semantics of our programs before they even exist, tools for reasoning about performance are less sophisticated and mostly rely on thinking really hard, pen and paper, or measuring after the fact. Modern compilers can sometimes realize on their own that something can be made faster without affecting the semantics of our program, and they will magically go ahead and optimize things for us. However, this only goes so far. We can't, for example, expect compilers to optimize programs that are conceptually inefficient by design, like doing complex arithmetic calculations with that beautiful but slow inductive representation of natural numbers from before.

It's worth noting that while it is easy to talk about performance in terms of time and speed, performance can be measured in other ways as well. Insofar as computers are concerned, for example, we can consider how much *space*, or memory, a particular program takes. But even beyond the computer, we can talk about how quickly something can be implemented, how many people need to be involved, or what kind of resources we need.

So yes, this is a book on economics too. It *must* be if we expect this type of programming to be realistic, to be relevant to civilization at all. But we are still just getting started, so let's mainly focus on getting our programs right for the time being. We'll come back to the matter of performance later.

Fifty-seven

A while ago we struggled with the Left. We learned that fmap wasn't able to do anything with the value contained in a Left constructor, it was only able to deal with the one on the Right. It had something to do with kinds. Essentially, the Functor typeclass expects its instances to be defined for type constructors of kind Type -> Type, where that input Type is necessarily the rightmost type parameter in our type constructor. So, for example, Either x, because of its kind Type -> Type, was a suitable candidate for a Functor instance. And what happened to the type of the fmap method in this instance? It became (a -> b) -> Either x a -> Either x b, a function that could only modify the payload on the Right. Nothing important happened on the Left.

What if we flip it, though? What if Either x y meant that x was the type of the payload on the Right, and y the type of the one on the Left? In the Functor instance for this brand new Either x, also of kind Type -> Type, we would still get fmap :: (a -> b) -> Either x a -> Either x b. That static x however, the one that stays the same, would now be talking of the Right payload, not the Left. And vice-versa, of course, which means that fmap would indeed allow us to modify the Left. The problem, however, is that we can't do that. Someone other than us came before, created Either the way we first met it, and shipped it with the Haskell language as such. We are stuck with it, forever, we have to embrace it as it is.

An alternative thing we can do is try to create a new datatype isomorphic to Either but with its type parameters flipped, so that its Functor instance works on the Left rather than the Right. Let's call it Fleither, short for "flip Either".

```
data Fleither a b = Fleft b | Flright a
```

Look how the b type parameter, the one our fmap method would allow us to modify, is now on the Fleft as we wanted.

```
instance Functor (Fleither x) where
  fmap = \g s -> case s of
                   Fleft b -> Fleft (g b)
                   Flright a -> Flright a
```

So clever. So much so that we neglected the Flright this time. How do we fmap a function over the Flright side of a Fleither? We can't, not with this approach. Changes to Fliright's payload are restricted for the same reasons that changes to Left's payload were restricted before. Let's throw Fleither away, we need something else.

Fifty-eight

When defining a new datatype like Either, Fleither or Pair, we don't always have to start from scratch, leaving all the payloads that show up in our fields polymorphic. Sometimes we can be a bit more concrete, and it works too.

```
data Flip b a = Flip (Either a b)
```

Here we are defining a new datatype named Flip, with two type parameters b and a. This datatype has a single constructor, also called Flip, which takes an entire Either a b as payload. That is, we construct a Flip by saying either Flip (Left ...) or Flip (Right ...). The interesting thing to notice is that the type parameters in Flip b a and Either a b are flipped, which would force a Functor instance for Flip b to target that a on the Left, rather than the b on the Right as fmapping over Either a b would normally do.

```
instance Functor (Flip x) where
  fmap = \g s -> case s of
                  Flip (Left a) -> Flip (Left (g a))
                  Flip (Right x) -> Flip (Right x)
```

There's nothing new here, we are just pattern-matching on the Flip constructor to extract its payload so that we can modify it if necessary. Of course, after modifying the payload on the Left constructor, we need to wrap the modified Either in the Flip constructor again. And similarly on the Right side. The type of fmap, which has now become (a -> b) -> Flip x a -> Flip x b, is demanding it. In other words, that s expression we are pattern-matching on has type Flip x a, and both expressions the right of the arrow -> have type Flip x b.

So how is Flip better than that failed Fleither from before? Well, it is still not ideal, so if that's what we are looking for we are not going to find it here. But look at this:

```
foo :: Either Natural Season -> Flip [Temperature] String
foo = \x -> fmap bar (Flip (fmap qux x))
```

Can you tell what this horrible yet didactic function does? What would be the result of foo (Left 3)? What about foo (Right Winter)? It doesn't matter what the *exact* result would be, but suffice to say that the bar function we assume here would transform that Natural number 3 on the Left side of the Either into a String somehow, and the qux function would transform the Season on the Right into a [Temperature]. Remember, the type level parameters to Flip appear deliberately in the opposite order than those in Either do, let's be aware of that. The interesting thing in this exercise is that we somehow managed to target both sides of an Either by relying solely on the behaviour of fmap. Unfortunately, we always end up with a `Flip` rather than an `Either` as output. For example, `foo (Left 5)` could result in the value `Flip (Left "bird")` rather than just `Left "bird"`. That's easy to fix, though. We just need to remove that `Flip` wrapper. After all, it doesn't serve us anymore once we are done `fmap`ping over it.

```
unFlip :: Flip a b -> Either b a
unFlip = \(Flip x) -> x
```

Other than the weird name `unFlip`, suggesting some kind of *un*doing, there should be nothing surprising here. We are just pattern matching on the `Flip` constructor to extract its payload before returning it. We know this `x` has the right type, an `Either` with its type parameters flipped, because that's what it says, literally, in the definition of this datatype. We can use a straightforward lambda expression to extract this payload because the `Flip a b` datatype has only one constructor. Otherwise, we would have needed to perform a case analysis to inspect all possible constructors. Long story short, `unFlip` will discard that `Flip` wrapper, so let's put it to use in our also awkwardly named function `foo`.

```
foo :: Either Natural Season -> Either String [Temperature]
foo = \x -> unFlip (fmap bar (Flip (fmap qux x)))
```

The implementation seems a bit convoluted, but remember that after we are done with things, we mostly just look at their types, not at the expressions that implement them. And here we just see that we go from one `Either` to another `Either`, not one detail about `Flip` *leaks* to the type of our function. Safeguarding the programmer from the ugly truth, hiding the bad stuff under the carpet, only ever looking at nice things: This is what software development is about. Just kidding, it's not. But still, one has to acknowledge the beauty of being able to prevent irrelevant implementation details, such as that `Flip` thing, from becoming a *cognitive burden* on whoever is trying to understand the purpose of this function. Kind of.

Fifty-nine

Granted, due to the lack of parametricity, our function `foo` is not the shinning light it's been made out to be. For all we know, without looking inside `foo`, this function could be simply ignoring its input and returning `Left "crocodile"` every time. Why not?

The problem, once again, is that we know *too much* about the payloads that go into that `Either`. We need to forget. How? Why, with parametricity, of course. So, rather than going from an `Either Natural Season` to a `Either String [Temperature]`, we will go from an `Either` of mysterious things to an `Either` of even more mysterious things. Let's make up some names.

```
foo :: Either a b -> Either c d
foo = \x -> unFlip (fmap ? (Flip (fmap ? x)))
```

The problem, now, is that we don't know what to write where we left those ? placeholders. Think about it. How will we modify a so that it becomes c, or how would we modify b so that it becomes d, if we know nothing about neither a, b, c nor d? Well, we wouldn't. But that's fine, that's what we wanted. It's so easy to forget why we are here sometimes. So what do we do? What does fmap do when it wants to modify the Right side of an Either, say, without knowing what's in it exactly? It simply defers the decision of how to modify that payload to the callers of this function, by asking them to provide the function that will modify the payload.

```
fmap :: (a -> b) -> Either x a -> Either x b
```

Let's copy this, then, but taking two functions rather than one. One for each side.

```
foo :: (a -> c) -> (b -> d) -> Either a b -> Either c d
foo = \l r x -> unFlip (fmap l (Flip (fmap r x)))
```

Sure, foo doesn't apply our fantasy bar nor qux on its own anymore, but that's alright, we could now say foo bar qux to achieve the same result. That's what we wanted, after all. Now, originally we named this thing foo because, this function being as ugly as it was, didn't deserve a better name. We punish ad-hoc code by not giving it a decent name, that's what we do. It keeps us from developing affection for it. But now, behold, foo is beauty, so let's give it a proper name.

```
bimap :: (a -> c) -> (b -> d) -> Either a b -> Either c d
```

It is a nice name, bimap. It evokes the idea that, somehow, we are mapping over *two* things. And indeed, that's exactly what we are doing.

Sixty

What about Pairs? We kind of forgot about them, but remember, just like Either is the most fundamental *sum type* out there, Pair is the most fundamental *product type*, so we need to pay attention to it as well. We can't forget it. Let's recall the definition of Pair.

```
data Pair a b = Pair a b
```

So, while Either a b has one of either a or b depending on whether we use the Left or Right constructor, Pair has both a and b values in it, always. Can we imagine mapping over both of them? Sure. Maybe we have a Pair Natural Season and we want to convert it to a Pair String (Maybe Temperature) for some reason. How do we do it? Well, let's just copy what bimap did, but this time for Pairs.

```
bimapPair :: (a -> c) -> (b -> d) -> Pair a b -> Pair c d
bimapPair = \f g (Pair a b) -> Pair (f a) (g b)
```

Alright, this works. We one of the functions to f, the other one to g, we pattern match on the a and b payloads of our Pair constructor, and after applying f and g to them, respectively, we put them back in a new Pair and we send them on their way. Beautiful.

Wait a minute. There's a pattern here. Let's compare bimap with bimapPair.

```
bimap     :: (a -> c) -> (b -> d) -> Either a b -> Either c d

bimapPair :: (a -> c) -> (b -> d) -> Pair   a b -> Pair   c d
```

Other than bimapEither allegedly being a more precise name for bimap, what do we see? The only thing that changes is our choice of Either or Pair, everything else in these types stays the same. We've seen this before, when somehow we managed to put all of (a -> b) -> List a -> List b, and (a -> b) -> Maybe a -> Maybe b and the like in the same bag. What did we reach out for? Functor. And what was Functor? A *typeclass* with a single method fmap that let us operate on different functor-like things homogeneously. Well, it turns out Functor is not the only interesting typeclass out there. We have others, like Bifunctor here:

```
class Bifunctor (f :: Type -> Type -> Type) where
    bimap :: (a -> c) -> (b -> d) -> f a b -> f c d
```

This should be straightforward, considering all we've learned so far. First, notice how we gave f, the bifunctor-like thing, an explicit kind Type -> Type -> Type. In instances of this typeclass, this f will become type constructors such as Either or Pair, taking *two* other Types as input. We kept the name bimap for the method in this typeclass. Why change something so beautiful? Hopefully we can see how replacing that f with concrete type constructors like Either and Pair leave us with a type of bimap *specialized* for types as concrete as Either a b, Pair c d, etc. Let's see some instances of this typeclass.

```
instance Bifunctor Pair where
    bimap = \f g (Pair a b) -> Pair (f a) (g b)
```

Nothing surprising in the Bifunctor instance for Pair. We just wrote the same implementation we had in our ad hoc bimapPair.

```
instance Bifunctor Either where
    bimap = \f g x -> case x of
                          Left a -> Left (f a)
                          Right b -> Right (g b)
```

Nothing surprising in this instance either. This time, however, we decided against re-using the implementation that used that Flip trick to target the Left. Why? Well, mainly because we can, but also because this way is more performant. Think about it. Here we are pattern-matching once, that's all, but in our Flip example we were indirectly pattern-matching once *each time* we used fmap, and we called fmap twice. So this implementation, in theory, should be at least twice as fast. Further optimizations made automatically for us by the compilers might render this argument moot, they might see the two fmaps, the Flip and the unFlip, and simplify all of that somehow. But still, it's important that we become familiar with this idea of considering performance matters early. Anyway, whether our performance optimizations have merit or not, this implementation looks very straightforward, so let's keep it.

With these instances in place, bimap (add 1) (add 10) will take us from Pair 2 5 to Pair 3 15, from Left 2 to Left 3, from Right 5 to Right 15, and the like.

Sixty-one

Here's a quick interlude, a curiosity. We just learned that the bimap method of the Bifunctor typeclass allows us to comfortably work on *both* the as and bs showing up in types like Either and Pair. And seeing how as and bs are all these types have inside, we should be able to tackle anything concerning Either, Pair and the like using just bimap. Right? Not so fast. Consider the swap function.

```
swap :: Pair a b -> Pair b a
swap = \(Pair a b) -> Pair b a
```

Like the identity function, swap is one of those functions for which there's only one possible implementation. It's *parametricity*, again, telling us what to do. Anyway, that's beside the point. We are trying to focus on something else here. We are trying to address how to implement swap in terms of bimap. It should be possible, if bimap is as powerful as we made it out to be.

```
swap :: Pair a b -> Pair b a
swap = \x -> bimap ? ? x
```

Well, well, well... it seems bimap is not as powerful after all. Let's recall the type of bimap.

```
bimap :: Bifunctor f => (a -> c) -> (b -> d) -> f a b -> f c d
```

In our case, the chosen f, our Bifunctor, is Pair. Let's specialize that.

```
bimap :: (a -> c) -> (b -> d) -> Pair a b -> Pair c d
```

Moreover, the expected output type is not Pair c d, but rather Pair b a. We can replace c with b and d with a throughout to convey this.

```
bimap :: (a -> b) -> (b -> a) -> Pair a b -> Pair b a
```

Alright. So, all we need to do is have the first function, the one that modifies the a value as it appears in the input Pair a b, return the b value that appears in that same input. And vice-versa in the case of the second function. Alas, we can't.

While in Pair a b we can say that a and b exist in some kind of context where they are each other's neighbour, bimap only allows us to address the values individually, forgetful of the neighbourhoood where they exist. We can't ask a who its neighbour is, we would need to be observing both a and b *at the same time* —that is, in the same function, as swap does— in order to answer that. The fact that we have a -> b and b -> a separately makes this impossible. So, no, we can't use bimap to implement swap, for any implementation of swap needs to know about the value next door.

Is this bad? No, not necessarily. Understanding and relying on the limitations of our abstractions is fundamental to our type of programming. Imagine a Pair MyMoney YourMoney. If we only ever have Bifunctor as a means to operate on this Pair, then we can be certain that our monies will never mix, and that is good. In time, we will discover more neighbourly abstractions.

Sixty-two

So what else is of kind Type -> Type -> Type? Why, functions of course. And by functions we mean the (->) type constructor, the one we shoehorned into the Functor typeclass by partially applying it to just one Type, rather than two. Here, however, the Bifunctor typeclass asks us for something of kind Type -> Type -> Type. So, proud of itself, feeling welcome, the function goes...

```
instance Bifunctor (->) where ...
```

... and then it dies. Just ponder upon the false beauty, the abyss, the type bimap gets for a function.

```
bimap :: (a -> c) -> (b -> d) -> (a -> b) -> (c -> d)
```

It says that given our function a -> b, the one we are attempting to modify, and two other functions a -> c and b -> d, it can transform it to a new function c -> d. We saw in fmap, just a different name for compose, how two functions b -> d and a -> b can come together to form a function a -> d, where that d is the result of applying b -> d to the b we obtain from applying a -> b to the a we will eventually receive as input. And if we put fmap and bimap side by side, we will see that things are not *that* different.

```
fmap  ::              (b -> d) -> (a -> b) -> (a -> d)
bimap :: (a -> c) -> (b -> d) -> (a -> b) -> (c -> d)
```

The mystery, the conundrum, must thus be in c, the one type that fmap doesn't mention at all.

If we look at what bimap returns, we see a function that receives a c as input, and somehow returns a d as output. Well, let's try and find the ways we could obtain a d. There is a b -> d function, which implies that if we somehow manage to get our hands on a b, then a d will easily follow. Alright, let's get a b then. There is a function a -> b that returns a b if we just give it an a. Well then, let's find a a. Oh, the disappointment. There's no a, and there is no way to obtain one either. All we have is a c, a bunch of functions we can't use, and a -> c laughing at us. How did we end up here? We need c -> a, not a -> c. Who flipped that arrow? We have the c already, it's the a the one we need, not the other way around. Just flip it. Flip it. Flip it! Ahh!

This is the place where many programmers have lost their minds. And understandably so, for this is a *fundamental* problem we are facing. But we are going to get through this, don't worry, and things will be much, much brighter on the other side. It will be alright.

Sixty-three

Let's have a quick interlude before we continue. We have been hiding something quite important about the types we write, and it's time we made this explicit.

Where do names come from? We learned that in the case of lambda expressions like \x -> ..., it is in our choice of the word x, which we write in between \ and ->, where we decide that x will be the name we will use within this lambda expression to refer to the input of this function. But function inputs are not the only things that we can name. Let's revisit our beloved identity function once again.

```
id :: a -> a
id = \x -> x
```

In this id example, we opted to bind the name x to refer to the input *expression*, and we opted to give the name a to the *type variable* describing the type of such input. And output, yes, yes. We know where the name x comes from, but we don't know where a comes from. We seem to be just using it out of the blue. And what if, next to this definition of id, we were to define a new expression none?

```
none :: Maybe a
none = Nothing
```

This is a perfectly valid expression. We know that Nothing is an acceptable constructor for a type Maybe a *for all* choices of a we may want to make. none :: Maybe Natural or none :: Maybe String, they are alright. But what about that a? Is it the same a as the one we wrote before in the type of id? And if we use id on a Natural number, does it mean that none must now always take the type Maybe Natural, since we seem to have specialized that a somehow? Not quite.

Similarly to what we saw before when we talked about name shadowing, where we said that each time we *bind* a new name to an input expression this name shadows an equal one that existed before, preventing them from being mistaken for each other within the *scope* of the function that binds the name, type variable names also have a similar mechanism for allocating names called *quantification*.

So far, the only type of quantification we've dealt with, and the only one we care about for the time being, is the *universal quantification* of type variables. And this type of quantification, being the most common, is also the default one that Haskell infers for us unless we explicitly ask for something else. That is, when we write a type like id's:

```
id :: a -> a
```

What's really happening, is that we are asking for this:

```
id :: forall (a :: Type). a -> a
```

The forall (a :: Type). part is *universally quantifying* the newly introduced name a to be something of kind Type. Meaning that within the *scope* where this quantification applies, which in this case is everything to the right of that dot ., any type among *all* the types that exist can take the place of this a. And, not knowing which type that will be, we can continue to use the name a to consistently refer to it. This is how type variable names come to existence. They are *quantified* somehow.

How does this solve the issue of the a in the type of id being different from the a in the type of none? Well, let's see the properly quantified type of none again.

```
none :: forall (a :: Type). Maybe a
```

Here, none is also universally quantifying a, meaning that within the scope of this quantification, a can be *any* Type at all, which implies that there is no need for this a to be equal to any other a that came before or after its time, much like what happens in lambda expressions and shadowing, where the new name we pick for the input expression doesn't relate nor conflict with other names, equal or not, at all.

Does this sound redundant and boring? That's perfect, it should. After all, we have been dealing with universal quantification ever since we saw our very first type, so there's nothing conceptually new for us here. Why does this matter, then? Well, because we humans sometimes forget why we are here at all, and explicitly writing our foralls will help us remember what we are dealing with. We will see this shortly.

Of course, we can also universally quantify more than one type variable at once, and their kinds can be something other than Type as well. Look, for example, at the explicitly quantified type for fmap.

```
fmap :: forall (f :: Type -> Type) (a :: Type) (b :: Type)
       . Functor f
      => (a -> b)
      -> f a
      -> f b
```

Other than some new syntax, there should be nothing surprising here. We learned a while ago that the f in Functor f needed to have kind Type -> Type, and here we are just making that explicit.

It's also possible to leave out the explicit kind information for a quantified name whenever it can be inferred from its usage. For example, we could have written the type of fmap without explicitly mentioning the kinds of a, b and f, and it would have been the same. Haskell infers that a and b have kind Type, and that f has kind Type -> Type.

```
fmap :: forall f a b
       . Functor f
      => (a -> b)
      -> f a
      -> f b
```

And, by the way, one last charming detail. Rather than writing forall, we can write ∀. It's the same, but looks arguably nicer. Unsurprisingly, we read ∀ out loud as "for all".

```
id :: ∀ x. x -> x
```

We'll make use of ∀ throughout this book, for typography is our treat and book real estate is at a premium.

Sixty-four

At the beginning of this book, for didactic purposes, we intentionally said something wrong. We said something about a function of type input -> output which eventually became id. Let's try and write a function with that type.

```
nope :: input -> output
```

Actually, let's add some explicit quantification. It will make things more obvious.

```
nope :: ∀ (input :: Type) (output :: Type)
       . input -> output
```

Before diving into the implementation of this function —that is, into the expression that defines it— let's read its type out loud. It says here that this function takes a Type as input, *any* Type among *all* the Types that exist, and returns yet another Type as output, again, *any* Type among *all* the Types that exist. We call these types input and output respectively. Alright, let's try.

```
nope :: ∀ (input :: Type) (output :: Type)
     . input -> output
nope = \x -> ¿
```

Taking in that input and binding it to a name like x so that we can refer to it afterwards is not a problem. It says in the type that input will be a Type, *any* Type among *all* the Types that exist, and we know that lambda expression name binding is able to cope with this. Any input we receive, notwithstanding its type, will be bound to x. There is nothing surprising there. The problem lies at the opposite end, to the right of the arrow symbol ->, where we are supposed to come up with an expression of type output which, as its universal quantification dictates, must be *any* Type. So do we just pick one, then? Perhaps Natural? I mean, if we are saying *any* Type is fine, we should be able to randomly pick one, right? No. Not at all. See what happens if we do that.

```
bad :: ∀ (input :: Type). input -> Natural
bad = \x -> 3
```

Sure, this function compiles and runs. It is a perfectly valid and boring function that ignores its input and returns a Natural number. But where is output? It doesn't show up in the function type anymore. The thing is, we said that output represented the idea of *any* Type, but 3 is not a value of *any* Type, rather, it is of *one* Type in particular, Natural, and by saying that a Natural is what we will return, we are implicitly saying that every other Type is what we won't. Thus, output disappears and a very specific Type, Natural, takes its place.

What our function would need to return in order to satisfy that fantasy input -> output type, literally, is an expression of *any* Type the caller of this function desires. Think about how silly that is for a second. Let's say the caller wants the name of a flower. Well, our function should be able to come up with one. Let's say our caller has a then-incurable disease and applies our function expecting a cure in return. Guess what? We must come up with that cure, too. But what about using x? After all, we already bound that name to an expression of type input which we agreed represents *any* among *all* Types, so why don't we just reuse x?

```
id :: ∀ (input :: Type). input -> input
id = \x -> x
```

Ciao bella identità, ci siamo rincontrati. No, id is *not* what we were looking for. The identity function is something else. It takes *any* Type as input, sure, but the Type that it takes, is also the one it returns. That is, whereas input and output had been quantified *separately* before, meaning the caller could have chosen an input different than the output if desired, here these two types have been equated, *unified* with each other. Whatever input is, whatever output was, they are now one and the same Type. Or the other way around. Remember, we have been here before, and names didn't matter back then either.

```
id :: ∀ (output :: Type). output -> output
id = \x -> x
```

Moreover, while moot at this point, because of the parametric polymorphism in id, we also know that the output *value* will be the same as the input value, which takes us even further away from our goal.

What we describe here with shame, a tragedy, we described before with pride at the beginning of this book. Why? Because like *l'identità*, who embraced her ignorance and by doing so she could reason all the way to meaning, we students must profit from our rapidly vanishing ignorance, too, while it lasts. Did the nuances of quantification matter sixty chapters ago? No, they did not. What mattered was the didactic vehicle we found, the segue into discovering something new. We'll have time to be old, know everything, and hinder our own learning process by not daring to be playful anymore. So learn, never stop learning. But dare too, and never stop playing, for it's often in the exploration, in dismantling the foundations and prejudices we embody, that we find new truths worth writing.

So no, we can't reasonably come up with an implementation for input -> output. *Unreasonably* however, because learning to tear down walls is as important as learning how to build them, we can, and we will do so now.

Sixty-five

The biggest mistake we humans make is to believe we are right when we are, in fact, wrong. Not because of how this wrong belief might affect *us*, but because of the damage it makes us inflict on others, willingly or not, and the havoc that will ensue from the wrong choices made by the people that follow suit, also vehemently believing the untrue.

So what can we programmers do to prevent this? How can we be certain we are not wrong? Well, we start from the assumption that we actually *are* wrong, and then we make sure we acknowledge any shortcomings in our model or reasoning process, lest we disregard or forget them later on. We achieve this by challenging ourselves, by actively trying to subvert our own software, our own understanding, our own beliefs. Only then, fully aware of any limitations, and *if* confident that our software stands up to scrutiny, we proceed to make the statement that is due.

To get comfortable with this subversive type of programming, we'll now uncover a fundamental shortcoming in Haskell's type system. Let's implement the impossible input -> output.

```
nope :: input -> output
nope = \x -> ?
```

Somehow, using only what we see here, we need to come up with a value of type output that nope can return. But how? It seems that x, of type input, is all we have to work with. Is it, though? Let's take a step back so that we don't miss the bigger picture. No, it is not. We also have *Haskell* to work with. And Haskell, like every other general purpose programming language out there, supports the idea of *general recursion*, meaning that nope itself can talk about nope. And what is nope if not a function that given a value of type input, just like our x, returns a value of type output like the one we so desperately need?

```
nope :: input -> output
nope = \x -> nope x
```

That's it. We did it. We wrote a well-typed expression that our stringent type-checker will happily accept and compile. An expression that, suddenly, everybody can use. An expression, a function, that given any input will return *any* output we ask of it. A function that promises to name every flower, find every cure, and do anything else we desire. With it, apparently, we have solved every question in the universe.

But of course, none of this is real. If we dare look at the abyss for a second, we'll see that nope 3, say, beta-reduces to nope 3, which itself reduces to nope 3, which again reduces to itself and forever continues to do so. That is all there is to nope. And while to the distracted reader this may sound a bit like what id is doing, it most definitely is not. When we say id 3, that application reduces to 3 and there's nothing more to reduce. But trying to reduce nope 3, on the other hand, brings us back to nope 3. Intuitively, nope never reduces to a smaller expression. Or, talking in terms of inductive recursion, it never gets closer to a *base case*. We saw something like this many chapters ago, when we discussed a broken fromNatural which got farther and farther away from its base case as time went by.

Nope, there are no useful answers ever coming out of nope. Like a dog trying to catch its tail, like going down Escher's stairs, all we get from nope are a promise, an ever lasting futile attempt, and an innocent caller forever waiting for this function to return. Theoretically, nope would use up all time and energy in the universe and still fail to deliver its promise. Technically speaking, we say that nope *diverges*. And that vertigo? That sense of infinity? That's what a paradox feels like. What happens in practice, however, if we ever use nope, is that our program *hangs*. It gets stuck for as long as the computer, unaware, continues to execute our program. Forcing the termination of our program, perhaps by rebooting our computer, mitigates this. *Have you tried turning it off and on again?* It really is fascinating how we can model and tame the infinite within a box, on a small piece of paper.

But it gets even more interesting, for even in this mud we can still use our tools to improve the implementation of nope. The first thing we can do is notice that nope is a function written in eta-expanded form \x -> nope x, which we could simplify by eta-reducing to just nope.

```
nope :: input -> output
nope = nope
```

The tautology is now even more obvious: nope is nope. Sure, quite helpful, thank you very much. However, something changed. Other than its type, nothing in the definition of nope suggests it is a function anymore. Has input -> output now become an overly specific type? If we say nope is a function, then nope = nope is simply stating that "a function is a function". But wouldn't this tautology also apply to types other than functions? Can't we say a Natural number is a Natural number? Can't we say a Maybe String is a Maybe String? Of course we can. We can say that *any* value of a particular Type is indeed a value of that Type, meaning that the type of nope could in principle be more general.

```
nope :: ∀ (x :: Type). x
nope = nope
```

Here, we are saying that nope is a value of *every* type out there. Do you want to add two Natural numbers? Try nope + nope. Do you need something to put in your Just? Maybe Just nope is your answer. And can we compose some nopes? Sure we can, try compose nope nope. But of course, none of these expressions accomplishes anything, even while they all type-check. So, what does this mean?

Eventually we will learn that there is a direct correspondence between types and logic, but for now, the gist of it is this: The types that we write down in Haskell correspond to *theorems* in logic —that is, statements we expect to be true— and the *expressions* that we write for those types are the *proofs* that the theorems are true. We briefly touched this topic before, when we talked about isomorphism and said that we could *prove* that two types a and b were isomorphic as long as we were able to implement two functions a -> b and b -> a with a particular relationship to each other. These two function types together are stating a theorem, they are stating that it's possible to convert between a and b and back as necessary, which, when combined with the further requirement that composing these two functions should result in the identity function —although that's beside the point— effectively *prove* that a and b are isomorphic. But let's forget about the intricacies of isomorphisms and look at more straightforward examples. If we manage to implement a function of type x -> y, then we are *proving* that we can obtain a y from an x. If we manage to define a value of some type z, then we are proving that at least one such value exists. And do we remember how we agreed that implementing a Bifunctor instance for functions was just *impossible*? How, in other words, we couldn't *prove* that functions were Bifunctors? Well, look again:

```
instance Bifunctor (->) where
  bimap = nope
```

This, I am sorry, type-checks. Of course it is nonsense, but still, it type-checks. What we are seeing here is that we can use nope to prove and implement *anything* we want, as absurd as the type, the premise, the theorem might be, and as useless and non-terminating, *diverging*, the implementation, the proof, will be.

In literature, nope is usually called *bottom* or ⊥, for reasons that will become apparent later on. In Haskell we call it undefined, which may or may not be a reasonable name depending on our expectations. That is, this monster comes with every Haskell distribution:

```
undefined :: ∀ (x :: Type). x
```

In practice, Haskell's undefined doesn't loop forever. Instead, it makes our program die right away, leaving a trail of forensic memorabilia about where the homicide took place. But that's just an optimization. Conceptually, the proverbial infinite loop is still somewhere in there.

Are we saying Haskell is fundamentally broken? Well, yes, but we need to put things in perspective and understand that what we are seeing here is merely an incarnation of the more general halting problem we described before, inherent to *every* general purpose programming language out there capable of computing *anything* that is computable. The existence of bottom, if you will, is a proof that given a program, *any* program, we can't reliably guarantee that it will ever terminate, that it will ever finish computing at some point in the future.

So we buck up, we *acknowledge* this fundamental pitfall in our model, in our reasoning framework, and move on. We accept that we can't readily trust programs written by others, obviously, and we *never* bottom our own. Particularly, we never use undefined, and we are careful not to loop ourselves into oblivion as the broken fromNatural did. In time, we'll find we have some tools for this.

But this exercise is not so much about undefined, Haskell, nor about general purpose programming languages. It's about understanding our limitations and who we are. We programmers are not in the business of lying, but in that of transparency and dealing with facts. Knowing about our shortcomings means we can continue our journey *aware* of the limb that we lack, the allergies we have, and still live a full life.

Does this make Haskell unsuitable for some kinds of problems? Of course it does. For example, Haskell would be a terrible theorem prover, as we would be able to prove anything we wanted with it, true or false. There are languages that sacrifice their generality for a bottomless existence, and in exchange they become perfect for this task instead. Other languages sacrifice types altogether for a stronger sense of kinship with the computer, who despite seeing but numbers, can still accomplish great things. Our toy lambda calculus was an example of this, and we'll see others as well. But all in all, Haskell is an excellent place to be.

So having subverted our system, shaken but wiser, on our merry way we go.

Sixty-six

Why is it that in input -> output, both types universally quantified, we struggle so much with output yet dealing with input is so straightforward? What is so fundamentally different between input and output? Let's explore that from a different angle.

We've grown accustomed to talking about how in a type like input -> output we have an input parameter or *argument* of type input and an *output* value of type output. And that is fine. But it turns out that, actually, *both* input and output can be seen as arguments of this function, arguments with different *positions*. This happens to be a richer way of describing the inputs and outputs of a function than saying, well, *inputs* and *outputs*. So let's learn.

We will describe nope :: input -> output by saying that, *from the perspective of* nope, input is an argument in *negative* position and output is an argument in *positive* position. This means that, if there is a need or possibility of doing so, nope would be responsible for *receiving* this input and for *producing* this output. This shouldn't sound surprising, for this is exactly what we've come to expect nope to do. An input goes in, an output goes out. And with these responsibilities in mind, it's quite easy to reason about why dealing with output in input -> output is so hard: It is nope's responsibility to produce an output, an impossible task, whereas coming up with an input is somebody else's responsibility, so we just don't worry about it.

But we can't really appreciate the utility of this nomenclature by just looking at nope, so let's look at compose instead.

```
compose :: (b -> c) -> (a -> b) -> (a -> c)
```

Which are compose's inputs? A quick glance suggests b -> c and a -> b, but let's not forget about the a that appears as an input of its own if we remove that redundant pair of parentheses to the right.

```
compose :: (b -> c) -> (a -> b) -> a -> c
```

But even then, b -> c and a -> b are themselves functions, so they clearly have inputs and outputs of their own as well. Should we add some of them to the list of inputs to compose? Should we talk about *the input of the input*? Not exactly, that would get confusing quite rapidly, and moreover, it would be wrong. We need a better vocabulary, so let's talk about arguments and their positions instead.

When we talk about argument positions, we always do so from the perspective of one particular type. In this case, we are doing it from the point of view of compose, so we will judge positions from the perspective of its type (b -> c) -> (a -> b) -> a -> c. *Perspective* is the most important thing to be aware of when reasoning about argument positions, so let's make sure we always keep track of it.

We said that the arguments in negative position are those which we *receive*. We already identified three of them: b -> c, a -> b and that standalone a. These are the types of three values to which we can refer by simply binding them to new names using a lambda expression.

```
compose :: (b -> c) -> (a -> b) -> a -> c
compose = \g f a -> g (f a)
```

We bound the first argument of type b -> c to the name g, the second argument, of type a -> b, to f, and the third argument, of type a, to the name a. The fact that we were able to somehow get a hold of these arguments, *receive* them, proves that these arguments were indeed in negative position as we were told. And what about the rightmost c in our type? Well, we know that we are *producing* a value of this type c as output where we say g (f a), so this c must necessarily be an argument in *positive* position. Let's add some annotations to keep track of what's positive + and what's negative -.

```
compose :: (b -> c) -> (a -> b) -> a -> c
            --------    --------    -   +
```

Is this all? Of course not. So far all we've done is say that b -> c, a -> b and a are negative arguments whereas c is a positive argument. We might as well have said they are *inputs* and *outputs*, respectively, and it would have been sufficient. So what else is there? Let's take it step by step. First, let's put those redundant rightmost parentheses back in the type of compose.

```
compose :: (b -> c) -> (a -> b) -> (a -> c)
```

Technically, this is *exactly* the same as before. Conceptually, however, rather than describing compose as the function that takes three values as input, with types b -> c, a -> b and a, and returns a value of type c as output, we now describe compose in a rather beautiful manner as the function that given two functions b -> c and a -> b, returns yet another function a -> c. And why does this matter? Because, conceptually, we now have two inputs rather than three, and because all of a -> c, the output we are producing, is in positive position now.

```
compose :: (b -> c) -> (a -> b) -> (a -> c)
           --------    --------    ++++++++
```

Is this correct? Are we able to produce an a -> c? Of course we are, we already did in \g f a -> g (f a). It might be more obvious if we write it a bit differently, though.

```
compose :: (b -> c) -> (a -> b) -> (a -> c)
compose = \g f -> (\a -> g (f a))
```

That rightmost \a -> g (f a) has the type a -> c.

So which statement about the position of our arguments is correct? Is a -> c positive, or is a negative and c positive? It turns out that both of these are true, and here we can finally start seeing the utility of characterizing arguments by their positions. When observed separately, those a and c have different positions, but when observed together as part of the same function a -> c, said function *itself* has a position of its own as well. Let's try to visualize all this information.

```
compose :: (b -> c) -> (a -> b) -> (a -> c)
           --------    --------    ++++++++
                                    -    +
```

There, it says that while a -> c is in positive position, the a and the c in it are in negative and positive positions of their own as well.

Let's look at a -> b now, an argument in negative position. What are the positions of a and b inside this function? Well, seeing how a is an input in that function, and how b is an output, presumably a is in negative position and b is in positive position, right? Not quite, not quite. We are forgetting our point of view, the *perspective* from where we are standing. Let's dig deeper.

From the point of view of compose, the entire function a -> b is in negative position because it's somehow provided by the caller as input, and all compose has to do is *receive* it. We see this clearly in the implementation, where we bind a -> b to the name f.

```
compose :: (b -> c) -> (a -> b) -> a -> c
compose = \g f a -> g (f a)
```

The rest of compose's type is saying that eventually we need to produce a c, and looking at what is available we can see that if we were able to obtain a b somehow, then we could apply g to that b and be done. How do we obtain that b? Why, we just apply f to the a that we have readily available.

But look carefully at what's happening. From the point of view of f, a is the input it receives, so it must be in negative position. But we don't care about f's point of view at all, so we ignore what f has to say and continue searching. The fact that f eventually receives an a at all implies that there is somebody else applying f to that a, which in turn implies that whoever calls f needs to *produce* an a for f to receive. And who calls f if not compose? So, from the point of view of compose, when it says f a, that a is in *positive* position. And this is despite that a *also* being in negative position from the point of view of compose when we bind it in \g f a -> .... So, contrary to our intuition, the a in a -> b, from the perspective of compose, is actually in *positive* position, for it's compose itself who is somehow coming up with an a to which it can apply f, even though that a also happens to be the one that was originally received by compose. Generally speaking, the fact that an a appears in negative position somewhere else in compose's type is irrelevant when determining that *this* occurrence of a in a -> b is indeed in positive position.

```
compose :: (b -> c) -> (a -> b) -> (a -> c)
           --------    --------    ++++++++
              +           -           +
```

And what about parametric polymorphism? Weren't *all* as appearing in a same type supposed to be the same? Why is one a positive while the other one is negative here? Apples and oranges. Parametric polymorphism talks about *types*, and the type a will indeed be the same throughout. But the position of arguments has nothing to do with types, it only has to do with who provides to whom.

So what happens to compose after it applies f a? It *receives*, in return, a b. And again, contrary to our intuition, from the point of view of compose, the output b of the function a -> b is in *negative* position because compose is *receiving* the b that has been provided, returned, by f.

```
compose :: (b -> c) -> (a -> b) -> (a -> c)
           --------    --------    ++++++++
              +           -    -       +
```

And once compose receives this b, it can *provide* it to g, of type b -> c, to finally receive a c in return. This proves that in b -> c, from the point of view of compose, b is an argument in positive position and c is one in negative position.

```
compose :: (b -> c) -> (a -> b) -> (a -> c)
           --------    --------    ++++++++
              +    -      +    -       -    +
```

This is certainly a richer vocabulary for talking about the inputs and the outputs of our functions. We can say that all the expressions in negative position are potentially inputs of our function, and likewise, that all those in positive position are its outputs.

And for completeness, let's see what happens if we keep adding those redundant "conceptual" parentheses to the right.

```
compose :: (b -> c) -> ((a -> b) -> (a -> c))
           --------    +++++++++++++++++++++++++
           --------    --------    +++++++++
              +    -       +    -      -    +
```

Nothing surprising there. We see the same result as we did when we put some parentheses around a -> c before. We are now talking about compose, the function that given a function of type b -> c as input, returns yet another function of type (a -> b) -> (a -> c) as output. So, b -> c is in negative position, and all of (a -> b) -> (a -> c) is in positive position. And what about going all the way, and putting some redundant parentheses around *everything*?

```
compose :: ((b -> c) -> ((a -> b) -> (a -> c)))
           +++++++++++++++++++++++++++++++++++++
           --------    +++++++++++++++++++++++
           --------    --------    +++++++++
              +    -       +    -      -    +
```

Ah, the most absurd example compose, out of thin air, receiving no input, *produces* a value of type (b -> c) -> ((a -> b) -> (a -> c)). A value that, thus, must be an argument in positive position. There is a pattern here, if you care to find it.

Sixty-seven

Understanding positions, developing an intuition for them, is quite necessary if we expect to excel at what we do. Not so much because of the matter of positions per se, but because of the closely related topic of *variance*, omnipresent in our field.

Let's continue looking at compose and its positions, but at the same time, let's not do it. Let's stop calling it compose, let's call it fmap instead.

```
fmap :: Functor f => (a -> b) -> f a -> f b
```

We will add some redundant parentheses around f a -> f b to emphasize what concerns us today. And while we are at it, let's leave out the Functor constraint, as we won't be paying attention to it for the time being.

```
fmap :: (a -> b) -> (f a -> f b)
```

In this light, we can describe fmap as taking a function a -> b and returning a function f a -> f b. A type that, when carefully arranged, reveals an interesting pattern.

```
fmap :: (  a ->   b)
     -> (f a -> f b)
```

Mathematics is the business of discovering beauty and structure in change. Thus, finding this type so charming, so organized, shouldn't surprise us at all.

What `fmap` is doing, in other words, is just putting an `f` around the a and the b. Well, *putting* is just a manner of speaking, for we know `f` a is an *input* to `fmap`, so it's not `fmap` but whoever applies it who puts this `f` around a. But still, conceptually, we are transforming a function from a to b, into a function from `f` a to `f` b. So much so, actually, that we often talk about `fmap` "*lifting* a function a -> b into `f`", rather than "`fmap`ping a function a -> b over `f` a".

But let's forget about `fmap` for a bit and compare these two functions a -> b and `f` a -> `f` b on their own merit. They both have an a somehow appearing as part of their input and a b somehow appearing as part of their output. In other words, directly or not, the as are part of arguments in negative positions and the bs are part of arguments in positive ones. That is, despite the extra `f`, the positions of the arguments in a -> b is preserved in `f` a -> `f` b. Lifting a -> b into `f` didn't change this. The name for this unsurprising phenomenon is *covariance*.

Sixty-eight

When we say `Functor f`, we are saying that `f` is a *covariant* functor, which essentially means that for each function a -> b, there is a corresponding function `f` a -> `f` b that is the result of *lifting* a -> b into `f`. The name "covariant", with the *co* prefix this time meaning *with* or *jointly*, evokes the idea that as the a in `f` a varies through a -> b, the entirety of `f` a varies to `f` b with it, in the same direction.

But as with many precious things in life, it's hard to appreciate *why* this is important, or at all interesting, until we've lost it. So let's lose it. Let's look at this matter from the other side, from its *dual*, a dual we find by just flipping arrows. So let's try that and see what happens.

```
contramap :: (  a ->    b)
          -> (g a <- g b)
```

A covariant functor `f`, we said, was one that could lift a function a -> b into a function `f` a -> `f` b. So, presumably, the *dual* of said covariant functor `f` —let's call it "g the *contravariant*"— is one that lifts a function a -> b into g a <- g b instead, arrow flipped. Conceptually, this is perfect. In practice, in Haskell, function arrows always go from left to right ->, never from right to left <-, so we need to take care of that detail. Let's write the function arrow in the expected direction, flipping the position of the arguments instead.

```
contramap :: (  a ->    b)
          -> (g b -> g a)
```

In Haskell, contramap exists as the sole method of the typeclass called
Contravariant, which is like Functor but makes our minds bend.

```
class Contravariant (g :: Type -> Type) where
  contramap :: (a -> b) -> (g b -> g a)
```

Bend how? Why, pick a g and see. Maybe perhaps? Sure, that's simple
enough and worked for us before as a Functor.

```
contramap :: (a -> b) -> (Maybe b -> Maybe a)
```

Here, contramap is saying that given a way to convert as into bs, we will be
granted a tool for converting Maybe bs into Maybe as. That is, a lifted
function with the positions of its arguments flipped, exactly what we
wanted. Let's try to implement this by just following the types, as we've
done many times before. Let's write the Contravariant instance for Maybe.

```
instance Contravariant Maybe where
  contramap = \f yb -> case yb of
                         Nothing -> Nothing
                         Just b -> ?
```

What a pickle. We know that f is of type a -> b, we know that yb is a Maybe
b, and we know that we must return a value of type Maybe a somehow.
Furthermore, we know that a and b could be anything, for they've been
universally quantified. There is an implicit ∀ in there, always remember
that. When yb is Nothing, we just return a new Nothing of the expected type
Maybe a. And when yb is Just b? Why, we *die* of course, for we need a way
to turn that b into an a that we can put in a new Just, and we have none.

But couldn't we just return Nothing? It is a perfectly valid expression of
type Maybe a, isn't it? Well, kind of. It type-checks, sure, but a vestigial
tingling sensation tells us it's *wrong*. Or, well, maybe it doesn't, but at least
we have laws that should help us judge. So let's use these laws to
understand why this behaviour would be described as *evil*, lest we hurt
ourselves later on. Here is the simplified version of the broken instance we
want to check, the one that *always* returns Nothing.

```
instance Contravariant Maybe where
  contramap = \_ _ -> Nothing
```

Like the *identity law* for fmap, the identity law for contramap says that
contramapping id over some value x should result in that same x.

```
contramap id x == x
```

A broken contramap for Maybe that *always* returns Nothing would blatantly
violate this law. Applying contramap id (Just 5), for example, would result
in Nothing rather than Just 5. This should be enough proof that ours

would be a broken `Contravariant` instance, but for completeness, let's take a look at the second `contramap` law as well.

Just like we have a *composition law* for `fmap`, we have one for `contramap` as well. It says that contramapping the composition of two functions f and g over some value x should achieve the same as applying, to that x, the composition of the contramapping of g with the contramapping of f.

```
contramap (compose f g) == compose (contramap g) (contramap f)
```

Isn't this the same as the composition law for `fmap`? Nice try, but take a closer look.

```
fmap (compose f g) == compose (fmap f) (fmap g)
```

```
contramap (compose f g) == compose (contramap g) (contramap f)
```

Whereas in the case of `fmap`, the order in which f and g appear at opposite sides of this equality is the same, it is *not* the same in the case of `contramap`. This shouldn't come as a big surprise, seeing how we already knew that `contramap` gives us a lifted function with its arguments in the *opposite* position. Intuitively, dealing with `Contravariant` values means we are going to be composing things backwards. Or, should we say, *forwards*? After all, `compose` initially shocked us with its daring, so-called *backwards* sense of direction, and now we are mostly just backpedaling on it.

But the important question is whether our broken `Contravariant` instance for `Maybe` violates this law. And the answer is, unsurprisingly I hope, *not at all.* Both sides of this equality *always* result in `Nothing`, so they are indeed equal. Many times they will be equally *wrong*, but that doesn't make them any less equal. And this is, technically, sufficient for our broken instance to satisfy this law. Thankfully we had that other law we could break.

So how do we `contramap` our `Maybe`s? Well, we don't. As handy as it would be to have a magical way of going from b to a when all we know is going from a to b, we just can't do that. Think how absurd it would be. If your imagination is lacking, just picture a being a tree and b being fire. So, to sum up, `Maybe`s are covariant functors, as witnessed by the existence of their `Functor` instance, but they are not contravariant functors at all, as witnessed by the impossibility of coming up with a lawful `Contravariant` instance for them. So once again, like in our previous `Bifunctor` conundrum, we find ourselves wanting for a function b -> a when all we have is a function a -> b. This time, however, we are prepared. We know that we keep ending up here because we are getting the position of our arguments, the variance of our functors, wrong. And we will fix that.

Wait. Covariant *functors*? Contravariant *functors*? That's right, *both* these things are functors. All that talk we had growing up about how functors are this or that? *A lie.* Well, not a lie, but rather, we hid the fact we were talking *only* about covariant functors. Now we know that there are other types of functors too. Unfortunately, the Haskell nomenclature doesn't help here. For example, one could argue that the `Functor` typeclass should have been called `Covariant` instead, or perhaps the `Contravariant` typeclass should be called `Cofunctor` and `contramap` should be renamed to `cofmap`, etc. Anyway, not important. As we continue learning more about functors we'll see that even these names fall short of the true nature of functors. There's more, yes, and it's beautiful. But despite their names, both `Contravariant` and `Functor` are typeclasses that describe, indeed, *functors*. And it shouldn't surprise us, considering how we saw in detail that except for their opposite argument positions, the types of `fmap` and `contramap`, as well as their laws, are exactly the same.

Sixty-nine

It is high time we learned a new function, and in the most exquisite tradition of this fine establishment, it will be a boring but beautiful one.

```
const :: a -> b -> a
```

But this time we'll leave it to you and parametricity to figure out what const does. Don't worry, you got this. You've been preparing for this moment all along.

Seventy

Let's make something very clear once and for all. We said it before, but here we say it again. Functors —that is, covariant `Functors`— are *not* containers, they are *not* little boxes.

Functors are type constructors that are *covariant* in at least one of its type parameters and abide by some laws. That's all. `Maybe` is a covariant `Functor` because in `Maybe a`, that `a` shows in positive position in `Maybe`'s `Just`, which in turn causes a function `a -> b` to always be lifted *covariantly* into `Maybe a -> Maybe b`.

```
data Maybe a = Nothing | Just a
```

Can't see how a is positive? Let's look at an example, then.

```
foo :: Maybe Natural
foo = Just 5
```

This 5, our a, is being *produced* here out of the blue. This proves that a is indeed an argument in positive position from the perspective of foo. Sure, from the perspective of the Just value constructor, the a it expects to *receive* is certainly in negative position, but a value constructor's perspective doesn't matter at all when judging the variance of a datatype. What matters, instead, is the perspective of the already constructed value.

In our case, it's the point of view of the `Maybe Natural` value itself, `foo`, not the `Just` constructor of type `Natural -> Maybe Natural`, what matters.

This truth might be hard to see because we've become accustomed to talking about the position of arguments of a *function*, yet here we see no function. *Or do we?* Think about what a function that takes *no input* and returns 5 as output would look like. Conceptually, no such thing could exist, for inputs are a function's *raison d'être*. No input, no function. So we can't fathom its looks, no. But still, without noticing, we just described 5 as the *output* of this hypothetical inputless function and had no trouble understanding what we meant, agreeing on what our goal was. So it's our lack of surprise, if you will, what suggests that 5 could in fact be in positive position.

But suggestions and perspectives are respectively insufficient and boring, so let's do something adequate and fun instead. Let's prove that an `a` on its own, just like the one we find inside the `Just` value constructor, is in positive position. We'll do so by showing how said `a` can be represented as `b -> a` instead, a proper function with an input where it's blatantly obvious that there is an `a` in positive position. And nevermind that `b`, it could be anything. So how do we show that a value of one type, here `a`, can be represented as a value of a different type, here `b -> a`? Well, generally we just prove that there is an isomorphism between these two representations. Unfortunately, while possible, doing so now would derail our learning process for unrelated reasons, so we'll just do half of the work instead. We'll just show that it is possible to go from `a` to `b -> a`, and we'll leave the conversion from `b -> a` to `a` for a later day. So, in concrete terms, we need to implement a function of type `a -> (b -> a)`. Let's do it.

```
hola :: a -> (b -> a)
```

So, `hola` is a function that given— Wait a minute... Can we drop those extra parentheses, please? Thanks.

```
const :: a -> b -> a
```

Hola indeed! `const`, we meet again. Let's write down `const`'s implementation this time, though, just to be certain we all share the same understanding.

```
const :: a -> b -> a
const = \a b -> a
```

I trust you arrived at the same place during your own exploration of `const`. But here it is, in case you didn't. It's possible you ended up with `\a _ -> a`, `\a -> \b -> a`, or similar instead. That's fine, it all means the same.

If you dare profit from your innocence and entertain the idea of an alternative implementation of `Maybe`, where rather than having a value of type `a` we have a function of type `b -> a` in it, there shall be no doubt whatsoever of the *covariance* of this datatype.

```haskell
data Maybe a = Nothing | Just (b -> a)
```

The experienced reader will notice that, for unrelated reasons, this Haskell code doesn't work. But experienced reader, *we are just playing*, so let's pretend it does.

And finally, here is foo again, accommodating the needs of our new representation.

```haskell
foo :: Maybe Natural
foo = Just (const 5)
```

And the b we just made up? Well, *who cares*? const certainly doesn't. It could be anything. As proof of this, let's implement a function that adds a 10 to the Natural number contained in one of these twisted new Maybes, if any.

```haskell
add_ten :: Maybe Natural -> Natural
add_ten = \yn -> case yn of
                   Just f -> 10 + f "potato"
                   Nothing -> 10
```

The f inside yn takes *anything* as input, ignores it, and returns a Natural number. The ignored thing is our made-up b, the one that showed up in our fake Just constructor. Here we decided to apply f to the String "potato", but we could have applied it to id, fmap or anything else instead, and it would have been fine. With this, the expression add_ten (Just 3), for example, equals 13, and add_ten Nothing equals 10. Of course, as the experienced programmer noticed, none of this works. That is, however, unimportant, for it has nothing to do with our current concern and our reasoning remains nonetheless untainted.

And this same reasoning applies to List, Either and any other Functor we may encounter. They are not little boxes, they are type constructors *covariant* in the rightmost of their type-parameters, that's all they are. The functor instance for functions makes it quite clear, as the covariance of b in a -> b can be seen from far, far away.

Seventy-one

So what about Bifunctor? *Two little boxes*, perhaps? No, definitely not. Let's recall the definition of the Bifunctor typeclass.

```haskell
class Bifunctor (f :: Type -> Type -> Type) where
    bimap :: (a -> c) -> (b -> d) -> f a b -> f c d
```

We have so many type variables here that, to the untrained eye, it's not obvious what's going on. So let's explore this together. Essentially, bimap is two fmaps in one. We already knew this, of course, considering how the whole reason for reaching out to Bifunctor in the first place was that we wanted to act on both the Left and the Right side of an Either at the same time. So yes, Bifunctor is two fmaps, two Functors in one. One that turns the a in f a b into c, and another one that turns the b in it into d. Together, they turn f a b into f c d.

```
bimap :: (a -> c) -> (b -> d) -> f a b -> f c d
 fmap :: (a -> c)              -> f a   -> f c
 fmap ::            (b -> d) -> f   b -> f   d
```

Of course, these examples don't type-check, for none of f a, f b, f c nor f d are Types. But, if we playfully pretend they are, we'll see the resemblance to fmap.

What this implies, among other things, is that the Bifunctor typeclass expects the two type parameters to f to be *covariant*. And yes, there are some laws that a successful Bifunctor needs to follow as well. An Either, for example, is covariant in both of its type parameters: Right has a payload exactly like Maybe's Just, thus covariant, and the only thing that's different about Left is its name. So yes, Either a b is covariant with a and b, which makes it a suitable Bifunctor candidate. In this regard, a Pair is no different, so a Pair is also a Bifunctor in principle. But a function? A function is most definitely *not* covariant in *both* its type parameters. Let's see why.

First, the easy part. We know that in a -> b, a change in b will be lifted *covariantly* into the function. That's what fmap tells us.

```
fmap :: (b -> c) -> (a -> b) -> (a -> c)
```

Seeing this in a prefix manner, and with some extra parentheses, might help.

```
fmap :: (        b ->        c)
     -> ((->) a b -> (->) a c)
```

That is, the unchanging (->) a part is our f here, our Functor. And we can see how, from the point of view of fmap, all the bs appear as arguments in negative position and all the cs appear in positive ones. The implementation of this function proves that functions are indeed covariant in their outputs. What about their inputs, though? Well, let's try to lift a function that changes the a in a -> b into c and see what happens.

```
foo :: (a -> c) -> (a -> b) -> (c -> b)
```

This type is similar to that of fmap, to that of compose, but not quite the same, so pay careful attention. We'll call it foo because this function doesn't fit in any of the typeclasses we have described so far. We'll figure out where to put this code later on. Remember: Here we are trying to modify the *input* of a -> b, not its output. So let's try to implement foo, and let's also drop those redundant rightmost parentheses while we are at it.

```
foo :: (a -> c) -> (a -> ɔ) -> c -> b
foo = \g f c -> ?
```

Ugh, trapped again. We have a function that knows how to convert an a
into the b that we need to return, and we have a c, but we have no way to
obtain an a to provide to a -> b. Once again we have a function a -> c
when what we actually need is c -> a. What is going on? Stop this torture
already, please.

Alright, let's arrange things differently to see more clearly, with prefix
notation, extra parentheses and such.

```
foo :: (      a    ->      c  )
    -> ((->) a b -> (->) c b)
```

This fool is trying to *covariantly* lift a function a -> c, intended to modify
the input of yet another function, and this can't happen. We know this
because if we squint hard enough and pretend that (->) _ b is our Functor
—notice the _ hole in there— then what we are looking at, really, is fmap.

```
fmap :: (  a ->    c)
    -> (f a -> f c)
```

But having failed, repeatedly, to implement this, we know this can't ever
happen. What if we flip the arrow, then? What if we try to get closer to
contramap rather than to fmap? All we have to do is change the position of
the lifted arguments. That is, rather than ending up with f a -> f c, we
should end up with f c -> f a.

```
contramap :: (  a ->    c)
         -> (f c -> f a)
```

And now that we have this, let's replace that f back with (->) _ b —again,
notice the hypothetical placeholder _ in there.

```
bar :: (      a    ->      c  )
    -> ((->) c b -> (->) a b)
```

We abandoned the name contramap because our type doesn't fit contramap's
shape anymore. But this is just a superficial issue related to the fantasy _
placeholder we just mentioned. It has to do with the fact that we are not
focusing on the y in x -> y, but rather on the x. We'll deal with this later,
it's not a fundamental issue at all. For now, just pretend bar is contramap.
Anyway, let's make things pretty again.

```
bar :: (a -> c) -> (c -> b) -> a -> b
```

We have an a, we have a way to go from a to c, and also a way to go from c
to b, which is what we ultimately wanted.

```
bar :: (a -> c) -> (c -> b) -> a -> b
bar = \f g a -> g (f a)
```

Or, another way of saying this, in a point-free manner:

```
bar :: (a -> c) -> (c -> b) -> (a -> b)
bar = \f g -> compose g f
```

That's right, all we had to do was flip the order of the input parameters to compose. Anyway, seeing how bar here is essentially contramap, this proves that functions are *contravariant* in their input parameter, the one we were focusing on.

So, no, functions are not Bifunctors because they don't have *two* covariant type parameters. They have one covariant parameter, its output, and another one that is *contravariant*, its input. Functions just don't fit bimap. Notice, however, that Bifunctor is also a terrible name, because just like how functors, semantically, can be either covariant or contravariant, the mathematical bifunctor can in principle have any of its type parameters be covariant or contravariant as well. Terrible naming indeed.

Alright, agreed, functions are not Bifunctors in Haskell's parlance. However, they are *bifunctors* nonetheless, in the true mathematical sense. Haskell calls a bifunctor that is able to accommodate a function's shape, where the first type parameter is contravariant and the second covariant, Profunctor. Haskell calls it Profunctor.

```
class Profunctor (p :: Type -> Type -> Type) where
  dimap :: (a -> c) -> (b -> d) -> p c b -> p a d
```

If we compare bimap and dimap side by side, we'll see they are essentially the same, except the as and cs appear in opposite positions.

```
bimap :: (a -> c) -> (b -> d) -> p a b -> p c d
dimap :: (a -> c) -> (b -> d) -> p c b -> p a d
```

The Profunctor instance for functions should be unsurprising. The trickiest part is seeing how our function type constructor (->) becomes that p. First we replace p with (->):

```
dimap :: (a -> c) -> (b -> d) -> (->) c b -> (->) a d
```

And then we move those prefix (->) type constructors to an infix position, adding some parentheses as necessary.

```
dimap :: (a -> c) -> (b -> d) -> (c -> b) -> (a -> d)
```

Beautiful. Although, let's remember that the rightmost parentheses are optional. Let's get rid of them, actually.

```
dimap :: (a -> c) -> (b -> d) -> (c -> b) -> a -> d
```

So now dimap can be described as the function that given an a, a way to obtain a c from that a, a way to turn that c into a b, and finally a way to convert that b into a d, returns said d.

```
instance Profunctor (->) where
    dimap = \f h g a -> h (g (f a))
```

Intuitively, dimap f h g a, where we've conveniently named f, g and h alphabetically, applies f, g and h to a, in that order. It's customary, however, to disregard that a and, in a higher-order point-free fashion, focus solely on the composition of f, g and h.

```
    dimap = \pre pos g -> compose pos (compose g pre)
```

This time we chose names pre and pos, rather than f and h, to highlight that the application of pre *precedes* the application g —our Profunctor, the function we are transforming— and that the application of pos is *posterior* to it. This is quite a common nomenclature. Of course, we could have composed pos and g first, and only then composed this with pre, as in compose (compose pos g) pre. It's all the same, for even if we haven't said so explicitly, function composition, just like adding or multiplying numbers, is *associative*.

So that's it. Functions are indeed bifunctors, yes, but the Profunctor kind, where their first type-parameter is contravariant and only the second is covariant.

Seventy-two

We know that a function a -> b is contravariant in a, yet we have neglected writing a Contravariant instance for it. Let's fix that.

It turns out that, in Haskell, saying that a -> b is contravariant in a is not quite straightforward for a very superficial, boring reason. That's why we haven't done it. We assumed in the previous chapter that our contravariant functor was (->) _ x, with that _ hole in it, but there's just no way to write that holey thing in Haskell, so we need some gymnastics to shoehorn (->) _ x into place. The main issue is the same we had with the Functor instance for Either, where we could only ever fmap over the Right because Functor expected a type constructor kinded Type -> Type as its f, and only Either x, with a fixed x on the Left, had that kind. In the case of functions, (->) x has that kind too, but the problem is that (->) x is fixing its contravariant parameter, the one contravariance ultimately cares about, to *always* be x, which ultimately defeats the whole purpose of contramap.

Ideally, what we would like to do is say that (<-) x, arrow flipped, is an instance of Contravariant where the x that has been fixed is the *output*, rather than the input of the function, leaving the contravariant parameter free for our mischiefs.

```
instance Contravariant ((<-) x) where
    contramap = \f g a -> g (f a)
```

In this case, contramap would take the type (a -> b) -> (x <- b) -> (x <- a) and all would be right in the world. However, we learned, there is no such thing as an arrow going from right to left <- in Haskell, so this won't work. Yet, we have something quite close.

```
data Op b a = Op (a -> b)
```

The Op datatype, standing for *opposite*, is essentially our flipped arrow <-. It's just a wrapper around a normal function with uglier looks. In both Op b a and a -> b we have a covariant b and a contravariant a, they just show up in opposite order as type-parameters. Other than there being a function inside Op, nothing here is new. We did the same thing a while ago with a datatype called Flip that flipped the sides of an Either.

```
data Flip b a = Flip (Either a b)
```

Look, Flip and Op are essentially the same. It's only their names and the type they wrap what changes.

```
data Op   b a = Op   ((->)   a b)
data Flip b a = Flip (Either a b)
```

Anyway, the important thing is that we can finally give functions a Contravariant instance through Op.

```
instance Contravariant (Op x) where
  contramap = \f (Op g) -> Op (\a -> g (f a))
```

The implementation is a bit noisy because once Op x is chosen as our Contravariant functor, the type of contramap becomes (a -> b) -> Op x b -> Op x a, which means we need to pattern match on the Op constructor to extract the b -> x function inside Op x b, and later apply the Op constructor to the function that will finally receive and transform the a, so that said function becomes Op's payload. But, if you put Op's contramap side by side with the wishful contramap for our fantasy leftwards arrow (<-), you'll see this in a more familiar light.

```
contramap = \f       g        a -> g (f a)
contramap = \f (Op g) -> Op (\a -> g (f a))
```

The Op stuff is just contortions to make Haskell happy. But in principle, we were right all along. We'll see more contravariant functors later on.

Seventy-three

But why do we care about functors, variance and profunctors so much? I mean, it took us *forever* to get here, so it better means something.

The beauty of Profunctor is that it generalizes the idea of a program beyond just functions. Suddenly, *anything* that has inputs and outputs can be manipulated, and that is *beautiful*.

```
dimap :: Profunctor p
      => (a -> c) -> (b -> d) -> p c b -> p a d
```

The type of dimap is saying that we can change *any* p that is a Profunctor
—that is, any p that is somehow a program— so that it accepts *any* a rather
than *any* c as input, and produces *any* d rather than *any* b as output. *Any*.

Moreover, we can do this in a predictable manner because the Profunctor
typeclass has *laws* making this whole thing a transparent and trustless
endeavour. Well, in the most beautiful anarchic sense anyway, for there's
no police to enforce such laws. We are talking about the usual functor
laws, of course. After all, a Profunctor is just two functors in one. So we
have an *identity* law and a *composition* law expected to be satisfied by every
Profunctor instance.

```
dimap id id == id
```

```
dimap (compose f g) (compose h i)
  == compose (dimap g h) (compose f i)
```

So what other things are Profunctor, beside functions? Ah, wouldn't we
like to know... We'll have to wait, I am afraid, for we moved so fast that
we've run out of problems to solve. Suffice it to say that not only have we
decoupled ourselves from the computer, the machine, but we've also
parted ways with the function which, we were told, was everything there
was. We are free now.

Seventy-four

So far, in this book, we have mostly learned about how to use types to
reason about our problems. Everything we learned was important and
fundamental somehow, and ultimately, has helped us develop the mindset
and principles we need. But unfortunately, we don't have much to show
for it yet. There's nothing mundane, nothing we can touch. There's so
much we know, yet we can barely write a program that does something.
So let's fix that now, let's write some earthly programs.

```
sum :: [Natural] -> Natural
sum = \xs -> case xs of
              [] -> 0
              x : rest -> x + (sum rest)
```

This function, sum, adds together all of the Natural numbers present in a
list. The type of sum doesn't say much, but if we look at the
implementation of this function we will see that it recursively adds each
element in the list. For example, applying sum to [2, 4, 3] would result in
9, for it reduces to 2 + sum [4, 3], which further reduces to 2 + (4 + sum
[3]) and then to 2 + (4 + (3 + sum [])) before finally becoming 2 + (4 +
(3 + 0)). If we add those numbers together, we end up with 9.

Returning 0 when we encounter an empty list [] makes sense if we consider that adding nothing together would result in, well, nothing, an idea often conveyed by 0.

And what about multiplying numbers together instead? Let's do that as well.

```
product :: [Natural] -> Natural
product = \xs -> case xs of
                 [] -> 1
                 x : rest -> x * (product rest)
```

product is quite similar to sum. All that changes is that we now multiply numbers with * rather than adding them with +, and that in the case of an empty list [] we return 1 rather than 0. The rest of this function is exactly the same as it was in sum. These new choices mean that, for example, product [2, 4, 3] will result in 24, for it reduces to 2 * product [4, 3] first, which then becomes 2 * (4 * product [3]), further reducing to 2 * (4 * (3 * product [])) to finally become 2 * (4 * (3 * 1)), which according to our calculator is 24. Those are really handy machines, use them.

The funny thing is the 1 there. It was easy to make sense of 0 in the case of sum because of its relation with the absence of things. If we have no numbers to add, we get 0 as result. Quite reasonable. However, in product we are saying that if we have no numbers to multiply, then we get 1 as result. Where does the 1 come from, if we have nothing to multiply at all? It is a bit funny, yes, I'll give you that. Yet, multiplying 2 * (4 * (3 * 1)) together really leads us to 24, which according to the abacus is exactly the number we expect as the result of multiplying all of 2, 4 and 3 together. So where's the trick? What's going on?

Seldom do we programmers, we mathematicians, care about things in isolation. It's in their relationships, in their belonging, that we find meaning. 0 is not interesting because of its familiar sense of absence, but because *adding* it to another number results again in that same number. 7 is 7, but 7 + 0 is also 7. It's in the relationship that 0 has with other numbers *through addition* that we find meaning. And the same is true for 1, which when *multiplied* by any other number results in that number once again. 7 is 7, but 7 * 1 is also 7. So close your eyes now, take a deep breath and look deeper. What do you see?

Identity, you are everywhere. You never leave, you never cease to amaze us. Yes, yes indeed, *adding zero* or *multiplying by one* are identities as well. We can write a function that adds zero, another that multiplies by one, call them id, and it would be alright. However, this time things are a bit different, and we are looking for something else. For one, we are assuming that the input to our identity function will be some kind of number, for we should be able to multiply it or add to it at times. This alone already violates the parametric polymorphism in ∀ x. x -> x, which says *nothing* about x, let alone it being a number. But beside this, we need to understand that it's not 1 nor 0 who are the identities, but instead, it's the combination of one these numbers with a particular operation, here * and +. So much so, actually, that as every other interesting thing in our journey, these numbers have special names. We say 0 is the *additive identity* and 1 is the *multiplicative identity*. We sometimes call them the *units* or *identity elements* of their respective operations, too. So, no, we can't really separate 0 from +, nor 1 from *, if we want to keep calling them identities.

But perhaps more importantly, at least insofar as sum and product are concerned, we need to acknowledge that being able to add and multiply numbers is necessary, notwithstanding whether these numbers are identity elements or not. When we arrived to 2 * (4 * (3 * 1)), for example, we were using * to multiply many numbers beside just 1, and that was fine. That is, while we certainly have a motivation for keeping the concepts of 1 and * together so that we can speak of them as an identity, we also have a motivation for keeping them separate, so that we can use * without involving 1 at times. In other words, we need to *pair* 1 and * without conflating them. So from now on, conceptually at least, we will pair elements with the operation for which they behave as identities, so that we are able to use them together or separately as we see fit. And these pairs, being as interesting as they are, will have a name too. We will call them *monoids*. We will explore them in detail later on, but for now let's just remember their name *monoid* and what it means.

Interestingly, what told sum and product apart brought them closer together as well, furthering our thesis that nothing mundane is as special as we are made to believe. The different behaviours of sum and product are simply due to their conscious choice of one of these monoids. But this is only half of the truth, of course, for if we leave monoids aside for a moment, the rest of sum and product are *exactly* the same, which should immediately catch our attention as something special too. So let's abstract away what's unique about sum and product, focusing on what's not.

```
foldr :: (Natural -> Natural -> Natural)
      -> Natural
      -> [Natural]
      -> Natural
foldr = \op unit xs ->
  case xs of
    [] -> unit
    x : rest -> op x (foldr op unit rest)
```

Things got a bit hairy. We'll clean this up soon enough, don't worry, but for now let's try to follow along. Fundamentally, only two things changed between product, sum and this new function foldr. Whereas before we mentioned 1 or 0 directly, respectively hardcoded inside the definitions of product and sum, here we are making it possible to receive this value as an input to this function. Its name, unit, should remind us of its purpose. Furthermore, we also abstract away the *operation* we perform on these numbers. We call it op, and it occupies the place that * and + did before. This time op appears in a prefix position, but that's a cosmetic matter that shouldn't surprise us. We know that a + b, say, is the same as (+) a b, so if (+) ever becomes that op, it will be alright.

The types that unit and op take can be inferred from their use, so don't let that scare you. First, let's keep in mind that foldr ultimately returns a value of type Natural. It says so on the tin. And, since unit is the value foldr will return when we encounter an empty list [], then unit must necessarily be of type Natural too. We use a similar reasoning for inferring the type of op, which we see being applied to two arguments. The first of these arguments, x, is an element found inside a list of Natural numbers, so it must be a value of type Natural itself. The second argument to op is the result of a recursive call to foldr which, we just said, results in a value of type Natural. And finally, not only must op take two Natural numbers as input, but it must also return yet another Natural, because like in unit's case, the return type of foldr mandates that the expression foldr returns always be Natural. And that's it. We have concluded that x must be of type Natural and op must be of type Natural -> Natural -> Natural, so these are the types of the expressions we see foldr take as new inputs, beside the actual list of Natural numbers to process.

With foldr in place, we can now redefine sum and product in a more concise and fundamental way by partially applying foldr, the function that somehow knows about the structure of a list, to the operations and units of the monoids that know how to combine the elements in that list.

```
sum :: [Natural] -> Natural
sum = foldr (+) 0

product :: [Natural] -> Natural
product = foldr (*) 1
```

So we found a new thing, foldr, that in a manner similar to fmap allows us to operate on lists for different purposes without repeating ourselves each time. And then we found those monoids too, which are somehow the essence of our programs. There's clearly something beautiful going on here.

Seventy-five

foldr is what we call the *right fold of a list*, and it is, essentially, a function that allows us to transform a list into almost anything else we can make out of its contents. If we recall, a list of values of type a can be either empty, as witnessed by the *nil* constructor [], or it can contain at least one value of said type a preceding the rest of the list, as witnessed by the *cons* constructor : which we usually use in infix form as a : rest. In order to construct a list, we must use these constructors somehow.

foldr is *exactly* the opposite. We could say that foldr is how we *destruct* a list. Think about how foldr op unit will use unit whenever it encounters [], and how it will use op on both a and rest whenever it encounters a : rest. A list that was constructed with [] will be "destructed" by foldr op unit by replacing [] with unit, and one that was constructed as a : rest will have this application of to a and rest replaced by op a (foldr op unit rest). And since [], : and those as are all there is to lists, we can say that foldr is effectively touching everything in the list, transforming it as it goes.

It might be easier to appreciate this visually, so let's try that. We said that sum [2, 4, 3] resulted in 2 + (4 + (3 + 0)), and in the past we also said that [2, 4, 3] could be written as 2 : (4 : (3 : [])). Let's put those two side by side and see what happens.

```
2 + (4 + (3 + 0))
2 : (4 : (3 : []))
```

Astonishing. Our unit, which in the case of sum is 0, is replacing the occurrence of [] in our list. And our op, which in this case is +, is replacing every application of :. In other words, conceptually, foldr is allowing us to go back in time to the moment when the list was constructed, and use the elements in that list, in the same order, to accomplish something other than creating a list. Quite beautiful, isn't it?

And at this point, curious as we are, we should stop and wonder what would happen if we pick [] to be our unit and : to be our op. Well, we don't need to do much thinking to figure it out, do we? We would be replacing every occurrence of [] with [], and every occurrence of : with :. In other words, everything would stay the same, nothing would change. What do we call that?

```
id :: [a] -> [a]
id = foldr (:) []
```

That's right, we can define the identity function for lists using foldr too. We said before that we could use foldr to convert lists into more or less anything we wanted, which suggests that it should be possible to convert them to themselves too, and our definition of id here is the proof of that. However, when we introduced foldr before, we did it in a context where we wanted to multiply and add numbers, so we focused on lists containing Natural numbers and on ultimately returning Natural numbers as well, so we can't really implement this id function with it. But all that Natural selection was unnecessary, and we could have made foldr a bit more polymorphic instead.

```
foldr :: ∀ a z. (a -> z -> z) -> z -> [a] -> z
foldr = \op unit xs -> case xs of
                        [] -> unit
                        x : rest -> op x (foldr op unit rest)
```

The implementation of foldr is exactly the same as before. All that's changed here is that instead of talking about Naturals in its type, we now talk about as and zs. Let's justify these choices. First, notice how a and z have been universally quantified separately, meaning that they could be *any* type, possibly different from each other. Second, notice how ultimately we are expecting to return a value of type z. Now, since the two things we ever return are our unit or a fully applied op, then both unit and the return value of op must have type z too.

```
foldr :: ∀ a z. (_ -> _ -> z) -> z -> [_] -> z
```

Then, we said that the elements in our list are of some type a, and the only place we ever deal with these elements directly is in our call to op, where an a is its first parameter.

```
foldr :: ∀ a z. (a -> _ -> z) -> z -> [a] -> z
```

The only thing remaining is the second parameter to op. But this is rather easy, too, since we can see how this parameter is a full application of op, which we already know returns a value of type z. So z it is.

```
foldr :: ∀ a z. (a -> z -> z) -> z -> [a] -> z
```

From as to z, this is where we wanted to be. And this new polymorphism, among other things, allows us to pick : of type a -> [a] -> [a] to be our op, and [] of type [a] to be our z, to finally have a foldr (:) [] of type [a] -> [a].

Folds, with their origami name reminding us that we are just giving a different shape to *all* the material present in a datatype, are the higher-order functions through which we can transform a datatype in its entirety to something else of our own choice. That's it.

Seventy-six

We talked about foldr being the *right* fold of a list. Intuitively, that means that when we write something like foldr (+) 0 [2, 4, 3], we end up with 2 + (4 + (3 + 0)), where the parentheses seem to be accumulating to the *right* side of this expression. But this is not the only way to fold a list, no. For example, we also have the *left fold* of a list, which we call foldl, and intuitively, accumulates parentheses on the left. That is, foldl (+) 0 [2,

4, 3] would result in ((2 + 4) + 3) + 0. Now, this difference is not particularly important if we are trying to add our numbers together, because whether the parentheses are to the left or to the right doesn't matter at all to our addition operator +. Both 2 + (4 + (3 + 0)) and ((2 + 4) + 3) + 0, if we try them in our calculator, will result in 9. However, many other times it does matter. For example, when we try to use [] and : as our unit and op. In this case, foldl (:) [] would in theory result in ((2 : 4) : 3) : []. However, that's nonsense that doesn't even type-check. Think about it. When we say 2 : 4, for example, we are trying to apply : to two values of type Natural. But we know that if the first input parameter to : is a Natural, then the second must necessarily be a list of them, of type [Natural]. So this just fails to type-check, meaning that our choice of foldl vs foldr matters here.

The reason why + and * succeed here is that they are *associative* operators, whereas : is not. We saw this before, but just as a reminder: A function f is associative if f a (f b c) and f (f a b) c are equal. Or, arranging things in an *infix* manner as in the case of +, we say that an operator + is associative if a + (b + c) equals (a + b) + c. The *cons* operator : is *not* associative, because a : (b : c) is not equal to (a : b) : c. Actually, this last example doesn't even type-check, so any discussion about associativity or equality is moot.

Anyway, associativity doesn't have much to do with folding, so let's just proceed to contemplate the implementation of foldl, the left fold of a list.

```
foldl :: (z -> a -> z) -> z -> [a] -> z
foldl = \op unit xs -> case xs of
                          [] -> unit
                          a : rest -> foldl op (op unit a) rest
```

Let's compare the types of foldr and foldl first.

```
foldr :: (a -> z -> z) -> z -> [a] -> z
foldl :: (z -> a -> z) -> z -> [a] -> z
```

The only difference between their types is in the first parameter to these functions, op, which in foldr takes the element of the list, of type a, as its first argument, whereas in foldl it takes it as its second argument.

The actual implementation of foldl is a bit trickier than that of foldr. This time, when faced with a : rest, foldl recursively calls itself as foldl op (op unit a) rest instead. That is, foldl is calling itself with op unit a as a new unit. Let's try to understand what's going on, because this is a bit surprising. Let's look at what the execution of foldl (+) 0 [2, 4, 3] looks like, step by step. That should clarify things for us.

First, foldl (+) 0 [2, 4, 3] reduces to foldl (+) (0 + 2) [4, 3]. This application then reduces to foldl (+) ((0 + 2) + 4) [3], which further reduces to `foldl (+) (((0 + 2) + 4) + 3) []`, finally returning all of `((0 + 2) + 4) + 3`, all of unit, as its output. Technically, this is a beautiful execution of our fold. Semantically, though, we find ourselves in a bit of a pickle. Suddenly, expressions like `0 + 2` or `((0 + 2) + 4) + 3` are taking the place of unit as we recurse into a next call to `foldl`. And while computers don't care about our name choices, we humans most certainly do. These things are not the *unit*, the *identity element* of our monoid, anymore. What's going on?

Lies. What's happening to the `unit`s of our `foldl` are *lies*. Well, kind of. When we introduced `foldr`, the relationship between its unit and the idea of the identity element of a monoid was so obvious that we decided to profit from this accident by taking the opportunity to explore it. You see, monoids are important, they bring *clarity* to everything they touch, and luckily for us they touch many, many things. And yes, we'll divert our attention to a monoid whenever we see one, and we'll see quite a few of them. So get used to the idea, profit from it, and learn to recognize those monoids. But no, monoids and folds don't have much to do with each other after all. In fact, to give a concrete example, `:` and `[]` are not a monoid, yet we were able to `foldr (:) []` just fine. So let's refine our perspective to understand the truth that's unfolding here.

Seventy-seven

Folding lists abstracts the idea of *iterating* over their elements. Rather than rewriting all the recursive machinery each time we need to transform a list, we keep around some higher-order order functions that do the boring bits for us. And we don't tackle the entire list at once. Instead, we address the elements of the list one by one. This allows us to focus, at each step, only on what is important. But we don't do this in isolation, no. If we did, we would be looking at something more like `fmap`, which allowed us to treat each element of our list *oblivious* of that element's relationship to the rest of the list. That's not a bad thing, though. Not knowing, actually, is what we love about `fmap` and functors. But folds are different, they are a more nosy, neighbourly abstraction. Let's take a look at `foldr` again.

```
foldr :: (a -> z -> z) -> z -> [a] -> z
foldr = \op acc xs -> case xs of
                      [] -> acc
                      a : rest -> op a (foldr op acc rest)
```

We renamed our unit parameter. We are calling it acc this time, short for *accumulator*. The intuition here is that `foldr`, as it goes about its origami business, *accumulates* in this acc what will eventually become the output of the whole execution of `foldr`. We see this situation, explicitly, when we find ourselves returning acc once we arrive to `[]`, the end of our list. Or the beginning, depending on the perspective you take. And it's actually these different *perspectives* that te.. foldl and foldr apart.

When we construct a list using [] and :, we tend to say that we *start* from an empty list [], and we build on top of this humble beginning by adding more and more elements to it using :. Eventually, we *end* up with a list of things we care about. This is the same perspective that foldr takes. foldr considers [] to be the beginning of the list, so that's where it starts working. It's easy to see this if we recall how we were taught in elementary school, in arithmetic, how we are supposed to tackle the things inside parentheses first. If we see 3 * (5 + 7), for example, we know the addition of 5 and 7 must happen before the multiplication by 3. The parentheses say so, that's their whole purpose. Well, look at what the output of foldr (+) 0 (2 : (4 : (3 : ([])))) looks like, this time with many redundant but still sensible parentheses. That is, look at 2 + (4 + (3 + (0))). The parentheses lean themselves to the right, they are deeper where [] was in our input list, and ultimately suggest that the elements that are closer to [] will be processed first.

Let's try and follow the execution of foldr step by step, and let's reduce any expressions that can be reduced, such as 5 + 1 into 6, as soon as we encounter them. As a high-level overview, here goes a list of all the reduction steps in this expression, written line by line, so that we can more easily visualize how we get from foldr (+) 0 [2, 4, 3] to 9.

```
foldr (+) 0 [2, 4, 3]
== 2 + (foldr (+) 0 [4, 3])
== 2 + (4 + (foldr (+) 0 [3]))
== 2 + (4 + (3 + (foldr (+) 0 [])))
== 2 + (4 + (3 + 0))
== 2 + (4 + 3)
== 2 + 7
== 9
```

As before, we start from foldr (+) 0 [2, 4, 3]. In a first iteration, as we process that leftmost 2, this will reduce to 2 + (foldr (+) 0 [4, 3]). Unfortunately, we can't really know the number that results from adding 2 to foldr (+) 0 [4, 3] just yet, for we won't know the numerical value of foldr (+) 0 [4, 3] until we reduce it further. So let's reduce it. foldr (+) 0 [4, 3] becomes 4 + (foldr (+) 0 [3]), and once again we have a number, 4, which can't be added just yet because we won't know the numerical value of the other operand, foldr (+) 0 [3], until we reduce it. So let's reduce it. We end up with 3 + (foldr (+) 0 []) this time, yet another addition we can't perform until we reduce foldr (+) 0 []. So we do, we reduce it, and finally get our hands in the rightmost numerical value in our fold, the deepest one inside our parentheses, 0, born out of the beginning of our list, [], out of *nil*, out of nothing. But don't forget that 0 was the number we've been passing as input to `foldr` all along, so there's no magic going on here. To sum up, we have gone all the way from 2 + (4 + (3 + (foldr (+) 0 []))) to 2 + (4 + (3 + 0)) now. Yes, this is boring, I know. Yawn through.

Alright, we have numbers now, so what do we do? Why, we *finally* start adding them, of course. We worked so much to get here, to this so-called "beginning" of the list, yet so far we haven't been able to perform *any* of the additions we wanted. We have all of 2, 4 and 3 waiting to add themselves to something. So let's start from what's closest to 0, to the inner parentheses. We had 3 + (foldr (+) 0 []) which we couldn't perform until we had a numerical value for `foldr (+) 0 []`, but now we do, it's 0, so let's tackle 3 + 0. We know this is 3, but let's not be so quick about it, let's try to understand what's going on. Remember, + here is what we called `op` at some point, and we said that `op` took two parameters as input. The first one being the list element that concerns us, which in our case is 3, and the second being the *accumulated* return value so far. Accumulated *how*? What does this mean?

Let's remind ourselves that we are looking at this particular scenario from the perspective of 3 now, from within the context of a constructor : being applied to this 3 and the `rest` of the list too, which in this case is the empty list []. And `rest` is the *only* part of our list for wich we were able to obtain *any* kind of numerical value so far, 0, by applying `foldr` to the *same* arguments + and 0. That is, 0 is part of our folding result somehow. But this result is not sufficient, for we still need to deal with the other elements in the list. So `foldr` just *accumulates* this intermediate result for the time being, and expects that somebody else *improves* it later on by somehow combining it with the other elements in our list. And who's the one doing the improving? It's `op` of course, by which we mean + in our case. `op` takes the current element being tackled as its first parameter and the accumulated result to be improved as its second argument. The type of `op` in `foldr` is a -> z -> z, where a is the type of each element inside the list we are folding and z is the type of `foldr`'s output altogether, recursively being fed to `op` so that it can be improved into a *new* z that somehow takes this a into account too. Improved how? Well, in our case, we just add the element in the list we are currently looking at, 3, to the result accumulated so far. And that's it, that's our "improvement". Now all of 3 : [] has been right-folded into 3. We have gone from 2 + (4 + (3 + 0)) to 2 + (4 + 3). And look at how the rightmost inner parentheses are gone too, enabling us to go back to adding that 4 that we couldn't add before.

And now we do the same process all over again, but this time focusing on 4, and with 3 rather than 0 as the initial value of our accumulator. `op`, that is +, is looking at the number 4 within the context of 4 : `rest`, and it knows that the parameters it's getting as inputs are 4 itself and the accumulated result of right-folding that rest, which we just said is 3. All + needs to do now is improve this result somehow, which, as before, it does by adding the element and the accumulated result together. So 2 + (4 + 3) becomes 2 + 7, and finally we do the same once again to improve 7, the result of right-folding 4 : (3 : []), by adding it together to the leftmost element of our list, 2, who has been waiting for us to finish this trip all along. Incredibly patient, that 2. And at last, we arrive to our longed numerical result, 9.

For comparison, here's how a reduction using foldl, rather than foldr, would happen.

```
foldl (+) 0 [2, 4, 3]
== foldl (+) (0 + 2) [4, 3]
== foldl (+) (2 + 4) [3]
== foldl (+) (6 + 3) []
== 9
```

foldr saw [] as the beginning of the list, which caused it to recurse its way from [2, 4, 3] to [] so that it could finally start operating on those list elements, starting from the rightmost 3, with the accumulator always being the rightmost input to op. But foldl, on the contrary, sees the list's most recently added element as its beginning, so it can start operating on lists elements right away, recursing with the accumulator as the leftmost input to op until the list has been exhausted, at which point the result accumulated so far is simply returned.

The result is the same 9 in both cases. foldr, however, did much more work to get there. Twice as much. So why would we ever use foldr? Well, for two reasons. First, because whenever op is not an associative operation, the results will definitely not be the same. We witnessed this when we picked the non-associative : as our operator. And, somewhat tangential, do you remember that thing we said about monoids bringing clarity to everything they touch? Well, this is another situation where they do. The operation of a monoid, like + or *, is *always* associative. There is no such thing as a monoid whose operation is not associative, which, at least for semantic purposes, takes the weight of picking between foldl and foldr off our shoulders. And second, foldr is necessary because it allows us to work with *infinitely long* lists, whereas foldl just can't cope with that. That's right. We haven't seen this yet, but Haskell has no problem at all in representing infinitely large datatypes, and by means of foldr we can operate on them within a finite amount of time and space. Yes, Haskell can do that.

What a ride. What a *boring* ride. Good thing we almost *never* have to follow through the execution of a program this way. This is the kind of stuff we let computers and type-checkers do for us. What matters is understanding the ideas and interesting concepts such as *folding*, developing an intuition for them, and learning how to tickle the computer in the right place so that it'll do them for us. Our ultimate goal is for our knowledge and skills to become latent, for that's how we get to spend time worrying about the important things instead. There will come a time in our journey when we won't *need* to worry about details unless we really want to. We will have decoupled meaning from execution, it will be our reward.

Seventy-eight

Here's another function for you to think about and implement on your own. And try to come up with a less distracting name for it, it will help.

```
pancake :: (a -> b -> c) -> (b -> a -> c)
```

Seventy-nine

Picture yourself facing a table with three plates on it, each one on top of another. The black plate is on top of the table, the red plate is on top of the black one, and the white plate is on top of the red. Table, black, red, white. Without being too clever nor showcasing any charming magician skills, how would you proceed if you wanted to rearrange this pile of plates so that the black plate is on top of the red, and the red on top of the white instead? You can only move one plate at a time.

We take the white plate, the one most recently added to the pile, and put it on the side. Now we have two piles of plates, rather than one. There's one with a red plate on top of a black plate, and another one with just a white plate. Next, we grab the red plate and put it on top of the white plate. Now we have a pile with the red plate on top of the white, and the original pile with just the black plate left in it. Finally, we take this black plate and put it on top of the red one. Now the original pile is empty, and the new pile has black on top of red, on top of white. We have reversed the order of the plates.

Linked lists are like these piles of plates. Or worse, perhaps. Imagine you can only look at this pile from above, so all you see is the plate at the top of the pile, the one that was added to it more recently, and you can't tell how many plates are below it. Maybe there's one, maybe there are none, or maybe there's infinitely many of them. You can't see what's beneath the *last* added plate without *first* taking it out of the pile. *Last in, first out.* That's what linked lists are.

In programming, we call structures like the linked list, which have this *last in, first out* property, "stacks". They behave like a pile, like a stack, of plates. That's why foldr, who insists on seeing [] as the first element of a list, often does twice as much work as foldl. It needs to move past *all* the plates, one by one, before doing any real work. Kind of. We will soon find this to be a lie too. But for now, it is our truth.

Eighty

Alright, let's fold some. Let's see if we can reverse the order of the elements in a list, the order of the plates in our pile, using some kind of fold. That is, let's see if we can turn [1, 2, 3] into [3, 2, 1] somehow. This could come handy, say, if we were running a plate business.

Like we learned, we achieve this by taking the plate that's at the top of the stack, the leftmost element of our list, and putting it on top of the new pile we are building instead. In Haskell parlance, the list we are building is our accumulator, and the leftmost element is the element to which foldl applies its accumulating function. So let's write that accumulating function for foldl, let's have it put the element on top of the accumulated list.

```
reverse :: [a] -> [a]
reverse = foldl (\acc a -> a : acc) []
```

That's it. Yes, reverse does the trick. Here's how the reduction of reverse [1, 2, 3] happens, written using the verbose infix constructor : to make a point.

```
reverse (1 : (2 : (3 : [])))
== foldl (\acc a -> a : acc) [] (1 : (2 : (3 : [])))
== foldl (\acc a -> a : acc) (1 : []) (2 : (3 : []))
== foldl (\acc a -> a : acc) (2 : (1 : [])) (3 : [])
== foldl (\acc a -> a : acc) (3 : (2 : (1 : []))) []
== 3 : (2 : (1 : []))
```

Elements flock from the input pile to the new pile acc as we peel, one by one, the : constructors. And, in case you didn't notice, \acc a -> a : acc is just flipping the order of the input parameters to :. If you think of : as a function of type a -> [a] -> [a], then \acc a -> a : acc is a function of type [a] -> a -> [a]. All that's happening in our chosen op is that the order of input parameters to : are being flipped. Does this remind you of something?

```
flip :: (a -> b -> c) -> (b -> a -> c)
flip = \f b a -> f a b
```

flip is a handy higher-order function we use quite often to achieve the same \acc a -> a : acc does, in a cleaner and more reusable package. Rather than saying \acc a -> a : acc, we can say flip (:). It's the same. Go ahead, prove that this is true using equational reasoning if you have doubts. We called this pancake before. And with this, we can define reverse as foldl (flip (:)) [] instead. Here's the same reduction as before, written differently just for show.

```
reverse [1, 2, 3]
== foldl (flip (:))          [] [1, 2, 3]
== foldl (flip (:))         [1] [2, 3]
== foldl (flip (:))      [2, 1] [3]
== foldl (flip (:)) [3, 2, 1] []
== [3, 2, 1]
```

An interesting property of reverse is that reversing the same list twice gives back the same list. Try it with the plates on your table, touch the identity, you'll see.

```
xs  ==  reverse (reverse xs)
```

Or, as we often say:

```
id  ==  compose reverse reverse
```

Eighty-one

Something else we'll often want to do is *concatenate* lists together. Say I have [1, 2] and you have [3, 4]. Concatenating these lists together should result in [1, 2, 3, 4]. Let's see if we can solve this using folds. After all, we were told we could fold our way into almost anything, so this should be possible too. We want to implement a function named concat of type [a] -> [a] -> [a], short for "concatenate", where among other examples, concat [1, 2] [3, 4] results in [1, 2, 3, 4].

```
concat [1, 2] [3, 4] == [1, 2, 3, 4]
```

But let's write this using : and [] instead, adding superfluous parentheses and arranging things differently so that Prägnanz can do its thing.

```
concat  (1 : (2 : ([]          )))
                 (3 : (4 : []))

    == (1 : (2 : (3 : (4 : [])))))
```

That's right. All that concat is doing is replacing the [] in its first argument, 1 : (2 : ([])) with the entirety of its second argument 3 : (4 : ([])). Everything else stays the same, all we do is replace [].

Replace [], replace [], couldn't foldr replace []? Yes, yes it could. In foldr op acc as, that acc ultimately replaces the [] in as. Well then, let's put this to use in concat. All we need to do is make sure the second argument to concat becomes that acc.

```
concat :: [a] -> [a] -> [a]
concat = \as bs -> foldr _ bs as
```

And what about that placeholder _ we left? Well, if you recall, we also said that, intuitively, the op in foldr op acc as replaced every occurrence of : in as. But in our case, the uses of : in as are just fine, so we don't really need to change them. Let's keep them where they are by replacing them with themselves.

```
concat :: [a] -> [a] -> [a]
concat = \as bs -> foldr (:) bs as
```

That's it. That's concat. Try it. Use equational reasoning to prove the concatenation of any list you want. Actually, it's quite interesting to see what happens when either as or bs are the empty list [], so maybe let's do that exercise here. Let's see it for bs first. That is, let's use equational reasoning to see how concat as [] reduces.

```
concat as []
== foldr (:) [] as
== as
```

Wait, what? Well, yes, we said many times before that foldr (:) [] was the same as id for lists, so here we are essentially asking for the identity of as, which is indeed as. And what about making [] our as instead?

```
concat [] bs
== foldr (:) bs []
== bs
```

We didn't say it explicitly before, but it turns out that \bs -> foldr (:) bs [] is *also* an identity function for lists. Think about it, foldr op acc [] has no more elements to process, so it simply returns acc, the result accumulated so far, unmodified. So if we make bs our acc, concat [] bs will effectively return that same bs given as input. In other words, the following is true.

```
xs  ==  foldr (:) [] xs  ==  foldr (:) xs []
```

So there we have concat, which can also be written in a more interesting manner using flip and a point-free style.

```
concat :: [a] -> [a] -> [a]
concat = flip (foldr (:))
```

Quite a surprise, isn't it?

Eighty-two

Almost every time we encounter something that behaves as an identity in the wild, chances are we are facing something interesting. And concat is no exception. So let's see what we have.

First, we know that concat [] xs and concat xs [] are the same as xs.

Second, we know, or at least we expect, that if we concatenate two lists [1, 2] and [3, 4] into [1, 2, 3, 4], and then we further concatenate that result with [5, 6], we should end up with [1, 2, 3, 4, 5, 6].

```
concat (concat [1, 2] [3, 4]) [5, 6]
== [1, 2, 3, 4, 5, 6]
```

But perhaps more importantly, we could also arrive to that *same* result by concatenating [1, 2] with the result of concatenating [3, 4] and [5, 6].

```
concat [1, 2] (concat [3, 4] [5, 6])
== [1, 2, 3, 4, 5, 6]
```

Generalizing this a bit, we can say for *any* three lists a, b and c, this is true:

```
concat (concat a b) c  ==  concat a (concat b c)
```

There is a name for this, and we know it. Perhaps it would be a bit easier to see if concat, rather than being a function expected to be used in prefix manner, was an operator intended for infix usage such as + or *. Let's have that infix operator, let it be ++. That is, *two* plus symbols.

```
(a ++ b) ++ c  ==  a ++ (b ++ c)
```

That's right. concat, which from now on we will call ++, is an *associative* operator just like + and *. But not only that. Just like addition by 0 behaved as an identity function, and how multiplication by 1 did the same, here, concatenation with the empty list [] behaves as the identity function too. In other words, the empty list [] is the *unit*, the *identity element*, for list concatenation.

```
a ++ []  ==  a
```

And whether we concatenate the empty list on the right or on the left doesn't matter. The result is the same.

```
[] ++ a  ==  a
```

List concatenation, in other words, just like the addition of numbers or their multiplication, all paired with their respective identity elements, seems to be a *monoid*.

Eighty-three

Intuitively, a monoid is a way of combining two values of a same type, where one of those values could behave as an identity, into a third value of that same type. For example, we combine 2 and 3 by multiplying them into 6, or we combine [1, 2] and [] by concatenating them into [1, 2].

That is, a monoid is two things: A *value* and a *function*. Or is it three things? A *type* x, a value of that type x, and a function of type x -> x -> x. Or perhaps four things? A type x, a value of that type x, and function of type x -> x -> x, let's call it f, where f a (f b c) and f (f a b) c are equal. That is, f is an *associative* function.

Actually, it's five things. It's a type x, an associative function of type x -> x -> x, let's call it f, and a value of that type x, let's call it unit, that behaves as the *identity* element for that function f, meaning that f unit a, f a unit and a are all equal. Yes, five is good for now. A *type*, a *value*, a *function* and *two laws*. One requiring that the function be *associative*, and the other one saying that our chosen value must behave as an *identity* element for that function.

We can't do much about those laws beside promising we won't violate them. So we stand up, look out the window, and as we marvel at nature, at the achievements of God and man, as the thought of a loved one fills our heart, we swear that our monoids will always be true, and that the burden of proving this will be ours. We can, however, do something about that value and that function. We know it's important to keep them separate, but it's also important that they always be close enough. Otherwise, if one's gone, we'd have a monoid no more. We need to pair them somehow.

Eighty-four

We know how to pair things. For example, without much ado, we can use Haskell's weird (tuple, syntax) to pair an associative function with its identity element.

```
addition :: (Natural, Natural -> Natural -> Natural)
addition = (0, (+))
```

Here we are saying that addition will be the earthly representation of our monoid. It's a pair, a tuple, made out of an associative function + and its identity element 0. But there's something else here, too. We grown-ups probably miss it because we are too easily blinded by what we want to see. But if we take a step back to contemplate the landscape, our inner child will notice that the most prominent figure here, actually, is the Natural type. We see it in the type of the identity element 0, and we see it all over the type of + too. In fact, maybe we see it too much, and this should be a sign that things could be better. For comparison, let's see how the definition of the list concatenation monoid would look like.

```
concatenation :: ∀ a. ([a], [a] -> [a] -> [a])
concatenation = ([], (++))
```

Superficially, things might look a bit messier, but essentially, all the type of concatenation is saying is that the contents of the list we want to concatenate are irrelevant. Our a, *universally quantified*, could be blackholes or rainbows. It wouldn't matter, for we'd be able to concatenate them nonetheless.

In concatenation we have the [a] all over the place, and in addition we have the same Natural everywhere too. Both monoids work just fine, but repeating the same thing over and over again is tiresome. Let's avoid this by using a custom product type, rather than Haskell's tuples.

```
data Monoid x = Monoid x (x -> x -> x)
```

Let's see Monoid in action. It will help us appreciate its merits.

```
addition :: Monoid Natural
addition = Monoid 0 (+)

multiplication :: Monoid Natural
multiplication = Monoid 1 (*)

concatenation :: ∀ a. Monoid [a]
concatenation = Monoid [] (++)
```

It looks neat. Rather than repeating our chosen type every time, we just repeat it once *inside* the definition of the Monoid datatype, where we have a value constructor also called Monoid with two fields in it: One with a value of the mystery type x mentioned as the type-parameter of the Monoid type constructor, and another one with a function making reference to that same x.

Moreover, whereas before we talked of how tuples containing things of a particular type and satisfying some laws could be seen as monoids, here we actually have a thing called "monoid". This is an improvement in the human communication front, for it's now easier to talk about these things out loud. But more importantly, it is a triumph on the type-checking front, seeing how the type-checker will now tell us that "a Monoid Natural was expected, not a String", or something along those lines, if we ever make a mistake worthy of that message. Interesting things deserve a name.

Alright, we have Monoid. What can we do with it? Well, for example, we said that while foldl and foldr don't really need a monoid to work, we can most certainly fold a list according to a monoid's intention if we have one. Let's try and wrap foldr, say, so that it takes a Monoid x rather than separately taking an initial accumulator and an accumulating function.

```
foldrMonoid :: Monoid x -> [x] -> x
foldrMonoid = \(Monoid unit f) -> foldr f unit
```

Quite an improvement, I'd say. We are partially applying foldr to the function and unit of our monoid so that all that's left to provide is the list to fold. Of course, foldrMonoid is less polymorphic than foldr, where the output type of the fold was not necessarily the same as that of the elements in the list, but that's an expected consequence of the type of our monoid function being x -> x -> x rather than x -> y -> y or y -> x -> y. In exchange for this sacrifice, we are lowering the number of values the caller of foldrMonoid needs to juggle at the same time, as well as the cognitive effort it takes to understand the type and behaviour of this function. So much so, that talking about foldr vs foldl is not necessary anymore. Why? Because monoids are *associative*. This means that, semantically, it doesn't matter whether we start accumulating from the left or from the right. The result will be the same anyway. In Haskell we call this function mconcat, an excellent name that doesn't make sense yet.

```
mconcat :: Monoid x -> [x] -> x
```

In practice, mconcat is often implemented with foldr because of its support for infinite lists, but that's unrelated to what we are discussing, so just pretend you didn't read this paragraph.

With mconcat in place, we can redefine our sum and product from before in a semantically more agreeable way.

```
sum :: [Natural] -> Natural
sum = mconcat multiplication

product :: [Natural] -> Natural
product = mconcat addition
```

Wait, what? There's something very wrong here.

Eighty-five

There's nothing wrong with our Monoid datatype. Yet, here we are, complaining that our product adds and our sum multiplies. What happened? They shared their types, that's what happened. Just like how when doing arithmetic we can mistake 2 with 4 because they are both values of type Natural, we can mistake multiplication with addition just as well, for they also share the same Monoid Natural type.

While in theory sufficient, and quite frankly rather beautiful, our approach didn't leverage the types as much as it could have and has thus become a weapon. Or, well, something that compiles yet does the unexpected. How can we fix this? How can we prevent multiplicative sums and additive products from existing at all?

It's the type-checker who decides what compiles and what doesn't, so all we need to do is speak its language, *types*, and ask for it. Luckily for us, we have a good starting point. Our monoids always refer to some type in them. Both addition and multiplication talk about Natural, say, and concatenation talks about [a]. But therein lies our problem.

It is true that Natural numbers can be added, yes, and it's also true that they can be multiplied. All that is true. But we can enrich these numbers a bit, with their intentions, and let the type-checker worry about any misplaced expectations.

```
data Sum = Sum Natural
```

This new datatype, Sum, is simply a wrapper around the Natural type.

When we construct a value of type Sum using the Sum value constructor, as in Sum 3, all we are doing is earmarking that Natural number 3 for a particular purpose. Its type is not just Natural anymore, so a function expecting a Natural number as input will reject it. Imagine we had a function called double that multiplied numbers by two.

```
double :: Natural -> Natural
double = \x -> x * 2
```

Applying double to the Natural number 3, as in double 3, would type-check just fine and eventually result in 6. But an application of double to Sum 3, as in double (Sum 3), will be rejected by the type-checker because it was a value of type Sum, not of type Natural as expected. We humans know that there is a Natural number hidden inside this value of type Sum, but the compiler doesn't. Or rather, it does, but it refuses to just ignore its belonging to a different species. Ultimately, all computers ever see are numbers. Sum 3? 432. Summer? 22. "telescope"? 324893. double? 47328902. But for all human intents and purposes, these numbers on their own are just nonsense. We have types because they convey different ideas that we don't want to get mixed up. For us, the Sum type conveys the idea that the Natural number is expected to be added through the addition monoid.

Sum is new vocabulary that both we and the type-checker understand. The only thing that's interesting about Sum, from the point of view of the type-checker, is that its name, its *identity* if you will, is different from that of every other one that came before, and from every other that will come after. This makes Sum unique, which makes it impossible to mistake a value of type Sum for a value of any other type. But the interpretation of Sum as a value to be combined through the addition monoid, on the other hand, is a choice we made at the level of expressions, and as such, the type-checker will forever remain unaware of it. The way we give meaning to Sum is by implementing functions that make use of it in the way we expect them to. So let's give it meaning by reimplementing addition in a way that takes Sum into account.

```
addition :: Monoid Sum
addition = Monoid (Sum 0) (\(Sum a) (Sum b) -> Sum (a + b))
```

We said Sum is essentially a wrapper around a Natural number. So here we are wrapping the Natural number 0 to be the identity element of this monoid, and in our associative function, before and after actually adding our numbers, we are unwrapping and re-wrapping these Sums. Let's see this in detail.

As we know, the type of addition, Monoid Sum, mandates that the type of its identity element be Sum, and that the type of its associative function be Sum -> Sum -> Sum. Creating a value of type Sum out of a Natural number is quite straightforward, all we need to do is *apply* the value constructor Sum to the

Natural number. And similarly, extracting the Natural number from inside the Sum expression is quite simple. All we have to do is pattern-match on Sum n to obtain that n. Actually, we can define a function that does this for us.

```
unSum :: Sum -> Natural
unSum = \(Sum n) -> n
```

It's interesting to notice that composing Sum and unSum results in an identity function for values of type type Sum.

```
compose Sum unSum :: Sum -> Sum
```

We can prove this is true by using equational reasoning. Actually, let's do it. It should be quick. Let's prove that compose Sum unSum (Sum n) equals Sum n.

```
compose Sum unSum (Sum n)
-- First we reduce compose.
== Sum (unSum (Sum n))
-- Then we inline unSum.
== Sum \(Sum n) -> n) (Sum n
-- Finally we beta-reduce.
== Sum n
```

And vice-versa, compose unSum Sum functions as an identity function for values of type Natural. You can do the equational reasoning to prove this yourself.

It's customary to call functions such as unSum, intended to just remove a wrapper like the one we named Sum, "*un*Something". Names matter to us humans, and talking about "wrapping" and "unwrapping" things simplifies communication.

And, by the way, we just accidentally proved that Sum and Natural are isomorphic. Using Sum and unSum we can convert between them without any loss of information. Isn't that nice? Not all accidents are bad.

Anyway, back to our addition monoid. As we know, in our Monoid, we are *receiving* values of type Sum wherever Sum appears as an argument in negative position, and we are responsible for *producing* values of type Sum everywhere Sum appears in a positive position.

```
data Monoid x = Monoid x (x -> x -> x)
                       +  +++++++++++++
                       +  -    -      +
```

In other words, we apply the Sum constructor wherever x, which in our case is Sum, appears in positive position. And we pattern-match on Sum n everywhere the x appears in negative position, in order to extract the Natural numbers inside these Sums and add them together before re-wrapping them in yet another Sum.

```
addition :: Monoid Sum
addition = Monoid (Sum 0) (\(Sum a) (Sum b) -> Sum (a + b))
```

Of course, rather than pattern-matching on the Sum constructor, we could
have used unSum instead.

```
addition :: Monoid Sum
addition = Monoid (Sum 0) (\a b -> Sum (unSum a + unSum b))
```

Here, a and b are of type Sum rather than Natural, but the whole thing
means the same anyway, so it doesn't really matter which version we
choose.

Alright, we have redefined addition to work exclusively with Sums. Now
let's have sum, the function supposed to add together all the Natural
numbers on a list, use it.

```
sum :: [Natural] -> Natural
sum = mconcat addition
```

Boom! If we try this, type-checking fails with an error saying something
about a value of type [Sum] being expected where a value of type [Natural]
was given. So not only does the *wrong* sum fail to compile, the one that
multiplied numbers rather than adding them, but also the *right* one fails,
the one which actually adds numbers. But while at first we might find this
depressing, a failure, what's really happening is that we have been
successful, yes we have.

Eighty-six

In Sum we created new vocabulary that the type-checker could understand.
Vocabulary that we could use to talk to the compiler and say "listen, my
friend, this is the only type of numbers we'll ever deal with". However, to
make effective use of this new vocabulary for our monoidal needs, we had
to change our addition monoid so that it worked with numeric values
wrapped in Sum, rather than with plain old Naturals. Now, we need to do
something similar in sum.

In sum we are blindly using mconcat, whose type is Monoid x -> [x] -> x,
and therein lies the problem. What is happening is that when we apply
mconcat to addition, of type Monoid Sum, the mystery x in our type becomes
Sum everywhere else. In other words, addition is forcing *every* value that
interacts with it to have the type Sum. That's great, that's what we wanted,
isn't it? This should be the type of sum, according to mconcat.

```
sum :: [Sum] -> Sum
```

This works just fine. However, we've trespassed. We've crossed a
forbidden boundary, and we shouldn't have. The users of sum, those who
just need to add numbers together, shouldn't have to concern themselves

with things such as Sum and monoids. Sure, internally, we are using this marvelous and unique vocabulary to implement sum, but we could as well have continued to use foldr (+) 0 as before, and from the point of view of the users of sum, it would have been equally fine. They don't care about how sum is made, all they care about is what sum accomplishes in the end.

This is the essence of *functional programming*. Our programs are essentially functions made out of functions. But at every step of the way, these functions have different and reasonable concerns. Just like how when running we don't need to think about our muscles contracting, about the impulses in our nerves, or about cells suddenly requiring more oxygen, functional programming is about worrying about the right thing at the right time. Never less, never more. In this new sum, we have our runner worrying about cells. That's not good.

We need to change sum so that rather than having the type [Sum] -> Sum, it can keep its previous type [Natural] -> Natural. That is, we need to modify the types of both the input and the output of sum. Alright, that's not a problem, we know how to do it.

```
dimap :: Profunctor p => (a -> c) -> (b -> d) -> p c b -> p a d
```

Remember Profunctors? Those were lovely. More specifically, specializing the type of dimap for our needs, we get something like this:

```
dimap :: ([Natural] -> [Sum])
     -> (Sum -> Natural)
     -> ([Sum] -> Sum)
     -> ([Natural] -> Natural)
```

Don't let the rather big type scare you, what's happening is really easy. Let's start from the very end. This use of dimap returns a value of type [Natural] -> Natural, which is *exactly* the type we want sum to have. Great. Then, we have an argument of type [Sum] -> Sum, which is exactly the type that mconcat addition has. Excellent. Let's start writing this down.

```
sum :: [Natural] -> Natural
sum = dimap _ _ (mconcat addition)
```

All we need now is a function of type [Natural] -> [Sum] to take the place of that first placeholder _, and another one of type Sum -> Natural to replace the second placeholder _.

We will use unSum as our value of type Sum -> Natural. There's not much to say about it beyond what we have already said. unSum is here and it has the right type, so let's just use it.

```
sum :: [Natural] -> Natural
sum = dimap _ unSum (mconcat addition)
```

What about the other placeholder? The function that takes its place is expected to convert every element in a list of Naturals into a value of type

Sum. We know how to convert *one* Natural into a Sum by simply applying the Sum value constructor, alright, but an *entire* list of them? Crazy stuff.

Just kidding. Remember lists are a Functor too, meaning we can use fmap, of type (a -> b) -> ([a] -> [b]), to lift the Sum value constructor of type Natural -> Sum into a function of type [Natural] -> [Sum], which is exactly what we need.

```
sum :: [Natural] -> Natural
sum = dimap (fmap Sum) unSum (mconcat addition)
```

Beautiful. Now sum has exactly the same type as before, but internally it relies on Sum and the addition monoid. Of course, we could have written this in a different way, but any difference would have been superficial. The fundamental issue of converting between Sum and Natural back and forth will always be present.

```
sum :: [Natural] -> Natural
sum = \xs -> unSum (mconcat addition (fmap Sum xs))
```

The experienced programmer might point out, concerned, that we are wasting computing power converting Naturals into Sums and back just to accommodate this, and in theory this is true. In practice, however, not only is Haskell beautiful, but also, as of the beginning of the 21st century, its compiler GHC is rather excellent as well. And among other things it can realize that Sum and Natural, despite their different types, are essentially the same. Knowing this, Sum will continue to exist at the type level, but it's existence as a term will have no cost whatsoever. Sometimes the compiler needs a couple of hints here and there to achieve this, but ultimately, it will happen. So, experienced programmer, don't worry too much about this.

Could we accidentally mix addition and a multiplication monoid of type, say, Monoid Product, where Product is to multiplication what Sum is to addition?

```
data Product = Product Natural

multiplication :: Monoid Product
multiplication =
  Monoid (Product 1)
          (\(Product a) (Product b) -> Product (a * b))
```

Yes, of course we could mix them up, but since we are not dealing directly with Naturals anymore, making that mistake would involve three accidents, rather than one, for we'd also need to mistake fmap Sum for fmap Product and unSum for unProduct. Chances are minimal.

Alright, we've accomplished something nice. In summary, we created these wrappers Sum and Product so that addition and multiplication, our monoids, couldn't be mistaken for each other. Of course we could get the implementation of addition or multiplication themselves wrong, if we mistook 0 for 5, say. But, quite frankly, that could be described as lack of responsibility. Ultimately, addition and multiplication are our sources of truth, the monoids themselves, so we need to make sure they do what we want. The computer can go quite far, but at some point we need to say "computer, *this* is what we want". And that's fine, because *we* are the programmers, *we* decide what ultimately happens. In our case, the definition of our monoid is the *only* place where we ever need to pay attention. *Just one place*, everything else builds on top of that. mconcat has been the only use case for our monoids so far, sure, but this was just an example. Monoids are *everywhere*, and in reality we use them all the time in a myriad of different ways beyond mconcat. So yes, it is scary, but it is also quite beautiful.

To sum up, we will always need to tell the computer what to do at least once, somewhere, for *we* are the programmers, not them.

Anyway, beyond the addition monoid, we don't have really another use for our Sum type. That is, we expect that there will be just *one* way to interpret Sum as a monoid, and never more. Actually, if we could somehow *prevent* a different monoidal interpretation for Sum, that would be even better. Can we do that? Can we have the Sum type itself, rather than addition, determine its behaviour as a monoid?

Eighty-seven

There's nothing wrong with our Monoid datatype. But there wasn't anything wrong with our representation of a monoid as a tuple either, and yet we somehow improved it. So let's improve this too. In Haskell, Monoid is not a datatype but a *typeclass*, just like Functor and friends.

```
class Monoid x where
  mempty :: x
  mappend :: x -> x -> x
```

How is this better? Let's see. First, and unrelated, notice that this typeclass is the first one we see having more than one method. This is fine. Just like product types can have many fields, typeclasses can have many methods too. Here, the Monoid typeclass has two methods named mempty and mappend. And if we compare this typeclass with the Monoid datatype we defined before, we'll see the resemblance is remarkable.

```
data Monoid x = Monoid x (x -> x -> x)
```

It seems that mempty is the name this typeclass is giving to the identity element of our monoid, and that mappend is the name it gives to the monoid's associative function. And, since there's nothing more to see, this must be all there is to the Monoid typeclass. So, without further ado, let's dive in. Let's make Sum be our x and implement an instance for this typeclass.

```
instance Monoid Sum where
   mempty = Sum 0
   mappend = \(Sum a) (Sum b) -> Sum (a + b)
```

It's not what we have gained from this instance what matters, but what we've lost. Which, actually, is also what we gained, so I guess it's both. Anyway, addition, the previous definition of our monoid, is gone. Poof. We don't need it anymore because its code lives in the Monoid instance for Sum now. And this is perfect, because while previously we *intended* the Sum type to determine the meaning of our monoid, in reality it was addition, who happened to mention Sum, who did. But now Sum must decide who it wants to be, for there can only be *one* instance of the Monoid typeclass for Sum, and whatever that one instance decides, will be the monoidal behaviour of Sum. It will be the only truth.

So let's profit from this. Let's see, for example, how the sum function, the one that adds Natural numbers, would be implemented using this function.

```
sum :: [Natural] -> Natural
sum = dimap (fmap Sum) unSum mconcat
```

The only thing that changed here, compared to our last definition of sum, is that we are *not* applying mconcat to addition anymore. So how does mconcat know that it needs to add the elements in our list? The answer is in mconcat itself.

```
mconcat :: Monoid x => [x] -> x
mconcat = foldr mappend mempty
```

What mconcat is saying is that given a list of elements of type x for which there exists a Monoid instance, such as our Sum, it will fold that list using mappend as the accumulating function and mempty as the accumulator. In other words, the value of type Monoid x we were explicitly passing around before, is now implicit. And all we need to do for this to work, essentially, is pick the right x. And how does sum pick this x? Both its use of the Sum value constructor and the unSum function *force* the Sum type to be this x. Anything else would fail to type-check.

So there we have it. A typeclass approach that requires us to have *one type per monoid*, making the type itself, rather than some expression in the wild, the source of truth. And moreover, it makes for a more comfortable experience, considering how we don't explicitly have to mention said gone monoid expression in our code anymore. For completeness, here's the code defining the rest of our monoids so far:

```
instance Monoid Product where
  mempty = Product 1
  mappend = \(Product a) (Product b) -> Product (a * b)

instance Monoid [a] where
  mempty = []
  mappend = (++)
```

And something else. Considering how monoids have laws, and how we
appreciate seeing laws in typeclasses so that we know all their instances
behave as expected, the Monoid typeclass in Haskell is accompanied by some
comments requiring compliance with two laws. First, an *identity law*,
whereby mempty is required to behave as an identity element:

```
mappend mempty x == x == mappend x mempty
```

And then an *associativity law*, in which mappend is required to be an
associative function:

```
mappend a (mappend b c) == mappend (mappend a b) c
```

That's all there is to Monoid in Haskell. For now, anyway. Of course there is
more.

Eighty-eight

Generally speaking, *to fold* is to iteratively convert the constituent parts of
a whole into a new whole, which could be something else entirely, by
means of some higher-order function. Kind of. For example, we iteratively
convert all the applications of : and [] in a list by replacing them with
values or function applications. But this idea of taking a whole and
converting it to something else part by part, in principle, sounds useful
beyond lists.

```
data Maybe a = Nothing | Just a
```

Let's see if we can come up with a function that folds a value of type Maybe
a somehow. We'll call it foldMaybe. Now, we know nothing about this
function except that it can be used to construct *anything* out of the
constituent parts of a Maybe a. So for now, let's just say the output of this
function will be some unknown, universally quantified type z.

```
foldMaybe :: ∀ z. ... -> z
```

Then, we need to deal with these "constituent parts" somehow, whatever
they might be. But we don't like being told what to do, so let's try to
discover this on our own. Let's just deal with "the whole", with the entire
Maybe a, at once.

```
foldMaybe :: ∀ a z. Maybe a -> z
foldMaybe = \ya -> case ya of
                     Nothing -> ⅋
                     Just a -> ⅋
```

We were reckless, we are stuck. `foldMaybe` can't construct values of some unknown type z out of thin air. This is a situation similar to the one we faced before when we dealt with `undefined`, also known as *bottom*. Yes, we are stuck here. Let's remind ourselves how `foldr` dealt with this, maybe we can find some inspiration there.

```
foldr :: ∀ a z. (a -> z -> z) -> z -> [a] -> z
```

We had different perspectives for understanding the behaviour of `foldr`, but the one that will help us now is the one that saw `foldr op acc` as replacing the empty list constructor [] with `acc`, and every use of `:` with an application of that `op`. In other words, `acc` of type z is what we use whenever the list is empty, and this `op` that somehow returns a z is what we use when we encounter an element of type a in our list. And, if we consider that a value of type `Maybe a` is essentially a one-element list, we can start copying some of these ideas.

First, we'll make sure `foldMaybe` is given a z that it can use whenever it encounters the `Nothing` constructor, analogous to the empty list constructor [].

```
foldMaybe :: ∀ a z. z -> Maybe a -> z
foldMaybe = \z ya -> case ya of
                       Nothing -> z
                       Just a -> ⅋
```

Then, we need something to deal with the "non-empty list" case. That is, with the `Just` constructor. In the case of `foldr`, this is a function of type a -> z -> z where the input of type a is the a in a `:` rest, and the input value of type z is the result of *recursively* folding the rest of the list. This makes sense for lists because they are a recursive datatype, but `Maybe`s are not, so perhaps we should just drop that input z altogether, and instead have a function a -> z that will convert the a inside a `Just` constructor when it sees one. Let's try that.

```
foldMaybe :: ∀ a z. (a -> z) -> z -> Maybe a -> z
foldMaybe = \f z ya -> case ya of
                         Nothing -> z
                         Just a -> f a
```

Yes, that's it. This is `foldMaybe`. Let's put it next to `foldr` so that we can compare their types.

```
foldMaybe :: (a ->      z) -> z -> Maybe a  -> z
foldr     :: (a -> z -> z) -> z ->    [a] -> z
```

The resemblance is quite apparent. Now, in Haskell, this function is *not* called `foldMaybe`. A bit confusingly, its name is `maybe`, all in lowercase, and the order of its input parameters also changes. But leaving aside those superficial matters, it is essentially the same thing.

```
maybe :: z -> (a -> z) -> Maybe a -> z
```

But why would we want to fold a Maybe, anyway? Couldn't we just perform a case analysis on the Maybe a value, pattern-match on the different constructors and manually deal with them as we've been doing so far? Well, yes, of course we could, but this is *functional* programming, so we appreciate having that same behaviour packaged in a function that we can readily apply and compose time and time again. We will use `maybe` shortly.

Eighty-nine

Can we fold an `Either a b` too? Of course we can. Actually, why don't you go ahead and implement this yourself?

```
either :: (a -> z) -> (b -> z) -> Either a b -> z
```

Generally speaking, the term "fold" is closely related to list-like things that have some kind of recursive structure, so saying that `maybe` and `either` are "folds" is perhaps not entirely accurate. Throughout our journey we'll see how these things can be generalized in different ways, but for now, "fold" is alright.

Anyway, luckily for us, the type of lowercase `either` fully determines its implementation. Unfortunately, this wasn't the case for either `foldr`, `foldl` nor `maybe`. For example, we could have implemented `maybe` as follows, it would have compiled, but it wouldn't have done quite what was expected of it:

```
maybe :: z -> (a -> z) -> Maybe a -> z
maybe = \z f ya -> case ya of
                     Nothing -> z
                     Just a -> z
```

That is, it wouldn't have used `f` at all. And how do we know what's expected of a well behaved `maybe`? Well, we don't. It really is up to the designer of this function to come up with a description for it and make sure its implementation matches that description. However, our fluffy intuition for what a "fold" is should suggest some sensible ideas. We could say, for example, that the following equality is expected to be true:

```
y  ==  maybe Nothing Just y
```

In other words, `maybe Nothing Just` is an identity function for values of type `Maybe a`, for all a. And when we consider this equality together with the parametric polymorphism of `maybe`, which says that it must return a z which can only be constructed by using one of the first two inputs to maybe, ruling out _ _ y -> y as the implementation of this identity function, we are safe.

So go ahead, implement either, and think about what makes it different from maybe in this regard.

Ninety

Another type we'll frequently encounter is Bool, which stands for *boolean*, and represents the idea of *truth* in logic. We use values of type Bool to denote whether something is *either* true *or* false. Either? Or? That sounds like a sum type, a coproduct.

```
data Bool = True | False
```

Indeed. In logic, things are *either* true *or* false, and Bool correctly represents that. For example, we could use a boolean as the answer to "is this list empty?".

```
isEmpty :: [a] -> Bool
isEmpty = \xs -> case xs of
                   [] -> True
                   _ -> False
```

isEmpty returns True when the given list is [], and False whenever it's not. Remember, that wildcard _ will successfully pattern-match against anything that has not successfully pattern-matched before.

There's not much more to say about Bool itself. It's just a rather boring datatype with two possible values.

Can we fold a Bool? Yes, of course we can. Remember, when we fold, we are trying to obtain a value of some type z by addressing each of the constituent parts of our datatype separately. In our case, those parts are True and False. So let's introduce a function called bool to do just that.

```
bool :: z -> z -> Bool -> z
```

We expect bool to replace True with one of the z inputs, and False with the other one. Which one, though? Which of the following implementations of bool is the correct one? And to make it a bit more interesting, let's say that if we pick the wrong one, we die.

```
\x y b -> case b of { True -> x; False -> y }
```

```
\x y b -> case b of { True -> y; False -> x }
```

This is one unfair game. It should be easy to understand what's happening if we compare the types of bool, maybe and either.

```
either :: (a -> z) -> (b -> z) -> Either a b -> z
maybe ::       z -> (b -> z) -> Maybe   b -> z
bool  ::       z ->      z -> Bool       -> z
```

Each of Either, Maybe and Bool have two constructors, and conceptually, they only differ from each other in their payloads. Either, for example, has payloads in both of its constructors. And this, combined with the parametric polymorphism in either mandating that a z be returned somehow, *forces* either to use a -> z when it encounters a Left constructor, and b -> z otherwise. either is beautiful.

```
either :: (a -> z) -> (b -> z) -> Either a b -> z
either = \f g e -> case e of
                     Left a -> f a
                     Right b -> g b
```

Similarly, maybe is still forced to use that standalone z when it encounters a Nothing constructor, for it has no b to which to apply b -> z. But on the other hand, when it encounters a Just constructor with a b in it, it can *choose* to apply b -> z or just return that standalone z, ignoring b altogether. And choice, in computers, sometimes means catastrophe. We are forced to clarify what happens in that case as part of our documentation. With bool things are even worse, because we have *two* standalone zs to pick from, and neither True nor False have payloads that could help us decide which one to use. We are at a loss, and must instead rely on documentation to speak the truth. One truth, its truth, which states that bool will return the first z when the given Bool is False, and the second z otherwise.

```
x == bool False True x
```

Thus spoke identity, with a very weak voice. This time, nothing mandated this choice.

Anyway, booleans are not too interesting to us now. What's interesting is what we can do when we have many of them, how we can *combine* them.

Ninety-one

In logic, *conjunction* is the operation through which two boolean values become one by agreeing on the truth of something. That is, the conjunction of two boolean values a and b is *itself* true, if both a *and* b are themselves true. Otherwise, it's false. In programming, we usually call this operation "and".

```
and :: Bool -> Bool -> Bool
and = \a b -> case a of
                False ->
                    case b of
                        False -> False
                        True -> False
                True ->
                    case b of
                        False -> False
                        True -> True
```

That's ugly. It gets the job done, sure, it only returns True if both a and b are True, but it is nonetheless ugly. Let's try to clean this up. First, let's realize that as soon as we discover that a is False, we can just return False without having to look at b.

```
and :: Bool -> Bool -> Bool
and = \a b -> case a of
                False -> False
                True ->
                    case b of
                        False -> False
                        True -> True
```

Second, notice how our analysis of b is saying that if b is False, then we return False, and if it is True, we return True. That is, we are saying that whenever a is True, we will return a new Bool with the same value as b. Well, in that case, we might as well return b itself.

```
and :: Bool -> Bool -> Bool
and = \a b -> case a of
                False -> False
                True -> b
```

Great, that's looking better. But could we use our bool fold instead? Sure, why not.

```
and :: Bool -> Bool -> Bool
and = \a b -> bool False b a
```

This is much nicer. If a is False we return False, otherwise we return b. But, actually, there is more. It turns out that the conjunction of two booleans, and, is a *commutative* operation, meaning that saying and a b or and b a leads to exactly the same result, which implies that *in theory* we are allowed to move around our as and bs to accommodate our needs.

```
and :: Bool -> Bool -> Bool
and = \a b -> bool False a b
```

What needs? The need to realize that in \a b -> bool False a b we are facing an eta-expanded form of \a -> bool False a, which in turn is an eta-expanded form of bool False. So what about eta-converting all of this, and leaving bool partially applied instead?

```
and :: Bool -> Bool -> Bool
and = bool False
```

Ah, functional programming. Nice. And look, all of these equalities hold.

```
and True  True   ==  True
and True  False  ==  False
and False True   ==  False
and False False  ==  False
```

This is true, and in general this is right. But in Haskell, this is wrong too.

Ninety-two

Contrary to almost every programming language out there, Haskell is a *lazy* one. And being lazy in Haskell, contrary to good human values, is often quite good.

A language being lazy, or more precisely the *evaluation strategy* of the language being lazy, means that an expression will only be evaluated when truly necessary, and never before. This is clever. Let's see how.

We are in our kitchen and we decide to cook something new. However, in order to cook this, we'll need some utensils and ingredients. We can't afford buying new utensils today, though, so we'll only start cooking if the utensils we already have are sufficient. Moreover, we remember that our pantry is almost empty, which means that we'll also need to go to the market to buy the ingredients that we need, and only if we can find them all will we be able to start cooking. So, we have two questions: Whether we have utensils, and whether we have the ingredients that we need. Only if we have all of this can we cook. In other words, our ability to cook depends on the conjunction of our having utensils and ingredients.

Being as lazy as we are, we take one thing at a time. We start by looking for our utensils because we dread going to the market, so if at all possible, we would like to avoid it. The kitchen is right here, so we look around and check whether we have everything we need, and eventually store in a value named we_have_utensils of type Bool our judgement about this fact. This is how we would proceed in real life, being the lazy cooks we are. *If* it is True that we_have_utensils, *then* we go to the market to find out whether we_have_ingredients or not. But if it is False that we we_have_utensils, then we can *avoid* that trip to the market altogether, for whether we_have_ingredients or not, we still won't be able to cook without utensils.

Haskell is lazy like that. When we ask for the conjunction of two Bool values, when we say and a b, we are asking for them to be considered and eventually conjoined in *that* order: First a, then b. That is, when we ask for the logical conjunction of we_have_utensils and we_have_ingredients, we want Haskell to save us from that trip to the market if possible by *first* evaluating whether we_have_utensils, and only if this is True, *then* force a trip to see whether we_have_ingredients or not. However, when we implemented and as bool False, we broke this.

```
we_can_cook :: Bool
we_can_cook = and we_have_utensils we_have_ingredients
```

Let's use equational reasoning to see how the evaluation of this application of and proceeds.

```
we_can_cook
== and we_have_utensils we_have_ingredients
== bool False we_have_utensils we_have_ingredients
== case we_have_ingredients of
     False -> False
     True -> we_have_utensils
```

In other words, our implementation of and as bool False is scrutinizing we_have_ingredients first, *before* considering whether we_have_utensils at all. This is exactly the opposite of what we wanted. This is wrong. So, yes, while on paper it is true that the logical conjunction of two boolean values is commutative, meaning that and a b and and b a shall result in exactly the same value, this model doesn't take into account the *cost* of obtaining the truths to conjoin. Mathematics is lucky like that, it needs not care about labour.

So we just change the order of the arguments, right? That is, we say and we_have_ingredients we_have_utensils rather than and we_have_utensils we_have_ingredients. Yes, sure, that works. However, booleans and this left-to-right way of dealing with them lazily have become so ingrained in programming, that flipping the order of these arguments in order to accommodate and encourage a sociopath and might not be a sane long-term choice. Can we fix it? Of course we can. All we have to do is go back to that time when and was behaving perfectly fine, before we decided to play with commutativity. We can accomplish this by reusing one of our previous definitions, or by using flip to swap the order of the input parameters in our current definition.

```
and :: Bool -> Bool -> Bool
and = flip (bool False)
```

I'll leave it up to you to prove, using equational reasoning, that this is indeed correct.

Ninety-three

So how do we *reason* about laziness? How do we tell people about and's lazy behaviour? How do encourage the conjoining of truths? Can we do that at all? We can, yes we can. Laziness can be an elusive concept, and it may take us a while to be able to comfortably work with it. But as anything worth talking about, we have precise vocabulary for it, so let's talk.

First, that which among friends we call "laziness", technically, is called "call-by-need evaluation strategy", a name that perfectly suggests how *need* will trigger evaluation. But here we are among friends, so let's continue talking about how lazy things are instead.

Second, while we previously suggested that everything in Haskell was an expression, and while that was "true", in reality, almost everything is actually a *thunk*. We juggle many truths here, and at different times some are more useful than others. In part, this abundance of truth is what makes programming and mathematics so complex and eternally exasperating. The secret to happiness, or at least to effectively functioning in this fractal truth landscape, seems to be in acknowledging that occasionally we'll have to content ourselves with the truth that addresses our immediate concerns, while remaining fully aware that we could be mistaken. Our empire could collapse at any time. There will always be a latent truth, a deeper one, comprehending more than what we've dared care for so far, quite likely proving us wrong. The silver lining is that we never stop learning: If we have more questions, we'll also have more answers. And then we'll have even more questions, of course, for that's the price of knowledge. It's humbling, it's fun, it's necessary.

Thunk, a broken tense form of the English verb "to think", made noun by the grace of this language's disregard for tradition, is the name we give to expressions that haven't been evaluated yet. In Haskell, when we say id 3, for example, this doesn't immediately become 3. Likewise, and True False doesn't immediately become False. Instead, these expressions become a *thunk*, a commitment to deliver a value of the expected type when requested and not before.

We want to defer computations for as long as possible, hoping, ironically, that some of them never run at all. This drastically changes the way we approach problems, the way we design solutions, because rather than worrying that the computer might be doing too much work, we can optimistically say things such as and we_have_utensils we_have_ingredients and trust that and won't make us go to the market unless we *really* have to. Without laziness, we just can't be so optimistic. Thunks help us achieve this. But moreover, as a separate virtue, *if* we actually need to evaluate a thunk for a particular purpose, then the profits of this labour need not end there. Once a thunk has been evaluated, once its ultimate value has been computed, it will be made available for others to exploit too without having to compute it all over again. Consider and we_have_ingredients we_have_ingredients. Are we *really* planning to go to the market *twice*? For what? To double check if we have the ingredients we need? That would be a waste of time, and Haskell, *thunks*, knows that. What happens instead is that we go to the market once, learn and commit to memory our understanding about the availability of ingredients, and the next time we are asked about it we just recall that answer, knowing that we have already *thought* about it before. Thus the etymology of the name.

We also implicitly suggested that Haskell's and was somehow special, different from and in other languages. And it's true, it is different, but not because it's special, rather because *it is not*. Most programming languages are *strict*, not lazy, meaning that when we say f a b in them, all of a and b are fully evaluated before f is called. There are no thunks nor any other means of automatically deferring computations for us. But Haskell is not like that, in Haskell f will be applied to its inputs a and b right away, even if a and b haven't been evaluated yet. *If* f ever needs to know more about a or b, *then* their thunks will be *automatically* evaluated and replaced with actual values. Generally speaking, it is not possible to implement functions like and in other languages in such a straightforward manner. However, logical conjunction is quite prominent in programming, and almost every programming language out there embraces the idea of market trip avoidance by automatically deferring the computation of the second input argument to and as an *optimization*, so as to prevent programs from doing more work than necessary. But how do they deal with this, if they are not lazy? Corruption, that's how. Mostly, they give booleans a special treatment. Whereas in Haskell Bool is a datatype we can define, in most other languages booleans and functions on them like and are baked into the language itself. In fact, being as baked in as they are, and having semantics different to the rest of the language, "functions" like and are rarely actual *functions*, they are some kind of kludge instead. Not in Haskell, though. We have kludges, sure, but not these.

Ninety-four

Let's build a list. An infinitely long list where every element is 0. Laziness makes this possible. Let's see how.

```
zeros :: [Natural]
zeros = 0 : zeros
```

The first thing to keep in mind is that just like how *defining* undefined, an expression that looped itself to oblivion, didn't make our programs crash, merely defining zeros doesn't make our program loop forever chasing its tail either. Essentially, zeros is a list of type [Natural] obtained by applying : to the Natural number 0 and to a recursive use of the name zeros. The type-checker accepts this because zeros has type [Natural], which is exactly what our use of : expected. Let's use equational reasoning to see how the 0s in zeros come to be.

```
zeros
-- We inline the definition of 'zeros'
== 0 : zeros
-- And then we do the same again and again ...
== 0 : (0 : zeros)
== 0 : (0 : (0 : zeros))
== 0 : (0 : (0 : (0 : zeros)))
== 0 : (0 : (0 : (0 : (0 : ...))))
```

Now, if we wanted to apply sum to zeros, say, the computation would
never end. Our program would diverge. This should be easy to see,
considering how sum is trying to come up with a *finite* number as the result
of adding together the elements of a list that happens to be *infinitely* long.
A futile endeavour, we know, but sum doesn't. Unfortunately, sum zeros
will type-check just fine. It's only when we attempt to *compute* sum zeros
that our program goes into an infinite loop, diverges, never halts, just like
undefined never did. This is yet another manifestation of the halting
problem that plagues general purpose programming languages. But of
course, divergence is not what we want infinite lists for. So let's go deeper,
let's try to understand the lazy nature of lists and see if we can profit from
it.

When we say zeros = 0 : zeros, we are creating a *thunk*. In fact, all Haskell
sees is something like this:

```
zeros :: [Natural]
zeros = <thunk>
```

Mind you, this is *not* proper Haskell. We are just using <thunk> to pinpoint
where Haskell sees a thunk.

To understand how this thunk becomes an actual value, we will need to
force its evaluation somehow and see what's underneath. Our most
unkind sum does that, but it is also too eager to see everything, too *strict*.
We need a gentler function this time, one that will kindly derive meaning
from our list without looking at it too hard, like isEmpty.

```
isEmpty :: [x] -> Bool
isEmpty = \xs -> case xs of
                    [] -> True
                    _  -> False
```

This function, isEmpty, doesn't really care about the contents of our list.
All it wants to do is scrutinize the outermost constructor, the one farthest
from [], and check whether it is [] or something else. Once we apply
isEmpty to zeros, we get our False result as expected, but something
happens to zeros as well:

```
zeros :: [Natural]
zeros = <thunk> : <thunk>
```

Our zeros is not *itself* a thunk anymore. Rather, it is now known to be an application of the : constructor to two other thunks. This is not a fact hidden in the promise of a thunk anymore, this is known. And it is important to realize that while this knowledge doesn't mention the 0 and zeros we originally gave as payloads to :, it is still enough knowledge for isEmpty to disambiguate whether we are talking about an empty list or not. And this is perfect, because if scrutinizing zeros had forced the evaluation of that rightmost thunk, isEmpty would have diverged for reasons we'll understand soon.

When a constructor application is the outermost expression inside a thunk, and we try to evaluate that thunk somehow, the constructor application becomes apparent immediately, but *not* its payloads. We say that the thunk, which held but mystery, has now been evaluated to its *weak head normal form*. And no, these are not words brought together just for their cadence, although one must wonder whether this was accident or design. *Weak head normal form* means that only the *head* of this thunk, the outermost expression in it, is guaranteed to have been evaluated. That's all it means. And being a *weak* head, of course, it won't force its constituent thunks to reveal themselves.

So there we have it. Gently, isEmpty was able to derive meaning from intent, from the promise of a list made through types, from the fact that it was : and not [] who revealed itself first. isEmpty doesn't need to know about the values in the list, so it doesn't peek, leaving the thunks unevaluated, preserving the list's laziness.

Funnily, however, our application of isEmpty to zeros created *yet another thunk* for a value of type Bool. So in reality, no evaluation has taken place so far in this chapter. It has all been a lie. Sure, we know to what extent isEmpty zeros *would* force the evaluation of zeros in order to become False if somebody asks for this boolean result, but nobody seems to be asking for this, so nothing is ever evaluated.

It might help to learn how programs *run* once they have been compiled. Essentially, every executable program has some sort of magical entry point, a function, that will *always* be evaluated in full so that the computer can find in it the orders it will blindly follow. So if we decide to use isEmpty zeros within this function, *this* would force a Bool out of our isEmpty zeros thunk, which would in turn force zeros to be evaluated to weak head normal form. But none of this matters for the time being, we are not in a rush to run anything anywhere, we are lazy, so let's go back to our thunks.

Ninety-five

Let's continue with the assumption that zeros has been evaluated to the extent it needed to be evaluated in order for isEmpty zeros to deliver its False result. That is, let's assume that zeros is known to be <thunk> :

```
<thunk>.
```

zeros is in *weak head normal form*, for *cons* : is known, yet acquiring this
knowledge didn't force the evaluation of any of the thunks mentioned as
its payload. But what about the expression isEmpty zeros itself, known to
be False? Is it also in weak head normal form? Surprisingly perhaps, *yes*,
yes it is, because discovering False didn't force the evaluation of any other
thunk within this constructor. Sure, there was never a thunk inside False
that could be evaluated to begin with, seeing how this constructor has no
room for carrying another expression as payload, but that's beside the
point. Nevertheless, this scenario of an expression being *fully* evaluated,
like False, with no thunks in it at all, is curious enough that it has its own
name. Very confusingly, we say that False is an expression in *normal form*,
and this is *in addition* to its being in weak head normal form.

In other words, in Haskell, the extent to which an expression has been
evaluated is one of two and a half alternatives. If the expression we are
considering is completely unevaluated, then it is a thunk. Otherwise, it is
an expression in weak head normal form. That's two. Now, expressions in
weak head normal form *may or may not* contain unevaluated sub-
expressions. If they don't, like False, then we say the expression is in
normal form too, *beside* being in weak head normal form. That's two and
a half, isn't it?

So, since our idea of infinity in zeros is hidden behind a thunk, as long as
we don't evaluate that thunk to normal form, we'll be alright. An infinity
in a box.

The problem with sum is that in order to deliver results, it *must* bring all of
zeros into normal form. For didactic purposes, let's look at how this
happens by going back and observing our first implementation of sum, the
one that didn't rely on monoids or folds. However, know that what we
are about to see is true of other versions of sum too.

```
sum :: [Natural] -> Natural
sum = \xs -> case xs of
               [] -> 0
               x : rest -> x + sum rest
```

As before, we start with zeros as an unevaluated expression, a thunk.

```
zeros :: [Natural]
zeros = <thunk>
```

We will assume something forces the evaluation of the thunk resulting
from saying sum zeros to weak head normal form. Which, in this case,
seeing as the result is a Natural number, will coincide with its normal form
too. That is, we are expecting the Natural number resulting from sum zeros
to be fully evaluated. Just a number, no thunks.

When we apply sum to zeros, which inside sum we call xs, the first thing that happens is that a case expression scrutinizes this xs in order to decide where to go next depending on whether we encounter [] or :. This forces xs to be evaluated to weak head normal form. Encountering [] means we are done, but in our case we know this will never happen because the list is infinite and there's no [] in it, so we pattern match on x : rest instead. Up until this point, zeros has been evaluated to <thunk> : <thunk>, just like in isEmpty zeros. However, we are not finished yet.

Next, we have an application of + to x and a recursive call to sum rest. Adding two numbers requires those numbers to be fully evaluated to normal form. If you recall, back when we were learning how to use foldr to implement sum, we found ourselves in a situation where we weren't able to perform our additions because we only knew about one of the two numbers we wanted to add, so we needed to wait until we could get an actual number to perform the addition. We were, in other words, waiting for these two parameters to be in normal form. Here, we are seeing this again. x + sum rest wants both x and sum rest to be in normal form, to be actual numbers free of thunks. Only then can x + sum rest itself become a numerical value.

While we are being terribly obsessive with the order of evaluation of our program, we should highlight that we don't really know which of these two arguments, x or sum rest, our function + will evaluate first. We can discover this by reading the documentation for + to see what it says, or with proper tinkering, we can discover it on our own. We will learn how to do that soon, but for now let's profit from our ignorance. Conceptually, it doesn't matter whether we evaluate x or sum rest first, because our program will diverge nonetheless. Eventually, both x and sum zeros will need to be actual numbers in normal form anyway. This can certainly happen for x, which is just 0, but it won't happen for sum rest. Why? Well, keep in mind that we are *already* trying to evaluate sum zeros, for which we are forced to evaluate sum rest. However, if we recall, our definition of zeros was 0 : zeros, so in this case rest is just another name for zeros. In other words, we are saying that evaluating sum zeros requires evaluating sum zeros. Thank you very much, that's not helpful at all. We are at a loss. At this point sum zeros depends on itself as much as undefined did. It diverges. Had we evaluated x before sum rest, we would have diverged nonetheless, but not before discovering the actual numeric value of x, 0, in vain. In fact, even if we ignored x altogether, we would still diverge. Let's see how.

Ninety-six

This is length, a function counting the number of elements in a list.

```
length :: [x] -> Natural
length = \xs -> case xs of
                  [] -> 0
                  _ : rest -> 1 + length rest
```

length doesn't pay attention to the actual elements of type x in this list, as witnessed by our use of an underscore _ in our pattern _ : rest. All length cares about is how many applications of : there are. That's all. Each time length encounters :, it adds 1 to the result of recursively applying length to the rest of the list. Using equational reasoning, it's easy to see how this works for a *finite* list like [9, 4, 6].

```
length [9, 4, 6]
== 1 + (length [4, 6])
== 1 + (1 + (length [6]))
== 1 + (1 + (1 + (length [])))
== 1 + (1 + (1 + 0))
== 1 + (1 + 1)
== 1 + 2
== 3
```

Indeed, 3 is the length of [9, 4, 6]. But what about the length of an *infinite* list like zeros? How do we calculate it? Well, we don't. Intuitively, expecting a finite number measuring the size of an infinite structure is kind of silly, isn't it?

If you look hard enough, you'll notice that the only difference between sum and length is that whereas sum adds together the elements of the list themselves, length adds a 1 instead, each time it encounters an element. That's all. In other words, it ignores the elements themselves, it never tries to evaluate them, they stay inside their thunks. Still, the use of 1 + length rest forces length rest itself to be evaluated to normal form, which diverges just like sum rest did, because given our definition of zeros as 0 : zeros, rest and zeros are really the same. So length rest means length zeros, which is exactly what we are failing to evaluate to normal form at the moment, so our program goes into an infinite loop.

Anyway, the fact that length doesn't try to evaluate the elements in the list implies that length, contrary to sum, is not trying to force the given list to normal form. Yet, both sum and length diverge. In other words, it doesn't take a normal form evaluation to have an infinite structure blow up, all it takes is being strict enough, so be careful.

length is essentially uncovering the structure of the given list without forcing the evaluation of the individual elements themselves. length will force each and every use of : and [] to their weak head normal form, but not the actual elements contained in them. We call this structure, devoid of its content, the *spine* of a list. Or, well, of any other non-list data type we might be dealing with.

In other words, evaluating the spine of a list means discovering whether its structure is [], [_, _], [_, _, _] or something else *without* necessarily evaluating the elements in them.

Ninety-seven

Being as strict as sum makes our programs diverge. Being a bit less strict but still strict enough, as length, still doesn't work. Being as lazy as isEmpty seems to do the trick, but that function doesn't do much. Let's see if we can find something more productive that can cope with infinite lists.

```
head :: [a] -> Maybe a
head = \case [] -> Nothing
             a : rest -> Just a
```

head is a function that returns Nothing if the given list is empty, or Just the first element of the list otherwise. Well, "first" might seem unfair, since we know that lists start from [], and using : we pile new elements on top of it. But here we really mean *first* in the *Last In, First Out* sense of the stack of plates. So, yes, head returns the first element of the list, if any. The name "head" might be confusing, considering we just talked about "weak head normal form" and the like, but this is just an accident: *that* head is unrelated to *this* head. Feel free to mentally rename this function to "first", "top", "outermost" or anything else that makes you happy.

Would head work with an infinite list like zeros? Let's see. head scrutinizes the given list so as to tell apart [] from :, just like isEmpty did. At this point, zeros has been evaluated to <thunk> : <thunk>. Next, head completely disregards rest, which in the case of zeros means just zeros. This is great. It means that we don't have to worry about the infinite zeros anymore, which suggests that *yes*, head works with infinite lists. And what about a? Well, we just put it inside another constructor, Just. This doesn't force the evaluation of a either. It simply preserves the evaluation state of a, thunk or not, inside a new constructor. But even if this did force the evaluation of a, head would still be able to cope with infinite lists, since it's never a in a : as who can make our programs diverge. Well, except when a is undefined or similar, but why would it be? head zeros, when evaluated, safely results in Just 0.

Ninety-eight

Alright, using head we are able to extract the first element of a list. But lists can have than one element in them, so in theory we should be able to write a function that *takes* the first few elements of a list, rather than just one of them. Let's write this function, let's call it take.

```
take :: Natural -> [x] -> [x]
take = \n xs -> case n of
                    0 -> []
                    _ ->
                        case xs of
                            [] -> []
                            x : rest -> x : take (n - 1) rest
```

Our implementation of take is rather ugly, aesthetically speaking. Very, actually. We'll take care of that soon, but for now let's content ourselves with the fact that it works as expected. take 3 zeros, say, returns a list with the three first elements of the list zeros. That is, [0, 0, 0]. And of course, it works for non-infinite lists too. For example, take 2 [5, 7, 4, 1] returns [5, 7]. If we ask for 0 elements, as in take 0 zeros, then we get an empty list [] as requested. Let's see how all of this works, and more importantly, *why*.

Actually, no. No, no, no. This is just too ugly. Let's clean it up before trying to make sense of it.

If you recall, we said that foldr was able to cope with infinite lists, so in theory we could rely on foldr to implement a fine looking take. And yes, yes we could, but the solution would be a bit tricky to follow, so maybe let's try something else first. In Haskell, the following two definitions of the id function are the same:

```
id :: x -> x
id = \x -> x

id :: x -> x
id x = x
```

That is, rather than binding the input value to the name x as part of a lambda expression, we do it on the left-hand side of the equals sign =. This is another *syntax* for defining functions, different from the \... -> ... syntax we've been using so far. But mind you, *conceptually*, from the point of view of the lambda calculus, id is still a lambda expression, a function, and will be treated as such. This is mostly just a *syntactic* difference, not a semantic one. And *why* do we have two ways of defining functions? Mostly because it can be convenient at times, we'll see. We call these alleged cosmetic improvements *syntactic sugar*. Haskell is sweet and sticky like that.

There are two interesting things about these equations. First, not only can they bind expressions to a name, they can also *pattern-match* on them. For example, let's redefine that isEmpty function from before using equations.

```
isEmpty :: [a] -> Bool
isEmpty [] = True
isEmpty _  = False
```

Ah, that's nice. We are saying that if the input to this function is the empty list [], we return True, and if it's anything else, as caught by the wildcard _, we return False. Just like in case expressions, the different patterns are tried from top to bottom, and the first one that matches gets to decide what happens next.

But perhaps more interesting, these equations allow us to bind and pattern-match more than one input at a time, something that we just can't do with case expressions. Well, not directly anyway. Let's profit from this by implementing a better looking version of take.

```
take :: Natural -> [x] -> [x]
take _ []        = []
take 0 _         = []
take n (x : rest) = x : take (n - 1) rest
```

Sweet, sweet sugar. What take is doing is quite clear now. If the given list is empty, we won't be able to select any elements from it, for there are none, so we just return another empty list []. The input list will be evaluated to weak head normal form as a consequence of attempting to pattern match against it. Incidentally, notice that because we completely ignored the first parameter to this function, we didn't force its evaluation. If it was a thunk, it would still be a thunk.

On the second take equation we consider the case of the given Natural being 0. If it is, then we are essentially being asked for a list with no elements in it, so we can return [] without even having to look at what's in our second input parameter. Of course, in order to compare this Natural number against 0, Haskell had to force it to normal form behind the scenes. So, by the time we are done trying to match against the pattern in this second equation, the Natural number will have been evaluated to normal form, too.

We could swap the order of these first two equations, too. This would have Haskell evaluate the Natural first, and the [x] afterwards, if at all necessary. Whether we do this or that depends on how we expect take to be used, and whether we expect the Natural or the [x] to be the the cheapest one to evaluate. Making a decision like this, in this take scenario, will involve a bit of wishful thinking for sure. We'll just keep it the way we wrote it, with [x] being evaluated first.

Finally, if none of the patterns in the first two equations matched, we move on to the final equation, in which we know that n is not 0 and the given list is not empty, because we already dealt with those cases before. So we proceed, confidently, to extract that x from the input of the list, to put it on the output list, and then to recursively call take asking for one less element from a list one element shorter.

Using equational reasoning, we can see how take 2 zeros reduces to [0,

0], for example.

```
take 2 zeros
== -- 3rd equation.
   0 : take 1 zeros
== -- 3rd equation.
   0 : (0 : take 0 zeros)
== -- 2nd equation.
   0 : (0 : [])
== -- Syntactic sugar.
   [0, 0]
```

Anyway, take consumes just enough from the infinite zeros so as to construct a list of length 2 as requested. After those two elements are found, it finalizes the list by sticking a [] at the end, rather than yet another recursive call to take as it had been doing so far.

Similarly, we can see how take 3 [4] has an early encounter with [] as we hit the first equation, even though 3 demanded more.

```
take 3 [4]
== -- 3rd equation.
   4 : take 2 []
== -- 1st equation.
   4 : []
== -- Syntactic sugar.
   [4]
```

Alright, take works. take takes a list, infinite or not, and consumes it to the extent that it needs to do so without diverging. This is already a useful achievement, but perhaps more importantly, contemplate how this lazy consumption of infinite structures allows us to keep the beautiful definition zeros as it is, reading like poetry, and take too, with its striking straightforwardery. But the real treat, actually, is how take *lazily produces* a list while *lazily consuming* another one.

In our third equation, where take returns x : take (n - 1) rest, what's really happening is that we are constructing a new thunk which when evaluated to weak head normal form will become x : <thunk>, where x preserves the evaluation state it had on the original list, thunk or not, and the thunk that shows up as the second parameter to : defers the recursive call to take. *Defers*, that's the important bit. So, really, no work has happened here either. take, on this equation, is taking a lazy list and lazily turning it into yet another lazy list. It's not until we actually evaluate the resulting lazy list that take does any work. So, for example, we could take the first seven hundred trillion elements of an infinite list, and this would be free as long as we don't actually try to do something concrete with these elements, such as adding them together with sum.

So why is it that take succeeds where sum fails? It really is quite simple. While sum is essentially folding a list with +, a function that will force the evaluation of its inputs to normal form before being able to produce an output, take, using :, can produce its output right away without having to force its payloads beyond their current evaluation state. That's all. This allows take to lazily produce a list, which it does by lazily consuming another one. This is the same laziness that allowed zeros to exist, together with a myriad of other beautiful constructions we'll see later on.

Ninety-nine

Just like in case expressions, we can rearrange the order of patterns in functions using equation syntax to some extent. However, we need to be careful about two things. First, laziness. To illustrate this, let's rewrite and using this equation syntax.

```
and :: Bool -> Bool -> Bool
and False False = False
and False True  = False
and True  False = False
and True  True  = True
```

This looks cute, but it is actually wrong. The problem is the same as before: and shouldn't be trying to scrutinize its second parameter until it has confirmed that the first one is True, because this forces it into normal form, which will in turn make us do that trip to the market, perhaps in vain. We need to avoid scrutinizing the second Bool until we are certain we need it.

```
and :: Bool -> Bool -> Bool
and False _     = False
and True  False = False
and True  True  = True
```

Much better. By the way, we are aligning things vertically because it looks nice, but it's not necessary at all. The position of the = sign is irrelevant, as well as the whitespace to its left and in between the function parameters. For example, this disgusting arrangement means exactly the same as above:

```
and :: Bool -> Bool -> Bool
and False          _       = False
and True False        = False
and          True    True = True
```

Anyway, this and works as expected. However, we are doing something silly. We force this second Bool into normal form, we scrutinize it, only in order to return a Bool with exactly the same value. Why not return that second Bool without even looking at it?

```
and :: Bool -> Bool -> Bool
and False _ = False
and True  x = x
```

Better. So, yes, laziness is something for us to keep in mind when deciding how and in which order we write our patterns and scrutinize our inputs.

The second thing we need to consider when arranging our patterns in equations or case expressions is whether they overlap with each other, which may accidentally cause a pattern to catch the wrong thing. For example, consider this rearrangement of the take equations.

```
take :: Natural -> [x] -> [x]
take _ []        = []
take n (x : rest) = x : take (n - 1) rest
take 0 _         = []
```

Compared to our previous take, we swapped the order of the last two equations. But now, the first and second equations alone will catch all the possible combinations of inputs to this function. Luckily for us, Haskell's most kind compiler GHC will let us know about this oversight during compile time. We can take a hint that this new version of take is doing something wrong by realizing that it will *never* force the evaluation of the given Natural value. Neither binding it to a name such as n, nor using n as one of the arguments in a function application like in n - 1, will force the evaluation of n. Actually, we could even get rid of n and that last equation, and things should behave exactly the same.

```
take :: [x] -> [x]
take []        = []
take (x : rest) = x : take rest
```

Do you recognize this new take? Pay close attention. That's right, it's the identity function once again. So be careful about the order in which you write your patterns, or you might end back where you started.

Moreover, in case you didn't notice, our broken take with its equations rearranged leads to the silly situation of trying to apply n - 1 even when n is 0, in which case the resulting value will not be a Natural number, for there's no natural number less than zero. Unfortunately, this type-checks and the problem goes completely unnoticed because the evaluation of this Natural never happens. Sure, take is failing in other ways too, so hopefully that makes our program collapse sooner rather than later, but we can't guarantee that. Nothing good can come out of this.

One hundred

There are two reasons why our broken take was trying to come up with these non-existent "negative natural numbers". One could be seen as an accident, but the other one was the irresponsibility that allowed the accident to take place.

The fundamental problem is in the type of -, the function that subtracts two `Natural` numbers and returns the result as yet another `Natural` number.

```
(-) :: Natural -> Natural -> Natural
```

If you recall, the smallest `Natural` number that can possibly exist is 0. However, the type of - allows us to say things like 3 - 4, where both 3 and 4 are perfectly acceptable `Natural` numbers, yet the expected result, also of type `Natural`, can't possibly exist. There is no such thing as a "negative one natural number", as 3 - 4 :: `Natural` would want us to believe. This is the problem our wronged take accidentally faced. But the culprit was not take. It was - who enabled the accident in the first place. So let's fix -, let's change its type. Or maybe let's keep - as it is, for reasons that will become apparent later on, but let's write a wrapper around it that will perform the subtraction safely. Let's call it safeSubtract, and then let's have take use it.

From where we are standing, the most evident solution would probably involve changing the return value to a `Maybe Natural`, so that `Nothing` is returned whenever the resulting `Natural` would be nonsense. That is, we want safeSubtract to have type `Natural -> Natural -> Maybe Natural`.

The implementation of safeSubtract is rather straightforward. All we need to do is ensure that b is not bigger than a, as this would result in a non-existent "negative natural number". Alright, for this task, let's use another infix function that comes with Haskell. It's (<), of type `Natural -> Natural -> Bool`, to be used as a < b and returning `True` whenever b is bigger than a.

```
safeSubtract :: Natural -> Natural -> Maybe Natural
safeSubtract a b = case a < b of
                      True -> Nothing
                      False -> Just (a - b)
```

So there we have it, safeSubtract will *never* return an invalid `Natural` number because it always checks that subtracting a and b makes sense before doing it. If it doesn't, it just returns `Nothing`. How would take make use of this function? Let's see. Let's start with a sane take.

```
take :: Natural -> [x] -> [x]
take _ []       = []
take 0 _        = []
take n (x : rest) = x : take (n - 1) rest
```

Now let's replace n - 1 with safeSubtract n 1.

```
take :: Natural -> [x] -> [x]
take _ []       = []
take 0 _        = []
take n (x : rest) = x : take (safeSubtract n 1) rest
```

But of course, this doesn't type-check anymore because take takes a `Natural` as its first argument, not a `Maybe Natural` as safeSubtract n 1 returns. So maybe we should pattern match on safeSubtract n 1 first, and only if we have a `Just` we recursively apply take.

```
take :: Natural -> [x] -> [x]
take _ []        = []
take 0 _         = []
take n (x : rest) = x : case safeSubtract n 1 of
                           Nothing -> []
                           Just m -> take m rest
```

Hey, look at that, it worked. And, actually, since this third equation is already dealing with the fact that n might be 0, perhaps we can drop that second equation altogether. Let's see.

```
take :: Natural -> [x] -> [x]
take _ [] = []
take n (x : rest) = x : case safeSubtract n 1 of
                           Nothing -> []
                           Just m -> take m rest
```

Hmm, no, that's not quite right. For example, consider take 0 [5, 3, 8].

```
   take 0 [5, 3, 8]
== 5 : case safeSubtract 0 1 of
          Nothing -> []
          Just m -> take m [3, 8]
== 5 : []
== [5]
```

That is, we are selecting the first element of our list *before* scrutinizing safeSubtract n 1 to see if we were asked for any element at all. Now, this is not necessarily a bad thing. Perhaps this is what we wanted take to do. It is important to tell apart "mistakes" like this one, where we are successfully repurposing a function, from actual silly ideas such as allowing "negative natural numbers" to exist. At some point we need to tell the computer what to do, and it's not the computer's responsibility to judge whether we asked for the right thing or not. Of course, as much as possible, once we know what it is that we want to solve, we want the type-checker to help us achieve this and nothing else. However, the type of take says *nothing* about its intended behaviour, so in return we don't get any help from the type-checker and make "mistakes". But don't let this scare you. It is possible to be precise enough in our types so that it becomes *impossible* to write a misbehaving take. Not yet, though. Maybe later.

Anyway, next take. Let's try to use safeSubtract *before* attempting to construct the list and see what happens.

```
take :: Natural -> [x] -> [x]
take _ [] = []
take n (x : rest) = case safeSubtract n 1 of
                       Nothing -> []
                       Just m -> x : take m rest
```

Ah, yes, this works as expected, just like our very first take. Is it *better*? Well, no, why would it be? Improving take was never our goal. Our goal was to make Natural subtraction safe, and *that* is all we did.

And by the way, perhaps surprisingly at first, we could have implemented safeSubtract as follows:

```
safeSubtract :: Natural -> Natural -> Maybe Natural
safeSubtract a b = bool (Just (a - b)) Nothing (a < b)
```

This works just fine because even if it *looks* like we are trying to subtract a - b right away, in reality, due to the laziness of bool, this subtraction doesn't even happen *unless* a < b is known to be False. Ultimately, this is why we cherish laziness. It's not so much about the lazy lists nor the other lazy structures, but about how expressive and ergonomic our programming experience becomes when we don't have to worry about *when* something will be executed. We write our expressions optimistically, and leave it up to Haskell to figure out if and when they deserve execution. It's so comfortable that we sometimes forget this even exists.

One hundred one

Adding two natural numbers results in another natural number. Multiplying two natural numbers results in another natural number. *Subtracting* two natural numbers, however, does *not* necessarily result in another natural number. Specifically, subtracting a - b results in a natural number *only if* a is greater than, or equal to, b. There is a pattern here, this is not an accident.

When an operation on one or more values of a particular type results in yet another value of that same type, we can say that this type is *closed under* this particular operation. For example, the natural numbers are closed under addition and multiplication, but they are *not* closed under subtraction. They lack this property.

Monoids too, for example, are closed under their mappend operation of type a -> a -> a, for some choice of a such as Sum or [x].

So what happens when we subtract Natural numbers, seeing as they are not closed under subtraction? We can address this question from many perspectives. For example, we can say that Nothing happens. That's what we did before. But, in reality, we "know" that subtracting 3 - 4, say, results in -1. So where does that -1 come from? The naturals find closure under subtraction in the *integer numbers*, which are the naturals together with their negative counterparts. In Haskell, integer numbers have type Integer. This is the type of -1.

So perhaps what we want, actually, is for safeSubtract to have type Natural -> Natural -> Either Integer Natural. That is, return a Natural if possible, or an Integer otherwise.

Really? I don't think so. We just said that the integer numbers include *all of the natural numbers* plus their negative counterparts. In other words, we are saying that the naturals are a *subset* of the integers, meaning that rather than having Either Integer Natural, artificially segregating the natural subset of the integers in this sum type, maybe we could just use Integer as the return value of safeSubtract, as Natural -> Natural -> Integer. We could, we could. However, since we are now returning Integers, we might as well take Integers as input so that we can do things such as safeSubtract -3 7. What we really want is a function of type Integer -> Integer -> Integer. That would work perfectly fine, because the integer numbers, contrary to the naturals, are closed under subtraction. Of course, if we now wanted to subtract Naturals through this function, we would need to convert them to Integers first. That's easy, though, considering how naturals are a *subset* of the integers, meaning that any Natural can be represented as an Integer too. In fact, Haskell comes with a function toInteger of type Natural -> Integer that does just this.

By the way, for very superficial parsing reasons, in Haskell we can't actually write safeSubtract -3 7. However, we can say safeSubtract (-3) 7, with those extra parentheses, or even safeSubtract (negate 3) 7, to achieve what we expect. Anyway, not important.

One hundred two

Actually, we can get rid of the name safeSubtract and use - directly to subtract Integer numbers. There's nothing that could be characterized as "unsafe" regarding the subtraction of integer numbers, so we might as well use the shorter infix name -, which we can do because the type of - is not Natural -> Natural -> Natural as we said before.

```
(-) :: Num x => x -> x -> x
```

The polymorphic type of - says that you can subtract values of types for which there is an instance of the Num typeclass, such as Integer and Natural. So yes, we can use - to subtract integer numbers too.

Num is one of the ugly kludges that we have in Haskell as of the third decade of the third millennium. We witnessed this before, when - promised to be able to subtract two natural numbers without leaving room for a potential failure, which we know was a lie. Luckily for us, we can mostly ignore Num and nonetheless be fine. Contrary to mathematical operators in many other languages, baked in as those booleans from before, Num and operators like - are just user written code that happens to be shipped together with the language, which we can choose not to use. We could, for example, define a closed - only for types where it makes sense, excluding Naturals. Will we ignore Num in this book? Not entirely, but we will learn how to do so if we so desire, of course.

Something we should highlight before not paying much more attention to Num throughout the rest of this book is that when we say 3 or -4, for example, the type of these expressions is neither Natural nor Integer, but rather Num x => x. For example, 3 is a value constructor meaning "the number three", which can be used to construct a value of *any* type that implements an instance for the Num typeclass. It is implemented as follows: Haskell magically converts your digits into an Integer, and then further magic applies fromInteger :: Num x => Integer -> x to that Integer, where fromInteger is one of the methods we get to implement as part of the Num typeclass. On the one hand, this is quite clever. On the other hand, in the Num instance for Natural numbers, say, we are forced to implement the method fromInteger with type Integer -> Natural. Really? What would be the answer to fromInteger (-4), say? There's no right answer. You see, a kludge. There are nice models for understanding what a number is, and we'll see some, but Num is not one of them. Num is why programmers cry.

One hundred three

One could argue that take should return Maybe [x] rather than [x], so that the caller gets a Nothing rather than a shorter list when the number of elements in the input list is less than requested. Indeed, one could argue that, but it would be in vain. Let's see why.

```
takey :: Natural -> [x] -> Maybe [x]
takey 0 _ = Just []
takey _ [] = Nothing
takey n (x : rest) = case takey (n - 1) rest of
                          Nothing -> Nothing
                          Just xs -> Just (x : xs)
```

Our new takey delivers the result it promises, and even deals with infinite lists just fine. So what's the problem? What's happening is that in takey's third equation, *before* making use of the lazy constructor :, we are *forcing* the evaluation of takey (n - 1) rest by scrutinizing it through a case expression. Otherwise, how would case know whether to return Just or Nothing without first learning whether Just or Nothing is the result of the recursive call to takey itself?

takey took the laziness take had away, for take never had to scrutinize its tail in order to construct a list with an x and a mere promise that there *could* be more xs, thus preserving the list's laziness. Remember, a list doesn't really need to know about its contents, not even when it's being constructed. It is only when we try to observe it beyond its weak head normal form that we start forcing secrets out of its thunks.

And why do we care about preserving laziness? Well, that's the wrong question. The right question is: Given that lists can be produced lazily, allowing lazy consumptions to be performed efficiently at a later time, if at all necessary, *why wouldn't we allow so*? We don't know what the future may call for, nor what this laziness may allow. We shouldn't arbitrarily restrict things just because we can. Not unless we can justify these constraints.

One hundred four

Many times, though, we *need* to be certain that our list has an exact number of elements in it. But even in those cases we may want to keep the core functionality of take as it is, and use it in combination with another function like length to ensure that the obtained list is of a particular length. Generally speaking, we'd rather avoid this because there are data structures other than lists that are more efficient at knowing their own lengths. Remember, a list must iterate itself completely to learn its length, it must uncover its *spine*, which is certainly a more expensive computation than just reading a number somewhere saying "oi, it'll be one-thousand seventy-five alright, the length". Alternatively, we could modify lists so that they carry this length information with them somehow. That would be interesting. We could, yes, but for now, let's just explore this whole idea using our boring and lazy linked lists.

```
takey :: Natural -> [x] -> Maybe [x]
takey n xs = case take n xs of
               ys ->
                 case length ys == n of
                   True -> Just ys
                   False -> Nothing
```

We are using the first case expression just to give take n xs a name, as we'll be using it twice afterwards. Whatever take n xs may be, we'll call it ys. Then, we are checking whether the length of ys equals n as requested. When this is True, we return ys wrapped in Just. Otherwise, Nothing.

We are using the infix function == for the first time here. This function returns a Boolean value indicating whether it is True or False that its two arguments, here length xs and n, are equal.

Scrutinizing the result of length ys == n to check whether this is True or False forces the evaluation of this expression to its normal form, which in turn forces length ys to normal form, which itself forces ys to reveal its spine, losing its structural laziness. This is where most of the laziness is gone. Its elements remain a mystery, though.

This == is actually defined as a method of a typeclass called Eq.

```
class Eq (x :: Type) where
  (==) :: x -> x -> Bool
  a == b = not (a /= b)

  (/=) :: x -> x -> Bool
  a /= b = not (a == b)

  {-# MINIMAL (==) | (/=) #-}
```

Ah, some new noise trying to distract us. Let's take this step by step. First, the high-level overview. Eq is a typeclass for Types with *two* methods in it, just like the Monoid class from before. One named == and the other /=. However, there's also a *pragma*, a special instruction that the authors of Eq wrote *for the compiler*, that says when *we* implement an Eq instance for a type of our choice, we don't really need to implement both methods. Rather, we can get by implementing only the minimum requirements. Namely, *either* == *or* /=. That is what the {-# MINIMAL (==) | (/=) #-} pragma is saying. Why, though?

Contrary to typeclasses we've seen before, both == and /= come with default definitions that will serve as the implementation of these methods for the instances of Eq in which we have chosen *not* to override some of these methods. In particular, a == b is defined to be not (a /= b), and vice-versa.

```
not :: Bool -> Bool
not True  = False
not False = True
```

not negates a boolean value. not True is False, not False is True. So what a == b = not (a /= b) is saying, concretely, is that comparing a and b for equality using a == b equals the *negation* of a /= b, which suggests that a /= b tells us whether a and b are *different*, rather than equal. And, indeed, if we look at the default definition of a /= b as not (a == b), we can see that this conveys the idea of a and b *not* being equal.

What's interesting about these two functions is that they *recursively* depend on each other. If we tried to evaluate a == b or a /= b as they are, without changing the definition of at least one of == or /= in the relevant Eq instance, our program would *diverge*, for a == b would depend on a /= b which would again depend on a == b, creating an infinite loop. This is why the authors added that MINIMAL pragma to the typeclass: It causes the compiler to complain if we fail to define at least one of == or /= explicitly.

As for why equality == and non-equality /= are both part of this Equality typeclass: It is mostly a matter of convenience. Sometimes it's easier or faster to decide that two things are equal, while other times it's easier to notice that they are different. We are studying *this* Eq because it gives us an opportunity to showcase the role of default method definitions and

MINIMAL pragmas in typeclasses, but conceptually, we could have Eq, == and /= defined this other way, and it would have been enough:

```
class Eq (x :: Type) where
    (==) :: x -> x -> Bool

(/=) :: Eq x => x -> x -> Bool
a /= b = not (a == b)
```

That is, we only kept the == method in the Eq typeclass —without a default definition, because there's not one definition that could possibly work for *all* choices of x— and we converted /= into a function that simply negates whatever == says. The Eq x constraint on /= requires that there be an Eq instance defined for the chosen x. Otherwise, without it, we wouldn't be able to say a == b within the definition of a /= b.

One hundred five

We used a strange syntax this time to define our infix operators, but there's nothing special going on beyond a high dose of syntactic sugar. All the following mean the same thing:

```
a == b = not (a /= b)

(==) = \a b -> not (a /= b)

(==) a b = not (a /= b)

(==) a = \b -> not (a /= b)
```

Well, operationally, there *might* be some difference between these definitions, but they do *mean* the same.

For completeness, one thing that's missing from the definition of the Eq typeclass are explicit *fixity* settings for == and /=. Without them, it will be hard for us to understand the precedence of these operators when used alongside other infix operators in the same expression. Somewhere in the source code where the Eq typeclass is defined we'll find some lines saying that both == and /= have an infix precedence of 4.

```
infix 4 ==
infix 4 /=
```

Notice that it doesn't say whether they associate to the left or to the right. That's fine, these operators are not associative anyway, so there's nothing to say.

One hundred six

Here's a funny story. That thing we learned about both == and /= being part of the Eq class? Not true anymore. While this book was being written,

Haskell changed. After a heated debate, the people that steer Haskell in one direction or another decided that it was better for == to be the sole method in the Eq typeclass, and for /= to be implemented as a normal function outside of Eq. So, things changed.

Was it a good change? Was it bad? Irrelevant. It *was*, and that's what matters to us. We need to get comfortable with the idea that software doesn't exist in a timeless vacuum. Software relates to ideas, to people, to technology. Software exists in the market. And over time, as those things change, so will relationships, and often, so will our programs. We rely on libraries and infrastructure made by others, we seldom build programs from scratch. A small change in one of those dependencies may cause our programs to stop compiling. Or even worse, they may behave differently yet continue to compile. We need to be aware of this and write software in such a way that if things were to change in the future, *when* things change in the future, we'll notice. We keep what's good, we reject what's bad, we adapt. We must.

As much as possible, we rely on the type system to let us know if there are changes to which we should be paying attention. Having programs fail to compile when they should is what we want. However, some situations will require us to rely on *tests* instead. Tests are programs that are not intended for public consumption, but for the programmers building things. We use tests to ensure that as we create our software, as our product evolves, as the environment where our program is expected to run changes, we can still deliver a working product. For example, were we building a program for watching movies, then we could write tests for making sure that we continue to support old movie formats over time.

We run these tests before committing any change to our product. If their expectations are not satisfied, they shall report the relevant details back to us so that we can fix whatever is broken. It's only after we find the tests to be enough in quality and quantity, and satisfied with our changes, that we commit said changes to the final product.

But tests are not only for backwards compatibility. We can, for example, have tests that run our functions against random inputs so as to check that we didn't accidentally forget to deal with some obscure scenario. We can write tests after having solved a problem in our programming, just to keep us from accidentally recreating that same problem in the future. We can also test that a particular function in our program acts in a particular way for a particular set of inputs. We can test many things. We'd rather not, however. See, we have a powerful type system that we can use not only for preventing bad programs from existing, but also for creating rich vocabularies for reasoning about problems and expressing their solutions. We want to stay here as much as possible. It's good. It's peaceful. Some things, however, will inevitably fall through the cracks. We've seen this.

And those things, we will test. We'd rather not, but we nevertheless will. We must.

One hundred seven

Enough distractions. Let's stop being lazy and go back to discussing how lazy other things are. Not us.

Much like take, fmap transforms a list *lazily*. This is quite easy to see if we look at an implementation of fmap for lists.

```
fmap :: (x -> y) -> [x] -> [y]
fmap _ [] = []
fmap f (x : xs) = f x : fmap f xs
```

When the given list is empty, there is nothing to transform, so we just return another empty list. But if there is at least one element x in the list, as witnessed by the : constructor, we return another list with f x as its head and fmap f xs as its tail. Yeah, among friends we call the second argument to : its "tail". Picture a snake with a head, and then a long, long tail. What's important to keep in mind here is that : is *not* forcing the evaluation of its payloads f x and fmap f xs. All this function ever outputs, if forced to weak head normal form at all, is <thunk> : <thunk>. That is, it defers the transformations to the list, preserving its original laziness if any. Promises, just promises.

The *type* of the list changes right away, though. We are always eager to know the types of expressions, even if they don't exist yet. This makes complete sense, considering how the type-checker knows *nothing* about the evaluation of expressions, so it couldn't possibly care about whether they have been evaluated or not. Types are types, expressions are expressions, and they have different concerns. Well, kind of. There will come a time when we'll blur that boundary too, but it won't be today.

So yes, fmap is also one of those functions that lazily produces a list while lazily consuming another one. It could choose *not* to do so, but what would be the point of doing that? It's only when somebody *needs* to observe one of the elements of the produced list that the application of f will be evaluated. Only when one *needs* to iterate over it will its tail be uncovered insofar as necessary.

Can we fmap infinite lists? Sure we can, why wouldn't we? Infinity is not an alien concept in Haskell, it's reasonable to expect things to work perfectly fine with infinite structures most of the time, unless something in their description suggests otherwise. For example, here is an infinitely long list of eights, obtained by replacing each element in the infinitely long zeros with an 8.

```
eights :: [Natural]
eights = fmap (const 8) zeros
```

We can use eights in any way we could use zeros.

One hundred eight

The natural numbers, we said, are the zero or a natural number plus one. *The zero or a natural number plus one.*

```
         the zero        or a natural number
            ↓                     ↓
naturals = 0 : fmap (\x -> 1 + x) naturals
                            ↑
                         plus one
```

This funny-looking code works just fine. Well, except for the pointy labels which are there just to make a point. Ignore them. naturals, of type [Natural], is indeed an *infinite* list of natural numbers. Every natural number that exists is somewhere in this list. That is, naturals gives us a never ending list starting with 0 : 1 : 2 : 3 : ....

Before trying to understand how naturals works, let's talk about unimportant superficial matters. In the definition of naturals, rather than saying \x -> 1 + x, we can say (+) 1 to make things a bit cleaner. This is just a partial application of +, in *prefix* form, to its first parameter 1. We've seen this before.

```
naturals :: [Natural]
naturals = 0 : fmap ((+) 1) naturals
```

This, I'm told, is a bit easier on the eyes. And as long as we are cleaning things up, we could also partially apply + to 1 in *infix* form instead, as (1 +). Notice how + is *not* isolated in its own pair of parentheses anymore. With this syntactic sugar, saying (1 +) x is the same as saying 1 + x or (+) 1 x. That is, the operand that appears partially applied to the left of + will remain on the left once + is fully applied to its two arguments. Alternatively, we can partially apply the operand on the right as (+ 1), which when used as (+ 1) x will preserve 1's rightful place as x + 1. In the case of addition, a *commutative* operation, it doesn't matter whether we say (+ 1) x or (1 +) x because the result is the same either way. But this syntactic sugar works for any infix operator, and the order of the arguments might make a difference for them. For example, compare (x :) xs and (: xs) x, both meaning x : xs. The order of the arguments to the list constructor : does matter.

```
naturals :: [Natural]
naturals = 0 : fmap (1 +) naturals
```

The zero or a natural number plus one. Cute. Now we can go back to important matters. We can use equational reasoning to see how naturals produces all of the Natural numbers. Let's do it step by step, it will be tricky this time. We start by inlining the definition of naturals.

```
0 : fmap (1 +) naturals
```

And then we inline naturals once more.

```
0 : fmap (1 +) (0 : fmap (1 +) naturals)
```

Now, in our most recent definition of fmap, we said that fmap f (x : xs) equals f x : fmap f xs. That is, we can apply f directly to the first element and move the fmap application to the tail of the list. We say fmap *distributes* the application of f over all the elements of the list. We can use this understanding to modify our outermost fmap (1 +) (0 : ...) so that it becomes (1 +) 0 : fmap (1 +) ..., or its less ugly version where we reduce (1 +) 0 to just 1.

```
0 : (1 : fmap (1 +) (fmap (1 +) naturals))
```

Excellent. Let's bring back memories, now. Do we remember the composition law of Functors, the one which said that fmap g (fmap f xs) is equal to fmap (compose g f) xs? We can use this law to clean things up. Rather than saying fmap (1 -) (fmap (1 +) ...), we can say fmap (compose (1 +) (1 +)) ..., thus removing one fmap application, making things simpler to observe.

```
0 : (1 : fmap (compose (1 +) (1 +)) naturals)
```

We know function composition all too well. We know that compose g f x applies f to x first, and g to this result afterwards. In our case, g and f both *add one*, meaning that compose (1 +) (1 +) x will add 1 to x, and then it will add another 1 to that result. In other words, compose (1 +) (1 +) x adds 2 to x. We will go ahead and replace all of compose (1 +) (1 +) with (2 +). You may want to write down the equational reasoning that shows that compose (1 +) (1 +) in fact equals (2 +). We won't do it here.

```
0 : (1 : fmap (2 +) naturals)
```

At this point we don't have much to do but inline the definition of naturals, so we do that.

```
0 : (1 : fmap (2 +) (0 : fmap (1 +) naturals))
```

As before, we can distribute the application of (2 +) in fmap (2 +) (0 : ...) as (2 +) 0 : fmap (2 +) ...). We do that, and reduce (2 +) 0 to just 2 as well.

```
0 : (1 : (2 : fmap (2 +) (fmap (1 +) naturals)))
```

And once again, we can apply the Functor composition law to simplify things a bit.

```
0 : (1 : (2 : fmap (3 +) naturals))
```

We could repeat this process over and over again, but I trust seeing the first handful of Naturals come to be is more than enough proof that naturals in fact contains an *infinite* amount of Natural numbers in it, *all* of them, starting from 0 and incrementally growing by one. There's nothing in our reasoning suggesting we will ever stop lazily producing new numbers.

One hundred nine

So far we've been using our brain, like cavemen, to understand *when* things are evaluated. We haven't had any help from the type system, we've been on our own. This is sad. There *are* ways to encode laziness in our types. For example, we can have a strict language, one that fully evaluates *everything* as soon as possible, but still allows us to mark some things as "lazy" when necessary. In this hypothetical language, we can imagine lazy lists being defined as follows:

```
data List a = Nil | Cons (Lazy a) (Lazy (List a))
```

In this language, constructor payloads are *strict* by default, but Lazy x defers the evaluation of x until somebody removes the Lazy wrapper somehow. For example, using a function evaluate of type Lazy x -> x provided by the language itself.

Our hypothetical language could also force the evaluation of a and b in an expression f a b *before* allowing f to even attempt to do anything with a and b. Unless, of course, a or b are wrapped in that Lazy type constructor, in which case their evaluation will be delayed until explicitly requested. For example, we could define and as follows, somewhat suggesting that the second input parameter won't be evaluated right away. Although it doesn't *really* say that, does it?

```
and :: Bool -> Lazy Bool -> Bool
```

So why don't we do this? Well, actually, some languages with a type system as rich as Haskell's do it. Haskell's lazy-by-default nature is the exception, rather than the rule. Although we'll be glad to know that at the flip of a switch we can make the Haskell code we write be strict by default if we want to. And we can implement this Lazy idea, too.

But just because we can do this, in Haskell or otherwise, it doesn't mean we *should*. The thing to keep in mind is that while types such as Maybe x convey the semantics about *what* something is, a type such as Lazy x talks about *when* x should be. We are conflating in the type system meaning with execution, which is exactly what we generally want to avoid. Should we *not* do this, then? Well, we are not saying that either. Maybe we should. Mandating a strict order of evaluation of our program, whether this shows up in the types or not, could make its computery performance easier to reason about. Do we *want* to reason and worry about this, though? Shouldn't the computer figure out *when* it is more convenient to do something, enabling us to instead focus on *what* our programs should do, blissfully oblivious of these operational details? So many questions, so very few answers.

It's unclear what's the right *default* approach, they each have their virtues and shortcomings. We'll talk about Haskell's shortly. But generally speaking, even if for a particular problem we'd prefer to have a different evaluation strategy, that doesn't mean that we need to change the *default* evaluation strategy of the entire language. We can usually override the defaults on a case by case basis. For example, we might want to define a strict list, one where evaluating the list to weak head normal form will force the evaluation of its constituent parts to weak head normal form too.

```
data List a = Nil | Cons !a !(List a)
```

You see those exclamation marks preceding the types of the payloads in Cons? Those exclamation marks, informally called "bangs", tell Haskell that when evaluating a value of type List a to weak head normal form, the payloads themselves will have to be evaluated to weak head normal form too. Otherwise, without the bangs, the payloads wouldn't be evaluated unless explicitly requested by whoever consumes the list. Of course, whether the a itself contains unevaluated expressions is a different matter that will need to be addressed too. This definition of List implies, among other things, that it is *not* possible to use Cons to define an infinite list in the same way we used : to define zeros or naturals, because as soon as something attempts to evaluate Cons to its weak head normal form, this will trigger yet another evaluation of itself, which will trigger another one, and another one, and another one, leading us to an infinite loop if the list is infinite.

Similarly, we can have functions force the evaluation of their input arguments to weak head normal form too, even when not necessary. For example, consider the const function from before:

```
const :: a -> b -> a
const a _ = a
```

This function completely ignores its second input parameter b, so evaluating const a b won't force b beyond its current evaluation level. If b is a thunk, say, it stays a thunk. This makes sense. Why would we evaluate b if we don't need to? Yet, if we add but a single *bang*, the story changes completely.

```
const :: a -> b -> a
const a !_ = a
```

This function *still* doesn't use its second input parameter. Yet, if we ever try to evaluate the result of const a b, then b will be evaluated to weak head normal form too. This function exists in Haskell, it is called seq, and it is not as useless as it seems. Its arguments, however, are in different order.

```
seq :: a -> b -> b
seq !_ b = b
```

One hundred ten

The quintessential example of why forcing the evaluation of some value to weak head normal form is necessary, even if not *really* necessary, is foldl. It turns out that the implementation we gave for it is broken in a curious way.

```
foldl :: (z -> a -> z) -> z -> [a] -> z
foldl = \f acc as -> case as of
                       [] -> acc
                       a : rest -> foldl f (f acc a) rest
```

The problem with this implementation is that on each iteration of foldl, our use of f acc a to improve the accumulator acc creates a new thunk that won't be evaluated until we reach the end of the list. In principle, this sounds appealing, but there is a problem that has to do with computer resources. Values occupy space in memory, values such as the inputs to f, one of which happens to be a thunk that the previous iteration created. So if new thunks, which themselves occupy memory, cling to these previous thunks that already occupied memory, Haskell won't have the opportunity to reclaim any of this memory until the result from the entire execution of foldl is evaluated. In other words, as foldl iterates over the list, it demands more and more memory. And if this goes on for long enough, our computer will run out of memory. All this for no significant gain, because contrary to foldr, foldl is incapable of producing an output as it lazily consumes a list. It must consume the list entirely before being able to produce something.

In other words, foldl is way too lazy in the way it produces its output. Ironically, it achieves this by being overly clingy while also being way too strict in the way it consumes its input, which prevents it from working with infinite lists at all. Way to go, foldl.

How do we fix it? Well, unsurprisingly I hope, all we need to do is *force* the evaluation of the thunk that f acc a creates before making a recursive call to foldl. Or, looking at it from a different perspective, we need to make sure that acc has been evaluated to weak head normal by the time we get our hands on it. In this way we can be certain that we never create a thunk that depends on yet another thunk that foldl itself created. How do we achieve this? As we did in seq, we just precede acc with a bang ! right where we bind its name. That's all.

```
foldl :: (z -> a -> z) -> z -> [a] -> z
foldl = \f !acc as -> case as of
                        [] -> acc
                        a : rest -> foldl f (f acc a) rest
```

With this one change we are preventing foldl from creating more thunks that it needs, thus reducing its memory consumption. This strict, unclingy version of foldl, is actually called foldl' in Haskell, and we should almost always use it instead of the way too lazy foldl, which is only useful if f happens to be lazy in its first argument, but it almost never is. The choice is usually between foldr and foldl', rarely is foldl worth considering. It's a shame the most beautiful name foldl is wasted. *Foldl prime* though —that's how we pronounce foldl'— has a rather heroic ring to it, so we'll take it.

In practice, it turns out that worrying about things being too lazy is more important than worrying about them being too strict. Mostly because when things are too strict, we experience divergence in some form right away. But with an overly lazy program, we may not immediately notice what may cause our system to collapse later on, once it has more data to process. There is a trick to avoiding this situation. Perhaps we shouldn't talk about it because, after all, this is not a book about tricks, but here it goes anyway just in case. When writing a function like foldl which uses some kind of accumulator, we *always* make that accumulator strict. We bang that acc. We'll have opportunities to repent and make things lazy later on if we realize they needed to be lazy after all, but by default we make them strict just in case. That's it. Our program continues to be lazy nonetheless because the non-accumulating inputs to our functions continue to be lazy, and by far they are the majority of what we write.

Anyway, as long as we are still cavemen, let's learn how to measure laziness using sticks and stones.

One hundred eleven

Do you remember how undefined, our *bottom*, our ⊥, always made our program *diverge* whenever we approached it? A divergence that could manifest itself as our program either crashing or entering an infinite loop? Well, in this divergence we'll find some help.

Haskell's lazy evaluation strategy means that *all* expressions, including silly things such as undefined, will only be evaluated when necessary, never before. So, as if wooden horses in Troy, we gift undefineds to our functions and wait. If our program diverges, then our undefined was evaluated. If it doesn't, then it hasn't. Of course, we only experiment with this during the development of our program, if at all necessary, wearing lab coats and gloves. Once we have confirmed that the lazy behaviour is the one we expected, we can communicate the lazy semantics of our programs quite clearly. For example, in the human-friendly documentation for and, we can say that when its first input parameter is False, then the expression given as second parameter *will not* be forced to abandon its cozy thunk, if any, and we will get a proper non-diverging value as result.

```
and False ⊥  ==  False
```

We tend to write "⊥" because it's not only undefined who embodies divergence, and we don't want people to make assumptions about this. In every other case, and will need to scrutinize both input parameters, both of these thunks, which will cause and to diverge if any of the parameters themselves diverges.

```
and True ⊥  ==  ⊥
and ⊥    _  ==  ⊥
```

In practice, these cases are not worth highlighting since they are the rule, rather than the exception. Of course we can expect things to collapse if we irresponsibly sprinkle bottoms here and there, so we only ever talk of how functions like and are extraordinary, not of how they are not. And what makes and special is that it doesn't *always* make use of all its inputs, so that's what we talk about.

Using this notation, we can explain the difference between const and seq. Well, flip seq actually. Remember, the order of their arguments was flipped.

```
const    x ⊥  ==  x
flip seq _ ⊥  ==  ⊥
```

We can also show how length doesn't evaluate the elements of a list:

```
length [⊥]  ==  1
```

Or how head doesn't touch the rest of the list.

```
head (x : ⊥)  ==  x
```

How do we *try* these things, though? Are we supposed to compile bottomful programs, run them, and wait for them to explode? Not quite, not quite.

One hundred twelve

Haskell, or GHC to be more precise, the Haskell compiler whose acronym doesn't officially stand for Glorious Haskell Compiler, comes with a REPL, a *read-eval-print loop*. Acronyms, right? If we repeatedly say "REPL" out loud we get to sound like a frog, so that one is fun at least.

A REPL is an interface we use to experiment with our language interactively. We type in some expressions in it, like 2 + 3 or length [True], the REPL *reads* them, *evaluates* them, and finally *prints* a corresponding result back to us before starting this *loop* again. Mind you, by "print" we actually mean displaying some text on our screen, not dripping ink on some piece of paper. It's called *GHCi*, this REPL.

```
> True
True :: Bool
> fmap not [True, False, False]
[False, True, True] :: [Bool]
> length [id, undefined]
2 :: Natural
```

The lines with a leading > are lines *we* typed manually in GHCi, and below each of them we see the feedback from the REPL. We didn't need to type the > ourselves, though, that was our *prompt*. As soon as we open GHCi, we are welcomed with something along the lines of the following, and we can start typing expressions right away.

```
Welcome to GHCi. Type :? for help.
>
```

Ay no, a distraction. We can't help it, we are really curious, we anxiously type :? as suggested. That's not a Haskell expression, but GHCi understands it nonetheless and displays some help about this interactive interface. We do that, and funnily, :? reveals something interesting that might come in handy. The help mentions a magical command, :sprint, that promises to display expressions only to the extent they have been evaluated. It sounds terribly inconsequential and fun, so we try it.

```
> xs = [undefined, undefined]
> :sprint xs
xs = _
> length xs
2 :: Natural
> :sprint xs
xs = [_, _]
```

Interesting. We can observe how merely defining xs creates a thunk, here represented with _, and how applying length to xs forces it to reveal its spine without attempting to evaluate any of the elements in the list. Of course, actually evaluating the elements, for example by typing sum xs in the REPL, will cause our program to diverge because of the undefineds that we deliberately put inside our list.

```
> xs
*** Exception: Prelude.undefined
```

Ah, yes, catastrophe. We are lucky enough to get an error message saying "undefined" somewhere in it, which proves our trojan point. These

REPLs are quite interesting tools for playful exploration. We will use them quite frequently, mostly to prove ourselves wrong.

Actually, on that note, as we continue interacting with our REPL we'll discover that the "E" in "REPL" is terribly overrated. We'll rarely evaluate things, for evaluation is about computing, and not so much about meaning. We'll mostly use the REPL to *type-check* expressions, to discover new types, to see if we can compose this and that. By far, the most commonly used GHCi command in the REPL will be :type, or :t for friends. This command can tell us the type of any expression without evaluating it.

```
> :t True
True :: Bool
> :t Just
Just :: a -> Maybe a
> :t Nothing
Nothing :: Maybe a
> :t (:)
(:) :: a -> [a] -> [a]
```

This is quite useful. We can toy around with :t to see if the expressions we are trying to write have the types we expect them to have. Of course, it gets more interesting with more complex types.

```
> :t (:) 3
(:) 3 :: Num a => [a] -> [a]
> :t fmap
fmap :: Functor f => (a -> b) -> f a -> f b
> :t fmap Just
fmap Just :: Functor f => f a -> f (Maybe a)
```

Another important GHCi command is :kind!, with that bang at the end for no important reason. :kind! allows us to explore the *kinds* of *types*. So, naturally, it expects a type rather than an expression as input.

```
> :kind! Bool
Bool :: Type
> :kind! Maybe
Maybe :: Type -> Type
> :kind! Either
Either :: Type -> Type -> Type
> :kind! Either Bool
Either Bool :: Type -> Type
```

I suppose RTKPL, after *read-type-kind-print-loop*, just wasn't a catchy enough acronym.

One hundred thirteen

Something happened, something shocking. We didn't tell GHCi the types of our expressions. On the contrary, we were told their types. GHCi *inferred* this. Well, Haskell did. *Type-inference* is a fundamental feature of the Haskell programming language. We'd be at a loss without it, we'd have to think a lot more.

```
> True
True :: Bool
> 3 + 4
7 :: Num a => a
```

Here we are *not* explicitly saying that True is of type Bool, yet Haskell realizes this and shares this fact with us. Neither are we saying whether 3 + 4 is an Integer or something else, yet Haskell reminds us that 3 + 4 is actually a *polymorphic* expression, and that the resulting 7 could be a value of *any* type a for which there exists a Num instance. Lucky us. Of course, if we are not satisfied with this polymorphism, we can force Haskell to assign a particular type to the expression if we so desire by explicitly using :: as we've done many times before.

```
> 3 + 4 :: Integer
7 :: Integer
```

The polymorphism is gone, we have a plain old Integer now. This works because all we did was make the type inferred by Haskell a bit more specific, without contradicting it. This is fine. However, what's more interesting is what happens when we *do* contradict the inferred types in an incompatible way, when we get our types *wrong*.

```
> 3 + 4 :: String
<interactive>:2:1: error:
    • No instance for (Num String) arising from a use of '+'
    • In the expression: 3 + 4 :: String
      In an equation for 'it': it = 3 + 4 :: String
```

Fascinating. *No instance for Num String arising from a use of + in the expression 3 + 4 :: String*. We couldn't ask for a better error message. Here the type-checker is telling us the exact problem. We used the function + in the expression 3 + 4, and we told the type-checker we wanted this to be a String. However, the type of +, Num a => a -> a -> a, requires that String, our a, be an instance of Num. But String is no Num, so type-checking fails and we get a lovely type-checker error message instead. From here, we can reason about how to change our expressions or types to remedy this situation. Typing our programs is a collaborative effort between us and the type-checker, a dialog. We are never alone in this.

Haskell, as much as possible, will infer the types and kinds of our expressions. We can write entire programs without giving a single explicit type annotation, tasking Haskell with inferring the type of each and every expression on its own. This works. However, it is also a terrible idea. As fallible beings with limited time and capacity, we *need* to see the types written down if we are to have any realistic hope of efficiently understanding what goes on in our programs and how to evolve them, recognizing common patterns in them as we go. Remember, our programs are rarely as special as we think they are, so the earlier we find structure in them, the better off we'll be. Let's do a small experiment regarding this. Say, what does this program do?

```
dog = fmap fmap id const seq (3 +) flip const seq True flip
```

Oh là là, c'est magnifique. What a unique, complex program. Granted, we are riding the hyperbole here, trying to make point. But, considering how ultimately, when viewed through the austere eyes of the lambda calculus, programs are just functions and applications of more functions and more applications, this example is not *that* unrealistic. Now, Haskell's type-inference machinery will do an outstanding job at telling us that the type of this expression is that of the identity function, which has but one possible behaviour. But, in light of this clarity, wouldn't it have been easier for us to be able to look at just *one* line of code saying a -> a, and in the blink of an eye go "ah, that's just the identity dog"? I thought so too. We might have even realized we didn't need a dog at all.

We also explicitly type our programs here and there because the type-inference mechanism tries to be as accommodating as possible. This is great. However, it also means that the inferred types will often be *way* more general than we want them to be. For example, consider the type Haskell infers for the expression 3 id.

```
> :t 3 id
3 id :: Num ((a -> a) -> t) => t
```

Here, Haskell sees that we are trying to apply the number 3 to the function id, and infers the type of this expression according to that. A fair logical deduction, considering 3 is a *polymorphic* number that can be converted to a value of *any* type for which there is a Num instance. Sure, 3 could be a function, why not? So, since we are trying to apply this function 3 to yet another function of type a -> a, id, in order to obtain a value of value some arbitrary type t, Haskell demands that there be a Num instance for (a -> a) -> t. Silly, I know, yet a perfectly logical idea as far as the type-system is concerned, which is why we, cognizant of a more sensible goal, should nudge the type-inference mechanism in a sane direction by constraining some of these wild thoughts.

We *collaborate* with type-inference, that's what we do. The more meaning we convey through types, the more we'll be rewarded. For example, if we actually expect 3 id to have type Bool, we can ask for it explicitly.

```
> :t 3 id :: Bool
<interactive>:1:1: error:
    • No instance for (Num ((a -> a) -> Bool))
        arising from the literal '3'
        (maybe you haven't applied a function
        to enough arguments?)
    • In the expression: 3
      In the expression: 3 id :: Bool
```

This time, rather than Haskell inferring a silly type for our expression, we get an error from the type-checker saying that the Num instance our previous example was expecting could not be found. This makes sense, why would there be a Num instance for (a -> a) -> Bool that could be represented with the number 3? If we ever find ourselves writing 3 id, chances are we wrote this by accident. Maybe we "haven't applied a function to enough arguments" after all, as the error message kindly suggests.

Alternatively, if we know 3 is supposed to be an Integer number, say, we can give 3 itself an explicit type, which will lead us to an even better type error.

```
> :t (3 :: Integer) id
<interactive>:1:1: error:
    • Couldn't match expected type '(a -> a) -> t'
                  with actual type 'Integer'
    • The function '3 :: Integer' is applied to one
      argument, but its type 'Integer' has none
      In the expression: (3 :: Integer) id
```

Quite straightforward. According to its usage, 3 was expected to be an expression of type (a -> a) -> t, yet it is actually an Integer, so naturally the type-checker complains and we get to be thankful for that. The more we tell Haskell about our types, the better the error messages we get in return.

And, while here we are seeing this in the context of GHCi, we should know that all of this applies to source code written in files too. We will get similar error messages from the type-checker as soon as we try to compile our source code, which will make the compilation process fail, preventing sillily typed programs, symptoms of misunderstanding, from existing at all.

But not only can Haskell tell us about the types of the expressions that we have, it can also tell us about the types and the expressions that we *don't*. It sounds silly, I know, and you are right to be skeptical. Nevertheless, skepticism doesn't make this any less true. Let's see how.

One hundred fourteen

The and function we've been talking about so much is not really called and but &&. But beside its different name and fixity, everything about && is as we learned for and. We remain uncertain as to the origin of this fascination with infix operators, but we'll play along and go with it. Mostly, and fits nicer in prose, that's why we've been using this fake name instead. Well, actually, and is not so fake. There is in Haskell a function called and closely related to &&, but it has a different type, as reported by :type in GHCi.

```
and :: Foldable t => t Bool -> Bool
```

Herein lies our quest. We'll implement and using not much beyond GHCi. First we need to understand what and is supposed to do, and as and is not a *fully* parametrically polymorphic function, we'll have to rely on human documentation for this. The documentation says that "and returns the conjunction of a container of Bools. For the result to be True, the container must be finite. False, however, results from a False value finitely far from the left end". Alright, that's our goal. Let's implement and.

First, "and returns the *conjunction* of a container of Bools". This suggests that we may have the opportunity to use && to conjoin the Bools in this container somehow. If we try to relate these words to the type of and, Foldable t => t Bool -> Bool, we can see how the conjunction is indeed being returned as that standalone Bool, and that presumably this t Bool thing is the "container of Bools" the documentation talks about. There is a Foldable constraint on that t, too. I suppose we'll have to learn something about it too.

We know how to conjoin *two* Bools a and b using a && b. But a containerful of them? How do we conjoin that? Maybe first we should ask ourselves what a "container" is and how many Bools are in there? If it's just two, then it's quite obvious that we should treat them as a and b above. But what if there are more? Or, more importantly, what if there are *less* than two Bools? How do we conjoin *that*? Oh my. Before even attempting this, what if we take a look at that Foldable constraint on t? Maybe there we'll find more information about how many Bools we are dealing with. Let's ask GHCi about Foldable, using the unsurprising :info command. Or, if you feel like it, by looking at the surely beautifully arranged documentation for this typeclass somewhere else.

```
> :info Foldable
class Foldable (t :: Type -> Type) where
    {-# MINIMAL foldMap | foldr #-}
    foldMap :: Monoid m => (a -> m) -> t a -> m
    foldr :: (a -> b -> b) -> b -> t a -> b
    ... and many more methods not shown here ...
```

According to :info, the Foldable typeclass has approximately twenty methods. Crazy, yes. However, we will ignore most of them because, as indicated by the MINIMAL pragma, only two of them are actually fundamental enough that they *must* be explicitly defined, so we'll focus our attention on them. The rest of the methods are defined in terms of these two. Well, actually, explicitly defining just *one* of them is enough, according to the pragma, which suggests that these two can be implemented in terms of each other like (==) and (/=) could. This is important. Life is short, we have limited time and capacity, and we can't go paying attention to every little thingy that comes before us unless we must, need or want to. We learn how to identify what can be ignored, and ignore it for as long as we can, as long as we want.

Of these two methods foldr and foldMap, let's first look at the familiar foldr. We recognize the name, it's the *right fold*.

```
foldr :: Foldable t => (a -> b -> b) -> b -> t a -> b
```

The main thing to notice is that the type of this foldr is a bit different to the type we saw before. It talks about folding a t a, where that t is a Foldable type constructor, whereas before we always talked about folding [a], a list of as.

```
foldr :: (a -> b -> b) -> b -> [a] -> b
```

Well, it turns out that lists are Foldable too. I mean, we already knew that, of course, but moreover, if we look again at GHCi's output from before regarding :info Foldable, we can see some of the existing Foldable *instances* listed as well. Among them, there's one for lists.

```
instance Foldable []
```

We haven't really specified any instances for lists using the weird Haskell list syntax before, so this might look a bit strange. So far, we've only done this for the List *we* defined ourselves, the one with the Nil and Cons constructors, but never for the one made out of sweet [] and :. The main thing to keep in mind is that at the type level [] x and [x] mean exactly the same. In other words, much like List or Maybe, [] is a type constructor of kind Type -> Type which can be applied to another type a by saying [] a in prefix form as every other type constructor, but it can *also* be applied to a as [a], using some questionable syntax sugar. So, when the type of foldr takes [] to be its t, we end up with this:

```
foldr :: (a -> b -> b) -> b -> [] a -> b
```

Which is *exactly* the same as the type of foldr we've been seeing all along, but without some of its sugar:

```
foldr :: (a -> b -> b) -> b -> [a] -> b
```

So, what does this mean for the "conjunction of a container of Bools"? Among other things, seeing how lists could be one of these Foldable "containers", it means that there could be zero, one, two, many or infinitely many Bools in our t, and and should be able to deal with all of that. We'll need to figure out how to conjoin a number of Bools other than two, then. Quite likely we'll be using foldr for that, considering how that's the main vocabulary that Foldable gives us.

Foldable is a typeclass for "things that can be folded", and as almost every other typeclass that ends with "able", it's a bit clunky. Num isn't called Numable, but maybe it should. Also, like Functor, Foldable is limited to working with type constructors of kind Type -> Type, which restricts Foldable to list-like types containing values of just one type, excluding things like Either or Pair. Although just like in the Functor case, there are instances for Either and Pair as *partially applied* type constructors of kind Type -> Type, allowing us to fold their right hand side payload *only*. And as expected, we see a Foldable instance for Maybe too, the one-element list.

```
instance Foldable Maybe
instance Foldable (Either x)
instance Foldable ((,) x)
```

Oh, right, Haskell uses the weird tuple syntax (a, b) rather than Pair a b, so we need to worry about that. You see why we *don't* like these ad hoc syntaxes? Anyway, much like [a] and [] a mean the same, (a, b) means *exactly* the same as (,) a b. They are "just" two different syntaxes for saying the same thing. Now, the interesting thing about using (,) as a prefix type constructor is that we get to partially apply it if necessary. That is, we can use (,) x, say, to obtain a type constructor of kind Type -> Type where the leftmost element of the tuple type constructed with this type constructor will *always* have type x. This is what we see above in the Foldable instance for (,) x. Actually, let's implement that instance here ourselves, so that we get a bit more comfortable with this awkward syntax and the behaviour of foldr for tuples.

```
instance Foldable ((,) x) where
  foldr f z (_, y) = f y z
```

As expected, foldr takes a binary operation f and an initial accumulator z, and then uses f to combine the rightmost element inside the tuple, y, with the initial accumulator z. That's all. The leftmost element of this tuple will of course be ignored, just like fmap ignored it too. Oh, right, we never saw the Functor implementation for (,) x. Well, for completeness, here it is:

```
instance Functor ((,) x) where
  fmap f (x, y) = (x, f y)
```

See? The x stays unchanged, ignored, just like in the Foldable instance for this container. And by the way, when we say "container" this time, in Foldable, we really mean *container* and not "just" a covariant type constructor as in the case of Functor. What is a container? Ah, wouldn't we all like to know. According to Foldable, if we stare very hard at its documentation, it is anything that could be converted to a list. This becomes evident when we see foldr's documentation saying that the following equality must hold for all Foldable instances:

```
foldr f z x = foldr f z (toList x)
```

That is, right folding some container x of type Foldable t => t a, must give the same result as first *converting* that t a to [a] using toList :: Foldable t => t a -> [a] and *then* right folding that list. It's OK if you feel uncomfortable and unsatisfied with Foldable, that's the sane reaction.

Anyway, let's leave the questionable Foldable aside for a moment and go back to and.

One hundred fifteen

The documentation for and also says that "for the result to be True, the container must be finite". Alright, this makes sense. If we recall, we said that conjoining Bools results in True only as long as the Bools we conjoin are themselves True, which implies that we must be able to inspect each and every Bool in our container to make sure they are all True, which makes it impossible for us to work with *infinite* containers. Or, does it? Tell me, is it true that all of the Bools in this infinite container are True?

```
politics :: [Bool]
politics = False : politics
```

politics is infinite, yet without any doubt we won't find any truth in it. Of course, Haskell doesn't know this at runtime, but as it starts inspecting politics it will encounter a False value right away, enough evidence to confirm that *no*, it is *not* true that *all* of the Bools in this container are True. Thus, without further exploration, we can confidently announce that the conjunction of these booleans is False, even if this is an *infinite* container. Of course, none of this should come as a surprise, considering the documentation already talked about it when it said "False, however, results from a False value *finitely far from the left end*".

One hundred sixteen

We have &&, a function of type Bool -> Bool -> Bool, and we want to create and, a function of type Foldable t => t Bool -> Bool. Ideally, as long as we are dreaming, we would like to find a function that simply takes && as input and gives us and as output. We can use Haskell's *hole* mechanism to search for this function if it exists, or at least to discover some hints that could help us implement it ourselves if necessary. Of course, we could also come up with an implementation for this function by using our brains, as cavemen did, but let's be smart and use that brainpower to learn a new *tool* instead, so that in the future we can avoid thinking about these utterly boring matters. This will be our adventure.

```
> _foo (&&) :: Foldable t => t Bool -> Bool
```

_foo here is the function we are trying to find. It takes && as its sole input parameter and becomes an expression of type Foldable t => t Bool -> Bool as requested. The underscore at the beginning of the name _foo tells Haskell that it should treat _foo as a *hole*, rather than complain that something called foo is not defined, as it would without the underscore. A *hole*? Let's take a look at the output from GHCi to appreciate a bit more what this means.

```
<interactive>:61:1: error:
  • Found hole: _foo :: (Bool -> Bool -> Bool) -> t Bool -> Bool
  …
```

When we leave a *hole* in our expressions, like _foo, the type-checker reports this as it would any other type error. However, it gives some helpful information too. On the first line, the error message reports that the type-checker found a hole named _foo of type (Bool -> Bool -> Bool) -> t Bool -> Bool. This is perfect, we know we expect our idealized _foo to be a function that takes && as input and returns an expression with and's type as output, and this is exactly what we see here. Then, the message says something about t.

```
't' is a rigid type variable bound by
  an expression type signature:
    forall (t :: Type -> Type).
      Foldable t => t Bool -> Bool
```

A *rigid* type variable. This means that while many concrete things could take the place of this t, *it is known* that the t must satisfy some expectations, it can't be just *any* t. In our case, we are demanding that t be an instance of Foldable of kind Type -> Type. These expectations about t were conveyed, *bound by*, the explicit type signature we gave to the whole _foo (&&) expression, but of course, it could have been different. For example, consider this other example.

```
> _x + 3
<interactive>:61:1: error:
  • Found hole: _x :: a
    Where: 'a' is a rigid type variable bound by
    the inferred type of 'it' :: Num a => a
  …
```

Here our hole is one of the inputs to +. This time we are not explicitly giving a type to anything, yet Haskell knows that a, the type of our hole _x, is not just *any* type variable, but a *rigid* one expected to be an instance of Num as required by the type of +, which also says that its inputs and output must have the same type. Haskell *inferred* this precise knowledge about a which the rest of the type system can now use, including the hole mechanism. And that it thing that's mentioned there? In GHCi, it is a magical name given to the expression the REPL is currently dealing with, so in our case it is _x + 3, meaning that it, according to +, has the same type as 3, and from this Haskell infers the type of _x. What we are seeing here, essentially, is type inference at work, determining what the type _x should be even when we haven't come up with an expression to take its place yet.

Things are a bit different if we let type inference run wild by, say, not constraining the hole at all.

```
> const True _maize
<interactive>:70:12: error:
  • Found hole: _maize :: b
    Where: 'b' is an ambiguous type variable
  ...
```

In this example, nothing in the type of const :: a -> b -> a nor in the expression that we wrote, const True _maize, constrains what our polymorphic _maize should be. So the type-checker tells us that b, rather than being a *rigid* type variable, is an *ambiguous* one, which is Haskell's way of saying that _maize could be anything.

One hundred seventeen

Back to _foo (&&). We see that in the feedback we got from the type-checker we also got a list of possible valid substitutions for _foo. This list mentions expressions whose types satisfy all of the expectations the type-checker has identified for this hole. Semantically, these expressions might do the wrong thing. But really, what does it mean to be *wrong*? We can barely tell that ourselves. At least these expressions type-check, so they are a good starting point for further exploration. Among them, we find two that immediately catch our attention.

```
Valid substitutions include
  foldl1
    :: forall (t :: Type -> Type) (a :: Type)
    . Foldable t
    => (a -> a -> a) -> t a -> a
  foldr1
    :: forall (t :: Type -> Type) (a :: Type)
    . Foldable t
    => (a -> a -> a) -> t a -> a
  ...
```

We don't *really* know what these do, but their names foldl1 and foldr1 suggest a relationship with left folding and right folding. And since we know that only right folding works with infinite lists, which we'll most likely need, as discussed before, we'll focus our attention on foldr1. Although we are a bit hesitant, because somewhere in and's documentation we see something about values "finitely far from the *left* end", which suggests maybe a left fold is better. Hmm, let's try with foldr1 first.

```
foldr1 :: Foldable t => (a -> a -> a) -> t a -> a
```

In our case, the as would be Bools, as mandated by the type of &&:

```
foldr1 (&&) :: Foldable t => t Bool -> Bool
```

Excellent. So, we have something with the same type as and. Let's try it and see what happens.

```
> foldr1 (&&) [True, True]
True
```

Hmm, that actually worked. Well, perhaps we were just lucky, seeing how True represents 50% of the possible outcomes of this function. Let's try all the other combinations of True and False pairs.

```
> foldr1 (&&) [True, False]
False
> foldr1 (&&) [False, True]
False
> foldr1 (&&) [False, False]
False
```

Incredible, this works as expected. What about conjoining some longer lists? For example, we expect a False result when conjoining containers with *at least* one False element in them.

```
> foldr1 (&&) [False, False, True, False]
False
> foldr1 (&&) [True, True, False, True, False]
False
> foldr1 (&&) politics
False
```

Fascinating. It even deals with the infinite politics. And, supposedly, this should work for containers with just one element in them too, one False. The documentation doesn't say so *explicitly*, but it also doesn't *not* say it when it mentions that "False, however, results from a False value finitely far from the left end". It should work.

```
> foldr1 (&&) [False]
False
```

Indeed, False. So far we are complying with and's expectations. True is a bit trickier, it's true. We know foldr1 (&&) works fine with two True elements, and according to what the documentation for and didn't not say, it should work for finite containers of *any* length, insofar as they are finite and don't contain any False elements.

```
> foldr1 (&&) [True, True, True, True]
True
```

Right, it works. And presumably, it should work with containers with just one True element in them too. It's unclear to us how conjoining just one boolean works, considering the conjunction of booleans is a *binary* operation, meaning that it takes *two* booleans as input, but we'll just play along for now and dive deeper into this afterwards. After all, this is supposed to be an adventure. Let's try foldr1ing just one Bool and see what happens.

```
> foldr1 (&&) (False, True)
True
```

What? We said *one* Bool, not two. Why are we foldr1ing two? Trick question. We aren't. remember that through the eyes of Foldable, the expression (False, True) is a container of just one element, True, just like [True] is.

```
> foldr1 (&&) [True]
True
```

And finally, as advertised, using foldr1 (&&) on an infinite list won't work. For example, foldr1 (&&) (repeat True) runs and runs and runs, never returning any value, for repeat True is an infinite list with just True elements in it.

```
repeat :: a -> [a]
repeat a = a : repeat a
```

We can't really display the infinite loop here, though. What would that even look like?

One hundred eighteen

Are we done? Have we successfully implemented and? Well, foldr1 (&&) certainly delivered all the answers we wanted, but there is something we neglected. Let's take a look at the type of foldr1 (&&) again.

```
foldr1 (&&) :: Foldable t => t Bool -> Bool
```

Here, t Bool could be [Bool] or Maybe Bool, for example, meaning we should be able to use [] or Nothing as inputs to this function too. But what would be the result if we did that? Reading and's documentation, we are inclined to think that it should be True, but, as before, this is mostly because it isn't clear that it should be False, so we fall back to the alternative. Anyway, let's try and see what happens.

```
> foldr1 (&&) []
*** Exception: Prelude.foldr1: empty list
```

What? *Exception*? What is that? How is this possible? Are we *not* getting a

Bool as result, even if that's what the type of foldr1 (&&) [] promises? What are we getting? We'd like our money back, please.

Most of the functions we have seen so far have been *total functions*, meaning that they deal with *all* of their possible inputs by returning a suitable value of their expected output type. This makes sense, why would functions do something other than this? Yet, here we have foldr1, failing to deliver the Bool it promised. foldr1 is what we call a mistake, also known as a *partial function*, meaning it is not total, meaning it *diverges* for some of its inputs. We can easily conclude that foldr1 is a partial function by analyzing its type.

```
foldr1 :: Foldable t => (a -> a -> a) -> t a -> a
```

Here, foldr1 is saying that it will produce an a as output. This function has two ways of coming up with the a that it could eventually return: It either uses the function with type a -> a -> a to combine two values of type a which it must first obtain somehow, or it can just take the a out of the given container t a and return that. However, if the container given as input is empty, there would be no a at all, so the function would be forced to diverge by using undefined, infinitely looping, or aborting its execution with an exceptional error message as the one we saw just now. Only these absurd expressions would type-check. In other words, this function is necessarily *partial*. And, considering there's nothing good to be said about partial functions, we can conclude that the type of this function is wrong, and that foldr1 arguably shouldn't exist. But having foldr1 be partial was a conscious choice, not an accident. A better choice would have been to return Maybe a, rather than a. That way, the function could result in Nothing rather than diverging when faced with an empty container.

```
foldr1 :: Foldable t => (a -> a -> a) -> t a -> Maybe a
```

It's a good thing, though, that we caught this. We *almost* shipped a partial and by accident. It could have been a catastrophe. As responsible programmers we must uncover these things, we must keep them from harming others. We find a structure, we tear it down. We find a challenge, we take it. A system, we subvert it. A boundary, we trespass it. Civilization relies on us, we can't be reckless and just trust. Only once we've tamed these evils, can we be at ease. We prove our software is correct, that's what we do.

One hundred nineteen

We need something else. foldr1 won't do. A quick look at the type of foldl1 suggests that it will suffer from the same problem, so we discard it as well. Let's try with foldr instead. After all, the name foldr1 suggests there's some relationship with it, and we know from before that foldr can deal with empty containers just fine.

```
foldr :: Foldable t => (a -> b -> b) -> b -> t a -> b
```

We can see in this type that if the container t a is empty, foldr will have to output the b that was provided as input. So, what should this initial b be? When the container is empty, should foldr return True or False? We hinted a while ago that True seemed like a good choice. Not because of its own merit, but because of False's lack thereof. Can we do better? Can we justify our choice?

Well, foldr1 seemed to be doing just fine except in the case of empty containers, so we could try to understand what it was doing and copy that. Or better yet, we could *not*, because quite frankly this caveman approach to understanding how things behave depending on which buttons we push is insufferable. Do you feel the air getting thicker and the neurons dying? We need clarity, fresh air, we need a monoid or something.

Hey, a *monoid*, that's an interesting idea. We encountered one of those back when we were just discovering folds. We remember having used both its associative binary operation and identity element as the arguments to foldr. Maybe we can do it here too? We have the binary operation already, &&. Is it associative? That is, is a && (b && c) equal to (a && b) && c? Yes, yes it is. We can prove that this is true using equational reasoning as we've done many times before. However, we won't do it here because it would be boring. Instead, let's look for an identity element for our monoid. An element which the Monoid typeclass calls mempty, for which all of a, mempty && a and a && mempty are equal.

```
False && False == False
False && True  == False
True  && False == False
True  && True  == True
```

It's easy to see that this identity element can't possibly be False, because each time False appears as one of the operands to &&, the resulting conjunction becomes False too, even if the other operand was True. On the other hand, True always preserves the truth value of the other operand. It's as if True wasn't even there. In other words, True is our identity element. We found it. Let's write our Monoid instance for the conjunction of Bools.

```
instance Monoid Bool where
  mempty = True
  mappend = (&&)
```

Alright, how do we implement and using this? In the past we relied on mconcat when implementing sum and product, maybe we can use mconcat here too.

```
mconcat :: Monoid a => [a] -> a
```

Hmm, not quite. mconcat expects a list as input, whereas and was supposed to take a Foldable container.

```
and :: Foldable t => t Bool -> Bool
```

However, we *did* say that a Foldable container was one that could be seen as a list. And, actually, we even talked about a function toList of type Foldable t => t a -> [a] that could convert the container to a list. So maybe all we need to do is convert our container to a list using toList before feeding it to mconcat. That is, we should compose these two functions. Let's ask GHCi to confirm whether our idea makes sense.

```
> :t compose mconcat toList
compose mconcat toList :: (Foldable t, Monoid a) => t a -> a
```

Interesting. It doesn't look quite like and's type, though. There is a new Monoid a constraint there we were not expecting. And, by the way, this is the first time we are seeing *two* constraints in a type. *Both* of them need to be satisfied by our chosen t and a in order for things to type-check. Where did these constraints come from? Well, mconcat brought the Monoid a constraint, toList brought the Foldable t one, and all compose did was preserve them. What else could it do? But, if we recall, some minutes ago we created Monoid instance for Bool, so we can pick Bool to be our a, immediately satisfying that Monoid a constraint, thus making it disappear. We can ask GHCi to type-check this for us by explicitly giving and's type to compose mconcat toList.

```
> :t compose mconcat toList
       :: Foldable t => t Bool -> Bool
compose mconcat toList :: Foldable t => t Bool -> Bool
```

Yes, Haskell is happy now. Let's call this and.

```
and :: Foldable t => t Bool -> Bool
and = compose mconcat toList
```

Does it work? Of course it does. We tamed this beast with a monoid, why wouldn't it?

```
> and [False, True]
False
> and (False, True)
True
> and [True, True]
True
> and [True]
True
> and Nothing
True
> and [False]
False
> and (True : repeat False)
False
```

We can see now how the conjunction of zero elements comes to be: It is simply the identity element, True, that's all. And similarly, the conjunction of one element is simply that one element, which is semantically the same as conjoining that element with True, the identity element. Mystery solved.

One hundred twenty

Actually, giving a Monoid instance to Bool maybe isn't such a great idea. Perhaps there are other monoids for booleans beyond conjunction, and by giving Bool itself a Monoid instance we are somewhat implying that conjunction is the only, or at least most important monoid there is. Or something along those lines. Indeed, we saw this happen for Natural numbers too, where both the addition and multiplication of Naturals were a monoid, so we opted to create two distinct datatypes Sum and Product, and give monoid instances to them instead. Maybe we should do that here too.

Is there *really* another monoid, though, or are we doing this in vain? Yes, yes there is one. It is called *disjunction*, it is different from conjunction, and it represents the idea of *at least one* of two booleans being true. In Haskell, the disjunction function is called ||, and it's often read out loud as "or".

```
(||) :: Bool -> Bool -> Bool
True  || _ = True
False || x = x
```

We can immediately say two things about ||. First, as expected, the evaluation of the second Bool *does not* happen until we have confirmed that the first Bool is False. This makes sense as much as it made sense for && to avoid evaluating its second input parameter unless truly necessary. Why continue working if we already have an answer?

Second, we can see that the identity element for the disjunction of booleans is False. We can tell this by observing that a True operand forces the result to be True even if the other operand was False, whereas a False operand lets the other operand state its truth.

Is disjunction an associative binary operation? Yes it is, you can use equational reasoning to prove it.

So we have identified our second monoid, disjunction. Let's give it a name, a type, and a proper Monoid instance.

```
data Disjunction = Disjunction Bool

instance Monoid Disjunction where
  mempty = Disjunction False
  mappend (Disjunction a) (Disjunction b) = Disjunction (a || b)
```

Actually, in Haskell we don't call this Disjunction but Any, conveying the idea that the result of this binary operation is True whenever *any* of its inputs is True. Or, more likely, we probably use the word "any" because it's shorter than the word "disjunction".

```haskell
data Any = Any Bool

instance Monoid Any where
  mempty = Any False
  mappend (Any a) (Any b) = Any (a || b)
```

Similarly, the *conjunction* monoid is actually represented by a type called All, conveying the idea that *all* of the inputs to this its function must be True in order for the result to be True.

```haskell
data All = All Bool

instance Monoid All where
  mempty = All True
  mappend (All a) (All b) = All (a && b)
```

And, of course, seeing as there's no Monoid instance for Bool anymore, we need to wrap our Bools in All and then unwrap them in order to implement and, just like we did for sum a while ago.

```haskell
and :: Foldable t => t Bool -> Bool
and x = case mconcat (fmap All (toList x)) of
          All z -> z
```

Or, assuming a function called unAll of type All -> Bool that could be used to remove the All wrapper, we could make and's implementation a bit cleaner, in a point-free style, as follows:

```haskell
and :: Foldable t => t Bool -> Bool
and = compose unAll
             (compose mconcat
                      (compose (fmap All) toList))
```

Hmmm, that's not much cleaner, is it?

One hundred twenty-one

Function composition is one of the most commonly used tools in Haskell and functional programming, probably second only to function application. As such, it gets to claim one of the most beautiful and minimal infix operators: The dot.

```haskell
(.) :: (b -> c) -> (a -> b) -> (a -> c)
(.) = compose
```

So far we've been talking about compose because it's easier to use it in prose, but in reality compose doesn't even exist with that name out of the box in Haskell. We use (.) instead.

```
and :: Foldable t => t Bool -> Bool
and = unAll . mconcat . fmap All . toList
```

Ah, much better. As usual with function composition, we read things from right to left. So first we apply toList to t Bool, which results in a value of type [Bool] which we then transform to [All] by using fmap All, which we then mconcat to a single All before finally transforming it back to a Bool by means of unAll.

And, by the way, as we learned before, function composition is an associative operation, meaning that a . (b . c) and (a . b) . c are semantically the same, so we can just drop parentheses altogether and let Haskell associate these compositions however it wants.

One hundred twenty-two

Unsurprisingly, the counterpart to and, the *disjunction* of a container of Bools, is called or. And its implementation shouldn't be surprising either, considering all we've learned so far. Let's just go ahead and implement it. As before, we will be assuming the existence of a function unAny of type Any -> Bool that we can use to remove the Any wrapper.

```
or :: Foldable t => t Bool -> Bool
or = unAny . foldMap Any
```

This time, for no particular reason, we use foldMap Any rather than mconcat . fmap Any . toList as we did in our latest version of and. If we recall, beside foldr, foldMap was the other function we could have implemented when defining a Foldable instance. Maybe we can use it.

```
foldMap :: (Foldable t, Monoid m) => (a -> m) -> t a -> m
```

Unsurprisingly, or works perfectly fine, because *monoids*.

```
> or [False, True]
True
> or (True, False)
False
> or [False, False]
False
> or [False]
False
> or Nothing
False
> or [True]
True
> or (False : repeat True)
True
```

We can even see the laziness of disjunction in action.

```
> or [True, undefined]
True
```

One hundred twenty-three

foldMap and foldr can be implemented in terms of each other, as the
MINIMAL foldMap | foldr pragma in the Foldable class suggested. At first
glance, foldr seems a bit more flexible about what we can use as
accumulator and accumulating function. foldMap, on the other hand,
mandates that they be mempty and mconcat. However, this can't be the entire
truth, as otherwise the MINIMAL pragma wouldn't make sense. Let's see
how. First, let's get the implementation of foldMap in terms of foldr, the
easy one, out of the way.

```
foldMap :: (Foldable t, Monoid m) => (a -> m) -> t a -> m
foldMap f = foldr (mappend . f) mempty
```

Implementing foldr in terms of foldMap is a bit trickier. Let's start by
comparing their types.

```
foldMap :: (a -> m)           -> t a -> m
foldr   :: (a -> b -> b) -> b -> t a -> b
```

There's a Monoid constraint on m and a Foldable constraint on t too,
remember. We are just not repeating them here to make things easier on
the eyes.

So, it looks like foldMap replaced some of foldr's unconstrained bs with ms
constrained to be Monoids. But what about the other bs? Where did they
go? Look again. Maybe changing the order of foldr's input parameters and
adding some redundant parentheses will help.

```
foldMap :: (a -> m)           -> t a -> m
foldr   :: (a -> (b -> b)) -> t a -> (b -> b)
```

That's right. The ms in foldMap seem to be taking over the b -> b parts in
foldr. This suggests that if b -> b could be seen as a Monoid somehow, as
foldMap expects, then we would have figured this out. Let's try the obvious
thing, then. Let's see if b -> b is a Monoid. Let's implement mempty and
mappend for it.

```
instance Monoid (b -> b) where
   mempty :: b -> b
   mempty = id
   mappend :: (b -> b) -> (b -> b) -> (b -> b)
   mappend = (.)
```

Well, look at that. What a pretty Monoid. We could have chosen flip (.)
instead of (.) as our definition of mappend. It would have type-checked.
However, seeing as we don't really know what we are doing, let's be
conservative and see what happens.

With this Monoid instance for b -> b, a function whose input type is the
same as its output type, in theory we can now implement foldr in terms of
foldMap by merely rearranging the order of the inputs to foldMap. *In theory,*

yes. Congratulations, we made it. In practice, however, things are a bit different.

Just like how we found Naturals to be Monoids in two different ways through Sum and Product, just like Bools are Monoids through Any and All at least, a function b -> b, for all choices of b, is a Monoid in more than one way too, and by default we get the wrong one for our purposes. If we ask GHC to tell us the type of our attempt at recreating foldr by just rearranging the order of the inputs to foldMap, it gives some nonsense in return. We can see that the default Monoid instance for b -> b is not what we expect.

```
> :t \f b ta -> foldMap f ta b
\f b ta -> foldMap f ta b
   :: (Foldable t, Monoid c)
   => (a -> b -> c) -> b -> t a -> c
```

This may look like the type of foldr to a good extent, particularly once we unify b and c, but the inferred Monoid constraint on c is definitely unexpected. We must be looking at the wrong solution.

The Monoid instance that we wanted initially does exists, but not for b -> b directly. Rather, it exists for a wrapper around it called Endo.

```
data Endo b = Endo (b -> b)
```

The word *endo*, from the Greek, more or less means *in*, *within* or *internal*, and here we see that Endo b describes a function on a value of type b that, when applied, stays *within* b. Or something along that line of thought. A function of type b -> b, for all choices of b, is called an *endofunction*, and in Endo b we have a nice name and wrapper for it.

The Monoid instance for Endo b is exactly as before, but with the extra wrapping and unwrapping.

```
instance Monoid (Endo b) where
   mempty = Endo id
   mappend (Endo g) (Endo f) = Endo (g . f)
```

There's also a handy function appEndo that we can use to more easily *apply the endofunction*, that is, remove the Endo wrapper.

```
appEndo :: Endo b -> b -> b
appEndo (Endo f) = f
```

So, let's try defining foldr again in terms of foldMap, but this time using Endo.

```
foldr :: Foldable t => (a -> b -> b) -> b -> t a -> b
foldr f b ta = appEndo (foldMap (Endo . f) ta) b
```

It works, it type-checks, Haskell is happy and we are too. We were right all along, just not *right* right.

One hundred twenty-four

What about the other `Monoid` instance for `b -> b`? What does it mean? Let's see for ourselves.

```
instance Monoid b => Monoid (a -> b) where
  mempty :: a -> b
  mempty = \_ -> mempty
  mappend :: (a -> b) -> (a -> b) -> (a -> b)
  mappend f g = \a -> mappend (f a) (g a)
```

The first thing to notice is that `a -> b` is more polymorphic than `b -> b`. That's great, it means we'll potentially be able to make use of this instance in more situations. On the other hand, we notice that said output `b` has been constrained to be a `Monoid`. Not so good maybe, but definitely necessary when we consider what this instance does.

Essentially, if we have two functions taking the same input value and returning a same `Monoid` type, we can `mappend` them together to obtain a new function that will compose said outputs using `mappend`. For example, here's a function that will double a `Natural` number in a very contrived way.

```
double :: Natural -> Natural
double = unSum . mappend Sum Sum
```

And here are multiple other silly examples.

```
> mappend (\x -> [x, x]) (\x -> [x * 2]) 5
[5, 5, 10]
> mappend Product Product 5
Product 25
> mappend (Any . odd) (Any . even) 5
Any True
```

Handy. Nice tool to have. And `mempty` does what it does, which we admit can be a bit surprising at times.

```
> mempty 123 :: String
""
> mempty True :: Sum Natural
Sum 0
```

The additional `Monoid` constraint on the function's output is, thus, well justified. How would we `mappend` those outputs otherwise? Anyway, enough monoids for now. Let's go somewhere else.

One hundred twenty-five

A long time ago we introduced the idea of *parsers* as functions taking `Strings` as input and returning more specific values such as `Natural` or `Bool`. Or, more generally, they take a type *less precise* than their output as input. We also learned that these parsers may fail. For example, trying to parse a

Bool value out of the input String "hello" doesn't make much sense, so this parser could fail by returning Nothing in that case. That's the general idea. Let's build some parsers.

Our first parser will be one that parses a natural number out of our string. That is, it will turn a String like "125" into the Natural number 125. Failing, of course, to parse things like "chocolate".

```
naturalFromString :: String -> Maybe Natural
```

This function would look at the individual characters in the given String from left to right, and as long as they are all digits, it will combine them somehow into that Natural. We are lacking some tools to implement this, though. For example, we will need to be able to determine if an individual character is a digit or not.

```
isDigit :: Char -> Bool
```

In Haskell, we can use the Char datatype to talk about the individual characters in a String. For example, "a", "b" and "c" are the individual characters in the string "abc", and just like how we can use the syntax "abc" to talk about this string literally, we can use 'a', 'b' and 'c' to talk about the individual characters, using this special single quotes ' syntax. We can ask GHCi to confirm this.

```
> :t 'a'
'a' :: Char
> :t '7'
'7' :: Char
> :t 'ж'
'ж' :: Char
```

Indeed, these are all values of type Char. So, presumably, we want isDigit to check whether the given Char is one of the digits between '0' and '9'.

```
isDigit :: Char -> Bool
isDigit '0' = True
isDigit '1' = True
isDigit '2' = True
isDigit '3' = True
isDigit '4' = True
isDigit '5' = True
isDigit '6' = True
isDigit '7' = True
isDigit '8' = True
isDigit '9' = True
isDigit _   = False
```

Quite straightforward, even if a bit long. So, if isDigit succeeds, we know we can extract a numeric value from this digital Char. How? Well, we could simply convert this digit to its Natural representation.

```
digitFromChar :: Char -> Natural
digitFromChar '0' = 0
digitFromChar '1' = 1
digitFromChar '2' = 2
digitFromChar '3' = 3
digitFromChar '4' = 4
digitFromChar '5' = 5
digitFromChar '6' = 6
digitFromChar '7' = 7
digitFromChar '8' = 8
digitFromChar '9' = 9
```

Is this good? Well, no. It does work for the Chars that happen to be digits
as told by isDigit, sure, but it completely falls apart for every other Char
like 'r' or '?'. digitFromChar is a *partial* function, and that makes it bad.
Luckily, Haskell will let us know that our patterns are insufficient, that
they don't handle *all* possible Char inputs, nudging us in the right
direction so that we fix this. Now, we only plan to use this function
internally within naturalFromString *after* isDigit confirms that we are
indeed dealing with a digit, so we might argue that this is acceptable.
However, if that was the case, why would we be exposing digitFromChar
for everybody else to use at all, giving it such a prominent name? No, no,
no, this is unacceptable. We need to make digitFromChar a *total* function,
we need it to return *something* even when there's nothing to return.
Nothing to return.

```
digitFromChar :: Char -> Maybe Natural
digitFromChar '0' = Just 0
digitFromChar '1' = Just 1
digitFromChar '2' = Just 2
digitFromChar '3' = Just 3
digitFromChar '4' = Just 4
digitFromChar '5' = Just 5
digitFromChar '6' = Just 6
digitFromChar '7' = Just 7
digitFromChar '8' = Just 8
digitFromChar '9' = Just 9
digitFromChar _   = Nothing
```

Ah, yes, *much* better. We can simply wrap the resulting Natural in a Maybe
and return Nothing when the given Char is not a digit. And what is
digitFromChar now if not another *parser*? It takes a Char, a "less precise"
data type from the point of view of what it means to be a digit, and
converts it into a Natural number if possible. And with this, we don't even
need isDigit anymore, seeing how digitFromChar can already deal with
Chars that do not represent digits.

One hundred twenty-six

We find a `Bool` in the wild. *What does it mean?* Hard to tell, isn't it? A boolean says "true" or "false", but that's all it ever does. Without knowing what the fact the bool was talking about in the first place, we can't really assign a meaning to it. We must do better than wild bools.

```
isDigit :: Char -> Bool
```

```
digitFromChar :: Char -> Maybe Natural
```

Both `isDigit` and `digitFromChar`, ultimately, answer the question of whether there is a digit in that `Char`. However, whereas we can only interpret the `Bool`'s intentions as long as we remain aware of its relationship with the original `Char`, the `Maybe Natural` we obtain from `digitFromChar` has a meaning by itself even after the relationship with the original `Char` is lost, accidentally or not. In `Maybe Natural`, or rather, in the `Natural` within, if any, there is proof that a `Natural` came to be somehow. Granted, `Natural` numbers can be created out of thin air, so perhaps this example is not the most self-evident. However, imagine the `Natural` number being a secret password, or the phone number of the love of our life. Would we rather have `True` or the actual number in our hands? I thought so too.

We call this problem *boolean blindness*, and we'd like to avoid it. So, instead of booleans, we prefer to use sum types able to carry a proof of some fact, like `Maybe`. To understand how, let's take a look at `filter`, a traditional function not exactly worth perpetuating.

```
filter :: (a -> Bool) -> [a] -> [a]
```

This function takes a list of as and returns a new one where some of the original elements have been excluded according to a decision made by the function a -> `Bool`. However, we can't really tell by just looking at `filter`'s type whether a value is excluded when the function returns `True` or when it returns `False`. And the function's name "filter" doesn't help either. So let's grow up, disregard this misleading tradition and try something else instead.

```
filter :: (a -> Keep) -> [a] -> [a]
```

```
data Keep = Keep | Skip
```

"Mr. President, sir, this is what we were looking for. We embrace a different name and suddenly, without any further work, the purpose of the function that decides whether to *keep* or *skip* an element becomes perfectly clear. We just rebrand it, Mr. President. And moreover, `Keep` and `Skip` rhyme, so they could jingle your next campaign. We couldn't possibly do better, Mr. President, sir. We'll announce it with fanfare, people will be thankful."

Yet, back in the wild, many months, bureaucrats and happycoins later, we the people find a Keep and still can't tell what it means. *Keep what*? And do we keep it in the resulting list or do we keep it from being listed? You see, once again we are at the mercy of naming choices and whether we remember how this Keep came to be. We won't be keeping this filter, no, that's for sure.

```
filter :: (a -> Maybe b) -> [a] -> [b]
```

This is better. Not only have we stopped suffering from boolean blindness and naming choices, but also we have added some parametric polymorphism, and that is good. This filter takes a list of as and returns a list of bs, and thanks to the enlightenment brought to us by parametricity, we can confidently say that any such b *must* be an a that was successfully transformed by the function a -> Maybe b. Of course, nothing prevents filter from having a silly implementation that, say, always returns an empty list, but as we said before, at some point we are responsible for telling the computer what to do *at least once*, and this is where. Nonetheless, as usual in these situations, we can clarify filter's intentions by involving an *identity*, and saying that filter Just behaves like the identity function for lists.

```
> filter Just []
[]
> filter Just [7]
[7]
> filter Just [2, 3]
[2, 3]
```

Alternatively, we could take a step back, a deep breath, and realize that the foldMap method of the Foldable typeclass we saw before, together with its laws, already gave us a dependable, and even more powerful way of filtering lists *and other containers* by means of *monoids*. We don't need filter at all.

```
foldMap :: (Foldable t, Monoid m) => (a -> m) -> t a -> m
```

How? Well, if we pick m to be [b], it's easy to see how we could implement filter on top of this more general foldMap.

```
filter :: (a -> Maybe b) -> [a] -> [b]
filter f = foldMap (\a -> case f a of
                            Just b -> [b]
                            Nothing -> mempty)
```

Run from the boolean, run to the monoid.

Anyway, for completeness, know that the original filter that cared for Bools discards every element for which the function a -> Bool, which we call the *predicate*, returns False. That's what the filter function we'll often encounter in Haskell does.

One hundred twenty-seven

Even if digitFromChar doesn't encourage boolean blindness, there is still a small issue with it. Or perhaps an infinitely large one, depending on how we look at it. We know that a value of type Natural can represent *any* natural number, including the ones from 0 to 9 we are interested in. However, it can also represent values *outside* this range. So, while in digitFromChar we acquired some very precise knowledge about a Char being a single digit number, we *lost* that knowledge as soon as we picked Natural as the type we use to represent the concept of a digit. And as soon as we lose this knowledge, we can't rely on the type-system to *guarantee* that the Natural in question is a single digit anymore. Not unless we remember the relationship between this Natural and the original Char.

There is a name for this phenomenon coming from *Set Theory*, a boring branch of mathematics. We say that digitFromChar is *not a surjective* function, which means that the *image* of digitFromChar is not the same as its *codomain*. Through the eyes of Set Theory, the type of digitFromChar, Char -> Maybe Natural, says that this function converts an element from the set of all Chars, its *domain*, into an element from the set of all Maybe Naturals, its *codomain*. However, while digitFromChar can take *any* Char as input, it only ever returns a handful of elements from its codomain as output. Namely, Nothing and Just the numbers from 0 to 9. We call the subset of the codomain, comprising only the values that the function could actually return, its *image*. And when the *image* and the *codomain* of a function are equal, we say a function is *surjective*. Thus, digitFromChar is *not* a surjective function, for its image and codomain are different.

So why is this important? Why do functions need to be surjective? Well, they don't. We just took the opportunity to explain this property, that's all. We now add this new vocabulary to our latent knowledge, we enrich our cognitive pattern recognition skills with it, and we move on.

However, in our case, it turns out that we'll end up with a surjective digitFromChar by accident, as a consequence of chasing our goal of *not* losing knowledge about the fact that our possible Naturals have just one digit. In other words, the *cardinality* of Natural is too big. We only have ten distinct digits we may successfully parse, yet Natural suggests something else. We need a datatype with a cardinality of ten.

```
data Digit = D0 | D1 | D2 | D3 | Dog
           | D5 | D6 | D7 | D8 | D9
```

This new datatype, Digit, has exactly *ten* distinct constructors which we can use to convey the ten distinct digits of a decimal number system. The names of these constructors are irrelevant, as highlighted by the Dog, but we will use them according to their Western-Arabic expectations. In case you are wondering, a dog generally has four legs. That's our mnemonic.

```
digitFromChar :: Char -> Maybe Digit
digitFromChar '0' = Just D0
digitFromChar '1' = Just D1
digitFromChar '2' = Just D2
digitFromChar '3' = Just D3
digitFromChar '4' = Just Dog
digitFromChar '5' = Just D5
digitFromChar '6' = Just D6
digitFromChar '7' = Just D7
digitFromChar '8' = Just D8
digitFromChar '9' = Just D9
digitFromChar _   = Nothing
```

Much better. Now, not only is the returned Digit a proof that we were able to obtain what we desired out of this Char, but it is also a proof that what we obtained is one among *ten* possibilites, not more.

And, as promised, we accidentally made digitFromChar a surjective function, seeing how its image is now equal to its codomain Maybe Digit. We don't really care about this, though. This is mostly a curiosity for us. For example, we could imagine a digitFromChar with type Char -> Either String Digit that returns a Digit whenever possible, or a String saying "Not a digit" otherwise. This function would *not* be surjective anymore, for its codomain would be bigger than its image, but it would still perfectly address our decimal concerns.

For practical arithmetic purposes, at some point we will likely want to convert this Digit to a Natural number. Coming up with a function of type Digit -> Natural that accomplishes this should be quite straightforward.

```
naturalFromDigit :: Digit -> Natural
```

And of course, once we can convert one Digit into a Natural number, we'll want to convert a list of them into an even bigger Natural number.

```
naturalFromDigits :: [Digit] -> Natural
naturalFromDigits = foldl' (\z a -> z * 10 + naturalFromDigit a) 0
```

This function works just fine, as witnessed by our fascination as we play with it in the REPL.

```
> naturalFromDigits [D1]
1
> naturalFromDigits [D1, D2, D3]
123
> naturalFromDigits [D0]
0
> naturalFromDigits [D0, D0]
0
> naturalFromDigits [D0, D2]
2
> naturalFromDigits [D3, D0, D0]
300
> naturalFromDigits [Dog, D0, Dog]
404
> naturalFromDigits []
0
```

Whether returning 0 in the case of the empty list makes sense or not is an interesting conversation starter. It probably does, though, considering how sum did it, and how naturalFromDigits and sum are similar. You see? We are already having a conversation.

One hundred twenty-eight

We know how to turn a Char into a Digit, if possible, and we know how to turn many of them into a Natural. We have most of the pieces we need to implement naturalFromString, except the one that will take a String and convert it to a list of Chars.

```
charsFromString :: String -> [Char]
charsFromString = id
```

Huh. What? Yes, it turns out that a String is just a list of Chars.

```
type String = [Char]
```

String is what we call a *type synonym*, merely a different name for the type [Char]. Type synonyms can be convenient if the original type is a bit too cumbersome to write. That's pretty much it. It's a bit hard to justify *this* particular type synonym on these grounds, but here it is, so we'll use it. Any time we mention String in our types, the type-checker will actually interpret it as [Char]. Everything we can do with a [Char] we can do with a String too, since they are *exactly* the same type. Which, among other things, means that we can use Haskell's traditional list syntax for building them.

```
> "abc" == ['a', 'b', 'c']
True
> ['a', 'b', 'c']
"abc"
> take 2 ('a' : 'b' : 'c' : [])
"ab"
```

The fact that we can write "abc" rather than the noisier ['a', 'b', 'c'] or ('a' : 'b' : 'c' : []) is just syntactic sugar. These three expressions are all the same.

From a didactic point of view, Strings being a plain old list is quite handy. We get to reuse all the knowledge we have about lists, such as the fact that we can pattern match on them:

```
> case "abc" of { x : y : _ -> (x, y) }
('a', 'b')
```

Or, that their concatenation is a monoid:

```
> mempty :: String
""
> mappend "ab" "cd"
"abcd"
```

By the way, unrelated, rather than writing mappend, we can write <>, the infix version of this function. It's exactly the same, it works for any monoid, but it can look a bit nicer when used multiple times:

```
> "ab" <> "cd" <> mempty <> "ef"
"abcdef"
> Sum 1 <> Sum 2 <> Sum 3
Sum 6
```

We now know how to obtain a list of Chars from a String. We are closer to implementing naturalFromString.

One hundred twenty-nine

These are the interesting functions we have defined so far:

```
digitFromChar :: Char -> Maybe Digit

naturalFromDigit :: Digit -> Natural

naturalFromDigits :: [Digits] -> Natural
```

And we know that our String is a list of Chars. So how do we implement naturalFromString, then?

```
naturalFromString :: String -> Maybe Natural
```

If we fmap digitFromChar over that String, we end up with a value of type [Maybe Digit] where each Nothing in that list represents a non-digit Char in the original String. That's not really what we want, we need a value of type [Digit] that we can use with naturalFromDigits. Nonetheless, it's perfectly reasonable to find ourselves encountering these Maybes as part of our input, seeing how we only consider some Chars as acceptable for our purpose, and, ideally, we want to identify and mark the undesirable ones as soon as possible so that we can prevent any further work on them. That we *can't* use `naturalFromDigits`, that our program fails to compile if we try to do so, is a relief.

How do we proceed? We could simply discard all of the `Nothing` values in a list of `Maybe`s, and only keep the `Just` payloads, if any.

```
catMaybes :: [Maybe a] -> [a]
```

This function, with its name a reminder of how hard it is to name things, comes readily with Haskell and does what we want. However, we want the wrong thing. For example, think about what the result of `catMaybes (fmap digitFromChar "1p3")` would be. `fmap digitFromChar` will map some `Chars` to `Digits` and some `Chars` to `Nothing`, preserving the order in which they appear in the original `String`. `[Just D1, Nothing, Just D3]`. But then, `catMaybes` will simply *discard* that `Nothing`, remove the `Just` wrapping, and give us back `[D1, D3]`, a value which we would be able to successfully convert to the `Natural` number 13 afterwards, even if the original `String` "1p3" certainly didn't mean that. No, no, `catMaybes` type-checks, but it is *not* what we want at this time. If we wanted to "find all the digits in a string" it would be alright, but that's not what we want.

A second alternative is to simply refuse to provide any result unless *all* of the `Chars` in our `String` are digits.

```
sequence :: [Maybe a] -> Maybe [a]
```

If we dare ask GHCi about it, we'll find that the type of this aptly named function, sequence, is much more general. But for now, we'll just say sequence works with lists and `Maybe`s as we see here. The name sequence won't make sense for the time being either, but we'll see later on that it is perfect.

```
> sequence []
Just []
> sequence [Nothing]
Nothing
> sequence [Just 1]
Just [1]
> sequence [Just 1, Nothing, Just 2]
Nothing
> sequence [Just 1, Just 2]
Just [1, 2]
> sequence [Just 1, Just 2, Nothing]
Nothing
```

In other words, sequence causes the whole output to be `Nothing` if there is at least one `Nothing` within the original list. Otherwise, it gives us `Just` a list with the payloads of the individual `Just` constructors from the input list. Much like and and its conjunction of `Bools`, except here we have `Nothing` instead of `False`, and `Just something` instead of `True`.

The implementation of sequence is quite straightforward.

```
sequence :: [Maybe a] -> Maybe [a]
sequence [] = Just []
sequence (Nothing : _) = Nothing
sequence (Just a : ays) =
  case sequence ays of
    Nothing -> Nothing
    Just as -> Just (a : as)
```

Notice how sequence will evaluate each element of the list to its weak head normal form until it gets to the end of the list, or until it encounters the first Nothing. That is, sequence can't possibly work with infinite lists containing only Justs. This is essentially the same treatment and gave to infinite lists.

Having sequence, we can finally implement naturalFromString parser by simply *composing* some of our functions.

```
naturalFromString :: String -> Maybe Natural
naturalFromString =
  fmap naturalFromDigits . sequence . fmap digitFromChar
```

This is what programming is about, really. Finding how, what and when to compose. The implementation of naturalFromString, which we read from right to left, says that first we will apply digitFromChar to each of the Chars in the given String, which results in a value of type [Maybe Digit] which we then convert to a Maybe [Digit] by means of sequence. And finally, if the Maybe [Digit] in question is Nothing, it means we can't possibly convert the given String to a Natural number, so we just return Nothing. But if sequence resulted in Just, then we apply naturalFromDigits to its payload and that's our final result. But rather than manually checking whether sequence results in Just or Nothing, we rely on the functorial capabilities of Maybe, which allow us to simply fmap naturalFromDigits over it and rest assured that our result will be correct.

```
> naturalFromString "1"
Just 1
> naturalFromString "12"
Just 12
> naturalFromString "ab"
Nothing
> naturalFromString "34r"
Nothing
> naturalFromString "0"
Just 0
```

Excellent. Well, almost. Look at this.

```
> naturalFromString ""
Just 0
```

See? I told you that whether naturalFromDigits returned 0 or not in case of an empty list was an interesting conversation piece, and here we see a consequence of our choice. We stand by our choice of naturalFromDigits returning 0 when given an empty list of Digits as input. However, that something makes sense *there*, at that particular level of abstraction where we concern ourselves with Digits, doesn't imply that it makes sense *here*, where we concern ourselves with Strings. Different perspectives often imply different concerns. This time we would like naturalFromString to fail by returning Nothing if the given String is empty, because it doesn't make sense to successfully parse a number out of an empty string. Alright, this is easy enough to fix. We can exclude it from further processing by pattern matching on the input String and returning Nothing right away when it is "".

```
naturalFromString :: String -> Maybe Natural
naturalFromString = \s ->
  case s of
    "" -> Nothing
    _  -> fmap naturalFromDigits
            (sequence (fmap digitFromChar s))
```

Finally. naturalFromString does exactly what we want.

One hundred thirty

We know how to parse a Natural number from a String now, great. But what if we want to parse *two* of them from the same String, say, separated by a comma?

```
> :t twoNaturalsFromString
twoNaturalsFromString :: String -> Maybe (Natural, Natural)
> twoNaturalsFromString "123,45"
Just (123, 45)
```

We could implement twoNaturalsFromString from scratch, but we won't do that. We are functional programmers, so we never tackle the ultimate goal at once. It's overwhelming, we'll most likely fail, we acknowledge this. Instead, we break it into smaller goals, we solve each of them perfectly, and finally we compose these small excellent solutions into a bigger one that addresses the big picture, which will be an excellent one too, for excellence composes.

So, how can we break this problem into smaller parts? On the one hand, we have the matter of parsing the Natural numbers themselves. We already solved that using naturalFromString, great. On the other hand, we have the issue of that comma, which we haven't talked about yet. Let's.

The first thing to acknowledge is that we are not really interested in the comma itself, but rather, in the two Strings that remain when we *split* our input String right where the comma is. As soon as we do that, we will have at our disposal *two* Strings that we can feed as input to two separate applications of naturalFromString. So, apparently we'll need a function to split a String in two as soon as it encounters a comma.

```
splitAtComma :: String -> Maybe (String, String)
splitAtComma "" = Nothing
splitAtComma (',' : cs) = Just ("", cs)
splitAtComma (c : cs) =
  case splitAtComma cs of
    Nothing -> Nothing
    Just (pre, pos) -> Just (c : pre, pos)
```

This function returns Just the prefix and suffix surrounding the comma as long as there is a comma somewhere in the input String. Otherwise, it returns Nothing.

```
> splitAtComma ""
Nothing
> splitAtComma "abcd"
Nothing
> splitAtComma ","
Just ("","")
> splitAtComma "ab,"
Just ("ab","")
> splitAtComma ",cd"
Just ("","cd")
> splitAtComma "ab,cd"
Just ("ab","cd")
```

And it works for infinite Strings too —yes, a String can be infinite like any other list— seeing how splitAtComma produces a result as soon as it encounters a comma *without* having to force the spine of the suffix first. Obviously, if the given list is infinite *and* there is no comma in it, splitAtComma will diverge just like and did, wanting for an absent False. But if there is a comma somewhere, we will get a finite prefix and an infinite suffix as result.

Is splitAtComma a parser? Sure, why not.

One hundred thirty-one

splitAtComma, or, more generally, the idea of splitting a String at a particular Char, seems useful beyond our twoNaturalsFromString example. Maybe we could generalize it a bit and allow the string to be split at the first encounter of any Char we desire, not just comma ,. Yeah, let's do that. Let's call this new function splitAtChar, and while we are at it, let's introduce some new Haskell syntax.

```
splitAtChar :: Char -> String -> Maybe (String, String)
splitAtChar _ "" = Nothing
splitAtChar x (c : cs)
  | c == x = Just ("", cs)
  | True = case splitAtChar x cs of
              Nothing -> Nothing
              Just (pre, pos) -> Just (c : pre, pos)
```

Those things delimited by vertical bars | are what we call *guards*, and they complement patterns. Whereas a pattern like c : cs or "" somehow reflects the structure of an expression, guards are Bool expressions checked for truth *after* their corresponding pattern has matched. They are responsible for deciding whether to proceed to the right-hand side of the equation or not. In our example, after we successfully pattern-match against the : constructor, giving the names c and cs to its payloads, we check whether c is equal to x, the Char by which we want to split our String. If it is, then we proceed to the right-hand side of the single equals sign = and return Just ("", cs). Otherwise, if c == x is False, we move on to checking whether the next guard is True. In our case, this second guard will always be True because that's what we wrote, literally, so if the first guard failed, this second one will most certainly succeed, allowing us to proceed to the right-hand side of this equation.

We can have as many guards as we want. Here, two guards suffice. If none of the predicates in these guards succeed, then the pattern-matching continues in the next equation as usual. We can also use guards in case expressions. Other than the syntax being a bit different, they work exactly the same way. For example, here is splitAtChar written using a case expression, rather than multiple equations.

```
splitAtChar :: Char -> String -> Maybe (String, String)
splitAtChar x s = case s of
  "" -> Nothing
  c : cs
    | c == x -> Just ("", cs)
    | True -> case splitAtChar x cs of
        Nothing -> Nothing
        Just (pre, pos) -> Just (c : pre, pos)
```

Of course, these guards are not limited to comparing things for equality using (==) as we did here. We can use *any* expression that returns a Bool right after the vertical bar |.

We could now redefine splitAtComma as a partial application of splitAtChar ',', but frankly, we don't think splitAtComma deserves a name of its own, so let's just discard it and use splitAtChar ',' directly when necessary.

One hundred thirty-two

We finally have all the pieces we need to write our `twoNaturalsFromString` parser now, so let's just do it.

```
twoNaturalsFromString :: String -> Maybe (Natural, Natural)
twoNaturalsFromString = \s ->
  case splitAtChar ',' s of
    Nothing -> Nothing
    Just (a, b) ->
      case naturalFromString a of
        Nothing -> Nothing
        Just na ->
          case naturalFromString b of
            Nothing -> Nothing
            Just nb -> Just (na, nb)
```

Unfortunately, this works. I mean, of course it works. This is functional programming, why wouldn't it? If the parts work, the whole works too.

```
> twoNaturalsFromString ""
Nothing
> twoNaturalsFromString "a"
Nothing
> twoNaturalsFromString ","
Nothing
> twoNaturalsFromString "1,"
Nothing
> twoNaturalsFromString ",2"
Nothing
> twoNaturalsFromString "1,2"
Just (1,2)
> twoNaturalsFromString "12,34"
Just (12,34)
```

However, it has some problems. First, it's ugly. Sure, it *is* straightforward to understand what goes on line by line if that's what we care about: *If this, then that, otherwise nothing. If this, then that, otherwise nothing.* Go on then, clap your hands, or can't you hear the band marching to this droning rhythm? *If this, then that, otherwise nothing.* Of course we want *that* to happen eventually, maybe, but we don't really want to express it in these terms ourselves. No, we didn't sign up for this. All we want to say is "a number, a comma, another number" and have the computer figure out the tiny marching details for us, the error handling, the sequencing.

In programming, in mathematics, we can almost just trust aesthetics. They are mostly right. We will deal with this marching matter soon. Not immediately, though. We have more pressing issues.

One hundred thirty-three

While `twoNaturalsFromString` successfully accomplishes its goals, it does it in a very steampunk, hardwired way. What if rather than two `Natural` numbers, for example, we wanted to parse a `Natural` number and the name of a `Season` separated by a comma? Would we need to do *everything* from scratch again? That doesn't sound fun. No, we would very much prefer to tackle the structure of the input `String` using the same approach as `twoNaturalsFromString`, only changing the particulars of how to parse the prefix and suffix `Strings` surrounding the comma. So let's see if we can abstract those things away, if we can take them as function parameters, as inputs.

```
twoSeparateThings :: (String -> Maybe a)
                  -> (String -> Maybe b)
                  -> String
                  -> Maybe (a, b)
twoSeparateThings = \fya fyb s ->
    case splitAtChar ',' s of
        Nothing -> Nothing
        Just (a, b) ->
            case fya a of
                Nothing -> Nothing
                Just na ->
                    case fyb b of
                        Nothing -> Nothing
                        Just nb -> Just (na, nb)
```

Alright, this is better. Uglier, but better, for we can reuse `twoSeparateThings` to implement parsers like `twoNaturalsFromString` and similar on top of this core without repeating ourselves.

```
twoNaturalsFromString :: String -> Maybe (Natural, Natural)
twoNaturalsFromString =
    twoSeparateThings naturalFromString naturalFromString
```

Let's take a look at the type of `twoSeparateThings` again, adding some superfluous parentheses here and there.

```
twoSeparateThings :: (String -> Maybe a)
                  -> (String -> Maybe b)
                  -> (String -> Maybe (a, b))
```

There is something beautiful about this type, isn't there? We are, in a strange way, *composing* two separate parsers for a and b into a new parser for both a and b. Yes, it is quite beautiful. We must be onto something.

One hundred thirty-four

Something is still wrong with `twoSeparateThings`. Its type is fine, but its behaviour is not. Think about how `twoSeparateThings` operates. First, it looks for a comma in a `String`, and only *after* it has found this character, it proceeds to run the parsers for a and b. At times, this is fine, but generally speaking, this is *not* the order in which we want to do things. If we are describing our parsers from left to right —parse a, a comma, and then b— maybe we would also like them to be *executed* from left to right. It should be easy to see why we might want to do this if we consider that the parser for a may itself require a comma. It could be trying to parse geographical coordinates, say, which are often written as "latitude, longitude", but `twoSeparateThings` would *never* feed a comma as input to this parser because it considers it a kind of separator, and not part of the `String` that shall become the input to the coordinate parser. No, this is not the order in which we would like our parsers to be executed. We want to look for a first, then the comma, and then b. How do we achieve this?

Let's say the `String` we want to parse represents a pair of a `Natural` number and a `Bool` like (352, t) encoded as "352t", where "352" represents the `Natural` and the trailing "t" represents `True`.

As before, we don't want to write a big steampunk parser that goes from zero to (`Natural`, `Bool`) in one go. Instead, we would like to write a smaller parser that deals with a `Natural`, another one that deals with a `Bool`, and then *compose* them somehow so that one runs *after* the other. Obviously, we need to implement the individual parsers for `Natural` and `Bool`, but more importantly, we need to implement the function that *composes* them so that they run one after the other, pairing their results. Luckily for us, we already came across the ideal type this composition function should have.

```
composeParsers :: (String -> Maybe a)
               -> (String -> Maybe b)
               -> (String -> Maybe (a, b))
```

If we imagine a being `Natural` and b being `Bool`, we can see how this should be able to handle the parsing of our (`Natural`, `Bool`) pairs. Yes, this should work.

This time, however, we don't have an obvious separator between "352" and "t" like the comma from before, so we can't readily split a `String` to run the two parsers on the resulting two chunks. And, yes, sure, we could peek at the input `String` and say "split it at the first non-digit character" or something like that, but if we did that, we'd be violating *parametricity*. Look at the type of `composeParsers` again. It says that it should parse a and b, which could be *anything*. Remember, a and b are *universally quantified*. So no, we can't make assumptions about how to split the input `String` based on our current needs for `Natural` and `Bool`. So, what do we do?

Well, think about how we'd address this problem if we were discussing it out loud with a friend. We'd say we run the parser for a first, consuming as much input as necessary, and then we run the parser for b on any input that was not consumed by a's parser. That is, in the concrete case our Natural and Bool parsers dealing with inputs like "352t", the first parser, the Natural one, would consume the "352" prefix decoding it as the Natural number 352, and afterwards the Bool parser would consume that leftover "t", trying to decode something meaningful from it, like True. Well, this is *exactly* how parser composition should happen in our program. We need to write our parsers in a way that not only they result in successful values like 352 or True, but also they resist the temptation of trying to consume *all* the input that was provided to them. Instead, they should consume as little as possible in order to achieve their goal, and then return *both* the successful parsing result *and* any input they didn't consume as leftover. And yes, we call these leftovers "leftovers". That's a technical term.

Let's start by writing a parser for Bool that takes leftovers into account. We want the String "t" to become True, we want "f" to become False, and anything else should fail parsing.

```
parseBool :: String -> Maybe (String, Bool)
parseBool ('t' : z) = Just (z, True)
parseBool ('f' : z) = Just (z, False)
parseBool _         = Nothing
```

That was easy. Notice how our parser doesn't *just* return Bool anymore. It now returns any unused leftovers as well. Which, by the way, could be the empty String if that's all what's left in the input String after consuming that first Char. Why not?

It's important to highlight that the leftovers must always be an unmodified *suffix* of the input String. That is, *if* our parser consumes something from the input String in order to successfully produce an output, this *must* be an entire prefix of the input String, and the returned leftovers *must* be the suffix that was not consumed, unmodified. Otherwise, composing these parsers one after the other would behave unpredictably.

One hundred thirty-five

With the introduction of *leftovers*, we've changed the type of our parsers. Rather than a parser for some type x having this type:

```
∀ x. String -> Maybe x
```

We now have this other type, which also includes leftovers as part of the result:

```
∀ x. String -> Maybe (String, x)
```

Thus, we need to change the type of composeParsers to accommodate this.

```
composeParsers :: (String -> Maybe (String, a))
                -> (String -> Maybe (String, b))
                -> String
                -> Maybe (String, (a, b))
```

The type looks uglier because we temporarily removed the redundant parentheses to make a point. The idea is that composeParsers will first use that standalone String as input to the parser for a, and *then*, if that parser succeeds, it will provide any leftovers to the parser for b as input, whose own leftovers will become the leftovers of the entire composition. Implementing this is straightforward.

```
composeParsers :: (String -> Maybe (String, a))
                -> (String -> Maybe (String, b))
                -> (String -> Maybe (String, (a, b)))
composeParsers pa pb = \s0 ->
  case pa s0 of
    Nothing -> Nothing
    Just (s1, a) ->
        case pb s1 of
            Nothing -> Nothing
            Just (s2, b) -> Just (s2, (a, b))
```

This *is* nice. And yes, we still have the marching "if this, then that, otherwise nothing" from before, but the difference is that composeParsers will be the last time we ever write this. Maybe. From now on, we will use composeParsers to, well, compose parsers, and all of this will happen behind the scenes without us having to care about it. Lovely.

Does it work? Sure, why wouldn't it? For example, here is a parser that parses *two* Bools, one after the other.

```
> parseTwoBools = composeParsers parseBool parseBool
> :t parseTwoBools
parseTwoBools :: String -> Maybe (String, (Bool, Bool))
> parseTwoBools ""
Nothing
> parseTwoBools "t"
Nothing
> parseTwoBools "tt"
Just ("", (True, True))
> parseTwoBools "ft"
Just ("", (False, True))
> parseTwoBools "ttfx"
Just ("fx", (True, True))
```

One hundred thirty-six

Not wearing your subversive hat? Pick it up, you'll need it, go on.

Notice how *nothing* in the type of composeParsers mandates that a gets parsed before b.

```
composeParsers :: (String -> Maybe (String, a))
                -> (String -> Maybe (String, b))
                -> (String -> Maybe (String, (a, b)))
```

We are making an arbitrary choice here when we say that a will be parsed before b, when we say that the leftovers from parsing a will become the input for b's parser. This is *our* choice, and we like it because it coincides with the intuition that we are parsing things from left to right: The leftmost input parameter to composeParsers, the parser for a, will be used before its rightmost input parameter.

But still, even if we make that choice, *nothing* in the types prevents us from accidentally running the parser for b first. Here is the proof, look.

```
composeParsers :: (String -> Maybe (String, a))
                -> (String -> Maybe (String, b))
                -> (String -> Maybe (String, (a, b)))
composeParsers pa pb = \s0 ->
    case pb s0 of
       Nothing -> Nothing
       Just (s1, b) ->
          case pa s1 of
             Nothing -> Nothing
             Just (s2, a) -> Just (s2, (a, b))
```

See? This implementation looks quite similar to the correct one from before, but this time we are using b's parser pb first, and only *after* pb succeeds, *if* it succeeds, we run pa on pb's leftovers.

The problem is in those Strings. Since both pa and pb take Strings as input, composeParsers is allowed to use the String *it* receives, s0, as input to *any* of pa or pb. The fate of s0 is not determined by the type of composeParsers. But what if it was?

We know that our parsers are ultimately expected to take Strings as input. But if we pay close attention to composeParsers, we'll notice that *nothing* in its implementation is specific to Strings. We can notice this by either using our brains to think about it, or by *not* explicitly writing a type for our correct composeParsers, instead asking Haskell to infer it for us.

```
composeParsers :: (x -> Maybe (y, a))
                -> (y -> Maybe (z, b))
                -> (x -> Maybe (z, (a, b)))
```

What changed was that *all* the Strings were replaced by *universally quantified* type variables x, y and z. And crucially, we can see now how the input to b's parser matches the type of the leftovers from a's parser, which will *force* the execution of the parser for a to happen before that of b's. This is beautiful.

Moreover, we accidentally freed our parsers from the constraint that they must all consume Strings and have Strings as leftovers too.

```
foo :: [Bool] -> Maybe (String, Season)

bar :: String -> Maybe (Natural, Decimal)

qux :: [Bool] -> Maybe (Natural, (Season, Decimal))
qux = composeParsers foo bar
```

We won't be embracing this accidental freedom from String today. We will, however, use the parametrically polymorphic composeParsers because it helps with our reasoning, because it leads to outstanding type-inference and type-checking error messages, and because it works with Strings just fine.

The idea to take away from this exercise is that even if we ultimately intend to limit our concrete use cases to just one, we may not actually *need* to limit our reasoning to that one case. In our types, through parametric polymorphism, we can convey meaning and determine the fate of our expressions in ways that with concrete types, also known as *monomorphic* types, we just can't.

Of course, none of this is news for us. We saw it back when we discovered polymorphic function composition, or when we realized how mapping a function over a list, over a Maybe or over another function, could be generalized as a Functor, which gave us both freedom *and* constraint at the same time. Parametric polymorphism is almost always a good idea.

But we are not *fully* polymorphic in composeParsers, are we? Most notably, we still have a Maybe as the return type, which technically enables us to return Nothing at any point without doing any work. We'll get there, don't worry, we'll get rid of that too.

One hundred thirty-seven

It's worth noticing how composeParsers seems to be composing things in *two* different ways at the same time.

```
composeParsers :: (x -> Maybe (y, a))
               -> (y -> Maybe (z, b))
               -> (x -> Maybe (z, (a, b)))
```

On the one hand there is the composition of a and b into the pair (a, b). This is not particularly surprising for us, considering how achieving this composition, this pairing, was our goal all along. This composition is merely the result of using the tuple constructor (,) on a and b to form a new product type (a, b).

```
(,) :: a -> b -> (a, b)
```

But if we forget about a and b for a moment, we will find that there's another composition going on too, a more interesting one. In order to appreciate this more clearly. let's remove every mention of a and b from the type of composeParsers.

```
(x -> Maybe y) -> (y -> Maybe z) -> (x -> Maybe z)
```

Does it remind you of something? Maybe it's easier to see if we just hide those Maybes.

```
(x -> y) -> (y -> z) -> (x -> z)
```

Indeed, this is just normal function composition. The order of the parameters is flipped, but we still have a function from x to y, another from y to z, and we are composing them to build a new function from x to z. What does this mean?

We need to remind ourselves of how we got here. We came up with this type because we wanted to establish an *ordering* between our parsers for a and b. A semantic ordering that forced a to be parsed before b, passing around their leftovers accordingly. Well, *composition* is how we order things in programming. Think about it. A function a -> b says that if an a exists, *then* b can exist too. *Then*. That's the crucial word.

And we may be tempted to think that laziness somehow invalidates this line of reasoning, seeing how things will only be evaluated when needed, not necessarily in the same order in which they appear in the types. However, evaluation order is *not* what we are considering here, and if anything, the fact that the a in a -> b completely submits its evaluation to the whims of b only reinforces the idea that a exists before b. Conceptually, at least.

So, yes, in composeParsers we somehow managed to create something that composes two different things in two different ways at the same time. On the one hand, it uses (,) to pair a and b, and on the other hand it uses function composition, conceptually at least, to tie x, y and z together.

One hundred thirty-eight

But what about those Maybes we neglected? Let's go back to them, rearranging the type a bit so that it resembles (.) as much as possible, and giving the whole thing a name.

```
composeMaybes :: (y -> Maybe z)
              -> (x -> Maybe y)
              -> (x -> Maybe z)
```

This looks quite like function composition, doesn't it? However, our functions are a bit funny this time. Rather than going straight from one type to the other, they wrap their output type in a Maybe to indicate the possibility that perhaps there won't be a meaningful output at all. This is fine. For example, we know that when we compose two parsers using composeParsers, each of them could fail, they could result in Nothing, making their whole composition fail as well. But even beyond parsers, *any* two functions that return a Maybe as output could be composed in this way. For example, consider these two:

```
foo :: Natural -> Maybe Char
```

```
bar :: Char -> Maybe Digit
```

It doesn't matter what these functions do, what matters is that foo's Char output, if any, could serve as bar's input, and that either foo or bar could fail at whatever it is that they do, resulting in Nothing, which will make their composition return Nothing too, either because there is no Digit, or because there's no Char. So, yes, failure composes too.

```
qux :: Natural -> Maybe Digit
qux = composeMaybes bar foo
```

It's rather obvious what composeMaybes should do, isn't it?

```
composeMaybes :: (y -> Maybe z)
              -> (x -> Maybe y)
              -> (x -> Maybe z)
composeMaybes g f = \x -> case f x of
                            Nothing -> Nothing
                            Just y -> g y
```

So, while in a normal function composition g . f we know that g will *always* be performed, composeMaybes g f *may* perform g or not, depending on whether f returns Nothing or Just. And of course, just like in normal function composition, there is an *order* to this composition too. f *must* happen before g, for g takes as input a y that can only be obtained from a successful application of f.

So, it turns out that composeParsers was actually performing *three* compositions at the same time. Tres. The parsing results, the inputs, and the failures, all of them were being composed in different ways. *Composition*, that's what matters. We humans have a tendency to focus on *things*, but in reality, it's the *relationship* between things what matters. And this book, in case you hadn't noticed, is about that.

But composition is terribly uninteresting unless we have *identities* to compose, too. Without them, say, we could have a broken implementation of composeMaybes that always results in Nothing. Identities prevent these silly things. In the case of composeMaybes, this identity is just Just.

```
composeMaybes Just f  ==  f
```

```
composeMaybes f Just  ==  f
```

In the case of function composition —like the one we kinda see in our leftover management— the identity is obviously id. We know this.

```
id . f  ==  f

f . id  ==  f
```

But what about the identity when *pairing* two things using (,), as we are doing with our parser outputs a and b?

```
(?, x)  ==  x

(x, ?)  ==  x
```

Ah, wouldn't we like to know? There *must* be an identity, otherwise we wouldn't be calling this product a "composition". We will learn about it, but not right away. First, we have something to wrap up.

One hundred thirty-nine

We know that we want *all* our parsers to have the type String -> Maybe (String, x), where x is something meaningful we hope to obtain from this String. For example, here's the implementation of our most recent parseBool using this type.

```
parseBool :: String -> Maybe (String, Bool)
parseBool = \s -> case s of
                    't' : z -> Just (z, True)
                    'f' : z -> Just (z, False)
                    _       -> Nothing
```

And we appreciate the type of functions like these so much that we want it to be its own *different* type.

```
data Parser x = Parser (String -> Maybe (String, x))
```

We have seen datatypes like this one before, having just one constructor and carrying a *function* as a payload. Op, for example, was one of them.

```
data Op a b = Op (b -> a)
```

They are very common when we want to give functions with a particular shape, serving a particular purpose, a distinct type so that we can more easily talk and reason about them, and at the same time distinguish them from other functions that would otherwise have the same type yet serve a different purpose. The idea is that rather than talking about "parseBool, the function that takes a String and returns a Maybe (String, Bool) where Nothing conveys parsing failure and Just conveys a successful parsing of a Bool accompanied by a leftover input String", we will assign a meaning to the Parser type, and then just talk about "parseBool, the Parser of Bool values".

```
parseBool :: Parser Bool
parseBool = Parser (\s -> case s of
                            't' : z -> Just (z, True)
                            'f' : z -> Just (z, False)
                            _       -> Nothing)
```

Two things changed in this definition of `parseBool` compared to the one from before. First, the type. Just look at it, `Parser Bool`, beautiful, all the noise is gone. Second, we are *wrapping* the entire `\s -> ...` function from before inside the `Parser` constructor. That is, this function becomes the sole payload of this datatype.

Remember, there is a type constructor called `Parser`, of kind `Type -> Type`, which we can apply to a `Type` like `Bool` to obtain yet another `Type`, `Parser Bool`. We can construct *values* of type `Parser Bool` —or `Parser whatever`, really— by using the value constructor *also* called `Parser`, of type `(String -> Maybe (String, Bool)) -> Parser Bool`, where `String -> Maybe (String, Bool)` is exactly like the type of our `\s -> ...` expression.

Here is a different `Parser` that does something else. A boring one that always fails to parse *any* x we ask of it.

```
fail :: ∀ x. Parser x
fail = Parser (const Nothing)
```

Does it type-check? Indeed it does, `const Nothing` has type `∀ a b. a -> Maybe b`, so if we pick `String` as our `a`, and `(String, x)` as our `b`, everything will click. It is not the most interesting `Parser`, but it is a `Parser` nonetheless.

So, what do we do with these `Parsers`? Well, the same thing we did with them back when they were still functions: We provide some input to them and see what happens. Doing this when parsers were functions was easy, all we had to do was apply them to an input `String`, that was all. But what can we do now? A `Parser` is not a function anymore. *Or is it?* Sure, at a very superficial level, a `Parser x` is *not* a function, but if we look beyond that `Parser` wrapper, all we'll find is a plain old boring function taking `String` as input and returning `Maybe (String, x)` as output. So, presumably, what we need to do is somehow remove the `Parser` wrapper and provide the `String` input *directly* to the underlying function.

We are not strangers to this approach, actually. For example, we saw something similar in `Sum`, remember?

```
data Sum = Sum Natural
```

You see, there is a `Natural` value *inside* the `Sum` value constructor, but in order to operate on it directly, we first need to remove the `Sum` wrapper. In fact, we introduced a function named `unSum` just for this.

```
unSum :: Sum -> Natural
unSum (Sum x) = x
```

Hmm, what would happen if we did the same thing to Parser?

```
unParser :: Parser x -> (String -> Maybe (String, x))
unParser (Parser f) = f
```

Of course. unParser gives us the function within the Parser value
constructor, which, being a function, can now be applied to its input
String by juxtaposition as usual. Traditionally, people call functions like
these, which when fully applied to all of its input arguments give the
impression of *running* the Parser or similar somehow, *run*Something,
rather than *un*Something. Maybe we should do that as well, out of respect
for tradition. And we will remove the redundant rightmost parentheses
while we are at it.

```
runParser :: Parser x -> String -> Maybe (String, x)
runParser (Parser f) = f
```

We can now talk about runParser, the function that runs the given Parser
on the given String, resulting in parsing leftovers and a parsing result, if
any.

```
> runParser parseBool "t"
Just ("", True)
> runParser parseBool "tf"
Just ("f", True)
> runParser parseBool "x"
Nothing
> runParser fail "hello"
Nothing
```

Excellent. The only important thing left to worry about is composeParsers,
which should now deal with values of type Parser rather than the String
-taking parsing functions from before.

```
composeParsers :: Parser a -> Parser b -> Parser (a, b)
composeParsers pa pb = Parser (\s0 ->
    case runParser pa s0 of
        Nothing -> Nothing
        Just (s1, a) ->
            case runParser pb s1 of
                Nothing -> Nothing
                Just (s2, b) -> Just (s2, (a, b)))
```

Look at that handsome type, all the noise is gone. All we did was replace
the previous parsing functions for values of type Parser, use runParser each
time we need to run a Parser on some input, and finally wrap the entire
expression in a Parser value constructor. The rest is the same as before.
Well, almost. We lost some of the parametricity we had, seeing how the
Parser datatype fixes the parsing input and leftover types to be Strings, so
we can't tell anymore the order in which of these Parsers are run just by

looking at the type of composeParsers. We now need to rely on documentation, routine and prayer. It's alright, though. Even though we made a big deal out of this before, that was mostly to make a point. In practice, this is a minor issue. In Haskell, tradition says things go from left to right, so that's the order in which we normally expect things to happen. Except when they go from right to left, which is about half of the time. One gets used to it.

```
> runParser (composeParsers parseBool parseBool) "tfx"
Just ("x", (True, False))
> runParser (composeParsers parseBool parseBool) "txf"
Nothing
```

However, if we cared enough we could bring back that lost polymorphism to the Parser type.

```
data Parser i o x = Parser (i -> Maybe (o, x))
```

In this type, i is the parser's input type, o is the parser's leftover type, and x is the actual parsing output. The implementation of composeParsers would be exactly the same, only its *type* would change.

```
composeParsers :: Parser x y a
               -> Parser y z b
               -> Parser x z (a, b)
```

We won't be doing any of this, though. We *could*, but to keep things "simple", we'll just embrace the left to right tradition.

One hundred forty

Let's take a small diversion. Why is a product type called a *product* type? Why is a *sum* type called that? We know some of this. In part it has to do with their *cardinality*, that is, with the number of values that can potentially be of these types. Or, as we say, the number of values that *inhabit* them. Let's start with something simple, let's start with Bool.

```
data Bool = True | False
```

The cardinality of Bool is *two* because Bool has only two inhabitants True and False. And what if we pair two Bools in a product type? What would be the cardinality of that pair?

```
(Bool, Bool)
```

It's *four*, of course. We can verify this by explicitly enumerating *all* of the possible values of type (Bool, Bool) —(True, False), (True, True), (False, False) and (False, True)— or by simply *multiplying* the cardinalities of the types that are part of this pair.

```
2 × 2  ==  4
```

Multiplication, product. From here *product* types take their name. But *four* is also the *sum* of *two* and *two*, isn't it?

```
2 + 2  ==  4
```

It is, and we can see this manifest itself as the cardinality of a *sum* type of Bools, each of them having a cardinality of *two*.

```
Either Bool Bool
```

Again, there are *four* values that could have type Either Bool Bool. They are Left False, Left True, Right False and Right True. From here, *sum* types take their name. Interestingly, the cardinalities of (Bool, Bool) and Either Bool Bool are equal.

```
2 + 2  ==  2 × 2
```

And surprisingly perhaps, this implies that (Bool, Bool) and Either Bool Bool are *isomorphic*. That is, we can convert between values of these two types back and forth without any information loss. Here is *one* proof of this.

```
fromEither :: Either Bool Bool -> (Bool, Bool)
fromEither (Left x) = (False, x)
fromEither (Right x) = (True, x)

fromPair :: (Bool, Bool) -> Either Bool Bool
fromPair (False, x) = Left x
fromPair (True, x) = Right x
```

Composing these functions, as expected of every isomorphism, results in identities.

```
fromEither . fromPair  ==  id

fromPair . fromEither  ==  id
```

In general, for any two datatypes having the same cardinality, we can always come up with a pair of functions that proves that an isomorphism exists between them. Or more than one, perhaps. For example, rather than using the Left constructor whenever the first Bool in the pair is False, we could have chosen to use the Right constructor:

```
fromEither2 :: Either Bool Bool -> (Bool, Bool)
fromEither2 (Right x) = (False, x)
fromEither2 (Left x) = (True, x)

fromPair2 :: (Bool, Bool) -> Either Bool Bool
fromPair2 (False, x) = Right x
fromPair2 (True, x) = Left x
```

These two functions form an isomorphism too. Is this a *better* isomorphism than the one given by our previous fromPair and fromEither? No, nor is it a worse one. It's just different.

One hundred forty-one

We said interesting things about functions seen *together* as isomorphisms, but we can also say interesting things about them *on their own*. Let's look at fromPair, for example. What can we say about it?

```
fromPair :: (Bool, Bool) -> Either Bool Bool
fromPair (False, x) = Left x
fromPair (True, x) = Right x
```

If we pay attention to the output of this function, we can see that it is a *surjective* one. Its *codomain*, comprising *all* the possible values of type Either Bool Bool, equals the *image* of this function, by which we mean the values of type Either Bool Bool that this function actually outputs. So, yes, fromPair is *surjective*.

But if we also include the input type in our reasoning, there's something else we can say about fromPair. Notice how each value in the *domain* of this function, (Bool, Bool), has a *distinct* corresponding value in its codomain. No two different input values result in a same output value, which ultimately leads the cardinality of its domain, *four*, to be equal to the cardinality of its image. We call this property *injectivity*, we say that fromPair is an *injective* function.

And yes, talking about the cardinality of the *image* of a function is fine. Generally speaking, *cardinality* is a property of *sets*, and it's mostly just a fancy name for talking about how many elements are in a set. If you have *five* friends, lucky you, then the cardinality of your set of friends is *five*.

For contrast, here is a *non-injective* function.

```
isEmpty :: [a] -> Bool
isEmpty [] = True
isEmpty _  = False
```

isEmpty is *not* injective, for some elements of its domain —namely, all the non-empty lists— map to a same element of its image, False. This is a surjective function, but it is not an injective one.

And why is injectivity good? Well, it's neither good nor bad, it's just a property. Like enjoying a particular song or having been born in June, it doesn't say much about us.

One hundred forty-two

What's interesting for us is that when a function is *both* injective *and* surjective, this function is said to be *bijective*, and all bijective functions have an *inverse* function which, essentially, undoes everything the bijective function did. We can also call these inverses *anti-functions*, but only if we like science fiction.

In our example, fromPair is a bijective function and fromEither is its inverse. Or, taking the opposite perspective, we can say that fromEither is our bijective function and fromPair its inverse. Generally, we just say these two functions are inverses of each other, and this implies their bijectivity.

And all of this for what? Because this is what an *isomorphism* is: a pair of functions that are the inverse of each other. That's it. At least insofar as types and functions are concerned.

One hundred forty-three

A *product* type multiplies the cardinalities of its parts, a *sum* type adds them. So what?

It turns out that there is a direct correspondence between our data types and algebraic expressions like 2 + 4 or 3 × a. That is, we can translate many algebraic expressions to a type and vice-versa. So much so, actually, that we call these things *algebraic data types*. In short, *ADT*.

For example, the Either sum type applied to two types a and b having a particular cardinality, corresponds to the *addition* of the two natural numbers representing those cardinalities. Say, if we acknowledge that the cardinality of Bool is *two*, and that the cardinality of Season is *four*, then the algebraic expression adding these two numbers corresponds, in Haskell, to a use of Either on the corresponding types Bool and Season.

```
2 + 4  ==  Either Bool Season
```

And indeed, if we were to count it manually, we'd find the cardinality of Either Bool Season to be *six*.

Notice that in our use of the symbol ==, we are *not* saying that these things are *equal*. They couldn't possibly be equal because we have numbers on one side and types on the other. All we are saying is that there is a *correspondence* between these algebraic expressions and these types.

But more generally, leaving the field of arithmetic and jumping deep into algebra, we can replace 2 and 4 with almost any other algebraic expressions a and b and the correspondence would still hold.

```
a + b  ==  Either a b
```

And similarly, we have *product* types, or pairs, corresponding to algebraic multiplication:

```
a × b  ==  (a, b)
```

Which of course we could further compose with algebraic addition:

```
a × (b + c)  ==  (a, Either b c)
```

And, if you recall from elementary school, multiplication had this interesting property of being *distributive* over addition, meaning that for all choices of a, b and c, the following two algebraic expressions are equal:

```
a × (b + c)  ==  (a × b) + (a × c)
```

Which, of course, can *both* be represented as types, since they are just additions and multiplications, which in Haskell correspond to Eithers and pairs.

```
(a, Either b c)  ≈  Either (a, b) (a, c)
```

Now, these two types are not exactly *equal*. If we pass a value of type (a, Either b c) to a function expecting a value of type Either (a, b) (a, c), our program will fail to type-check. But they are *isomorphic*, as signaled by our usage of the symbol "≈" above, and as proved by this pair of inverse functions.

```
foo :: (a, Either b c) -> Either (a, b) (a, c)
foo (a, Left b)  = Left  (a, b)
foo (a, Right c) = Right (a, c)

bar :: Either (a, b) (a, c) -> (a, Either b c)
bar (Left  (a, b)) = (a, Left b)
bar (Right (a, c)) = (a, Right c)
```

But the correspondence between algebraic data types and algebra doesn't end here. Look at what happens when we start mixing numbers with variables.

```
1 + a  ==  Maybe a
```

Maybe? How? Well, let's recall its definition and see for ourselves.

```
data Maybe a = Nothing | Just a
```

That is, a value of type Maybe a can be constructed by using Nothing *or* by applying Just to a value of type a, of which we have as many inhabitants as its cardinality says we have. So, if we take Maybe Bool for example, we would have *three* ways of constructing a value of this type, namely Nothing, Just False and Just True. In other words, Maybe Bool corresponds to the number 3.

```
3  ==  1 + 2  ==  Maybe Bool
```

Using this same line of reasoning, we can come up with an algebraic datatype of *any* cardinality we want. For example, a while ago we came up with a Digit type because we needed something of cardinality *ten*. But isn't *ten* just another natural number that can be expressed as additions and products of some other numbers? Sure it is, and we can use an approach not unlike equational-reasoning to explore this, playing fast and loose with the fact that we are intertwining algebraic expressions, numbers and types.

```
Digits
  == 10
  == 2 × 5
  == Bool × 5
  == Bool × (Bool + 3)
  == Bool × (Bool + (1 + 2))
  == Bool × (Bool + (1 + Bool))
  == Bool × (Bool + Maybe Bool)
  == (Bool, Either Bool (Maybe Bool))
```

And yes, we could have factorized 10 in a different way, eventually arriving at a different datatype. Different, but still *isomorphic*.

Are Digit and (Bool, Either Bool (Maybe Bool)) isomorphic? They better be. If we are able to come up with a *bijective* function from Digit to (Bool, Either Bool (Maybe Bool)) we will have proved this, seeing how every bijection has an inverse, and together they form an *isomorphism*.

```
naturalFromDigit :: Digit -> (Bool, Either Bool (Maybe Bool))
naturalFromDigit D0  = (False, Left  False)
naturalFromDigit D1  = (False, Left  True)
naturalFromDigit D2  = (False, Right Nothing)
naturalFromDigit D3  = (False, Right (Just False))
naturalFromDigit Dog = (False, Right (Just True))  -- Woof!
naturalFromDigit D5  = (True,  Left  False)
naturalFromDigit D6  = (True,  Left  True)
naturalFromDigit D7  = (True,  Right Nothing)
naturalFromDigit D8  = (True,  Right (Just False))
naturalFromDigit D9  = (True,  Right (Just True))
```

Is naturalFromDigit *injective*? Yes, it takes each element from its domain to a distinct element of its image. Is it a *surjective* function? Yes, its image and codomain are the same. Is naturalFromDigit a *bijective* function? Yes, for it is both injective and surjective. Does naturalFromDigit have an *inverse*? Yes, every bijective function is invertible, although quite often the inversion is an exercise for the reader. Are Digit and (Bool, Either Bool (Maybe Bool)) *isomorphic*? Yes, for there are two functions, naturalFromDigit and its inverse, that are by definition inverses of each other. Is it the *only* isomorphism between these types? No, we made many arbitrary choices when we came up with our mapping between these types, and we could have chosen differently.

So, why do we come up with types like Digit if Eithers, pairs, Bools and the like seem to be enough? *Humans*, that's why. Internally, computers and programming languages use simple sum and product types like Either and pair a lot. Digit, *ten*, Either Bool (Maybe Bool), 3 + 7, 10, it doesn't matter what we call it, computers don't care. But we humans, we care about names. The words "digit" or "season" *mean* something to us, so we come up with technically irrelevant stuff like names in order to improve our quality of life, and that's fine.

One hundred forty-four

What about *the one*? That is, *one*, literally.

```
1
```

It's quite straightforward, actually. It barely deserves any attention.

```
data Unit = Unit
```

See? One constructor, cardinality *one*, done. Next topic.

```
> :t Unit
Unit :: Unit
```

One hundred forty-five

Just kidding. Of course there's more to Unit than its definition. For example, we recently said that the algebraic expression 1 + a corresponds to the Haskell datatype Maybe a. But that's not the sole truth, is it? If we are saying that Unit corresponds to *one*, and that Either corresponds to addition, then what's wrong with this?

```
1 + a  ==  Either Unit a
```

Nothing, there's nothing wrong with that. Indeed, Maybe a and Either Unit a are isomorphic. Here's the proof, this time using the maybe and either functions rather than pattern-matching, for fun.

```
fromMaybe :: Maybe a -> Either Unit a
fromMaybe = maybe (Left Unit) Right

fromEither :: Either Unit a -> Just a
fromEither = either (const Nothing) Just
```

Now, we know that addition is a *commutative* operation, meaning that even if we swap the order of the operands, the result stays the same.

```
1 + a  ==  a + 1
```

This is indeed true in algebra. However, in Haskell, the corresponding types are not exactly *equal*, just *isomorphic*.

```
Either Unit a  ≈  Either a Unit
```

That is, we can't just use a value of type Either Unit a where a value of type Either a Unit or Maybe a are expected, even if algebra says these are equal.

```
Bool  ≈  Either Unit Unit
```

Isomorphic, but not equal.

One hundred forty-six

What about multiplying by one?

```
1 × a  ==  (Unit, a)
```

Right, of course, multiplication corresponds to product types, to pairs, so we just pair Unit, *one*, with our chosen a, and that's our answer. As before, however, if we try to exploit the *commutativity* of multiplication, we end up with just an *isomorphism* on the Haskell side of things, not an equality. That is, while in algebra 1 × a and a × 1 are equal, in Haskell types they are just isomorphic.

```
(Unit, a)  ≈  (a, Unit)
```

Why do we insist on isomorphisms so much? Well, because something interesting is about to happen. Look.

```
a × 1  ==  a  == 1 × a
```

Back when we were learning about multiplication of natural numbers, we learned that *one* was the identity element or *unit* of multiplication, meaning that multiplying a number by this unit leads back to that same number. Or, more generally, in algebraic terms, multiplying *any* variable a by the multiplicative identity, *one*, leads back to that same a. And this, of course, has its correspondence in types too.

```
(Unit, a)  ≈  a  ≈  (a, Unit)
```

This isomorphism is important because equality won't be enough where we are going. At times we will end up with (Unit, a) when, conceptually, a is all we care about.

It's interesting to reason about *why* a and a pair of a and Unit are conceptually equal, isomorphic, even if (a, Unit) clearly has more information than a, and converting between (a, Unit) and a clearly loses some information. Namely, Unit.

```
byeUnit :: (Unit, a) -> a
byeUnit (Unit, a) = a
```

See? Unit is gone. Poof.

Unit differs from every other type in that it has only *one* inhabitant, so if we know that a value has the type Unit, then we also know that this value *must* be the constructor also named Unit, simply because there is no alternative. So if we are asked to come up with the inverse function of byeUnit, of type a -> (Unit, a), we can simply create the Unit that we need, right there on the spot.

```
hiUnit :: a -> (Unit, a)
hiUnit a = (Unit, a)
```

There is simply no other possible implementation of a function with this type. So byeUnit didn't *really* lose any information about Unit, because there was no information to be lost. Unit can come and go as necessary. It has no meaning, it's there simply to fill the gaps.

One hundred forty-seven

In types, *equality* is a very strong statement. Generally speaking, a type is only equal to itself.

The idea of types, in Haskell, comes from a nice branch of mathematics unsurprisingly called *Type Theory*, or more precisely, *constructive* Type Theory, which, as its name indicates, *constructs* things. In particular, it constructs *truths*.

When we define a datatype like Bool, we are saying "this is Bool, nothing else is Bool, and these are its inhabitants True and False". From that point on Bool, True and False exist, and they are different from anything else that came before them. *This is the new truth*, and from this construction follows that Bool is only equal to itself.

But nothing in this construction talks about how Bool could be considered equal to an isomorphic type, so it is up to us to prove, if necessary, that any non-Bool type we might be interested in can be seen as equal to Bool by means of an isomorphism. In concrete terms, this means explicitly applying a function to convert between Bool and this other type in the desired direction. Anything less than that will fail to type-check.

But if Bool and Either Unit Unit are obviously isomorphic, or more generally, if any two datatypes with the same cardinality are, why doesn't the compiler magically come up with an isomorphism between them and use that to reason about Bool and Either Unit Unit as equals? It would be nice, wouldn't it? They *mean* the same, don't they? Well, you tell me.

```
data Hire = DoHireHim | DoNotHireHim

data Fire = DoFireHim | DoNotFireHim
```

According to their cardinality, these are isomorphic too. Perhaps we'd like to consider them as equals? No, of course not. To *hire* and to *fire* mean completely different things to us, even while technically, yes, they are isomorphic.

Hire and Fire are equal in an *extensional* manner where we only consider structural matters about types such as their cardinalities, but they are *not* equal in an *intensional* manner by which we assign meaning to these types beyond what the type-system sees.

So, *no*, it would not be such a great idea to allow the type-system to *automatically* guess whether two types should be treated as equal based simply on their extensional properties. One could envision a type system where we *explicitly* allowed some types to be treated as equal, such as (Unit, a) and a, even if ultimately they are not. This is closely related to the idea of *univalence*, which we won't be covering here, but is different from equality and lies at the core of modern type theories that among other things attempt to tackle this matter.

One hundred forty-eight

A while ago we said that the multiplication of natural numbers, with *one* as its identity element, was a *monoid*. And we are saying now that multiplication corresponds to product types, and that *one* corresponds to Unit. Are we implying that product types are monoids too? Why, of course we are.

Let's see. What is a monoid? On the one hand, there is an associative binary operation taking two values of a same type. Do we have this? Kinda.

```
(,) :: a -> b -> (a, b)
```

a and b are most certainly *not* the same type. They *could* be, sure, but their separate universal quantifications say this is not necessary. But even if a and b were the same type c, say, the output type (c, c) would still be different to the inputs of type c.

```
(,) :: c -> c -> (c, c)
```

So, what? Monoid no more? Not necessarily. Let's compare the type of (,) with that of (*), the function that multiplies Natural numbers —or any Num, really, but we are only considering Naturals just to keep things simple.

```
(*) :: Natural -> Natural -> Natural
```

If we were to describe this function out loud, we would say that (*) is the function that takes two *values* of type Natural and outputs yet another *value* of type Natural. That is, we are talking about the things we do to *values* of type Naturals.

But (,) is not so much about the *values* it receives as input —which, as mandated by parametricity, remain completely untouched anyway— but about the *types* of its inputs and output. We are talking about product *types*, not product values. So, not unlike (*) concerns itself with *values* of a same *type*, (,) concerns itself with *types* of a particular *kind*.

(,) is the binary operation of our monoid, yes, but rather than taking two values of a same *type* into another value of that same type, it takes two types of a same *kind* into another value of that same kind. But if we are now talking about types and kinds rather than values and types, we better look at *type constructors*, not value level functions. In other words, our binary operation is not so much the value constructor (,) of type a -> b -> (a, b), but the *type constructor* (,) of kind Type -> Type -> Type.

```
> :kind! (,)
(,) :: Type -> Type -> Type
```

Now, this takes us to a very different place. In particular, we won't be able to use the Monoid typeclass we explored before because that was designed to operate on *values*, not on types. But that's fine, we are not actually planning to implement anything, we are simply exploring some ideas in our mind, beyond Haskell.

So we have (,), a binary operation on *types* of kind Type. Great. Is it associative? Well, if we are claiming this is a monoid, it must be. That is, the type (a, (b, c)) must be equal to the type ((a, b), c). Now, these types are most definitely not equal, but they are *isomorphic*, as proved by the following bijective function which isn't enough for the type-checker to automatically treat one of these types as equal to the other, but it is enough for us humans to acknowledge the associativity of (,).

```
assoc :: (a, (b, c)) -> ((a, b), c)
assoc (a, (b, c)) = ((a, b), c)
```

So we have an associative type level function (,) operating on *types* of kind Type. Do we have an identity element, a unit, of that same kind Type? This should be a type that, when combined with another type x, should result in x once again. Well of course we have it, it's the Unit type, corresponding to the number *one*, the identity element for the product of numbers.

```
> :kind! Unit
Unit :: Type
```

Again, while not *equal*, the following types are all isomorphic, which is good enough for us.

```
(Unit, a)  ≈  a  ≈  (a, Unit)
```

So, yes, we can reason about the product of *types* as monoids in the same way we could reason about the product of *numbers* as monoids.

And in doing this we kinda left Haskell behind, which brings our attention to the fact that many of the ideas we are learning, showing up here as values and types, have a place outside Haskell too. It's just that in Haskell they can *run*, and that's what makes this place so interesting.

And by the way, if we really wanted, we could have these type level monoids in Haskell too. To a good extent anyway, *up to isomorphisms*. Not today, though, for we are learning something else instead.

One hundred forty-nine

What about *addition*? We saw that there was a clear correspondence between the addition of numbers and *sum types*. And, seeing how the addition of numbers is a monoid, as we saw many times before, presumably *sum types* are a monoid too. And sure enough, they are.

Just like (,), Either is a type constructor of kind Type -> Type -> Type. Is it an associative one? Yes, it is, as witnessed by this invertible function.

```
assoc :: Either a (Either b c) -> Either (Either a b) c
assoc (Left a) = Left (Left a)
assoc (Right (Left b)) = Left (Right b)
assoc (Right (Right c)) = Right c
```

But what about our identity element, which according to the addition of natural numbers was *zero*? Well, it's just a type with *zero* inhabitants, which in Haskell we call Void, presumably to convey a vast sense of emptiness.

```
data Void
```

And *no*, we didn't forget anything. We are saying that Void is a new datatype, but since it has *no* inhabitants, we are listing *no* constructors for it, and this is fine.

So how do we construct a value of type Void? Well, we don't. That's the beauty of it. Void exists as a type, but it does *not* exist as a value, which among other things allows us to claim that *yes*, Void is the identity element for the addition of types, that is, for Either.

```
Either Void x  ≈  x  ≈  Either x Void
```

How? All we have to do is find the relevant isomorphisms. Going from x to Either x Void is straightforward, we just use x as payload for the Left constructor and ignore the Right side completely.

```
toEither :: x -> Either x Void
toEither = \x -> Left x
```

And similarly, we would use the Right constructor if we were trying to construct an Either Void x. But what about the inverse of this function? What about the one that goes *from* Either x Void to x?

```
fromEither :: Either x Void -> x
fromEither (Left x) = x
fromEither (Right ？) = ？
```

It's clear that if we receive a Left constructor with an x as its payload we must simply return that x. But what if we receive a Right? What do we do with the Void value that comes as its payload? How do we obtain the x that we must return from that Void value?

Read that again, go on. *A Void value*? Impossible, there's no such thing as a Void value, that's absurd. And that's the clever trick. How would we construct a value of type Either x Void using the Right constructor if we wanted to do so? Well, we wouldn't. We just can't do it. In that case, the type of the Right constructor would be Void -> Either x Void, which implies that in order to obtain a value of type Either x Void we first need to provide an *impossible* value of type Void, which renders the Right constructor completely unusable, meaning that a value of type Either x Void can only ever be Left x. That is, much like how we could determine that an expression of type Unit must *always* be the value Unit, a value of type Either x Void must *always* be Left x. And, of course, the opposite is true for Either Void x, where only Right x can exist.

So how do we implement fromEither, then? The answer will disappoint you.

```
fromEither :: Either x Void -> x
fromEither (Left x) = x
fromEither (Right _) = undefined
```

You see, that second equation will *never* happen, so we can happily make our function bottom out there, confident that this code will *never* be evaluated.

The Void type corresponds to the idea of *zero*, the identity element for the addition of types, for Either. And, indeed, once we acknowledge the relevant isomorphisms, we find in sum types a monoid too.

One hundred fifty

What about other common algebraic operations such as *subtraction* and *division*? Well, no luck there. It doesn't make sense to subtract types, nor to divide them, so there's no correspondence between those operations and our algebraic datatypes.

Think about it. What would Digit *minus* Bool, allegedly corresponding to 10 - 2, be? Sure, we could come up with a completely new type with cardinality *eight* that doesn't mention Digit at all, but this would be just an ad hoc transformation of the type Digit, not the result of applying a type constructor akin to Either or (,) to both Digit and Bool in order to get a *new* datatype of the desired cardinality. No, there's no direct correspondence between *subtraction* and algebraic datatypes. And similarly, there's no correspondence with *division*. Generally speaking, there's no correspondence between types and algebraic operations that could make the cardinalities of our types smaller. Types can only grow, not shrink. Well, technically, they could go down to *zero* if we pair them with Void.

```
(Bool, Void)
```

See? This type has *zero* inhabitants, for we can never create the value of type Void we need in order to construct a value of type (Bool, Void). This corresponds to the idea that multiplying by *zero* equals *zero*. But other than this one corner case, types don't shrink, they only ever grow, so shrinking operations such as division and subtraction do not have a correspondence with operation on types. Having said that, look at this datatype:

```
data List a = Nil | Cons a (List a)
```

A List is made out of products, sums and *recursion*. What can we do with it? Is there a correspondence between recursion and algebraic expressions? Let's see.

Let's start by writing down the definition of List as an algebraic equation, pretending that list is the name of a mathematical function, and translating the right hand side of this definition to additions and multiplications as we've been doing so far.

```
list(a) = 1 + a × list(a)
```

Now, this is just an algebraic equation, so as with any other equation we can move things around preserving the equality, ideally with a goal in mind. In our case, since we are trying to determine what list(a) means, we want to move things around so that list(a) remains on one side of the equation, and on the other side we have an expression *not* involving list(a) at all. In algebraic parlance, we want to *solve* this equation for list(a). You might have learned how to do this in high-school, but you may also have forgotten, so here's how. Be warned, this is boring.

The basic idea is that if we perform a same operation, *any* operation, on *both* sides of the equation at the same time, the equality between these sides is maintained. The trick is that carefully selected operations will cancel out with others that were previously present in the equation, and things will seem to "move around". Let's try. Let's subtract a × list(a) from both sides of our equation.

```
list(a) - a × list(a) = 1 + a × list(a) - a × list(a)
```

Now, on the right-hand side of our equation, we are *adding* a × list(a) only to *subtract* it afterwards, taking us back to where we were before adding it in the first place, which prompts us to wonder whether we could perhaps avoid all this trouble if we just *don't* add nor subtract a × list(a) at all. In other words, this addition and this subtraction cancel each other out. Let's get rid of them.

```
list(a) - a × list(a) = 1
```

Alright. Also, unrelated, we know that *one* is the multiplicative identity, meaning that multiplying by it doesn't change the value of our expressions. So, just to make a point, let's multiply our leftmost list(a) by *one*.

```
1 × list(a) - a × list(a) = 1
```

On the left side of this equation, list(a) is multiplying *both* 1 and a in this subtraction, so we can factor this out as a multiplication of list(a) by 1 - a. Remember, multiplication distributes over addition and subtraction, and by factoring out that multiplication we are "undistributing" it.

```
list(a) × (1 - a) = 1
```

Next, we can try to divide both sides of this equation by 1 - a.

```
(list(a) × (1 - a)) / (1 - a) = 1 / (1 - a)
```

And finally, seeing how on the left-side, 1 - a appears as part of a multiplication and a division that cancel each other out, we can just remove them.

```
list(a) = 1 / (1 - a)
```

Great. We arrived somewhere. This is saying that list(a), the algebraic expression that corresponds to our Haskell datatype List a, equals the algebraic expression 1 / (1 - a). And now what? How do we translate *that* back to Haskell? We only have divisions and subtractions here, but we know that we have *no* correspondence between those operations and Haskell types. What's going on? Have we lost once again?

It turns out that an expression like 1 / (1 - a) can be seen through the eyes of geometry as an *infinite* geometric series with this shape:

```
1 / (1 - a) = 1 + a + a² + a³ + ...
```

And so on, infinitely. But look, we only have additions and powers there, and powers are just multiplications repeated over and over again.

```
1 / (1 - a) = 1 + a + a × a + a × a × a + ...
```

This is fine. We know how to translate additions and multiplications to types:

```
Either Unit
      (Either a
              (Either (a, a)
                      (Either (a, a, a)
                              ... infinitely many more ...
```

In other words, we are saying that a list of as is *either* Unit, which corresponds to the empty list, *or* one a, corresponding to the one-element list, *or* a pair of as, corresponding to the two-element lists, *or* three as, corresponding to a list with three as in it, and so on, infinitely. Of course we can't list infinitely many constructors, which is why we use *recursion* to define lists instead. But yes, even recursive datatypes have corresponding algebraic counterparts.

One hundred fifty-one

We know how to translate the algebraic expression y^3 to types. It's just y multiplied by itself, and by itself again, which in types we can easily express as a product type of *three* ys such as (y, y, y). But what about something like y^x? How do we multiply y by itself x times? How many times are "x times" anyway? Have we lost? Not quite. It turns out that algebraic exponentiation corresponds to *function* types. In particular, y^x corresponds to a function x -> y whose cardinality equals the numeric value of y^x. Let's look at an example, 3^2, which in Haskell corresponds to a function taking as input a type with cardinality *two*, such as Bool, and returning as output a type with cardinality *three*, such as Maybe Bool.

```
Bool -> Maybe Bool
```

What we are claiming is that there are 3^2 inhabitants of this type. That is, there are *nine* different scenarios we could write that take a Bool as input and return Maybe Bool as output. And indeed, here they are, all nine of them implemented as partial applications of the bool function.

```
f1, f2, f3, f4, f5, f6, f7, f8, f9
  :: Bool -> Maybe Bool
f1 = bool Nothing       Nothing
f2 = bool Nothing       (Just False)
f3 = bool Nothing       (Just True)
f4 = bool (Just False)  Nothing
f5 = bool (Just False)  (Just False)
f6 = bool (Just False)  (Just True)
f7 = bool (Just True)   Nothing
f8 = bool (Just True)   (Just False)
f9 = bool (Just True)   (Just True)
```

And yes, those first two lines saying f1, ..., f9 :: Bool -> Maybe Bool are valid Haskell syntax too. When we have many expressions with a same type, we can write down their names separated by commas and assign a type to all of them at the same time using this syntax.

We also claimed before that types with the same cardinality are isomorphic to each other. Is this still true for functions? Let's see, let's try to find an isomorphism with another type having the same cardinality as Bool -> Maybe Bool, which is *nine*.

```
data X = X1 | X2   X3 | X4 | X5 | X6 | X7 | X8 | X9
```

Sure, X could work. It has *nine* inhabitants. All we need now is a bijection between X and Bool -> Maybe Bool, as well as its inverse. Let's try converting values of type X to values of type Bool -> Maybe Bool first.

```
fromX :: X -> (Bool -> Maybe Bool)
fromX X1 = f1
fromX X2 = f2
fromX X3 = f3
fromX X4 = f4
fromX X5 = f5
fromX X6 = f6
fromX X7 = f7
fromX X8 = f8
fromX X9 = f9
```

No surprises here. We simply map every distinct input to a distinct output function among the ones we defined before, making sure that the image of fromX equals its codomain, which indeed it does. But what about the *inverse* of this function?

```
toX :: (Bool -> Maybe Bool) -> X
```

We haven't quite said it yet, but functions *can't* be compared for equality using (==), nor can they be pattern-matched against —not in Haskell, anyway— so we can't really inspect how this function was defined and decide which X to return based on that. Once a function exists, all we can do with it is apply it. Alright then, let's see if we can solve this by applying it. And *yes*, applying a function could be a more computationally intensive process than just pattern-matching against constructors, so in principle, exploiting our isomorphism in this direction could be a more demanding operation for our computer. Nonetheless, computationally intensive or not, the result would be correct.

How do we do this? We have a function Bool -> Maybe Bool, we know all we can do is apply it, and we know that there are *two* such Bools to which we could apply it, True and False. So, to which of them do we apply it? Well, we are trying to come up with a *bijective* function, which among other things implies that the image of this function and its codomain must be the same. That is, this function must be *surjective*. But see what happens if we apply the function to, say, False.

```
toX :: (Bool -> Maybe Bool) -> X
toX = \f -> case f False of
                Nothing    -> ?
                Just False -> ?
                Just True  -> ?
```

Even though the cardinality of our codomain X is *nine*, the cardinality of our image is just *three*. For this reason, this is *not* a surjective function, let alone a bijective one. And of course, applying our function to True would lead to the same result, since we'd still only have *three* different patterns to match against f True, leading to the cardinality of toX's image to be just *three*. No, what we need to do is apply f to *both* False and True, making sure we deal with all the *nine* possible combinations of their results.

```
toX :: (Bool -> Maybe Bool) -> X
toX = \f -> case (f False, f True) of
                (Nothing,    Nothing)    -> X1
                (Nothing,    Just False) -> X2
                (Nothing,    Just True)  -> X3
                (Just False, Nothing)    -> X4
                (Just False, Just False) -> X5
                (Just False, Just True)  -> X6
                (Just True,  Nothing)    -> X7
                (Just True,  Just False) -> X8
                (Just True,  Just True)  -> X9
```

This is fine. We can apply f as many times as we want. Or in general *any* function, not just f. Other than our computer doing arguably unnecessary work, nothing bad can come out of applying functions. Anyway, we have restored the surjectivity and bijectivity of toX, demonstrating at last the existence of an isomorphism between X and Bool -> Maybe Bool, two types with the same cardinality.

```
toX . fromX = id

fromX . toX = id
```

So, *yes*, functions are isomorphic to other types with the same cardinality. But of course, functions embody *mystery* and computation too, which is why we use them instead of their isomorphic types. In other words, this exploration about a function's isomorphism was mostly a curiosity, and we can forget all about it now.

One hundred fifty-two

Wait, wait, wait. Sorry, false alarm. Don't forget about functions and exponentiations just yet. There's actually something else we should bring our attention to, first. Two things, actually, two *fascinating* situations. The first one is x^1, which in algebra equals x, and in Haskell translates to an isomorphism between Unit -> x and x.

```
Unit -> x  ≈  x
```

Is this true? Let's see if we can come up with a bijective function and its inverse. First, let's go from Unit -> x to x.

```
foo :: (Unit -> x) -> x
foo f = f Unit
```

That was easy. We have a function that consumes a Unit in order to produce an x, and we know we can always make a Unit out of thin air. So we just make one, apply f to it, and return the resulting x. What about the inverse of foo?

```
oof :: x -> (Unit -> x)
oof = const
```

Even easier. oof is a function that takes an x, and returns a function that ignores any Unit it receives and outputs the original x. We could have pattern matched against Unit, of course, but const, of type ∀ a b. a -> (b -> a), works too. Remember, those rightmost parentheses are redundant. Curious, right?

And what about x^0, which algebra says equals *one*, which in Haskell corresponds to Void -> x being isomorphic to Unit?

```
Void -> x  ≈  Unit
```

Strange, isn't it? Let's see if we can implement a bijective function from Void -> x to Unit.

```
bar :: (Void -> x) -> Unit
bar _ = Unit
```

That was easy. Values of type Unit can be created out of thin air, so we just create one and return it, why not? In any case, even if we wanted, we wouldn't be able to use that function Void -> x because there's just no Void to which we could apply it, so all we can do with that function is ignore it.

bar is clearly a *surjective* function, since its image matches its codomain, but is it an *injective* one? That is, does bar map each distinct value of type Void -> x into a distinct value of type Unit? Well, yes it does, because it turns out that there's only one such value.

```
absurd :: Void -> x
absurd = const undefined
```

Absurd, I know. This function says "give me a Void, and in return I'll give you any x you want". Tempting, right? Except we can never give absurd a Void, so we have no way of proving that absurd is just bluffing. This bluff, nevertheless, inhabits Void -> x. And it is its *only* inhabitant, seeing how *bluffing* is all a function with this type could ever do. So yes, our function bar above is an injective one. It takes each distinct value of type Void -> x, of which there is just one, absurd, into a different value of type Unit.

What about the inverse of bar, the function that converts Unit into a function of type Void -> x? Well, we know there's only one such Unit, and we just learned that there's only one such Void -> x, so the implementation is rather straightforward, even if absurd.

```
rab :: Unit -> (Void -> x)
rab Unit = absurd
```

See? Fascinating. Silly, but fascinating nonetheless.

And now we can finally forget about all this. For real this time. Poof. The correspondences between algebraic expressions and types continue, sure, but *we* will stop here. All we wanted to say, really, is that the product of types is a *monoid* with Unit as its identity element, and we already said that a couple of chapters ago, before taking our long diversion—a word that in Spanish, by the way, means *fun*.

One hundred fifty-three

The reason why the product of types being a *monoid* matters is this:

```
composeParsers :: Parser a -> Parser b -> Parser (a, b)
```

Remember composeParsers? It was the function which took two Parsers, one for a and one for b, and composed them into a new Parser that after parsing a and b, in that order, produced both of them as result in a pair (a, b). Internally, it made sure the leftovers from parsing a were fed into the b's Parser as input, and that parsing failures from the individual Parsers were composed too.

And what does composeParsers have to do with monoids? Well, *everything*, for composeParsers is a monoid too. Just look at it side by side with another well-known monoid.

```
composeParsers :: Parser a -> Parser b -> Parser (a, b)
(,)            ::        a ->        b ->        (a, b)
```

composeParsers behaves just like (,), the product of types, which we know is a monoid with Unit as its identity element. At least insofar as the as and the bs are concerned, that is.

Monoids bring understanding. We'll soon distill Parsers and myriad monoids more, allegedly complex beasts, into a very simple idea. This is the way of the monoid. We've seen *many* of them already, we just didn't know they were monoids too.

But composeParsers and (,) are not *exactly* the same monoid, the product of types. If we were to pair two Parsers for a and b as a product type with (,), we'd end up with (Parser a, Parser b), not Parser (a, b). Of course, we could always create a function that takes one such (Parser a, Parser b) and returns a Parser (a, b) to mitigate this fact.

```
foo :: (Parser a, Parser b) -> Parser (a, b)
foo (pa, pb) = composeParsers pa pb
```

Wait, what? All this function is doing is taking Parser a and Parser b out of the tuple, and giving them to composeParsers as separate inputs. Other than the fact that the input Parsers come inside a pair, this function does exactly the same as composeParsers. What's going on?

One hundred fifty-four

Any two functions (x, y) -> z and x -> y -> z are *isomorphic* to each other. This has nothing to do with parsers nor monoids, but with that delicious Indian flavour.

```
curry :: ((x, y) -> z) -> (x -> y -> z)
curry f = \x y -> f (x, y)
```

Just kidding. This has nothing to do with cuisine. Apologies for the temptation. But yes, this function is historically called curry. We wrote down its implementation for completeness, but parametricity says this is the *only* possible implementation of this function, so we could have skipped it.

We say that x -> y -> z is the *curried* form of the function (x, y) -> z. And, of course, since we called these two types *isomorphic*, there must be an inverse to this function, too.

```
uncurry :: (x -> y -> z) -> ((x, y) -> z)
uncurry f = \(x, y) -> f x y
```

Again, a straightforward and unique implementation. We say that (x, y) -> z is the *uncurried* form of x -> y -> z. Or we don't, actually. Nobody talks like that.

```
curry . uncurry  ==  id
```

```
uncurry . curry  ==  id
```

Now, the reason why this isomorphism is of particular interest to us programmers is because it proves that there's really no difference between a function taking *two* input values as separate parameters, or as just *one* parameter that happens to be a pair containing those values. More generally, a function with *any* number of input parameters is isomorphic to a function taking just one product type with that many values as its sole input.

```
curry
  :: ((a, b) -> c)
  -> (a -> b -> c)

curry . curry
  :: (((a, b), c) -> d)
  -> (a -> b -> c -> d)

curry . curry . curry . uncurry . uncurry . uncurry
  :: (a -> b -> c -> d -> e)
  -> (a -> b -> c -> d -> e)
```

Technically speaking, we know that functions only ever take *one* input parameter at most. We have those invisible rightmost parentheses in our function types, remember?

```
curry :: ((a, b) -> c) -> (a -> (b -> c))
```

What's happening, conceptually at least, is that curry transforms a function taking two inputs values as a product type into two *functions*, each of them taking just one input value individually.

In other words, we've accidentally proved that product types are *unnecessary* for computation, seeing how we can always take the values inside a product type as separate input parameters instead. Are product types convenient? Sure, but they are unnecessary too. Indeed, we didn't say anything about pairing back when we were exploring the lambda calculus, which supposedly can do everything a computer can using nothing but functions, so there must be some truth to this. We'll learn more about this later on, when we build our own programming language.

One hundred fifty-five

Alright then, composeParsers is a monoid somehow related to product types, to (,), but not quite the same thing. What kind of monoid is it, then? Let's see if we can uncover the truth by just trying to *prove* that this is indeed a monoid. As a starting point, we have a binary operation, composeParsers.

```
composeParsers :: Parser a -> Parser b -> Parser (a, b)
```

Is it an *associative* binary operation? That is, are composeParsers pa (composeParsers pb pc) and composeParsers (composeParsers pa pb) pc the same? Well, they are most certainly not *equal*, because while one results in a value of type Parser (a, (b, c)), the other one results in a value of type Parser ((a, b), c). But they are *isomorphic*. Both in the fact that the pairs (a, (b, c)) and ((a, b), c) are isomorphic themselves, and that in both cases pa will consume its input first, followed by pb, and by pc at last. So, yes, composeParsers is associative *up to isomorphism*.

```
composeParsers pa (composeParsers pb pc)
  ≈ composeParsers (composeParsers pa pb) pc
```

We can prove this isomorphism by implementing a bijective function that simply rearranges the *output* produced by one of the Parsers.

```
parserAssoc :: Parser (a, (b, c)) -> Parser ((a, b), c)
parserAssoc p = Parser (\s0 ->
  case runParser p s0 of
    Nothing -> Nothing
    Just (s1, (a, (b, c))) -> Just (s1, ((a, b), c)))
```

Proving that composeParsers is associative, however, is not just a matter of rearranging the pairs that show up as *output* of this Parser. If it was just that, then the following nonsense would prove associativity too:

```
government :: Parser (a, (b, c)) -> Parser ((a, b), c)
government = const (Parser (const Nothing))
```

While the type of government suggests it achieves something useful, the execution of government proves otherwise. government exists by taking away the accomplishments of the given `Parser (a, (b, c))` and replacing them with grand promises that turn out to be `Nothing`. Defined in a fascinating manner so as to put readers in awe, government doesn't prove associativity at all.

The associativity of `composeParser` must happen in two places: on the a, b and c from the `Parser`'s output, of course, but also *inside* the `Parser`, which must *compose* in a predictable manner, as mandated by the second half of a monoid, its *identity element*.

One hundred fifty-six

`Unit` is the identity element for product types. And, seeing how `composeParsers` and `(,)` are so closely related, we'll probably also have to involve `Unit` somehow.

```
(Unit, a)  ≈  a  ≈  (a, Unit)
```

If, once again, we observe how `(,)` and `composeParsers` compare, we'll find a hint.

```
composeParsers :: Parser a -> Parser b -> Parser (a, b)
(,)            ::        a ->        b ->        (a, b)
```

`composeParsers` expects our as and bs to be *outputs* from these `Parsers`, not just standalone values, so we have to take that into consideration. That is, just like how applying `(,)` to a value of type `Unit` and a value of type `a` results in a value isomorphic to that original `a`, applying `composeParsers` to a value of type `Parser Unit` and a value of type `Parser a` should result in a value isomorphic to that original `Parser a`.

Where do we get a value of type `Parser Unit`, though? Can't we just invent one?

```
unit :: Parser Unit
unit = Parser (\s -> Just (s, Unit))
```

Nice. Here, `unit`, all in lowercase letters, is a `Parser` that doesn't consume any input yet it always *succeeds* to parse a `Unit`. On the one hand, this is fine, because we know that `Units` can be created out of thin air. But on the other hand, we are departing from the idea that a `Parser` is expected to derive meaning from an input `String`. In our case, `unit` is ignoring its input `String` completely, and the `Unit` it returns is certainly no "meaning" derived from it.

Is `unit` a silly idea, then? No, it's not, and we'll soon explain why. But first, let's just assume it's right, and while we have `unit` at hand, let's finally state the *identity law* that `composeParsers` must satisfy:

```
composeParsers unit pa  ≈  pa

composeParsers pa unit  ≈  pa
```

That is, composing *any* Parser with unit must result in that same Parser right away. Of course, we know that if pa has type Parser a, then composeParsers unit pa, say, will have the *different* type Parser (Unit, a). But we are grown ups now, and we understand that while these two expressions don't even have the same type, they could still be isomorphic. There's hope.

But, is there *truth*? We need truth. Warriors would prove this to be true using equational reasoning. We, subject to the physical constraints of these pages, but more importantly, to our love for these words, will just spell it out. *Yes*, it is true that these things are isomorphic.

First, because unit doesn't consume any of its input, making it available as leftovers to whomever comes next, untouched. unit, in this regard, is identity. Second, because unit never fails at parsing. unit, in this regard, is Just. And third, because the output unit provides, Unit, adds no meaning of its own. unit, in this regard, is Unit.

unit is the composition of three units, three identities, in one. This is how we know isomorphisms between Parser a, Parser (Unit, a) and Parser (a, Unit) exist. Composing something with an identity has never led to anything but itself.

In contrast, would the following be an identity for our monoid?

```
not_the_unit :: Parser Unit
not_the_unit = Parser (const Nothing)
```

No, of course not, it says so on the tin. But if history has taught us anything, it's that people tend to misunderstand blatantly obvious things. People will see the type and assume the wrong thing. So let's dig deeper. If not_the_unit was our identity element, our unit, composing it with a Parser a would lead to a Parser (Unit, a) isomorphic to Parser a. Which, among other things, implies that the newly composed Parser (Unit, a) would succeed in parsing any input which the original Parser a would. Alas, it would not, for whether the original Parser a succeeds or not, not_the_unit would cause the newly composed Parser (Unit, a) to fail.

To sum up, we now know that composeParsers is the associative operation of our monoid and we know that unit, all in lowercase, is its identity element. What we don't know is how can Unit be a sensible output from a Parser that is supposed to derive meaning from Strings. Unit means nothing.

One hundred fifty-seven

We have been describing a Parser x as something that given a String produces something of type x, and we have focused our attention on the fact that this x is somehow "meaning" that was trapped in that String. But what if we've been paying attention to the wrong thing?

A Parser *produces* a value of type x, that's what fundamentally matters. And as such, it can produce *any* x it sees fit, like Unit or the number 7. That a Parser also happens to be able to derive meaning from String is tangential. It's important, sure, that's the whole reason why the Parser type even exists, but the importance of a Parser "parsing" rather than doing something else is less relevant than we may think. The uniqueness of "what it means to be a Parser" is comparable to what it means to be a saxophone. Does it sound different than a piano? Sure it does, but ultimately, they both *sound*, and that's what matters to us programmers. We are composers. We care about harmony and rhythm, about music as a whole, not about the individual trumpets, cymbals or guitars. We care about whether something *produces* an x or not. How that x comes to be is irrelevant.

And *production* is something we learned to associate with *positive* argument positions a long time ago. But, more importantly, with the closely related topic of Functors. Well, it turns out that a Parser, despite its apparent uniqueness, is mostly just another boring Functor. And we love boredom, it's fun. I mean Functors, we love Functors.

We can prove that a Parser is indeed a Functor by implementing a Functor instance for it, which boils down to implementing fmap, the function that *covariantly* lifts a function of type a -> b into a function of type Parser a -> Parser b.

```
instance Functor Parser where
  fmap :: (a -> b) -> (Parser a -> Parser b)
  fmap f = \pa -> Parser (\s0 ->
    case runParser pa s0 of
      Nothing -> Nothing
      Just (s1, a) -> Just (s1, f a))
```

Easy enough. And, by the way, specifying the type of fmap in this instance, or more generally, of any method in any instance, is unnecessary. We do it here just for clarity. And yes, we said this before, but we are repeating it here also for clarity. The type of fmap is fully determined from the instance head, where we said that Parser is who we are giving this instance for.

In fmap, we are creating a new Parser b that, when run, will actually provide its entire input String to the original Parser a, reuse any leftovers from it as its own, and preserve any parsing failure too. That is, this new Parser b, insofar as "what it means to be a Parser", behaves *exactly* like the original Parser a did. However, rather than eventually producing a value of type a, it produces a value of type b obtained by applying the function a -> b to the a produced by Parser a.

And as if by magic, among other examples, we can now have a Parser that parses "t" as False and "f" as True by simply fmapping not over the value of type Bool that results from parseBool.

```
parseNegatedBool :: Parser Bool
parseNegatedBool = fmap not parseBool
```

We can try parseNegatedBool in GHCi to verify that it accomplishes what we expect.

```
> runParser parseNegatedBool "tfx"
Just ("fx", False)
> runParser parseNegatedBool "x"
Nothing
```

In the first example, where parseBool would have produced True as its parsing result, parseNegatedBool produces False instead, the result of applying not to that original True. We also see how fmapping over a Parser doesn't affect its leftover management in any way, for "fx" is the same leftover String that parseBool would have returned. In the second example, we can see how parsing failures are preserved too.

And, of course, like with every other Functor, we must ensure that the functor laws, which talk about identity and composition, hold.

```
fmap id  ==  id
```

```
fmap (g . f)  ==  fmap g . fmap f
```

Feel free to grab your pen and paper to prove, using equational reasoning, that indeed they hold.

Now, knowing that Parser is a Functor, we can do something silly.

```
regime :: Parser Unit
regime = fmap (const Unit) parseBool
```

You see, there used to be some truth in the output of parseBool. But now, silenced by the regime, the truth is gone and the Unit took its place. Nevertheless, regime still sequesters the relevant String, the evidence that led to the stolen Bool.

```
> runParser regime "teverything is fine"
Just ("everything is fine", Unit)
> runParser regime "feverything is fine"
Just ("everything is fine", Unit)
```

███ ███ the ████ ██ ██ regime ██ ██ ████ ██████ shows that it is perfectly possible ██ ██ ███ to have a Parser's output be completely unrelated to the String it consumes.

In `Parser x`, the `Parser` and the `x` have lives of their own. In the grand scheme of things, it's only a coincidence that from time to time the `x` and the input `String` are related. `fmap` is how we deal with the `x` output, and functions poking inside the guts of a `Parser`, like `composeParser`, are how we deal with `Parsers` themselves.

So, the regime does something as a `Parser`, but as a `Unit` it's rather moot. Can we have the opposite? Can we have a `Parser` that doesn't do anything interesting as a `Parser` but produces a meaningful output nonetheless? Sure we can, and the process is almost the same.

```
problem :: Parser Bool
problem = fmap (const False) unit
```

What changes is that we take `unit` as our starting point, the `Parser` that does nothing. All we need to do is use `fmap` to modify the `Unit` value which `unit` normally produces. In our case, we are replacing it with `False`.

```
> runParser problem "help"
Just ("help", False)
```

See? ██ ██ ██ ██ ██ no█ ██ ██ problem ██ everything ██ ██ ██ is ██ ██ ██ ██ fine.

One hundred fifty-eight

Before proceeding, let's explore a different way of defining `fmap` for `Parser` that highlights even more the fact that `Functor` doesn't care at all whether we are parsing a `String` or sending a rocket to outer space. Let's rewrite `fmap` in a way that doesn't explicitly pay attention to the insides of a `Parser` at all.

```
fmap :: (a -> b) -> Parser a -> Parser b
fmap f = Parser . fmap (fmap (fmap f)) . runParser
```

This silly looking definition is quite enlightening, even if cryptic too. We'll read it together to understand how, but first let's remind ourselves of the definition of the `Parser` datatype.

```
data Parser x = Parser (String -> Maybe (String, x))
```

The type of `fmap` implies three interesting things in our implementation. First, that the type of `f` is `a -> b`. Second, that the type of the `Parser` constructor we are ultimately using to construct a `Parser b` has type `(String -> Maybe (String, b)) -> Parser b`. And third, that `runParser`, the function we use to remove the `Parser` "wrapper", has type `Parser a -> (String -> Maybe (String, a))`.

In other words, in our allegedly cryptic point-free definition, in between the applications of `runParser` and `Parser`, we are converting a function of type `String -> Maybe (String, a))` into a function of type `String -> Maybe`

(String, b)), where all that's supposed to change is the a into a b. So, how do we do that?

We start by noticing that the a is inside a tuple, a pair. And pairs, we know, are Functors too. More precisely, in our case, (,) String is a Functor, meaning that fmapping a function a -> b over a pair (String, a) will lead to a pair (String, b), which is exactly what we need. However, this pair is trapped inside a Maybe (String, a), meaning that perhaps there's no pair at all. That is, we would only be able to apply fmap f over the tuple when the Maybe happens to be constructed with Just.

Luckily for us, Maybe is a Functor too, meaning that using fmap we could lift fmap f, our function of type (String, a) -> (String, b), into a function of type Maybe (String, a) -> Maybe (String, b) which would apply said fmap f only when the Maybe is a Just containing a (String, a).

But we are not done yet, because in between runParser and Parser we are transforming a *function* of type String -> Maybe (String, a), not just a Maybe (String, a). That is, what we are actually trying to modify is the *output* of a function. Well, this is what function composition is for, isn't it? And function composition goes by many names, one of them being fmap. In other words, using fmap once again, we could lift our function fmap (fmap f) of type Maybe (String, a) -> Maybe (String, b) into a function of type (String -> Maybe (String, a)) -> (String -> Maybe (String, b)), which is exactly what we need.

So, much like unit was the composition of three different identity-like things, the Functor instance for Parser is just the composition of three other Functors, of three fmaps.

If we ask GHCi what the type of said composition is, the answer is rather beautiful.

```
fmap . fmap . fmap
  :: (Functor f, Functor g, Functor h)
  => (a -> b)
  -> f (g (h a))
  -> f (g (h b))
```

All f, g and h are Functors. If we pick f to be (->) String, g to be Maybe, and h to be (,) String, then we end up with this:

```
fmap . fmap . fmap
  :: (a -> b)
  -> (->) String (Maybe ((,) String a))
  -> (->) String (Maybe ((,) String b))
```

Which looks horrible in prefix notation, but if we rearrange it to its infix form, we see this is *exactly* what we were looking for.

```
fmap . fmap . fmap
  :: (a -> b)
  -> (String -> Maybe (String, a))
  -> (String -> Maybe (String, b))
```

And we know that String -> Maybe (String, x) is just a different type for what we called Parser x, so all that remains is to take care of the arguably superfluous Parser wrapper, which is what the applications of runParser and Parser accomplish.

So, you see, it doesn't really matter what our Parser does. Our implementation of its Functor instance, which deliberately ignored inputs, leftovers and failure altogether, proves that.

One hundred fifty-nine

What's important, now, is that Parser *is* a Functor, which means we can peek and poke its output any way we see fit without taking its guts apart each time, without even mentioning we are dealing with a Parser at all. For example, if we imagine a function foo of type Bool -> String, then fmap foo would have this type:

```
fmap foo :: Functor f => f Bool -> f String
```

fmap foo is the function that given a Functor f producing values of type Bool, returns that same f, but this time producing values of type String rather than Bool. *Any* Functor, including Parser.

```
> :t fmap foo (Just True)
:: Maybe String
> :t fmap foo [True, False]
:: [String]
> :t fmap foo id
:: Bool -> String
> :t fmap foo parseBool
:: Parser String
```

The cases where we fmap foo over the function id and parseBool are particularly interesting because they are modifying the *output* of the function and the Parser, respectively, even if *no input* has been received by them yet. Which, depending on where you are coming from, could be unsurprising and expected, or completely shocking.

One hundred sixty

So, like many other beautiful things in life, Parser is a Functor. And while it doesn't really fit the Monoid typeclass, Parser is a beautiful monoid too. But moreover, seeing as this convergence of beauty is unlikely to be an isolated event, we give the situation a name, somewhat hoping we have more of them. We call a monoid known to be a Functor, a Functor known to be a monoid, a *monoidal functor*. And this name has a home, too.

```
class Functor f => Monoidal f where
  unit :: f Unit
  pair :: f a -> f b -> f (a, b)
```

Intuitively, pair takes *two* separate values a and b, both wrapped in the same f, and returns *one* value (a, b) still wrapped in that f. That's pretty much all the intuition we'll need in practice. Read this again, go on. We take *two* things and turn them into *one*.

The Monoidal typeclass says two things. First, as every other typeclass, it says that reasonable definitions of unit and pair exist for types f for which there is a Monoidal instance. Parser is one such f, with unit just as we defined it before, and pair being just a different name for what so far we've been calling composeParser. Let's give the full definition of the Monoidal instance for Parser.

```
instance Monoidal Parser where
  unit :: Parser Unit
  unit = Parser (\s -> Just (s, Unit))

  pair :: Parser a -> Parser b -> Parser (a, b)
  pair pa pb = Parser (\s0 ->
    case runParser pa s0 of
      Nothing -> Nothing
      Just (s1, a) ->
        case runParser pb s1 of
          Nothing -> Nothing
          Just (s2, b) -> Just (s2, (a, b)))
```

But the Monoidal typeclass also says something else, using a new syntax that we haven't seen before. Monoidal f says that f *must* be a Functor too.

```
class Functor f => Monoidal f ...
```

If we try to implement a Monoidal instance for some type f that is *not* a Functor, then type-checking will fail. We say that Functor is a *superclass* of the Monoidal typeclass, *constraining* the Monoidal instances that can exist. So, yes, *all* Monoidal functors are Functors. The opposite, however, is not true. Not all Functors are Monoidal functors. We'll see examples of this later on.

Generally speaking, a typeclass X can have one or more superclasses A, B, etc., specified in the following way:

```
class (A, B, ...) => X ... where ...
```

But this doesn't *just* mean that A and B are constraints that need to be satisfied in order for some instance of X to exist. It also means that whenever X itself is satisfied, then A and B are satisfied too. We can observe this useful feature with a small GHCi example:

```
> :t fmap id unit
fmap id unit :: Monoidal f => f Unit
```

When we ask GHCi about the type of the expression fmap id unit, it says
its type is Monoidal f => f Unit, even though we are clearly using both unit
from the Monoidal typeclass, and fmap from the Functor typeclass. If
satisfying a constraint like Monoidal f didn't *imply* that the superclasses of
Monoidal f are satisfied too, then the type of fmap id unit would need to
mention both Monoidal f and Functor f as their constraints.

```
fmap id unit :: (Monoidal f, Functor f) => f Unit
```

But this is kind of silly. It is correct, sure, but saying Functor f doesn't
really add any new meaning to Monoidal f, does it? Generally speaking, a
typeclass constraint being satisfied implies that *all* of the superclasses of
that typeclass are satisfied too.

And, by the way, if while learning this you are reminded of logical
implication but baffled at the fat arrow => going in the opposite direction,
then you are right, it goes in the opposite direction. If, on the other hand,
you know nothing about logical implication, then ignore this paragraph.

Finally, as with any other sensible typeclass worth talking about, we have
laws that any instances of this typeclass must abide by. First, we have a law
requiring that pair be an associative function. Up to isomorphism, of
course.

```
pair a (pair b c)  ≈  pair (pair a b) c
```

Second, we have a law requiring that unit behaves as the identity element
for pair.

```
pair unit a  ≈  a  ≈  pair a unit
```

Technically, this law can be split into two. There is a *left identity* law
where Unit appears on the left.

```
pair unit a  ≈  a
```

And a *right identity* law where Unit appears on the right.

```
pair a unit ≈ a
```

We don't care much about that difference for the time being. Anyway,
these are essentially the monoid laws. Nothing new for us here. There is,
however, an additional law.

```
pair (fmap f ma) (fmap g mb) == bimap f g (pair ma mb)
```

This law, called the *naturality* law, says that fmapping f and g separately
over two monoidal functors ma and mb before pairing them, must be equal
to fmapping f and g together over pair ma mb, something we can easily
accomplish using bimap. Luckily, we don't really need to think about this
law. Thanks to parametricity, it will *always* be automatically satisfied by all

One hundred sixty-one

Parser is a monoidal functor. So what? What can we do with this knowledge that we couldn't do before? Technically speaking, *nothing*. We already knew that Parsers could be composed in this way, we just didn't know *why*. But we must acknowledge that we only arrived to that composition by chance, after many failed attempts. We could have avoided that pain and gone straight to the right answer had we known what to ask for. Well, now we know. Is it too late? No, not at all.

Parser is just one among *many* monoidal functors, so all the pain we went through to understand the composition of Parsers was not in vain, for we shall be able reuse this knowledge with any other monoidal functor. That we couldn't readily recognize the structure in what we believed to be unique didn't make those structures any less true. Going forward, it'll be our loss if we don't try to identify them as soon as possible.

The basic question we should ask ourselves, as soon as we recognize something as Functor, is whether it makes sense to combine any two such Functorial values into one. If it does, then chances are we have a monoidal functor, which will allow us to reason about this combination, this composition, in a predictable way.

Let's take another Functor, then, to see how this works. Let's take Maybe. Do we think it makes sense to try and combine a value of type Maybe a with a value of type Maybe b, into a value of type Maybe (a, b)? Maybe.

```
pair :: Maybe a -> Maybe b -> Maybe (a, b)
```

Maybe x conveys the idea of a value of type x maybe being there, maybe not. pairing two such values, presumably, could convey the idea of *both* those values maybe being there, maybe not.

```
pair :: Maybe a -> Maybe b -> Maybe (a, b)
pair Nothing  _          = Nothing
pair _        Nothing    = Nothing
pair (Just a) (Just b) = Just (a, b)
```

Naturally, if any of the input Maybes is Nothing, there'd be no "*both* a and b" we could pair, so we simply return Nothing in those situations. Otherwise, if we have both Just a and Just b, then we can have Just (a, b) too. Is pair associative? Certainly. Up to isomorphism, of course. You can prove it yourself if you are bored, you know how.

What about unit, the value of type Maybe Unit which, when paired with a Maybe x, results in a value isomorphic to that Maybe x? Well, considering there are only two inhabitants of this type, we can try both of them and see what happens. If said Maybe Unit was Nothing, then pairing it with *anything* would lead to Nothing as result, which is fine whenever our Maybe x is Nothing, but terribly wrong otherwise. It ain't Nothing, no. It must be the other inhabitant then, Just Unit. And indeed, if we pair Just Unit with Just x, the resulting Just (Unit, x) would be isomorphic to Just x. Or the other way around if we switch the order of the input parameters to pair. And if we pair this Just Unit unit with Nothing, we get Nothing as output, which is the same value we put in. So, *yes*, Just Unit is our unit.

```
unit :: Maybe Unit
unit = Just Unit
```

Of course, pair and unit should actually be defined inside the Monoidal instance for Maybe, but we know that already.

```
instance Monoidal Maybe where
  unit = ...
  pair = ...
```

Alright, Maybe is a monoidal functor. We didn't ask for this, but we got it nonetheless. Why is this useful? It's not too hard to come up with a toy example that justifies the *need* for this. In fact, we can reuse that cooking example from before, where we wanted to be sure we had *both* utensils and ingredients before we could cook.

```
utensils :: Maybe Utensils

ingredients :: Maybe Ingredients

cooking_stuff :: Maybe (Utensils, Ingredients)
cooking_stuff = pair utensils ingredients
```

And even if we haven't really focused our attention on it, just like in the case of Boolean conjunction using &&, the *laziness* in our definition of pair will make sure we don't go looking for ingredients until *after* we've made sure we have Utensils nearby. Laziness is one of those things most easily noticed when ripped away from us. Cherish it, it'll prove terribly useful as we advance. We'd be at a disadvantage without it.

We'll probably need drinks for dinner too.

```
drinks :: Maybe Drinks

everything :: Maybe ((Utensils, Ingredients), Drinks)
everything = pair cooking_stuff drinks
```

And so on, and so on. We can keep on pairing like that for as long as we want. What have we achieved? Well, how many times have we checked so far whether we had the things we wanted or not? *Zero.* How many times would we *need* to check if we wanted to start our dinner? *One,* we'd just have to check if everything is Just or Nothing, even though there are *three* things that could potentially be missing. That's something.

So, you see, `Parser` and `Maybe` accomplish two completely different things. One parses `Strings` and the other helps us organize dinners. Yet, we can reason about both of them in the same `Monoidal` way. Types can be whatever they want to be in order to achieve whatever they need to achieve. And despite their uniqueness, through well understood structures like the `Functor`, the `Monoid` or the `Whatnotoid`, they can *belong*. They will interact, they will compose.

One hundred sixty-two

Let's tackle our next `Monoidal` functor, but this time let's do something different. Let's *not* think about what pairing these things even means. Let's just do it. Let's ride the parametric polymorphism and see where it takes us.

```
instance Monoidal ((->) x) where
```

That's right, `(->) x` shall be our `Functor` today. That is, the function that takes x as input. Where, of course, x is a universally quantified type-variable that could be any `Type` we want.

Just like functions and values universally quantify type variables automatically unless we ask for something different, `instance` definitions do so too. In reality, this is what's happening behind the scenes:

```
instance ∀ (x :: Type). Monoidal ((->) x) where
```

We don't usually write the explicit quantification, but there's nothing wrong in doing so. In fact, since it makes things more explicit, some may argue it's better.

Anyway, `(->) x` is our chosen `Functor`, our chosen type constructor, so these are the types that Haskell will mandate for the methods in our `Monoidal` instance:

```
unit :: (->) x Unit
```

```
pair :: (->) x a -> (->) x b -> (->) x (a, b)
```

They look horrible written in this prefix form, but we can easily rearrange them to their infix form to see things more clearly.

```
unit :: x -> Unit
```

```
pair :: (x -> a) -> (x -> b) -> (x -> (a, b))
```

What do you see? It doesn't matter. We said we'd chase parametricity, so that's what we'll do. Let's leave the sightseeing for the tourists.

```
unit :: x -> Unit
unit = \_ -> Unit
```

This is the only implementation `unit` can have. Is it a bit silly? Sure.

Nonetheless, it's the only behaviour that makes sense. If unit must return a Unit, then it will be forced to ignore any value it is given as input. What about pair?

```
pair :: (x -> a) -> (x -> b) -> (x -> (a, b))
pair f g = \x -> (f x, g x)
```

Again, the parametric polymorphism in this type forces pair to do this. There's nothing else it could do that would type-check. pair takes two x-consuming functions f and g, and returns a new x-consuming function that returns, in a tuple, the outputs from applying *both* f and g to that x.

What is this good for? Well, imagine we have two functions that we want to apply to a *same* input:

```
double :: Natural -> Natural
double = \x -> x * 2

triple :: Natural -> Natural
triple = \x -> x * 3
```

We can pair them into a *single* function that, given a Natural number, returns *both* the double and the triple of that number at the same time.

```
double_and_triple :: Natural -> (Natural, Natural)
double_and_triple = pair double triple
```

In this case the types of our inputs and outputs are the same, but that's just a coincidence. We could as well have paired a function Natural -> Bool with a function Natural -> String to obtain a function of type Natural -> (Bool, String), and it would have been fine. Anyway, let's try our function in GHCi.

```
> double_and_triple 5
(10, 15)
```

Great, it works. And, of course, we can pair this further with any other function that takes a Natural as input. Like even, say, the function that takes a Natural number and returns True if the number is even, or False otherwise.

```
> pair double_and_triple even 5
((10, 15), False)
> pair even (pair triple double) 2
(True, (6, 4))
```

Is unit the appropriate identity element for our associative binary operation pair? It better be, for there's no way to implement unit differently.

```
> double 5
10
> pair unit double 5
(Unit, 10)
> pair double unit 5
(10, Unit)
```

All isomorphic results. Great.

In summary, we took a well understood idea, that of *monoidal functors*, and tried to see if our thing, the function, fits that idea. And it did, which led to an "accidental" discovery of a terribly useful new way of composing functions, one we hadn't even thought about before. Isn't this beautiful? Hadn't we tried this, chances are we'd be forever stuck applying a myriad of different functions to the *same* input values time and time again.

Structures are here. It's our loss if we refuse to acknowledge them.

This way of composing functions also goes by the name of *fan-out*, a hard to forget name if you keep in mind that when stuff hits the *fan*, it spreads *out* all over the place. This is such a common way of composing functions that Haskell dedicates an infix operator to it.

```
(&&&) :: (x -> a) -> (x -> b) -> (x -> (a, b))
(&&&) = pair
```

Actually, the type of this function is a bit more general than this because it's designed to work with function-like things, rather than just functions. Remember Profunctor? Not that, but similar. Anyway, we'll pretend it only works with functions for the time being.

```
> (double &&& triple) 10
(20, 30)
> (double &&& triple &&& quadruple) 10
(20, (30, 40))
```

We can see how &&& associates to the *right*. This doesn't mean much, however, considering how &&& must be an associative function as mandated by the way of the monoid. It could have associated to the left and the result would have been the same. Up to isomorphism, that is.

One hundred sixty-three

Another Functor, Either x. Let's see if this one is Monoidal too.

```
instance Monoidal (Either x) where
  unit :: Either x Unit
  pair :: Either x a -> Either x b -> Either x (a, b)
```

We'll see the implementations of pair and unit soon, but first, let's highlight something. Notice how x, the type of the eventual payload of a Left constructor, if any, is fixed to be x everywhere. This is not unlike the x in (->) x, which was fixed to be x everywhere too.

```
pair :: (x -> a) -> (x -> b) -> (x -> (a, b))
```

However, unlike that situation, where pair f g created a new function that *consumed* an x which it could then provide to both f and g, here, if pair needs to produce a Left x for some reason, then that x will need to come from one of the input Eithers. From which one, though? Both of them could have an x in them.

```
pair :: Either x a -> Either x b -> Either x (a, b)
```

This situation can be understood by simply remembering that while Either x y is *covariant* in x, x -> y is *contravariant* in it. Meaning that, quite often, what works for Either doesn't work for (->) and vice-versa.

So, what? Isn't Either x a Monoidal functor after all? It can be, but in order to be one it must make a choice.

```
pair :: Either x a -> Either x b -> Either x (a, b)
pair (Left x)  _          = Left x
pair _         (Left x)   = Left x
pair (Right a) (Right b)  = Right (a, b)
```

Obviously, just like we saw in the Monoidal instance for Maybe, we can only construct a pair (a, b) if both a and b are present at all. That is, if both our inputs values have been constructed using Right. There's no choice to be made there, it's trivial.

pair makes a choice when it decides what to do when both its inputs happen to be Lefties, a situation dealt with by the second equation above. pair chooses that it will be the payload from the *leftmost* Left the one that will be reproduced in the resulting Either x (a, b). The leftmost, not the rightmost. This is a choice. By swapping the order of the last two equations, we could have chosen to perpetuate the rightmost Left payload instead. However, seeing as we expect things to be evaluated from left to right as they appear in our code, returning the leftmost Left we encounter, arguably, makes the most sense.

However, that's not the entire truth. Later on we'll learn there are stronger forces at play here mandating that this be the case. The leftmost Left shall always be prioritized. So, in reality, there was no choice to be made here. Sorry, false alarm. The take-away from this experience is that when we don't know the whole truth, we may be tempted to make uninformed choices that could be, in fact, wrong. This is what dangerous humans do. Please, don't do that. Being aware of one's lack of understanding is good. We ask ourselves not how we are right, but how we are wrong.

And what about unit? Right, we almost forgot. There isn't much we can say about it that we haven't said about the unit for Maybe.

```
unit :: Either x Unit
unit = Right Unit
```

Only `Right Unit` can behave as the identity element in our `Monoidal` instance for `Either`, for exactly the same reasons that only `Just Unit` could behave as one in `Maybe`'s. Similar beasts, `Either` and `Maybe`. So similar, that we'll frequently benefit from these two fully parametric functions, `hush` and `note`, able to convert between them.

```
hush :: Either x a -> Maybe a
```

```
note :: x -> Maybe a -> Either x a
```

One hundred sixty-four

What does the `Either` monoidal functor mean?

```
head :: [a] -> Maybe a
head (a : _) = Just a
head _ = Nothing
```

`head` only ever "fails" when the input list is empty, and it communicates this failure by returning `Nothing`. In other words, we embraced the type constructor `Maybe` to communicate the possibility of failure.

But what if we had a function that could fail for more than one reason? In that case, `Nothing` wouldn't be enough anymore. Let's go back to cooking.

```
utensils :: Maybe Utensils
```

```
ingredients :: Maybe Ingredients
```

These values, despite their bad names misplacing expectations, perfectly convey the presence or absence of `Utensils` or `Ingredients` individually. However, once we pair them, some of that clarity is lost.

```
pair utensils ingredients
   :: Maybe (Utensils, Ingredients)
```

This pairing still perfectly conveys the idea of *both* the `Utensils` and the `Ingredients` being there or not. However, if either one or both of them are missing, all we get as result is `Nothing`, without any knowledge of who's to blame for our failed attempt at cooking. Using `Either`, however, we can achieve a bit more.

```
data WhyNot = NoUtensils | NoIngredients
```

If as the result of our pairing we had `Either WhyNot (Utensils, Ingredients)` rather than `Maybe (Utensils, Ingredients)`, then whenever our pairing didn't result in a `Righteous` pair of `Utensils` and `Ingredients`, we would be able to peek at the `WhyNot` value on the `Left` to understand why not.

```
pair (note NoUtensils utensils)
     (note NoIngredients ingredients)
   :: Either WhyNot (Utensils, Ingredients)
```

Using note, the function that enriches a value of type `Maybe y` by converting it to a value of type `Either x y`, we modify both utensils and ingredients so that if they turn out to be `Nothing`, we convert them to a `Left` with a value of type `WhyNot` explaining what went wrong.

And notice that while it makes sense to use `WhyNot` in this pairing, it wouldn't make much sense to use it in a standalone utensils, say.

```
utensils :: Either WhyNot Utensils
utensils = Left NoIngredients
```

This new type says that utensils could potentially report the absence of ingredients, something which doesn't make sense at all. What does utensils know about ingredients? Generally speaking, we should make our errors as precise and useful to our audience as possible, so that nonsense like this doesn't happen. Otherwise, the burden of understanding what utensils really meant passes on to whomever needs to use these Utensils.

```
case utensils of
   Right u -> ... We have some utensils. Nice. ...
   Left NoUtensils -> ... Too bad. We have no utensils. ...
   Left NoIngredients -> ... What is this nonsense? ...
```

We make our errors as small and precise as it reasonably makes sense, and we let our callers grow them as they see fit, for example, by using note. The bigger our errors are, the smaller the meaning they convey.

Anyway, back to our pairing of noted Maybes. If there are no utensils, we get `Left NoUtensils` as result. If there are no ingredients, we get `Left NoIngredients`. And if there are neither utensils nor ingredients, we get `Left NoUtensils` as result. Unsurprisingly, I hope.

pairing two `Eithers`, we learned before, leans to the leftmost `Left` in case two `Left`s are provided as input. So, if ultimately we are trying to evaluate `pair (Left NoUtensils) (Left NoIngredients)`, the result will be `Left NoUtensils`. Well, actually, pair won't even check whether its second input argument is `Left` or `Right` at all. It'll just return the leftmost `Left` as soon as it encounters it. On the one hand this is great, because we evaluate nothing beyond the leftmost `Left`, preserving laziness. But on the other hand, it could be handy to know whether *both* the Utensils and the Ingredients are missing.

Well, we can't have a cake and eat it too. We either look at *both* `Eithers` in order to judge their contents, accumulating the error reports from *all* the `Left`s we discover, *or* we bail out as soon as we encounter the first `Left`, without having to evaluate any other `Eithers` beside it. With `Either`, we only get the latter behaviour. We *could* have a different `Monoidal` functor that did something else. However, this alternative behaviour will have to wait. First, we need to address some parsing issues.

One hundred sixty-five

The composition of Parsers shares an unfortunate trait with the composition of Maybes, where Nothing becomes the result whenever a Nothing is among the Maybes we attempt to pair, yet we can't tell *which* of the Maybes in question is the Nothing leading to this result. When pairing Parsers, if one of them fails, we are not able to identify which one either.

```
> :t pair parseBool parseDigit
:: Parser (Bool, Digit)
> runParser (pair parseBool parseDigit) someString
Nothing
```

Why did the Parser fail? We'll never know unless we manually inspect someString and try to deduce it ourselves, something completely unmanageable when rather than two Parsers we find ourselves composing hundreds of them. Remember, as soon as we can compose two things, we can compose infinitely many of them. Composition não tem fim.

So, how do we solve this? We know how. Rather than using Maybe for conveying failure, we need to use Either. We need to change the definition of our Parser datatype.

```
data Parser x = Parser (String -> Either String (String, x))
```

We've chosen to put a String on the Left side of our Either. The idea is that a Parser can now cry something interesting like Left "Expected 't' or 'f'" as it fails, rather than just saying Nothing.

```
parseBool :: Parser Bool
parseBool = Parser (\s0 -> case s0 of
  't':s1 -> Right (s1, True)
  'f':s1 -> Right (s1, False)
  _      -> Left "Expected 't' or 'f'")
```

But a lonely Parser is terribly uninteresting. We better modify the Functor and Monoidal instances, taking the new changes into account, so that our Parser can make friends.

```
fmap :: (a -> b) -> Parser a -> Parser b
fmap f = Parser . fmap (fmap (fmap f)) . runParser
```

Wait, this fmap has exactly the same implementation as before. Why, yes, both Either String and Maybe are Functors, so we don't really need to change anything here. The fmap that was lifting a function (String, a) -> (String, b) into Maybe (String, a) -> Maybe (String, b) before is now lifting it to Either String (String, a) -> Either String (String, b), but other than that nothing changes. What about pair?

```
pair :: Parser a -> Parser b -> Parser (a, b)
pair pa pb = Parser (\s0 ->
    case runParser pa s0 of
        Left e -> Left e
        Right (s1, a) ->
            case runParser pb s1 of
                Left e -> Left e
                Right (s2, b) -> Right (s2, (a, b)))
```

Compared to our previous implementation, all that changed is that rather than dealing with Nothings and Justs, we deal with Lefts and Rights everywhere. Now, assuming we added a proper error message to parseDigit too, our previous example would lead to a much better experience.

```
> runParser (pair parseBool parseDigit) "x3"
Left "Expected 't' or 'f'"
> runParser (pair parseBool parseDigit) "tx"
Left "Expected a digit '0'..'9'"
> runParser (pair parseBool parseDigit) "t3"
Right ("", (True, D3))
```

Obviously, we changed the type of runParser too.

Writing beautiful and informative error messages is a delicate matter. We'll get better at it as we move forward.

One hundred sixty-six

In a sense, picking String as our error type goes against our previous recommendation that the type we chose to convey errors should be as small and precise as possible. However, we need to take the context where this error exists into consideration. Small and precise error messages mainly make sense if they can help a program trying to recover from that error automatically.

```
utensils :: Maybe Utensils
utensils = case our_utensils of
                Just x -> Just x
                Nothing -> neighbour's_utensils
```

In this example, if we have our own Utensils available, we'll go ahead and use them. But if we don't, we'll ask some neighbour for their Utensils, if any. We are, in other words, *recovering* from failing to have our own utensils, a well known situation, by having the computer *automatically* do something else for us.

But if we consider the use case for Parsers, which could fail parsing for a myriad of different reasons, none of them which we expect to automatically recover from, then a String makes a bit more sense. The errors reported by our Parsers are exclusively intended for human consumption, not for computers. How would a program *automatically* recover from an error saying that a particular character in our input is not a digit? Would the program just go and replace the character in question with a digit? Which digit? That doesn't make much sense. What if the input was right, actually, and it was the Parser who was expecting the wrong thing?

If a parsing error happens, *any* error, it's either because the input is malformed or because the Parser is wrong, and fixing any of those situations requires human intervention. So the best we can do, really, is to try and be precise about *where* the parsing error happened. Even more so if we consider that the number of Chars in our input String could very well be in the millions.

Location. All we care about this time is *location*. We are not interested in designing our errors so that they can be mechanically analyzed, but rather, so that they convey useful information about *where* parsing failed, and what the Parser was trying to accomplish there.

```
> runParser (pair parseBool parseBool) someString
Left "Expected 't' or 'f'"
```

Without looking inside someString, can we tell which of the two parseBools failed to parse its input? Of course we can't. Any of them could have failed, and since both of them generate exactly the same error report, we can't really tell their failures apart. In a sense, we recreated the uninformative Nothing from before. How do we fix that?

Well, how do we *want* to fix it? There are at least two obvious solutions. One of them could say at which absolute character position in the input String we encountered the error. Say, at character position 0, 3 or 43028. This solution manages to be both very precise and very uninformative at the same time. Alternatively, we could say something along the lines "failed to parse the user's birthdate month number", which gives a higher level overview of *what* the Parser was trying to accomplish when it failed, without telling us exactly where the crime happened. Luckily, we don't have to chose between these alternatives. They complement each other, and we can have them both.

We'll tackle this in the same way we tackle every other problem. We'll correctly solve the individual small problems we identified, and then we'll just compose these correct solutions into a bigger and still correct solution. Correctness composes.

One hundred sixty-seven

Let's tackle the matter of the absolute position of a problematic Char in the input String first. The idea is that, when a Parser fails, we want to see *both* an error message and an absolute character position indicating *where* the parsing failed.

```
> runParser (pair parseBool parseBool) "xy"
Left (0, "Expected 't' or 'f'")
> runParser (pair parseBool parseBool) "ty"
Left (1, "Expected 't' or 'f'")
```

In the first example above, it's the leftmost parseBool in the pair the one that *first* fails to parse the input "xy". And seeing as 'x' is the Char that's causing the whole thing to fail, and how said 'x' appears as the very first Char in the input String, then we report the Natural number 0 as the absolute position of the first conflicting Char in the input String. Remember, we always start counting from *zero*, not from one.

In the second example, the first parseBool in the pair successfully parses its input, but the second one fails. Accordingly, the output from our use of runParser reports that the Char at position 1 was the conflicting one. That is, the second Char, the 'y'.

Alright, let's do this. First, *types*. We are counting things starting from *zero*, so we'll probably need a Natural number somewhere. In fact, we already know we want one of those as part of runParser's output, so maybe let's start by changing that type.

```
runParser :: Parser x
          -> String
          -> Either (Natural, String) (String, x)
```

Great. We are saying that if parsing fails, as communicated by a Left result carrying a (Natural, String) payload, then said Natural shall tell us at which *absolute* position in the input String parsing failed, and next to this Natural we get a Stringy error message too, describing what went wrong. The Right hand side of the Either is as before. In order to implement this new runParser, however, we'll need to massage our Parser type a bit.

```
data Parser x = Parser
        (String -> Either (Natural, String) (String, x))
```

Well, well, well... Things are getting a bit insane, aren't they? That's fine. Ours is an exercise in taming complexity, and we are deliberately trying to show how the solution to these so-called complex problems boils down to finding the right types and getting their composition right.

This new type, just like the runParser function, says that if our parsing function is going to fail, then it better provide information about *where* the failure happened. The rest is the same as before. Let's rewrite parseBool to take this into account.

```
parseBool :: Parser Bool
parseBool = Parser (\s0 -> case s0 of
  't':s1 -> Right (s1, True)
  'f':s1 -> Right (s1, False)
  _      -> Left (0, "Expected 't' or 'f'"))
```

Since parseBool only ever deals with the first Char of its input String, it will always report 0 as the failing position when it receives some unexpected input. Of course, we could easily imagine a different Parser that took more than the first Char of the input String into consideration. For example, a Parser expecting the String "blue" as input but getting the String "bland" instead, could report 2 as the position of the first unexpected Char, "a".

Anyway, seeing as we changed the type of Parser, we'll need to modify pair in its Monoidal instance so as to take this new type into account.

```
pair :: Parser a -> Parser b -> Parser (a, b)
pair pa pb = Parser (\s0 ->
  case runParser pa s0 of
    Left (n, e) -> Left (n, e)
    Right (s1, a) ->
        case runParser pb s1 of
            Left (n, e) -> Left (n, e)
            Right (s2, b) -> Right (s2, (a, b)))
```

The Functor instance stays the same as before because all we did was change the payload on the Left, which fmap completely ignores. Let's see if this works, then.

```
> runParser (pair parseBool parseBool) "xy"
Left (0, "Expected 't' or 'f'")
> runParser (pair parseBool parseBool) "ty"
Left (0, "Expected 't' or 'f'")
```

What? This ain't right. The 0 reported in the first example is fine, yes, because the Char at position 0 in the input String "xy" is certainly wrong. But in the second example, the error also reports 0 as the failing position, even if we know that this time the Char at position 1, the 'y', is the culprit. What's going on? Well, not *composition*, that's for sure.

One hundred sixty-eight

The problem we are seeing is that when individual Parsers such as parseBool fail, they report the *relative* position where they fail with respect to their own input, not with respect to the *entire* input that was initially provided to runParser by us in GHCi. That is, these errors do not report the *absolute* position of the conflicting Chars in the input String.

This makes sense, however. How would parseBool know *where* it is being executed in this initial input String? In pair pa pb, that second parser pb doesn't know how much input pa consumed before pb's time to parse, so there's no way it can calculate an absolute position that takes that into account. Should it, though? Should pb know about pa's diet? No, it should not. It *could*, but that would put an additional burden on pb, who'd now need to worry about other Parser's affairs too. No, that won't work. There is, however, somebody who is *already* worrying about both pa and pb, so perhaps we can address this matter there. Namely, pair.

If pair knew how much input pa consumed in order to successfully produce its output, then it would be able to *add* that amount to the conflicting Char position reported by pb, resulting in a new position that takes both pa's and pb's affairs into account, without imposing any burden on neither pa nor pb. Well, almost no burden. We do need to change our Parser datatype so that also *successful* executions of our parsing function return how much input they consumed. pair will need to take this number into account. But, quite frankly, this is a very nice feature that sooner or later we would have wanted anyway. *Of course* we will appreciate knowing how much input was consumed in order to produce a successful result, perhaps even more so than how much of it was consumed in order to produce a failure. Let's change our Parser datatype to accommodate this.

```
data Parser x = Parser (String -> Either (Natural, String)
                                         (Natural, String, x))
```

Now, our Either type reports on *both* sides, as a Natural value, a number indicating how many Chars were successfully consumed by the parsing function in order to produce the result it did, be it Left or Right. For example, this is what parseBool would look like now.

```
parseBool :: Parser Bool
parseBool = Parser (\s0 -> case s0 of
  't':s1 -> Right (1, s1, True)
  'f':s1 -> Right (1, s1, False)
  _      -> Left (0, "Expected 't' or 'f'"))
```

Indeed, the Righteous scenarios now report how much input they consumed. The Left situation is the same as before. In any case, all of these scenarios still talk about the *relative* positions of the Chars they consume as they appear in the input String *this* Parser receives. parseBool still doesn't care about how much input other Parsers consume.

Notice that, seeing as both the Left and the Right must produce a Natural number now, conveying essentially the same meaning in both cases, we could have chosen to abstract that common Natural away in an isomorphic product type like (Natural, Either String (String, x)). This is not unlike algebra, where a × b + a × c equals a × (b + c). We'll encounter this situation time and time again in our travels. Anyway, we picked a representation already, so let's stay with it.

The real magic happens inside pair.

```
pair :: Parser a -> Parser b -> Parser (a, b)
pair pa pb = Parser (\s0 ->
  case runParser pa s0 of
    Left e -> Left e
    Right (n0, s1, a) ->
      case runParser pb s1 of
        Left (n1, msg) -> Left (n0 + n1, msg)
        Right (n1, s2, b) ->
          Right (n0 + n1, s2, (a, b)))
```

Well, that's a lot. Or perhaps barely anything, if we consider how many
things are going on in this composition, all of which we are taming in a
few lines of rather mechanical code. pair deals with inputs, with different
types of output, with leftovers, with failure, with error messages, with the
length of consumed input, with the position where a parsing failure
occurred, and with making bigger Parsers out of smaller Parsers. All of
that while respecting the monoidal functor laws, which makes this whole
affair terribly predictable. We *wish* taming all of this complexity was as
"complicated" as these few lines. Oh, wait, it is. Nice.

Mind you, we are learning, so our implementation has deliberately been
made rather crude and primitive in the sense that it deals with *all* of these
concerns at the same time, in the same place. But in reality, these different
concerns could be treated separately, and pair would simply *compose* all
those treatments into one. After all, we *know* that a Parser is made out of
Functors, and we've profited from this knowledge before when we
implemented Parser's own fmap as fmap . fmap . fmap. Wouldn't it be nice
if we could implement pair like that, too? Indeed. And we can. And we
will. But not yet.

Compared to the previous implementation of pair, two things changed.
One is that all occurrences of the Right constructor now carry a *three*
-element tuple of type (Natural, String, x) in it, where that Natural
conveys the number of Chars successfully consumed by the Parser. The
other thing that changed is that in the Eithers we return *after* both Parsers
have run, the Natural number in question is now the *addition* of the
number of Chars consumed by both Parsers. That change is *all* we need to
do in order to make sure that the relative location of any parsing failure, as
well as the relative number of Chars a successful parsing consumes, convey
an *absolute* amount with respect to the entire input String initially
provided to runParser.

Finally, assuming we take care of updating our definitions of fmap and
runParser to take the new insides of the Parser datatype into account,
rather straightforward things to do, we can check whether we
accomplished what we wanted.

```
> runParser (pair parseBool parseBool) "xyz"
Left (0, "Expected 't' or 'f'")
> runParser (pair parseBool parseBool) "tyz"
Left (1, "Expected 't' or 'f'")
> runParser (pair parseBool parseBool) "tfz"
Right (2, "z", (True, False))
> runParser
    (pair parseBool (pair parseBool parseBool)) "tft"
Right (3, "", (True, (False, True)))
```

Beautiful, isn't it? The take-away from this experiment in taming complexity, in case it wasn't made clear by the last hundred chapters or so, is that it's not really *things* what matters, but the *relationships* between things. Parsers can *be*, and that's fine, but what's interesting is how Parsers can be *together*. And we have absolute control over that, *we* decide what it means to compose things.

One hundred sixty-nine

There's a barely noticeable issue in our latest implementation of pair that we should fix.

When we said n0 + n1, we created a *thunk* that, when evaluated, would result in the Natural value arising from the addition of n0 and n1. However, what if n0 or n1 were thunks *themselves*, as they would be if they were the result of previous uses of pair? In that case, we'd be accumulating unevaluated thunks in memory until the very last moment when somebody decides to evaluate them. This is not so bad in small examples like ours, but if we acknowledge that pairing thousands of Parsers is a reasonable thing to do, then a thousand unnecessary thunks would occupy a significant amount of memory. We better get rid of them.

```
pair :: Parser a -> Parser b -> Parser (a, b)
pair pa pb = Parser (\s0 ->
  case runParser pa s0 of
    Left e -> Left e
    Right (!n0, s1, a) ->
      case runParser pb s1 of
        Left (!n1, msg) -> Left (n0 + n1, msg)
        Right (!n1, s2, b) ->
          Right (n0 + n1, s2, (a, b)))
```

That's it. All we did is put a bang ! in front of the n0 and the n1s that are eventually used in the expression n0 + n1. This new code keeps the expression n0 + n1 *itself* a thunk, but this time the n0 and the n1 in it are guaranteed to have been evaluated by the time they are used in this addition. By doing this we prevent creating more thunks than we actually need.

Are we abandoning laziness by doing this? Not really. By the time we scrutinize the Either resulting from runParser in order to disambiguate whether is Left or Right, the parsing function has already consumed as many Chars from the input String as it needed in its attempt to successfully parse something, so it clearly knows the number of Chars in question.

This issue almost went unnoticed, didn't it? Remember what we said back when we were studying laziness: *Always* make your accumulators strict. Just do it. And what is this Natural value, which accumulates the number of Chars consumed so far, if not an accumulator?

One hundred seventy

It's a bit disappointing having to worry about counting those Chars even when we *successfully* parse something, isn't it? Well, that's kinda our fault, for we had parseBcol do too many things at once. All we wanted to say, really, is that if the first Char is either 't' or 'f', then we would be able to turn it into True and False, otherwise parsing should fail with some message of our choice. We don't really want to worry about consumed input length nor leftovers.

```
boolFromChar :: Char -> Either String Bool
boolFromChar = \c -> case c of
  't' -> Right True
  'f' -> Right False
  _   -> Left "Expected 't' or 'f'"
```

We want something like boolFromChar above. Yes, that's neat. Can we have that? Not exactly, because it's not a Parser. But if we had a function able to convert a value of its type Char -> Either String Bool, or more generally Char -> Either String x, to a value of type Parser x able to deal with all the leftover and position nonsense inside, then it could work. Yes, let's build that.

```
parse1 :: (Char -> Either String x) -> Parser x
parse1 f = Parser (\s0 ->
  case s0 of
    [] -> Left (0, "Not enough input")
    c : s1 -> case f c of
      Left e -> Left (0, e)
      Right x -> Right (1, s1, x))
```

parse1 is a higher-order function dealing with all that nasty Parser business, but deferring any decision about whether the leading Char in the input String is good or not to the given function f. The name parse1 reminds us of how hard it is to name things, but hopefully it also evokes the idea that we are trying to *parse* exactly *one* Char from our input String. Inside parse1, there's nothing we haven't seen before. Having this nice tool, we can redefine parseBool in a rather neat manner.

```
parseBool :: Parser Bool
parseBool = parse1 (\c -> case c of
  't' -> Right True
  'f' -> Right False
  _   -> Left "Expected 't' or 'f'")
```

Excellent. And, by the way, completely unrelated, every time we find ourselves writing a lambda expression which, after binding an input to a name, immediately proceeds to perform a case analysis on it like this:

```
\x -> case x of A -> ...
                B | foo -> ...
                  | bar -> ...
```

We can replace it with what Haskell calls a "lambda case" expression, which essentially replaces that \x -> case x of part with just \case.

```
\case A -> ...
      B | foo -> ...
        | bar -> ...
```

Mind you, this is just *syntactic sugar* for what we already know. We are simply avoiding the unnecessary introduction of a name, thus keeping ourselves from having to worry about it. Additionally, we are saving ourselves from typing a handful of characters. That's all. But, as it sometimes happens with syntactic sugar, it makes things beautiful. And what would life be without beauty? We live, we die, and if we did things right, at best *beauty* is what we leave behind.

This use of the word case reminds us that some words like case, forall or instance are magically reserved by Haskell for some very specific purposes, and not for anything else. If we try to implement a function, say, using these words, the compiler will yell at us. These words are forbidden, like ███████, ██ or ███████████. It's just a syntactic matter, not a semantic one.

```
class :: instance -> instance
class :: \case -> case
```

Anyway, with the help of \case, our parseBool becomes a bit more pleasing to contemplate.

```
parseBool :: Parser Bool
parseBool = parse1 (\case
  't' -> Right True
  'f' -> Right False
  _   -> Left "Expected 't' or 'f'")
```

One hundred seventy-one

With parse1, we can flawlessly parse *one* Char without having to worry about Parser details. With pair we can flawlessly compose Parsers, without having to worry about Parser details. So, in theory, we have all we need in order to flawlessly parse as many Chars as we want, again, without having to worry about Parser details.

Not quite, not quite. Eventually we'll find ourselves wanting to compose Parsers in a way different than how pair does it. Or two more ways, actually. At that point, *yes,* we'll have finished. Remember, we control composition, and it's perfectly fine to have things compose in many different ways. We know that *functions,* for example, compose in at least two ways.

```
fmap :: (b -> c) -> (a -> b) -> (a -> c)

pair :: (x -> y) -> (x -> z) -> (x -> (y, z))
```

Nevertheless, we have enough tools now that we can start building interesting and bigger Parsers using these tools *only,* without having to peek inside the Parser datatype again. That's the rule. We can't peek inside Parser. What tools do we have, concretely?

```
fmap :: Functor f => (a -> b) -> (f a -> f b)

unit :: Monoidal f => f Unit

pair :: Monoidal f => f a -> f b -> f (a, b)

parse1 :: (Char -> Either String a) -> Parser a
```

And, by the way, notice how only parse1 works exclusively with Parsers. The rest of our tools require just an f satisfying some constraint. That is, unless what we build relies *explicitly* on parse1, it should work for any such f.

Let's start by implementing a Parser that deals with an entire String, rather than an individual Char.

```
expect :: String -> Parser Unit
```

The idea is that expect "blue" succeeds if "blue" is at the beginning of the input String, otherwise parsing fails. expect "blue" *expects* "blue" to be there, not something else. expect "" always succeeds, obviously, for the empty String will always "be" there, not occupying any space.

Perhaps surprisingly, this Parser returns a meaningless Unit as output if successful, emphasizing the fact that we are not interested in deriving any *meaning* from the String eventually provided to runParser, but rather, that we are ensuring that the well-known String we provide as input to the expect function, "blue" in our example, is present in the raw String being parsed. That's where this name comes from, *expect.* We are just *expecting* something to be there, that's all. In other words, we won't be using expect for the output it could produce, but for the *side effects* it has as a Parser.

That is, we are interested in the current absolute position inside the `Parser` being advanced, the leftovers being passed around, and the failures from parsing being composed nicely, but we are not interested in this `Parser` producing any *output*. Not in the functorial sense of the word, anyway.

```
> runParser (expect "blue") "blues"
Right (4, "s", Unit)
> runParser (expect "blue") "blank"
Left (2, "Expected 'u'")
```

So, how do we write this? Well, we don't, not yet. expect would be doing too many things at once, and we need to start getting more comfortable with the idea of correctly solving the *small* issues first, composing the small solutions into a bigger solution afterwards.

If a `String` is essentially a list of `Char`s, why don't we implement a function `expect1` first, of type `Char -> Parser Unit`, doing what expect would do for an entire `String`, but just for one `Char`? That is, `expect1` would make sure that the leading `Char` in the `String` being parsed is the expected one. Once we have `expect1` working, we could in theory just pair many of them, one for each `Char` in the `String` given to expect, and this should accomplish our goal.

```
expect1 :: Char -> Parser Unit
expect1 a = parse1 (\b ->
  case a == b of
     True -> Right Unit
     False -> Left ("Expected " <> show a <>
                    " got " <> show b))
```

`expect1 'x'`, say, only succeeds if "x" is the `Char` at the beginning of the `String` being parsed. Otherwise, it fails with a useful error message.

```
> runParser (expect1 'x') "xf"
Right (1, "f", Unit)
> runParser (expect1 'x') "yf"
Left (0, "Expected 'x' got 'y'")
```

Excellent.

One hundred seventy-two

We introduced a new function in expect1, one named show.

```
> :t show
show :: Show a => a -> String
```

show can be used to convert values of types for which there is a `Show` instance into a `String` representation of said value. Mainly for internal display purposes like ours. In expect1 we are trying to construct a `String` to use as an error message, and we are using `<>`, of type `String -> String -> String` here, to concatenate the smaller `String`s that are part of our message.

But the a and the b which we want to mention in our message are not Strings, so we need to convert them first. That's when show comes handy.

Technically, since we know that String is just a type synonym for [Char], we could have concatenated [a], say, rather than show a. However, the latter renders the Char more carefully, surrounding it with single quotes ' and making sure that special Chars like '\n', more or less corresponding to the "enter" key on our keyboard meaning "render a line break here", don't actually cause the line to break but is instead simply displayed as '\n'. show is not really about rendering things so that they are pretty, but about making the details about the value in question very clear to whomever is reading show's output. So, yes, we prefer using show in our error messages rather than having to worry about these little details.

show is actually a method of that typeclass called Show.

```
class Show a where
  show :: a -> String
```

We can implement Show instances for datatypes we define ourselves quite easily.

```
data Foo = Bar | Qux Integer

instance Show Foo where
  show = \case Bar -> "Bar"
               Qux i -> "Qux " <> show i
```

See? Easy. Recursive, even. We can now show our values of type Foo.

```
> show Bar
"Bar"
> show (Qux 8)
"Qux 8"
```

Actually, we can just *not* write show and GHCi will automatically apply show to our values and print the resulting String.

```
> Bar
Bar
> Qux 8
Qux 8
```

Among other things, Show is how GHCi knows how to render on our screen the expressions it evaluates. In fact, look at what happens if we try to evaluate something for which there is *no* Show instance, like functions.

```
> id
<interactive>:89:1: error:
    • No instance for (Show (a -> a)) arising from ...
```

Yes, we get a type-checker error saying that we can't print the expression because there's no Show instance for its type.

Also, tangentially, notice how what GHCi prints looks different depending on whether we say Bar or show Bar. This is because whereas the type of Bar is Foo, the type of show Bar is String, and the respective Show instances for these types, used by GHCi, will do different things.

Anyway, these Show instances are so straightforward that Haskell can automatically *derive* their implementation for us most of the time. For example, rather than manually writing the Show instance for our Foo type above, we could have said this:

```
data Foo = Bar | Qux Integer
  deriving (Show)
```

That line saying deriving (Show) tells Haskell to magically implement a Show instance for us.

```
> show (Qux 8)
"Qux 8"
```

See? Still working. There are actually *many* typeclasses like Show which Haskell can automatically derive for us, we'll talk about them as they appear. Some we've encountered already. Like Eq, the typeclass that gives us (==) and (/=). We can ask Haskell to derive it too.

```
data Foo = Bar | Qux Integer
  deriving (Show, Eq)
```

With this, we can now compare values of type Foo for equality using (==), something that previously would have resulted in a type-checking error.

```
> Qux 4 == Qux 7
False
```

Anyway, back to our parsers.

One hundred seventy-three

We have expect1, great. How do we build the expect we described before with it? Straightforwardly.

```
expect :: String -> Parser Unit
expect = \case
  "" -> unit
  x : rest -> fmap (const Unit)
                   (pair (expect1 x) (expect rest))
```

If expect is asked to parse the empty String, which we agreed should always succeed because, conceptually, the empty String "is always there", then we just use unit as our value of type Parser Unit. We know that unit will never fail nor affect our parsing in any way, and this is exactly what we need.

Otherwise, if we are given a non-empty String, as determined by a successful pattern match on the cons : constructor, then we construct a

Parser for the first Char, x, using expect1, and pair it with a *recursive* application of expect to the rest of the String. However, the type of pair (expect1 x) (expect rest) is Parser (Unit, Unit), not Parser Unit as desired, so we have to fmap the function const Unit over it in order to convert that (Unit, Unit) output to just Unit. That's all. Does it work?

```
> runParser (expect "") "hello"
Right (0, "hello", Unit)
> runParser (expect "h") "hello"
Right (1, "ello", Unit)
> runParser (expect "hello") "hello"
Right (5, "", Unit)
> runParser (expect "bread") "hello"
Left (0, "Expected 'b' got 'h'")
> runParser (expect "her") "hello"
Left (2, "Expected 'r' got 'l'")
```

Beautiful. Something interesting about the way in which we've defined our Parser is that, not only will the parsing of the entire String stop as soon as we encounter an unexpected Char, but also, due to laziness, the Parsers being paired themselves will not be constructed *at all* until it's been confirmed that indeed they will need to be executed. This has nothing to do with expect but with the lazy semantics of pair. We can easily observe this in GHCi using undefined, our trojan horse.

```
> runParser (pair (expect1 'a') undefined) "zt"
Left (0, "Expected 'a' got 'z'")
```

In this example, expect1 'a', the leftmost Parser in our pair, fails, so the rightmost Parser, our tricky undefined, is not evaluated *at all*. This is clever. Not only was the execution of this rightmost Parser prevented, but *zero* resources were spent on forcing it out of its thunk. This is very handy for us when constructing potentially very large Parsers out of many small Parsers. Because of laziness, the large Parsers won't really exist as a process until their existence has been proven to be necessary by the actual input we are dealing with at runtime. Very beautiful.

Anyway, *lists*. Strings are a list of Chars. And if we are saying that we are performing unit on every occurrence of [], and pair on every occurrence of :, then maybe we could express all of this as a *fold* instead.

```
expect :: String -> Parser Unit
expect = foldr (\x z -> fmap (const Unit) (pair (expect1 x) z))
               unit
```

Indeed. In this partial application of foldr, we are saying that any occurrence of [] will be replaced by unit, and that any occurrences of : will be replaced by the function where x is the Char currently in the spotlight, and z is analogous to the recursive call made to expect rest in our previous example. And, since foldr is lazy in the right way, the lazy semantics of this implementation of expect are still the same as before.

One hundred seventy-four

Let's build a more interesting Parser now. One that parses a natural number like "123" into a Natural value. Obviously, we'll start by writing the Parser that parses just one Digit, which we'll most certainly need. We've even seen this Parser before in our examples, we named it parseDigit. We'll figure out how to make a Parser for Naturals out of it later on.

We had a function digitFromChar before, let's see if we can reuse that.

```
digitFromChar :: Char -> Maybe Digit
```

Building a Parser Digit out of digitFromChar is straightforward.

```
parseDigit :: Parser Digit
parseDigit = parse1 (note "Expected a digit 0..9" . digitFromChar)
```

digitFromChar takes a Char and returns a Maybe Digit, so we compose this function with note in order to pimp that return type into an Either String Digit suitable for parse1. This works just fine, of course.

```
> runParser parseDigit "8"
Right (1, "", D8)
> runParser parseDigit "v"
Left (0, "Expected a digit 0..9")
```

How do we go from parsing just *one* Digit, to parsing a Natural number made out of many of them? We actually implemented a function doing just that a while ago, but it had type String -> Maybe Natural rather than Parser Natural. It relied on a function called sequence with type [Maybe x] -> Maybe [x]. Maybe we can use sequence again?

Back then, we understood sequence as a function that, assuming all the Maybes in a given list were Just, then the result would also be Just. Otherwise it would be Nothing. And that was true. However, what we didn't know is that this "Nothing vs Just" dichotomy was just a consequence of pair's behaviour, and that sequence works for *any* Monoidal functor, not just Maybe.

```
sequence :: Monoidal f => [f x] -> f [x]
```

It might not be immediately obvious, but sequence is essentially pair in disguise.

```
pair :: Monoidal f => f a -> f b -> f (a, b)
```

Let's do the math. First, let's uncurry pair.

```
uncurry pair :: Monoidal f => (f a, f b) -> f (a, b)
```

Second, let's acknowledge that a tuple is essentially a two-element list.

However, unlike the elements in a tuple, which can have different types, all the elements in a list must be of the same type. We better make our as and bs be the same type, then.

```
uncurry pair :: Monoidal f => (f x, f x) -> f (x, x)
```

If we pick that x to be Either a b, say, then we'd still be able to pair as and bs somehow, by putting one of them in the Left and the other in the Right, so this change is not a big deal for our exploration.

What remains now is to make sure we work with lists of xs, rather than with pairs of xs. And as soon as we have that, we can build our sequence out of pair, unit, and some other minor things.

```
sequence :: Monoidal f => [f x] -> f [x]
sequence = \case [] -> fmap (const []) unit
                 fx : rest -> fmap (\(x, xs) -> x : xs)
                                   (pair fx (sequence rest))
```

If sequence gets an empty list [] as input, that means there's nothing to be done, so we just create a decorative f [] out of the blue by fmapping const [] over the innocent unit.

Hmm. Funny. We keep fmapping const whatever over unit time and time again. Maybe we should introduce a function that does just that, and thus avoid some of the noise.

```
pure :: Monoidal f => a -> f a
pure = \a -> fmap (const a) unit
```

A straightforward function, pure, but we seem to be reaching out for it quite frequently, so we better give it a name. The word "pure" should remind us that while the output type, f a, says that f could be doing some potentially fallible work in order to produce that a, like trying to parse some input, in reality, said a is being produced in a *pure* manner where f behaves as some kind of *identity* doing nothing interesting at all. And, actually, look at what happens if we forget about the f for the moment, and just worry about the a.

```
id :: a -> a
```

Oh my, oh my, how exciting. pure is but an oddly shaped *identity* function. Naturally, this tells us we are in the right place. We'll study this pure transformation in depth later on, but for now, we'll just use it to simplify the implementation of sequence a bit.

```
sequence :: Monoidal f => [f x] -> f [x]
sequence = \case [] -> pure []
                 fx : rest -> fmap (\(x, xs) -> x : xs)
                                   (pair fx (sequence rest))
```

Great. Let's move on. On the second pattern, when there's at least one such value of type f x in that list, we pair it with a recursive call to sequence rest. This pairing, however, results in a value of type f (x, [x]), not f [x] as we wanted. But, by fmapping the function \(x, xs) -> x : xs of type (x, [x]) -> [x] over this result —a function which we could also have written as uncurry (:)— we can convert our Parser's output to the expected type.

If we wanted to use sequence to parse *three* Digits, say, we could.

```
parseThreeDigits :: Parser [Digit, Digit, Digit]
parseThreeDigits = sequence (replicate 3 parseDigit)
```

replicate n x, which we haven't seen before, is a function that repeats n times the value x in a list. That is, replicate 3 parserDigit equals [parseDigit, parseDigit, parseDigit]. Then, we simply provide this list to sequence. According to GHCi, it works just fine.

```
> runParser parseThreeDigits "357"
Right (3, "", [D3, D5, D7])
```

Of course, a parsing failure on *any* of these Parsers would cause the entire thing to fail.

```
> runParser parseThreeDigits "4x6"
Left (1, "Expected a digit 0..9")
```

If sequence sounds too similar to our recent expect, it is because expect is a essentially just a special case of sequence.

```
expect :: String -> Parser Unit
expect = fmap (const Unit) . sequence . fmap expect1
```

This definition of expect first fmaps expect1 over the input String, which is just a list of Chars, to create a value of type [Parser Unit]. This list is a suitable input to sequence, which will turn it into a Parser [Unit], whose output of type [Unit] we are not really interested in, yet we must convert to Unit in order to keep the compiler happy. So we fmap the function const Unit over it to finally get our Parser Unit.

To sum up, while pair allows us to compose *two* values produced by some Monoidal functor, sequence allows us to compose *zero or more* of them. However, this wouldn't be possible without unit or pure giving us a sensible answer to the baffling question of what it means to sequence zero things. sequence brings "successful" things together in a way that reminds us of how and, the conjunction of *many* booleans, does it. Which, by the way, suggests that another way to look at pair is to think of it as the conjunction of just *two* functorial values. If both of these values are "true", by which we mean that none of them fail in the sense Left, Nothing or a failing Parser would, then pair gives us a "true" value as result.

One hundred seventy-five

Alright, we can parse Digits, and even *lists* of Digits. We should be able to parse Natural numbers now. All we have to do, presumably, is fmap our old function naturalFromDigits, of type [Digit] -> Natural, over our parsing result.

```
> runParser (fmap naturalFromDigits parseThreeDigits) "325"
Right (3, "", 325)
```

Amazing. What about parsing a smaller number, like 7?

```
> runParser (fmap naturalFromDigits parseThreeDigits) "7"
Left (0, "Expected a digit 0..9")
```

Hmm, this ain't right. What about a bigger number, like 12345?

```
> runParser (fmap naturalFromDigits parseThreeDigits) "12345"
Right (3, "45", 123)
```

What? Why are we only getting 123 as a Natural result, and "45" as leftovers? Well, the function is called parse*Three*Digits, isn't it? So this behaviour is exactly what we asked for. We parse *three* digits at a time, not less, not more, and we make a Natural number out of them.

But this isn't *really* what we want. We want to be able to parse Natural numbers of arbitrary lengths. We want parseNatural, a Parser of Natural numbers, with a behaviour like this:

```
> runParser parseNatural ""
Left (0, "Expected a digit 0..9")
> runParser parseNatural '3"
Right (1, "", 3)
> runParser parseNatural '4159"
Right (4, "", 4159)
> runParser parseNatural "42hello"
Right (2, "hello", 42)
```

Unfortunately, it turns out *we can't* implement a parseNatural like this unless we poke inside the Parser datatype again. Let's think about *why*.

The first example, where we try to parse a Natural number out of the empty String, obviously fails because there are no digits we can parse in this empty String. The second and third examples successfully parse the *full* input String into two different Natural numbers 3 and 4159. However, notice how the String representations of these numbers have *different* lengths. This is different from what parseThreeDigits does when it expects precisely *three* Chars. Moreover, even the last example manages to successfully parse the number 42 out of "42hello", leaving "hello" as leftover for whomever comes next. Yes, this is what we want.

```
> runParser (pair parseNatural (expect "he")) "42hello"
Right (4, "llo", (42, Unit))
```

As it goes about its business, parseNatural most certainly asks itself whether the fine Char currently in the spotlight is a digit *or* not, and makes a decision about whether to continue parsing *or* not based on that. *Or.* We said *or*. A digit *or* not. But *or* means *disjunction*, it means Either, it means *sum*. It doesn't mean conjunction, product nor tuples. It does not mean sequence, it does not mean pair.

And we are stuck, because while we'd most certainly be able to come up with a fascinating implementation of the dubitative parseNatural if we were to open up the Parser constructor and poke inside, those are *not* the rules of the game. We said we'd only build Parsers using fmap, unit, pair and parse1, *not* the Parser constructor. We are stuck. We need to *compose* differently.

One hundred seventy-six

Let's open up Parser again, but only so that we can create a new way of composing it.

What are we trying to accomplish? We have said, over and over again, that pair and unit have a correspondence to the product of types, a *monoid* with (,) as its associative binary operation and Unit as its identity element. That's where the "monoidal functor" name comes from: Monoidal recreates the monoidal features of product types for Functorial values.

```
(,)  ::                    a ->   b ->   (a, b)
pair :: Monoidal f => f a -> f b -> f (a, b)

Unit ::                 Unit
unit :: Monoidal f => Unit
```

The name Monoidal is a bit unfortunate, however, because it somewhat implies that the *product* of types is the only monoid for which we'd want to find a correspondence in Functors, which is most certainly false. What about the *sum* of types? Isn't it true that composing types together using Either as an associative binary operation is a monoid with Void as its identity element? Of course it's true, and we'd like to have a typeclass for that correspondence, too.

This typeclass will have two methods. One shall correspond to the Void type, the identity element for the sum of types, conveying the idea of *zero*. And the other shall correspond to the Either type constructor, the associative binary operation for this monoid.

```
Either :: a -> b -> Either a b
```

Either takes two types a and b, and returns yet another type Either a b. So, mimicking the correspondence between the tuple type constructor (,) and pair, we'll need a function where the a and b given as input, as well as the

Either a b returned as output, are all wrapped in some Functor f.

```
alt :: Functor f => f a -> f b -> f (Either a b)
```

We'll call the function alt, suggesting that Parser a and Parser b are different *alternatives*. This *or* that. The correspondence between Either and alt, regarding their shape at least, is unquestionable.

But alt is not enough. Our monoid needs an identity element too. In the case of Either, it is Void. For us, it'll be Void too, but much like unit wrapped Unit in some Functor f, we'll wrap Void in an f too. Let's call it void, in lowercase.

```
void :: Functor f => f Void
```

And since we like a good pun, and moreover we are obliged to rebel against the misnamed Monoidal, we'll name our typeclass Monoidalt and chuckle for a while. We'll learn, eventually, that neither Monoidal nor Monoidalt are the names these things go by, so any discussion about their naming would be moot and temporal. Let's have fun.

```
class Functor f => Monoidalt f where
  alt :: f a -> f b -> f (Either a b)
  void :: f Void
```

Excellent. We have types. And, obviously, since we are deriving these ideas from a monoid, we have *laws* too. First, left and right identity laws.

```
alt void a  ≈  a
```

```
alt a void  ≈  a
```

Then, an associativity law.

```
alt a (alt b c)  ≈  alt (alt a b) c
```

And finally, a naturality law, which we won't ever need to prove manually because of parametricity.

```
alt (fmap f ma) (fmap g mb) == bimap f g (alt ma mb)
```

These laws are exactly the same as the Monoidal ones, except we have replaced all mentions of pair with alt, and unit with void.

Alright then, let's implement a Monoidalt instance for Parser. Let's focus on the alt method first. alt takes a Parser a and a Parser b as input, and returns a Parser producing a value Either a b that tells us whether it was the Parser for a or the one for b which succeeded. The hope is that we will be able to say "parse either this *or* that", and it should work.

Let's see a motivating example first. Let's compose two Parsers using alt. One of them will be expect 'friend", of type Parser Unit, which only succeeds if the String being parsed starts with "friend". The other Parser will be parseBool, of type Parser Bool, converting a leading 't' in the input

String into True, or a leading 'f' into False. Composing these two Parsers with alt will give us a value of type Parser (Either Unit Bool), rather than Parser (Unit, Bool) as pair would.

```
expect "friend" :: Parser Unit

parseBool :: Parser Bool

alt (expect "friend") parseBool :: Parser (Either Unit Bool)
```

First we'll try to parse "friend", which is exactly the minimum input our leftmost parser expect "friend" needs to succeed. alt, seeing as expect "friend" succeeds, proceeds to wrap its Unit output in a Left constructor that fits the expected Either Unit Bool type.

```
> runParser (alt (expect "friend") parseBool) "friend"
Right (6, "", Left Unit)
```

Excellent. It worked. And notice how in this situation, the rightmost parseBool was completely ignored by alt. The idea is that alt will only proceed to evaluate the rightmost Parser *if* the leftmost fails. For example, expect "friend" would fail to parse an input like "foe". However, if we use alt to compose it with parseBool, which would successfully parse False from "foe", we will have created a Parser that *because* expect "friend" fails, will try doing something else, and it will succeed at that.

```
> runParser (expect "friend") "foe"
Left (1, "Expected 'r' got 'o'")
> runParser parseBool "foe"
Right (1, "oe", False)
> runParser (alt (expect "friend") parseBool) "foe"
Right (1, "oe", Right False)
```

Of course, the Bool output from parseBool is wrapped inside a Right constructor so as to accommodate the type Either Unit Bool, but that's a minor detail. We can see that the False output is still there on the Right, much like how Unit was on the Left before.

It's interesting to notice that while expect "friend" won't successfully parse "foe", the very first 'f' in "foe" *will* be successfully parsed by the expect1 'f' that's used somewhere inside the definition of expect "friend". This is communicated to us quite clearly by the error we get from failing to parse "foe" using expect "friend":

```
> runParser (expect "friend") "foe"
Left (1, "Expected 'r' got 'o'")
```

The error arises when trying to parse the 'o' in position 1, not before. Yet, despite the leading 'f' having already been consumed by then, alt, seeing how the leftmost Parser failed, will provide the *entire* input String "foe" to its rightmost Parser parseBool, which will successfully turn that 'f' into

False. Not just "oe".

This is an interesting property of alt, and interesting properties have names. We call it *backtracking*. We say that the Parser constructed by alt pa pb will "backtrack" its input in full before providing it to pb, if pb needs to run at all, despite how much pa may have consumed. Both pa and pb will receive the *same* input.

Funnily, even if alt wanted to provide just "oe" to parseBool after a failed attempt by expect "friend" to parse "foe", it would *not* be able to do so, for we were careful enough when designing the Parser datatype. We made sure that any unsuccessful parsing, as conveyed by a Left output from the parsing function, does *not* mention any leftovers at all. Only a successful parsing result communicates its leftovers. Clever, right? I'm sure there's a lesson here about not putting things in your output that were already part of your input, unless truly necessary.

What if *both* Parsers fail, though?

```
> runParser (alt (expect "friend") parseBool) "water"
Left (0, "Expected 't' or 'f'")
```

The whole composition fails, of course, what else could happen? But notice how the error reported by alt is coming from the rightmost Parser, parseBool. In other words, alt pa pb is biased towards preserving pa's success, but pb's failure.

Alright, let's implement alt for Parsers.

```
alt :: Parser a -> Parser b -> Parser (Either a b)
alt pa pb = Parser (\s0 ->
  case runParser pa s0 of
    Right (n, s1, a) -> Right (n, s1, Left a)
    Left _ -> case runParser pb s0 of
      Left x -> Left x
      Right (n, s1, b) -> Right (n, s1, Right b))
```

There's not much going on here. All that's happening is that while pair pa pb only executes pb when pa succeeds, alt pa pb only executes pb when pa *fails*, respectively wrapping as and bs in Left and Right. That's all.

And just like in pair, the fact that alt doesn't touch pb unless truly necessary means that any laziness in pb is fully preserved. We can observe this by composing undefined, our trojan horse, to the right of a successful Parser.

```
> runParser (alt parseBool undefined) "t"
Right (1, "", Left True)
```

Had alt evaluated that undefined, our program would have crashed. However, that only happens when the leftmost Parser fails.

```
> runParser (alt parseBool undefined) "x"
*** Exception: Prelude.undefined
```

What about void, of type `Parser Void`? This `Parser` clearly must fail, for there's no way we can produce a value of type `Void` as output. But failing is easy, we just return a `Left` with a dummy error message, and report *zero* `Chars` of consumed input.

```
void :: Parser Void
void = Parser (const (Left (0, "void")))
```

Does it work? Well, void is supposed to *fail* by design, so if that's what you mean by "work", then sure, it works.

```
> runParser void "t"
Left (0, "void")
> runParser (alt void parseBool) "t"
Right (1, "", Right True)
> runParser (alt parseBool void) "t"
Right (1, "", Left True)
```

pairing with void obviously fails, since this corresponds to the idea of multiplying by *zero*, which results in *zero*.

```
> runParser (pair void parseBool) "t"
Left (0, "void")
> runParser (pair parseBool void) "t"
Left (1, "void")
```

In other words, whereas unit always succeeds by doing nothing, void always *fails* by doing nothing. And its output value of type `Void`, a value *impossible* to construct, is a testament to this.

One hundred seventy-seven

Alright, we have `Monoidalt`, alt and void now. Can we finally get `parseNatural`? That's what we wanted, right?

```
parseNatural :: Parser Natural
parseNatural = fmap naturalFromDigits (some parseDigit)
```

Look at how beautiful it is. We parse some digits, and then we simply `fmap` `naturalFromDigits` over them. That's it. The magic, of course, happens inside some.

```
some :: (Monoidal f, Monoidalt f) => f a -> f [a]
```

some is a function that, given some f a where the f has *both* `Monoidal` and `Monoidalt` instances, it will return a list with *at least one* a in it. We can try `some parseDigit` on its own to see how this works.

```
> runParser (some parseDigit) ""
Left (0, "Not enough input")
> runParser (some parseDigit) "1"
Right (1, "", [D1])
> runParser (some parseDigit) "123"
Right (3, "", [D1, D2, D3])
> runParser (some parseDigit) "12y4"
Right (2, "y4", [D1, D2])
```

That is, as long as *one* execution of parseDigit is successful, some parseDigit will succeed. The return type of some is a bit disappointing, seeing as [a], as far as the type system is concerned, could be an empty list, but in fact this will never be the case. We'll come back to this later. For now, let's implement some. Actually, let's implement optional first, a smaller function that some will benefit from.

```
optional :: (Monoidal f, Monoidalt f) => f a -> f (Maybe a)
```

While alt fa fb executes fb if fa fails, optional fa *never* fails. Instead, it replaces failures in fa with a "successful" Nothing result. In a sense, optional demotes failures in f to failures in Maybe. We can try optional in GHCi and see how this works in practice.

```
> runParser (optional parseBool) "f"
Right (1, "", Just False)
> runParser (optional parseBool) "hello"
Right (0, "hello", Nothing)
```

See? In both cases runParser returns Right, even when parseBool could not parse "hello" into a Bool. In that case, optional replaced that parsing failure with a successful Nothing result. The implementation of optional is quite straightforward:

```
optional :: (Monoidal f, Monoidalt f) => f a -> f (Maybe a)
optional fa = fmap (either Just (const Nothing)) (alt fa unit)
```

We simply compose fa with unit using alt, meaning that alt fa unit will *always* succeed even if fa fails, for unit itself never fails, so at worst alt fa unit will successfully produce Right Unit. Finally, we use fmap to convert the Either a Unit produced by alt fa unit to the isomorphic and more ergonomic Maybe a. We see both Monoidal and Monoidalt constraints on this function because we are using *both* unit and alt, methods belonging to these different typeclasses. Having optional, defining some is just a small step.

```
some :: (Monoidalt f, Monoidal f) => f a -> f [a]
some fa = fmap (\(a, yas) -> a : maybe [] id yas)
              (pair fa (optional (some fa)))
```

some fa requires *at least one* execution of fa to succeed, but it will collect the results of as many successful executions of fa as it can before finishing its work. From this idea follows that some must pair one execution of fa with an optional execution of a recursive call to some fa. That's all. If that first fa fails, then the whole thing fails, but as long as it succeeds, then whether any following executions of fa succeed or not is irrelevant, some will succeed nonetheless. Finally, since pair fa (optional (some fa)) has type f (a, Maybe [a]), we fmap over it a function that converts all of that to [a]. And we can see here that, indeed, this returned list will never be empty, even if the type [a] says otherwise. We'll fix that later on.

One hundred seventy-eight

There aren't nearly as many Functors for which we can provide implementations of alt and void compared to those for which we can implement pair and unit, considering alt only makes sense for Functors that have some concept of "success" and "failure" like Parser does. Maybe is one of those few.

```
instance Monoidalt Maybe where
  void = Nothing

  alt (Just a) _ = Just (Left a)
  alt _ (Just b) = Just (Right b)
  alt _ _        = Nothing
```

For Maybe, Nothing means failure and Just success.

```
> alt Nothing Nothing
Nothing
> alt (Just True) Nothing
Just (Left True)
> alt (Just True) (Just False)
Just (Left True)
> alt Nothing (Just False)
Just (Right False)
```

And, as expected, the rightmost Maybe is not evaluated unless strictly necessary. Laziness is preserved.

```
> alt (Just True) undefined
Just (Left True)
```

Using alt with Maybe allows us to check whether we have one thing, *or* the other, *or* Nothing at all.

One hundred seventy-nine

Monoidal is where the fun's at. What other things can we sequence? We tried Maybe and Parser already. Let's try Either String next. To keep things neat, let's say all of the expressions below are of type Either String [Natural].

```
> sequence []
Right []
> sequence [Left "red"]
Left "red"
> sequence [Left "red", Left "blue"]
Left "red"
> sequence [Left "red", undefined, Right 4]
Left "red"
> sequence [Right 3, Left "red"]
Left "red"
> sequence [Right 2]
Right [2]
> sequence [Right 4, Right 1, Right 8]
Right [4, 1, 8]
```

Essentially, Either String —or Either whatever, really— behaves just like Maybe.

What about functions next? Let's see what happens if we pick (->) Natural to be our Monoidal functor, and assume we'll be using sequence on lists containing values of type Natural -> Bool. Let's try with the function named even first, of type Natural -> Bool, returning True if the given number is even or False otherwise.

```
> even 4
True
> even 3
False
> :type sequence [even]
sequence [even] :: Natural -> [Bool]
> sequence [even] 4
[True]
> sequence [even] 3
[False]
```

Interesting. What about putting *more* functions of type Natural -> Bool in that list?

```
> sequence [even, odd] 4
[True, False]
> sequence [even, odd] 3
[False, True]
> sequence [even, odd, \x -> x == 5] 3
[False, True, False]
> sequence [even, odd, \x -> x == 5, \x -> x < 10] 3
[False, True, False, True]
```

Fascinating. Like pair and &&&, sequence still *fans-out*, but it does so on a larger scale.

```
> sequence [(+ 1), (* 10), \x -> x * x] 3
[4, 30, 9]
```

What other Monoidal functor do we know? What about lists? They are a

Functor, right? Wait, wait, wait, we never discussed this. Are we saying lists are Monoidal functors? How?

One hundred eighty

For something to be a Monoidal functor, it needs to implement unit, pair and satisfy a bunch of laws. Well then, if we can do that for lists, we'll have discovered a new Monoidal functor. Can't wait to see what it does!

```
instance Monoidal [] where
  unit :: [Unit]
  pair :: [a] -> [b] -> [(a, b)]
```

Interesting. It's unclear what unit should do. Should it be an empty list? Hmm, perhaps not, seeing how that would be akin to unit being Nothing in the case of Maybe, which we already know is incorrect. Should it be a list with one Unit in it? What about seven Units in it? What about it being an *infinite* lists of Units? Interesting questions. But, considering this function is called unit, evoking the idea of *one*, let's try our luck with a list containing exactly *one* Unit in it and see what happens. If necessary, we'll have the opportunity to repent and change our decision later on.

```
unit :: [Unit]
unit = [Unit]
```

What about pair? The type says it takes a list of as and a list of bs, and it turns them into a single list of as and bs. So, presumably, it just pairs the individual elements of these lists in order.

```
pair :: [a] -> [b] -> [(a, b)]
pair (a : as) (b : bs) = (a, b) : pair as bs
pair _        _        = []
```

Like in the case of Maybe, we only seem to be able to produce a list containing tuples of as and bs when the lists given as input are themselves not empty. Do we have a monoidal functor? Well, let's check if our chosen unit and pair satisfy the monoidal functor laws. First, for any list x, pair x unit, x and pair unit x should all be isomorphic. Let's pick some lists and try. Let's try with the empty list first.

```
> pair [] unit
[]
> pair unit []
[]
```

Alright, this seems fine. Let's try a list with *one* element next.

```
> pair [4] unit
[(4, Unit)]
> pair unit [4]
[(Unit, 4)]
```

Excellent. These results are isomorphic to [4]. What about a list with *two* elements in it?

```
> pair [5, 7] unit
[(5, Unit)]
> pair unit [5, 7]
[(Unit, 5)]
```

What? That's not right. Where did the 7 go? The identity law for monoidal functors has been violated. Did we pick the wrong unit? Actually, no. It's just that we accidentally implemented the wrong pair function. The function we implemented is commonly known as zip, not pair.

```
zip :: [a] -> [b] -> [(a, b)]
zip (a : as) (b : bs) = (a, b) : zip as bs
zip _        _        = []
```

zip is useful when you have two lists of the same length, and you want to pair up the individual elements of those lists in the same order they appear on those lists.

```
> zip [1, 2, 3] [100, 200, 300]
[(1, 100), (2, 200), (3, 300)]
```

However, when one of the given lists is shorter than the other one, all the "extra" elements from the longer lists are simply dropped from the result.

```
> zip [1, 2, 3] []
[]
> zip [1, 2, 3] [100]
[(1, 100)]
> zip [1, 2, 3] [100, 200, 300, 400, 500, 600]
[(1, 100), (2, 200), (3, 300)]
```

What failed when we tested for the identity law is that we were trying to zip a list of length *one*, unit, with a longer list, and this caused all the elements beyond the first one to be discarded.

Could we have pair use longer lists, then? We could, yes, we could. But to spice things up, let's try something else. Let's have pair obtain the *cartesian product* of these lists instead, and see what happens. That is, pair shall pair each element of one list exactly once with each element of the other list.

```
> pair [1, 2, 3] [True, False]
[(1, True), (1, False),
 (2, True), (2, False),
 (3, True), (3, False)]
```

You see, all of 1, 2 and 3 are paired with *both* True and False. And, symmetrically, all of True and False are paired with 1, 2, and 3. If we were to pair with unit, the result would be isomorphic to the given list.

```
> pair [1, 2, 3] unit
[(1, Unit), (2, Unit), (3, Unit)]
> pair unit [1, 2, 3]
[(Unit, 1), (Unit, 2), (Unit, 3)]
```

And pairing with the empty list, obviously leads to an empty result.

```
> pair [5, 6, 7] []
[]
> pair [] unit
[]
```

Let's implement pair, then. We said we want to pair *each* element of one list, to *each* element of the other list. Well, fmap is how we address "each" element of a list, so presumably we are saying we'll somehow need to fmap twice, once for each list.

```
pair :: [a] -> [b] -> [(a, b)]
pair = \as bs -> concat (fmap (\a -> fmap (\b -> (a, b)) bs) as)
```

Indeed. Let's pay attention to the fmap (\a -> ...) as part first. This is saying that each element of the list named as, the first parameter to pair, will be transformed somehow using the function given as first parameter to this fmap. This function, however, has *another* use of fmap in it. \a -> fmap (\b -> (a, b)) bs says that each element in bs will be paired up with the a coming from as that's currently in the spotlight. If, for example, as was [1, 2, 3] and bs was [True, False], then this code would first select 1 to be a and True to be b, then it would keep a as is but move on to picking False as b, and only then, after bs has been exhausted, would 2 become a and the process would continue until as is exhausted.

However, if we consider that the type of the function we are fmapping over as is a -> [(a, b)], and that generally speaking fmap for lists means *lifting* a function of type x -> y into a function of type [x] -> [y], then lifting a function type a -> [(a, b)] will give us something of type [a] -> [[(a, b)]]. But do you see those double brackets around (a, b)? We don't want them. This is where concat comes in, the function that given a list of lists, concatenates them together into a single one.

```
> concat []
[]
> concat [[]]
[]
> concat [[], []]
[]
> concat [[], [1, 2], [3]]
[1, 2, 3]
> concat [[1, 2], [], [3, 4], [5]]
[1, 2, 3, 4, 5]
```

So, yes, lists are Monoidal functors too, and pairing them gives us the *cartesian product* of the given lists. I leave it up to you to prove, using equational reasoning, that the Monoidal functor laws hold for our chosen unit and pair. Here's the full definition of the Monoidal instance for lists.

```
instance Monoidal [] where
  unit :: [Unit]
  unit = [Unit]
  pair :: [a] -> [b] -> [(a, b)]
  pair = \as bs -> concat (fmap (\a -> fmap (\b -> (a, b)) bs) as)
```

What could sequence do for lists?

```
sequence :: Monoidal f => [f a] -> f [a]
```

This question becomes ever more intriguing when we realize that our f will be the type constructor []:

```
sequence :: [[a]] -> [[a]]
```

Well, seeing how pair gives us the cartesian product of *two* lists, and how sequence is essentially pair but for *zero or more* lists rather than two, sequence will presumably give us the cartesian product of *all* the input lists.

```
> sequence []
[]
> sequence [[]]
[]
> sequence [[], []]
[]
> sequence [[1, 2], [3, 4, 5]]
[[1, 3], [1, 4], [1, 5], [2, 3], [2, 4], [2, 5]]
> sequence [[1, 2], []]
[]
> sequence [[1, 2], [], [3, 4, 5]]
[]
> sequence [[1, 2], [3], [4, 5, 6]]
[[1, 3, 4], [1, 3, 5], [1, 3, 6],
 [2, 3, 4], [2, 3, 5], [2, 3, 6]]
```

Beautiful. Notice how the number of lists we obtain as result of using sequence equals the multiplication of the lengths of the lists given to it:

```
> sequence [[1, 2], [3, 4, 5]]
[[1, 3], [1, 4], [1, 5], [2, 3], [2, 4], [2, 5]]
```

The length of [1, 2] is *two*, the length of [3, 4, 5] is *three*, thus the length of the list returned by sequence [[1, 2], [3, 4, 5]] is *three* times *two*, *six*. Following this reasoning, we can see how including an empty list [], whose length is *zero*, among the input to sequence, would lead to an empty list as result. Multiplying by *zero* results in *zero*.

```
> sequence [[1, 2], [], [3, 4, 5]]
[]
```

Another interesting property of the list resulting from sequence is that, provided there are no empty lists among the input, all of the individual lists returned by sequence will have the same length, which will equal the number of lists that were provided to sequence as input. For example, if we provide *two* lists as input to sequence, in return we get a list of lists of length *two*:

```
> sequence [[1, 2], [3, 4, 5]]
[[1, 3], [1, 4], [1, 5], [2, 3], [2, 4], [2, 5]]
```

Here we provide *four*:

```
> sequence [[1], [2, 3], [4], [5]]
[[1, 2, 4, 5], [1, 3, 4, 5]]
```

And here, just *one*:

```
> sequence [[1, 2, 3]]
[[1], [2], [3]]
```

sequence is fun. René would enjoy it.

```
> fmap product (sequence [[1, 2], [3, 4, 5]])
[3, 4, 5, 6, 8, 10]
```

One hundred eighty-one

Is every Functor a Monoidal functor? Most are, but some are not. Take tuples, for example. We know that (,) x is a Functor where fmap can be used to modify the *second* element of a tuple.

```
> fmap even ("hello", 8)
("hello", True)
```

If we wanted to give a Monoidal instance for tuples, then one of the things we'd have to implement is a unit having this type:

```
unit :: ∀ x. (x, Unit)
```

But that type says *for all* x. How do we create, out of the blue, a tuple with a value of some arbitrary type x as its first element? We don't. We can't. That's what undefined tried to do, remember? And what about pair?

```
pair :: ∀ x a b. (x, a) -> (x, b) -> (x, (a, b))
```

Well, we'd need to combine the two xs that we get as input in a way that respects the Monoidal laws. But how would we do that if we don't know what x is? Let alone how to compose it.

So there we have it, (,) x is a non-Monoidal Functor.

Or is it? Sure, if we don't know how to create an x out of the blue, nor how to compose two of them into a third one, we are at a loss. But what if, say, we knew x was a Monoid?

```
instance Monoid x => Monoidal ((,) x) where
  unit = (mempty, Unit)
  pair (xa, a) (xb, b) = (mappend xa xb, (a, b))
```

This is perfectly valid Haskell. We *can't* change the types of the unit nor pair methods, so we can't add a Monoid constraint on x *there*, on the method types. But we can say that for (,) x to be an instance of the Monoidal typeclass, then x must be a Monoid too. And once the Monoid constraint on x is satisfied, we can simply use mempty to come up with a default value for x in unit, and mappend to compose the two xs in the case of pair.

String, that is [Char], is a Monoid, so unit can create a value of type (String, Unit) just fine.

```
> unit :: (String, Unit)
("", Unit)
```

And we can pair values of type (String, Unit), too.

```
> pair ("jelly", True) ("fish", 8)
("jellyfish", (True, 8))
```

But, if we try to do any of this with tuples where their first element type is not a Monoid, like (Void, Unit), say, we get an error from the type-checker complaining about the missing Monoid instance.

```
> unit :: (Void, Unit)
<interactive>:31:1: error:
    • No instance for (Monoid Void)
      arising from a use of 'unit'
```

So, yes, tuples are Monoidal functors, but only if their first element is a Monoid. So, no, not all Functors are Monoidal functors.

One hundred eighty-two

We've said time and time again that as long as we know how to compose *two* things in a monoidal kind of way, we can compose as many of them as we need, sometimes infinitely many. And this is true for both pair and alt.

Imagine we had a product type named Foo with four fields in it.

```
data Foo = Foo Natural Bool Digit Bool
```

Could we use pair to, say, construct a Parser for Foo? Of course.

```
parseFoo :: Parser Foo
parseFoo = fmap (\(a, (b, (c, d))) -> Foo a b c d)
                (pair parseNatural
                      (pair parseBool
                            (pair parseDigit
                                  parseBool)))
```

This works as expected.

```
> runParser parseFoo "1234f8t"
Right (7, "", Foo 1234 False D8 True)
```

However, hopefully we can agree that the implementation of parseFoo is
rather inconvenient. It is simple, yes, we just use pair as many times as
necessary and fmap a rather straightforward function over the parsing
result in order to construct a Foo, but it is not very ergonomic. Imagine if
Foo had *ten* more fields. Then, the implementation of parseFoo would be
much longer and distracting.

What about sum types? What if we wanted to parse a datatype like Season,
having *four* constructors?

```
parseSeason :: Parser Season
parseSeason =
  fmap (\case Left Unit -> Winter
             Right (Left Unit) -> Spring
             Right (Right (Left Unit)) -> Summer
             Right (Right (Right Unit)) -> Fall)
      (alt (expect "winter")
          (alt (expect "spring")
              (alt (expect "summer")
                  (expect "fall"))))
```

That's... unfortunate. There is beauty in alt's simplicity, but that doesn't
necessarily make alt an *ergonomic* solution. Half of parseSeason's
implementation is spent simply rearranging alt's output using fmap, and
doing this time and time again gets frustrating rather quickly. But there is
an alternative.

```
(<|>) :: Monoidalt f => f x -> f x -> f x
fa <|> fb = fmap (either id id) (alt fa fb)
```

This function, (<|>), intended to be used as an associative infix operator, is
essentially just alt. However, rather than taking functorial values of
different types f a and f b as input, eventually returning an f (Either a b)
as output, it takes inputs of a same type f x, making it possible to forego
the produced Either output, and instead simply return an f producing the
leftmost x that succeeded. And with this, we can implement parseSeason in
very concise way.

As always, let's start by solving the smaller problems first. Let's start by
implementing Parsers for each Season *individually*.

```
parseWinter :: Parser Season
parseWinter = fmap (const Winter) (expect "winter")
```

parseWinter shares its type with parseSeason from before, but rather than
worrying about all four possible Seasons, it worries only about Winter.
How? It expects "winter" to lead the input String being parsed, and if it
does, then this Parser's boring Unit output will be replaced with Winter by
simply fmapping const Winter over it.

```
> runParser parseWinter "winter"
Right (6, "", Winter)
```

And we can do the same for the rest of the Season values.

```
parseSpring :: Parser Season
parseSpring = fmap (const Spring) (expect "spring")

parseSummer :: Parser Season
parseSummer = fmap (const Summer) (expect "summer")

parseFall :: Parser Season
parseFall = fmap (const Fall) (expect "fall")
```

And with this, we have constructed the *four* Parsers that could, in principle, add up to the totality of what it means to parse a Season. All we need to do now is compose them into a bigger Parser. And seeing how the Parser we ultimately want, parseSeason, has the same type as the four Parsers we just defined, Parser Season, we can go ahead and use <|> to do it.

```
parseSeason :: Parser Season
parseSeason = parseWinter <|> parseSpring <|>
              parseSummer <|> parseFall
```

As confirmed by GHCi, this works as expected.

```
> runParser parseSeason "winter"
Right (6, "", Winter)
> runParser parseSeason "spring"
Right (6, "", Spring)
> runParser parseSeason "summer"
Right (6, "", Summer)
> runParser parseSeason "fall"
Right (4, "", Fall)
```

We've made alt a bit more ergonomic, that's all. Interestingly, as we said, <|> is an associative binary operation, and whenever we have one of those, we should ask ourselves whether there is an identity element too. Because if there is, then we'll have a monoid. Obviously, knowing that alt formed a monoid, and how <|> is just alt in disguise, we expect there will be an identity element too.

However, considering the type of (<|>) is Monoidalt f => f x -> f x -> f x, *for all* x, then our identity element will have to be of type ∀ x. Monoidalt f => f x. *For all* x, not just Void as it's been so far. Well, that's easy.

```
empty :: Monoidalt f => f x
empty = fmap absurd void
```

Remember absurd? It was that function of type ∀ x. Void -> x saying "give me a Void and I'll give you *anything* you want", a bluff. And, sure enough, the type of void, Monoidalt f => f Void, says there *could* be a Void someday.

So, taking inspiration from politicians, all we need to do is keep lying, `fmap` the absurd over the void, lies on top of lies, into what will be our deepest lie yet, for there will *never* be a Void, thus there will never be an x, yet there will be a *zero*, the identity element for our monoid. We call it `empty`, and together with `<|>`, they form the `Alternative` typeclass, which is how our exploratory `Monoidalt`, alt and void manifest themselves in Haskell.

```
class Functor f => Alternative f where
    empty :: f x
    (<|>) :: f x -> f x -> f x
```

`Monoidalt` doesn't come with Haskell out of the box. Nor `alt`, nor `void`. All we have is `Alternative` and its methods. Beyond this book, we'll probably never write a `Monoidalt` instance for `Parser` nor any other monoidal functor. The `Monoidalt` typeclass just isn't there in Haskell by default. We'd have to write our own, and nobody would care. So, we write `Alternative` instances instead.

```
instance Alternative Parser where
    empty :: Parser x
    empty = Parser (const (Left (0, "empty")))
    (<|>) :: Parser x -> Parser x -> Parser x
    pa <|> pb = Parser (\s0 ->
      case runParser pa s0 of
          Right r -> Right r
          Left _ -> runParser pb s0)
```

The implementation of the `Alternative` instance is even simpler than that of `Monoidalt`, since we don't need to worry about massaging a and b into `Either a b`. We can keep our efforts to a minimum and do what we said we wanted to do, which is run `pb` only if `pa` fails. That's all. The implementation for `empty`, despite having a different type and a different error message, is exactly the same as `void`'s. The type of `empty` says there *could* be an x, but `empty` will *never* produce such an x. No promises are broken.

Of course, once we have `empty` and `<|>`, if for some reason we want `void` and `alt` back, we can define them in terms of `empty` and `<|>`. Implementing `void` is easy, seeing as it's just a less polymorphic version of `empty`. We don't need to do anything.

```
void :: Alternative f => f Void
void = empty
```

And `alt` is not much more complex than this, really. Essentially, we need to do something similar to what we did when parsing Seasons using `<|>`, but for `Either` rather than for `Season`.

```
alt :: Alternative f => f a -> f b -> f (Either a b)
alt fa fb = fmap Left fa <|> fmap Right fb
```

That's it. The `Alternative` instance for `Maybe` is quite straightforward too:

```
instance Alternative Maybe where
    empty :: Maybe x
    empty = Nothing
    (<|>) :: Maybe x -> Maybe x -> Maybe x
    Just x <|> _        = Just x
    _      <|> Just x = Just x
    _      <|> _        = Nothing
```

That's the story of how `Monoidalt` ceased to be.

One hundred eighty-three

Infix operators are, generally speaking, uncomfortable to use outside
textbook scenarios. Mainly because it's hard to tell how adjacent
expressions using these operators associate unless we put parentheses
around them. Luckily, this is not a textbook, so we'll take this into
account.

Consider for example 2 + 4 / 5 * 3 / 4 / 5 - 2 / 2 * 3. Even if we have
been doing arithmetic all our lives, chances are we'll need to stop for a
while in order to contemplate and understand how the expressions we just
wrote, using what we call an *infix operator soup*, associate with each other.

In Haskell we compose *many* expressions, and the fewer infix operators
we use, the easier it will be to understand what's going on at a glance. Even
more so knowing that these operators may, at times, be partially applied.
Perhaps it won't be necessarily easier on *our* eyes as we write these
expressions for the first time, but it will certainly be easier for those who
come to read them afterwards. Which, by the way, includes our future
selves too.

With this in mind, we'll present an alternative and enlightening way of
defining parseSeason, say, avoiding the use of <|>.

```
parseSeason :: Parser Season
parseSeason = asum [parseWinter, parseSpring,
                    parseSummer, parseFall]
```

Having seen the previous definition of parseSeason using <|>, we should be
able to decipher what asum is doing by just thinking about it. asum will try
as many Parsers as it needs from the given list, from left to right, until one
of them successfully produces a value of type Season. If none of them do,
then the whole Parser constructed by asum fails. The name asum shall
remind us of the correspondence between Alternative and the *sum* of
types, which we acknowledged through alt and void. The definition of
asum is very straightforward.

```
asum :: Monoidal f => [f a] -> f a
asum = foldr (<|>) empty
```

If you recall, one of our intuitions for `foldr` is that all occurrences of `:` in the list are replaced by the given binary operation, and `[]` is replaced by the given initial accumulator. In our case, we replace them with `(<|>)` and `empty`.

The list provided to `asum` in our `parseSeason` example could have been written using `:` and `[]`, rather than Haskell's `[funny, list, syntax]`:

```
parseSeason = asum (parseWinter : parseSpring :
                    parseSummer : parseFall  : [])
```

Replacing those `:` and `[]` for `<|>` and `empty`, as `asum` would, takes us back to where we started:

```
parseSeason = parseWinter <|> parseSpring <|>
              parseSummer <|> parseFall <|> empty
```

Well, not *exactly*. We have a superfluous `empty` at the end. That's not too bad, seeing as this doesn't change the *meaning* of our parser. It does, however, change the error message with which `parseSeason` fails, seeing how if `parseSeason` fails, it will be `empty`'s fault and not `parseFall`'s. Perhaps we don't want that. Who knows what kind of important message is there on that error? Let's write `asum` differently so that we preserve the error from the last element of the list, if any.

```
asum :: Monoidal f => [f a] -> f a
asum []         = empty
asum [fa]       = fa
asum (fa : rest) = fa <|> asum rest
```

Quite straightforward. All we did is prevent any further recursion into `asum` whenever we receive a list with a *single* element as input, as matched by the second pattern above. This effectively means `empty` will only happen if `asum` is applied to `[]` by an outside caller, because `asum` will never recursively apply itself to the empty list `[]`, making sure any errors arising from that last `f a` in the input list, if any, are preserved.

It's interesting to highlight that just like there is a direct correspondence between `sequence` and the conjunction of booleans, in the sense that `sequence` requires that *all* the functorial values given to it succeed or "be true", there is also a direct correspondence between `asum` and the *disjunction* of booleans, seeing how `asum` expects *at least one* of them to succeed. At least one, *any* one. `sequence` and `asum` are the functorial versions of `and` and `or`.

Here are some examples of `and` and `sequence`, assuming we are dealing with functorial values of type `Maybe Natural`.

```
> and []
True
> sequence []
Just []
> and [True, False]
False
> sequence [Just 3, Nothing]
Nothing
> and [True, True]
True
> sequence [Just 3, Just 4]
Just [3, 4]
```

And here are or and asum, for comparison:

```
> or []
False
> asum []
Nothing
> or [False, False]
False
> asum [Nothing, Nothing]
Nothing
> or [False, True]
True
> asum [Nothing, Just 4]
Just 4
```

One hundred eighty-four

What about Monoidal, unit and pair? Can we get rid of them, too? Is that something we want to do? Why, yes, just look at how awkward parseFoo is otherwise.

```
parseFoo :: Parser Foo
parseFoo = fmap (\(a, (b, (c, d))) -> Foo a b c d)
               (pair parseNatural
                  (pair parseBool
                     (pair parseDigit
                         parseBool)))
```

The implementation of parseFoo is simple, relying only on fmap and pair, but it is also rather noisy. The problem, just like alt's before, is that pair can only deal with *two* Parsers at a time, but Foo is constructed by combining the outputs of *four* Parsers, which inevitably calls for some massaging of said outputs using fmap. What can we do about it?

```
data Foo = Foo Natural Bool Digit Bool
```

First, let's acknowledge that the fact that Foo is a datatype has little to do with what we want to achieve. What is interesting to us is that the Foo value constructor is being *applied* to multiple values, each of them produced by a different Parser.

```
Foo :: Natural -> Bool -> Digit -> Bool -> Foo
```

Instead of Foo, we could have been dealing with a normal function and the problem would still be there. Let's take replicate, for example, the function we learned about a while ago.

```
replicate :: Natural -> x -> [x]
replicate 0 _ = []
replicate n x = x : replicate (n - 1) x
```

This function outputs a list that simply repeats a given value as many times as requested.

```
> replicate 3 'z' :: [Char]
"zzz"
> replicate 0 True :: [Bool]
[]
```

Now, let's say we want to construct a Parser that parses a String like "3z" into "zzz", or "15t" into "ttttttttttttttt". That is, it treats a Natural number followed by a Char as the inputs to replicate. We can easily achieve this by relying on fmap and pair.

```
parseNChar :: Parser String
parseNChar = fmap (\(n, c) -> replicate n c)
                  (pair parseNatural parseChar)
```

Having parseNatural and parseChar readily available, the implementation of this new Parser is quite straightforward. Keep in mind that String is a type synonym for [Char], so replicate 3 'z', say, truly is a String.

```
> runParser parseNChar "3f"
Right (2, "", "fff")
> runParser parseNChar "22g"
Right (3, "", "gggggggggggggggggggggg")
> runParser parseNChar "0s"
Right (2, "", "")
> runParser parseNChar "145"
Left (3, "Not enough input")
```

Easy enough. And, by the way, parseChar does the obvious thing:

```
parseChar :: Parser Char
parseChar = parse1 Right
```

If we compare parseFoo and parseNChar, we'll see they are implemented exactly in the same way. First they pair as many Parsers as they need, and then, using fmap, they modify the output produced by those pairs. It doesn't matter whether the output is modified using a value constructor like Foo or a normal function like replicate, the approach is always the same.

Alright, this works, but can we make it nicer somehow? Can we do something like what <|> did for alt, but for pair? Maybe we can somehow say "use the outputs from these n Parsers as input to this function taking n parameters". Or, more generally, any Monoidal functor f, not just Parser. Yes, that would be nice.

Let's start with a function taking *one* input parameter, not any arbitrary n. That is, given a function a -> z and a value of type f a, we should be able to apply that function to the a trapped in the Monoidal functor f so that f a becomes f z.

```
foo :: Monoidal f => (a -> z) -> f a -> f z
```

Wait, that's just fmap, isn't it?

```
fmap :: Functor f => (a -> z) -> f a -> f z
```

The only difference is that fmap doesn't really require a Monoidal constraint on f. A mere Functor constraint, like the one Monoidal implies, suffices. This makes sense, considering the main reason why Monoidal exists is pair, which allows us to bring *two* functorial values together, something fmap doesn't need to do. We don't really need to have a Monoidal version of fmap. Well, not yet anyway.

What about using a function taking *two* input parameters with fmap? Initially, it would seem we *can't* do that because fmap's first argument must be a function taking just *one* parameter as input. However, isn't it true that *all* functions take just one parameter as input, and that the fact that we talk about a function taking more than one of them is merely a convenience? Indeed. A function of type a -> b -> c, say, is really a function of type a -> (b -> c). We tend to drop those parentheses because they are redundant, but insofar as the type-checker is concerned, the parentheses are always there. So, going back to fmap, if in a -> z we pick b -> c to be our z, then this is the type fmap would get:

```
fmap :: Functor f => (a -> b -> c) -> f a -> f (b -> c)
```

Conceptually, fmapping a function of type a -> b -> c over some f a allows us to partially apply that function to its first argument. The partially applied function, however, now of type b -> c, gets trapped in that same f. This means that we are not able to further fmap it over an f b, for the simple reason that fmap expects a *function* of type b -> c as its first input, not an f (b -> c). fmap can help us apply a function to its *first* argument, but that's as far as it will go. This is why we have ap, too.

```
ap :: Monoidal f => f (b -> c) -> f b -> f c
```

Yes, ap is exactly what we need. fmap allowed us to apply that first a -> b -> c to f a, and ap will help us further apply f (b -> c) to an f b so that we can finally obtain an f c. It's interesting to look at the types of fmap and ap side by side.

```
fmap :: Functor  f =>   (x -> y) -> f x -> f y
ap   :: Monoidal f => f (x -> y) -> f x -> f y
```

The resemblance is striking. Both `fmap` and `ap` ultimately accomplish the same thing, they both apply the function x -> y to the x in f x, returning the resulting y inside yet another f. The only difference between them, really, is that while in `fmap` we are dealing with just *one* functorial value as input, f x, in `ap` we are dealing with *two* of them, f (x -> y) and f x. And that's where the `Monoidal` constraint on f comes from, for we know that in order to compose these two fs, we'll need to pair them.

```
ap :: Monoidal f => f (x -> y) -> f x -> f y
ap fg fx = fmap (\(g, x) -> g x) (pair fg fx)
```

The implementation of `ap` in terms of `pair` is trivial, seeing as we just need to keep doing what we've been doing so far. First, we apply `pair` to fg and fx, which results in a value of type f (x -> y, x), and finally we `fmap` a function applying the first element of that tuple to the second. That's all. Having `fmap` and `ap`, we can redefine our most recent `parseNChar` in this other way:

```
parseNChar :: Parser String
parseNChar = ap (fmap replicate parseNatural) parseChar
```

And notice that, just like in `pair foo bar`, there is an implicit default left-to-right ordering. That is, by default, `foo` will be performed and scrutinized before `bar`. In `ap foo bar` we have the same default implicit ordering. In our concrete example, this means that `parseNatural` will happen *before* `parseChar`, which is exactly what we want.

Is this implementation of `parseNChar` better? Well, it's not immediately obvious that it is, but let's consider what would happen when dealing with a function taking more than two parameters. Say, four, like our value constructor Foo.

```
Foo :: Natural -> Bool -> Digit -> Bool -> Foo
```

If we were to apply `fmap Foo` to a value of type f Natural, we'd get a partially applied Foo *inside* f as result. This is exactly the same thing that happened in our previous example when we `fmap`ped `replicate` over a value of type f Natural.

```
fmap :: Functor f
     => f (Natural -> Bool -> Digit -> Bool -> Foo)
     -> f Natural
     -> f (Bool -> Digit -> Bool -> Foo)
```

Now, the returned function of type Bool -> Digit -> Bool -> Foo, wrapped in f and allegedly taking *three* arguments, is really taking just *one*, the first Bool, and returning another function of type Digit -> Bool -> Foo as output. As usual, adding redundant parentheses highlights this.

```
fmap :: Functor f
     => f (Natural -> (Bool -> (Digit -> (Bool -> Foo))))
     -> f Natural
     -> f (Bool -> (Digit -> (Bool -> Foo)))
```

So, if we used that returned value as the first argument to ap, it would make ap take this type:

```
ap :: Monoidal f
    => f (Bool -> Digit -> Bool -> Foo)
    -> f Bool
    -> f (Digit -> Bool -> Foo)
```

If we think of fmap as allowing us to partially apply a function to its first parameter, then we can think of ap as allowing us to do the same when the function itself is wrapped in some functor f.

The resulting f (Digit -> Bool -> Foo) can now be applied to some f Digit in order to obtain an f (Bool -> Foo) which, when further applied to an f Bool, will finally give us the f Foo we wanted all along.

```
parseFoo :: Parser Foo
parseFoo = ap (ap (ap (fmap Foo parseNatural)
                       parseBool)
                   parseDigit)
               parseBool
```

But wait, isn't this uglier than before? Kinda, even despite that flamboyant vertical alignment. This ugliness comes from the fact that ap is designed to be used as an *infix* operator, not as a function in prefix position. In Haskell, ap is actually called <*>

```
(<*>) :: Monoidal f => f (a -> b) -> f a -> f b
```

And with <*>, this is how we'd write parseFoo:

```
parseFoo :: Parser Foo
parseFoo = fmap Foo parseNatural
                <*> parseBool
                <*> parseDigit
                <*> parseBool
```

Ah, yes, much better. Look at that. We apply Foo to the outputs of successful executions of parseNatural, parseBool, parseDigit and parseBool, in that order, to build a Parser for a value of type Foo.

But there's actually one additional change we can make to make this even nicer. It turns out there is an *infix* version of fmap, too.

```
(<$>) :: Functor f => (a -> b) -> f a -> f b
(<$>) = fmap
```

And with it, we can write parseFoo in an arguably nicer way by replacing fmap Foo parseNatural with Foo <$> parseNatural.

```
parseFoo :: Parser Foo
parseFoo = Foo <$> parseNatural
               <*> parseBool
               <*> parseDigit
               <*> parseBool
```

Look at that. All we read now is Foo and the name of the four Parsers that concern us. The rest of the noise is gone, replaced with <$> and <*>, two different ways of applying functions. Ah, the beauty of designing composition. Here is parseNChar, written in this style:

```
parseNChar :: Parser String
parseNChar = replicate <$> parseNatural <*> parseChar
```

Much better. We call this the *applicative* style of programming.

One hundred eighty-five

We could argue that this *applicative* style is merely syntactic sugar, and in a sense it would be a fair assessment. However, in reality, there's no special syntax going on here. These are just normal functions defined by us being used in infix form. It just so happens that these functions *compose* their inputs in such a beautiful way that we get something sweet and stylish in return.

It's interesting to notice that all of this only works because <*> associates to the left, just like normal function application using juxtaposition, meaning that the implicit parentheses around our expressions are placed in the same way we explicitly placed them in our previous example that used ap.

```
parseFoo :: Parser Foo
parseFoo = ((((Foo <$> parseNatural)
                <*> parseBool)
                <*> parseDigit)
                <*> parseBool)
```

If <*> associated to the right, then none of this would work and we'd need to go back to pair. Which, by the way, reminds us that just like we were able to implement <*> in terms of pair, assuming we had come up with <*> first, we could just as easily have defined pair in terms of <*> instead.

```
pair :: Monoidal f => f a -> f b -> f (a, b)
pair fa fb = (,) <$> fa <*> fb
```

But we can go even further, actually, seeing how <*> and pair only differ in the way they combine the values produced by their functorial inputs.

```
pair  :: Monoidal f => f a        -> f b -> f (a, b)
(<*>) :: Monoidal f => f (x -> y) -> f x -> f y
```

Whereas pair combines the outputs of the given fs by putting them in a tuple, <*> combines them by applying one to the other. That's all the difference there is between them, really. And, well, seeing as that's all the difference, maybe we can generalize both pair and <*> into a single function that takes this combining function as input as well, beside the two functorial values.

```
liftA2 :: Monoidal f => (a -> b -> c) -> f a -> f b -> f c
```

Do you recognize this function? Of course not, we've never seen it before. But do you recognize the *essence* of what this function does? Perhaps if we add some redundant parentheses and align things a bit...

```
liftA2 :: Monoidal f
       => ( a ->   b ->   c)
       -> (f a -> f b -> f c)
```

Prägnanz. liftA2, as its name suggests, *lifts* a function that takes *two* input parameters, a function of type a -> b -> c, into a function of type f a -> f b -> f c. This is not unlike fmap, which lifts a function of type a -> b, a function taking *one* input parameter, into a function of type f a -> f b.

```
fmap   :: Functor f
       => ( a ->   b)
       -> (f a -> f b)

liftA2 :: Monoidal f
       => ( a ->   b ->   c)
       -> (f a -> f b -> f c)
```

With liftA2, we can implement both pair and <*> in a rather straightforward manner by simply *lifting* the appropriate combining function. By lifting (,), of type a -> b -> (a, b), we get pair.

```
pair :: Monoidal f => f a -> f b -> f (a, b)
pair = liftA2 (,)
```

And by lifting \g a -> g a, of type (a -> b) -> a -> b, we get <*>.

```
(<*>) :: Monoidal f => f (a -> b) -> f a -> f b
(<*>) = liftA2 (\g a -> g a)
```

Completely unrelated but fun, <*> can be implemented as liftA2 id, too. Notice how \g a -> g a is merely an eta-expanded form of \g -> g, which is just another way of saying id. Fun.

```
(<*>) :: Monoidal f => f (a -> b) -> f a -> f b
(<*>) = liftA2 id
```

All of pair, <*> and liftA2 can be implemented in terms of each other. For example, here is liftA2 implemented in terms of pair.

```
liftA2 :: Monoical f => (a -> b -> c) -> f a -> f b -> f c
liftA2 g fa fb = fmap (\(a, b) -> g a b) (pair fa fb)
```

And here is liftA2 implemented in terms of <*>.

```
liftA2 :: Monoidal f => (a -> b -> c) -> f a -> f b -> f c
liftA2 g fa fb = g <$> fa <*> fb
```

Why do we have *three* ways of saying the same thing, then? Why do we have pair, <*> and liftA2 if they more or less accomplish the same things and can be defined in terms of each other? Mainly, it has to do with their purpose. <*>, being mostly concerned with *operational* matters, is an excellent tool for writing stylish code, but it doesn't really highlight the relationship between product types and monoidal functors in the way pair does it, which is exactly what makes pair superior for *didactic* purposes. liftA2, on the other hand, highlights that both pair and <*> *choose* to combine two functorial outputs in a certain way, but ultimately, that choice is irrelevant. liftA2 brings together the *side effects* of performing f a and f b, with little regard for how a and b are eventually combined.

Here's another take on parseNChar, this time *lifting* replicate, of type Natural -> x -> [x], to a function of type f Natural -> f x -> f [x] using liftA2, and subsequently applying it to parseNatural and parseChar.

```
parseNChar :: Parser String
parseNChar = liftA2 replicate parseNatural parseChar
```

Is this better? Well, it doesn't matter. It exists, that's the point. And, for completeness, we also have liftA3, liftA4 and many other liftA*n* functions that do what we expect. They are, of course, defined in terms of the more fundamental pair, <*> or liftA2. Here's liftA4, for example.

```
liftA4 :: Monoidal f
       => (  a ->   b ->   c ->   d ->   e)
       -> (f a -> f b -> f c -> f d -> f e)
liftA4 g = \fa fb fc fd ->
  g <$> fa <*> fb <*> fc <*> fd
```

And with it, we could redefine parseFoo as follows:

```
parseFoo :: Parser Foo
parseFoo = liftA4 Foo parseNatural
                      parseBool
                      parseDigit
                      parseBool
```

Is this something we'd want to do? Perhaps not, but the point is that if we *wanted*, we would be able to do so. And no, there is no liftA*n* for an arbitrary *n*. There *could* be, but type inference for functions of an arbitrary number of input parameters is a very tricky business, so for now we'll stick to concrete choices of that *n*.

One hundred eighty-six

So, which is better? pair, <*> or liftA2? We like to think of liftA2 as the most fundamental among these three functions, considering how it defers the choice of how to combine the functorial outputs to its caller. And intuitively, that's fine. Strictly speaking, however, liftA2 and <*> are as fundamental as each other. pair, however, is not. We won't go into details regarding *why* just yet, but suffice it to say that neither pair nor Monoidal nor unit come with Haskell out of the box, and instead we get to define Applicative.

```
class Functor f => Applicative f where
  pure :: a -> f a

  (<*>) :: f (a -> b) -> f a -> f b
  fg <*> fa = liftA2 id

  liftA2 :: (a -> b -> c) -> f a -> f b -> f c
  liftA2 g fa fb = g <$> fa <*> fb

  {-# MINIMAL pure, ((<*>) | liftA2) #-}
```

Applicative, land of the *applicative style*, where the *A* in liftA2 comes from.

Conceptually, monoidal functors exist, and everything we learned about them is true. But in our day-to-day Haskell, we just don't have a typeclass called Monoidal, thus we never use unit nor pair. And once something has been stripped from our vocabulary, it has been stripped from our thoughts as well. In Haskell we use Applicative, pure, <*> and liftA2, and reason about our functors in their terms instead.

According to the MINIMAL pragma, an instance of the Applicative typeclass *must* define at least pure, which is exactly the same pure we encountered before, and either <*> or liftA2, which, just like == and /= in the Eq typeclass, have default implementations in terms of each other. The reason why we get to define *both* <*> and liftA2 is merely one of performance. If for some reason our implementation of liftA2 can be made more performant than the default one in terms of <*>, or vice-versa, then we have the opportunity to do that. But we don't really *need* to. Defining just one of <*> or liftA2 is enough.

Let's implement the Applicative instance for lists.

```
instance Applicative [] where
  pure :: a -> [a]
  pure = \a -> [a]
  (<*>) :: [a -> b] -> [a] -> [b]
  fg <*> fa = concat (fmap (\g -> fmap g fa) fg)
```

Excellent. We've chosen to implement <*> this time, rather than liftA2. The implementation is *almost* the same as it was for pair before. We still combine the elements of the list in a cartesian manner, but rather than combining them with (,), we combine them by applying one to the other.

```
> pure 8
[8]
> (*) <$> [1, 2, 3] <*> [10, 100, 1000]
[10, 100, 1000, 20, 200, 2000, 30, 300, 3000]
> liftA2 (+) [1, 2, 3] [10, 100, 1000]
[11, 101, 1001, 12, 102, 1002, 13, 103, 1003]
```

If we use (,) as our combination function, then we get the same behaviour as we did with pair. But of course, we already knew that.

```
> (,) <$> [1, 2, 3] <*> [10, 100, 1000]
[(1, 10), (1, 100), (1, 1000),
 (2, 10), (2, 100), (2, 1000),
 (3, 10), (3, 100), (3, 1000)]
```

Nothing changed. Still the same as pair.

One hundred eighty-seven

Obviously, there are laws to Applicative. They are, essentially, the monoidal functor laws in disguise. But there are some extra bits too, and generally they look more complex because they lean towards the more operational side of things.

We don't really need to learn these laws by heart, we already have an intuition for what they achieve. Nevertheless, we'll quickly go through them one by one so as to get more comfortable with this *applicative* style.

First, there is a requirement that if both <*> and liftA2 are explicitly defined, which we may want to do for performance reasons, then said definitions must behave the same way as the default ones specified in terms of each other.

Second, there is an *identity* law. There's always an identity law.

```
pure id <*> fa  =  fa
```

Here, id has type a -> a and fa has type f a. This law is analogous to the identity law for Functors, which said that fmap id fa must be equal to fa.

Then, there is a *composition* law.

```
pure (.) <*> fh <*> fg <*> fa  =  fh <*> (fg <*> fa)
```

Here, fh has type b -> c, fg has type a -> b, and fa has type f a. This composition law is analogous to its non-functorial version, which says that (.) h g a, which can also be written as (h . g) a, is equal to h (g a).

As we can see, and as the *applicative* name suggests, these laws are focused on how to deal with *applying* functions wrapped in functors, rather than just pairing them as we did before.

Then there is a *homomorphism* law, which, assuming a function g of type a -> b and a value a of type a, says that there's no difference between wrapping these things in pure before or after their application.

```
pure g <*> pure a  ==  pure (g a)
```

And finally, there is an *interchange* law:

```
fg <*> pure a  ==  pure (\g -> g a) <*> fg
```

This law says that using pure to the left or to the right of <*> is the same. We can *interchange* the position of the arguments to <*> whenever one of them is constructed with pure, and there won't be any surprising consequences.

One hundred eighty-eight

Even though they don't say so explicitly, the Applicative laws imply that fmap g fa will always be equal to pure g <*> fa. This suggests that we don't really need to define fmap when we have already defined pure and <*>. I mean, of course we do, but we can use pure g <*> fa as our fmap implementation, and that will suffice.

As an example, let's create a new datatype and let's make it both a Functor and an Applicative functor. Let's make it the *simplest* possible one.

```
data Identity a = Identity a
```

The simplest indeed. Identity type merely wraps a value of type a in its Identity constructor. It doesn't add any meaning to a.

```
> :type True
True :: Bool
> :type Identity True
Identity True :: Identity Bool
```

Identity is rather boring, yet, it is a Functor. Its fmap merely applies a given function to the value inside the Identity constructor.

```
instance Functor Identity where
  fmap g (Identity a) = Identity (g a)
```

And it is also an Applicative functor where pure simply wraps a given value in Identity, and <*> applies a function inside an Identity to a value inside another Identity almost as if Identity wasn't there.

```
instance Applicative Identity where
  pure = Identity
  Identity g <*> Identity a = Identity (g a)
```

But, seeing how Identity is an Applicative functor, we could have given a much simpler implementation of fmap that doesn't mention Identity at all.

```
instance Functor Identity where
  fmap g fa = pure g <*> fa
```

Granted, it's not immediately obvious that this implementation is simpler than the one before. However, if we consider that this same trick applies to *all* Applicatives, then it becomes a bit more tempting. For example, using this same trick, we could have avoided rolling our own implementation of fmap for Maybe.

```
instance Functor Maybe where
  fmap g fa = pure g <*> fa
```

This works for *any* Applicative functor. For our convenience, this implementation of fmap in terms of pure and <*> comes prepackaged with Haskell, it's called liftA.

```
liftA :: Applicative f => (a -> b) -> f a -> f b
liftA g fa = pure g <*> fa
```

With it, we can simply define fmap to be liftA.

```
instance Functor Maybe where
  fmap = liftA
```

This might feel a bit awkward, seeing how Functor is a superclass of Applicative, implying that a Functor instance for f must exist in order for f's Applicative instance to exist. Yet here we are, relying on features from said Applicative instance in order to define the Functor instance. Awkward, yes, but reasonable. Haskell can deal with mutual recursion just fine. Frankly, this is not unlike any other two mutually recursive functions.

```
even :: Natural -> Bool
even 0 = True
even n = odd (n - 1)

odd :: Natural -> Bool
odd 0 = False
odd n = even (n - 1)
```

Here, even depends on odd, and odd in turn depends on even. Nevertheless, this is fine. Escher would be proud. Or disappointed. After all, both even and odd eventually terminate.

Of course, as in *any* situation involving general recursion, we are exposing ourselves to the possibility of creating a diverging program. If we plan to use liftA as our implementation of fmap, then we should *not* mention that same fmap inside our definitions of pure, <*> nor liftA2, considering how fmap itself would refer to some of these, sending us into an infinite loop. But this is fine, because as we'll see later on, we'll never *need* to mention fmap inside our Applicative instances.

Perhaps surprisingly, if we wanted to implement liftA in terms of pair, we would *not* be able to do so without referring to that same fmap. This, among other things, is why Haskell doesn't use pair as its foundation for Applicative, and why we consider liftA2 and <*> to be strictly superior to pair for *practical* purposes. Ah, real life, you always get your way.

And, by the way, that Identity functor we just implemented? *Very* important.

One hundred eighty-nine

Monoidal is gone, so we need to reimplement some things in terms of Applicative instead. Luckily for us, these things will become much more straightforward as a consequence of this change, seeing how Applicative was designed with ergonomics in mind.

For example, this is how we'd rewrite sequence, the function that requires *all* the functorial values in a given list to succeed, accumulating their outputs from left to right.

```
sequence :: Applicative f => [f a] -> f [a]
sequence = foldr (liftA2 (:)) (pure [])
```

Yes, quite straightforward. Or, maybe not so much yet. As we know, intuitively, foldr replaces occurrences of : with the given binary operation, and [] with the given initial accumulator. Do you see what's happening? sequence, given a list of functorial values, replaces [] with its functorial version pure [], and : with a lifted version of :, of type f a -> f [a] -> f [a]. In other words, sequence is merely making the list constructors : and [] functorial somehow. sequence is beautiful, and it will be even more beautiful later on.

Another example. This is how we'd implement optional, taking both Applicative and Alternative into account:

```
optional :: Alternative f => f a -> f (Maybe a)
optional fa = fmap Just fa <|> pure Nothing
```

Excellent. And we'll need to rewrite some, too. Which, actually, we can do without even mentioning optional this time, seeing how the implementation is now even simpler without it.

```
some :: Alternative f => f a -> f [a]
some fa = liftA2 (:) fa (some fa <|> pure [])
```

Fascinating. The relationship between Applicative and Alternative is rather interesting. In Haskell, surprisingly I hope, Applicative is a superclass of the Alternative typeclass, which looks more or less like this:

```
class Applicative f => Alternative f where
  empty :: f a
  (<|>) :: f a -> f a -> f a
```

This is rather disappointing. Strictly speaking, there is no reason for Applicative to be a superclass of Alternative. Yet, history has somehow brought us here. Applicative is a superclass of Alternative mostly just because.

An Alternative is said to be "a monoid on Applicative functors", and from this colloquial definition follows that, indeed, every Alternative functor must be an Applicative functor too. We know, though, that this is not really necessary. We defined Alternative and Monoidalt before without mentioning either Applicative nor Monoidal, just Functor, and it was fine. Moreover, if we dare search for the Alternative laws, we'll find almost none, which highlights even more how the concerns of this typeclass are mostly of an operational nature, and not so much about maths. Nevertheless, for most practical purposes, we can reason about Alternative as we'd reason about sum types, *this or that*, and we'll be alright.

For us, this is mostly an exercise in thickening our skin. We'll encounter many lawless and uncomfortable situations like this one going forward, and there's a lot of value in learning how to muddle through in spite of them, aware of them.

On a positive note, an Alternative f constraint being satisfied implies an Applicative f constraint being satisfied too, which at least will help us reduce the number of constraints we have to write down when using features coming from both Alternative and Applicative. Like when implementing optional, say.

One hundred ninety

We haven't really implemented any Applicative instances beside the one for lists. We should do so. Or, should we?

```
(<*>) :: [a -> b] -> [a] -> [b]
fg <*> fa = concat (fmap (\g -> fmap g fa) fg)
```

Mind you, when talking about liftA a while ago, we said that we *shouldn't* be using fmap inside our definition of <*> if we wanted to be able to use liftA as the definition for fmap in order to avoid duplicating our work. Yet here we are, using fmap. Don't worry, we'll fix this eventually.

Anyway, if we look hard at this definition of <*>, we'll see that the *only* thing in there that talks about lists is concat, of type [[b]] -> [b]. The rest of this definition relies solely on fmap and function application, suggesting that in principle this could work with *any* Functor. Could it? Should it? Well, let's see.

First, let's abstract away concat, let's take it in as an input parameter which we'll call join, reminding us that concat joins many lists into a single one.

```
mkAp :: ([[b]] -> [b]) -> ([a -> b] -> [a] -> [b])
mkAp join = \fg fa -> join (fmap (\g -> fmap g fa) fg)
```

By partially applying this mkAp to a suitable joining function like concat, we will get a suitable implementation of <*> in return, a function also known as ap. We would be "making an ap", that's where the funny name comes from. The mk prefix is a rather common convention for this sort of thing.

```
(<*>) :: [a -> b] -> [a] -> [b]
(<*>) = mkAp concat
```

Boring so far? Boring so far. But look, here comes the fun part. Seeing how mkAp doesn't talk about lists anymore, but about Functors, its type can be made more polymorphic.

```
mkAp :: Functor f
     => (f (f b) -> f b)
     -> (f (a -> b) -> f a -> f b)
```

All that seems to matter is that f be a Functor, and that we be able to provide a joining function of type f (f b) -> f b. In the case of lists, concat of type [[b]] -> [b] fits this shape, so everything clicks together and we get a working implementation of <*> in return.

But what about other Functors? Maybe, for example, can be seen as a list of zero or one elements. Could mkAp be used to implement <*> for Maybe? Maybe. We'd need to find a joining function for Maybes, though. Perhaps we can get inspiration from concat. Let's remind ourselves of its definition.

```
concat :: [[x]] -> [x]
concat = \xss -> case xss of
                   [] -> []
                   xs : xss' ->
                     case xs of
                       [] -> concat xss'
                       x : xs' -> x : concat (xs' : xss')
```

This code is exaggeratedly verbose, we don't often write code like this, but we are doing it here because it showcases the exact *order* in which the execution of concat happens, which is what concerns us at the moment.

```
> concat [[], [2, 3], [], [4], [5, 6, 7], []]
[2, 3, 4, 5, 6, 7]
```

concat takes a list of lists of values of type x and joins them all together into a single list of values of type x. It achieves this by *first* inspecting the *outer* list, here called xss, and *then*, *if* that list is not empty, it proceeds to do whatever it needs to do with the leftmost *inner* list, here called xs, *before* recursively calling itself to do any *further* work that may be necessary. *First, outer, then, if, inner, before, further.* These are words conveying order, time, decision, consequence. concat observes the outermost list somehow, and depending on what it encounters, decides whether and how to proceed with the inner list.

Order is important because, if we recall, when we say fg <*> fa, unless somebody explicitly asks otherwise, we want fg to be evaluated before fa. And indeed, this is what we see in mkAp.

```
mkAp :: Functor f
    => (f (f b) -> f b)
    -> (f (a -> b) -> f a -> f b)
mkAp join = \fg fa -> join (fmap (\g -> fmap g fa) fg)
```

By fmapping over fg first, we make its f, rather than fa's, be the outermost f
that join will encounter. Why does it matter? In the case of lists, when
pairing [1, 2] and [3, 4], say, this ordering means that we get [(1, 3), (1,
4), (2, 3), (2, 4)] as result, rather than [(1, 3), (2, 3), (1, 4), (2, 4)].
This may not seem too important here, but if we consider other functors
like Parser, where pair fa fb most certainly does *not* mean the same as pair
fb fa, seeing how the order of these parameters will determine which of fa
and fb consumes input first, then we must acknowledge that order can be
important, and unless asked to do otherwise, we must respect it.

Wait, are we saying mkAp could work for Parsers too? I'm glad to see us
paying attention. Perhaps. But in any case, let's try Maybe first.

Seeing how Maybe is essentially a list of zero or one elements, we should be
able to replicate what concat does, but for Maybes rather than lists. Let's call
our joining function joinMaybe.

```
joinMaybe :: Maybe (Maybe x) -> Maybe x
joinMaybe (Just a) = a
joinMaybe Nothing  = Nothing
```

That was easy. First we scrutinize the outer Maybe, and if it's Just, we
simply output the Maybe x payload in that Just. Otherwise, Nothing. In
concat we further scrutinized the inner list because the recursive nature of
lists forced us to do so, but this time it is not necessary and we can simply
return the value of type Maybe x that we already have at hand. Yet, despite
this simplification, in joinMaybe we can still see the ideas of time and
dependence, in the sense that we wouldn't be able to do anything with the
inner Maybe until *after* we had judged the outer one to be Just.

```
> joinMaybe Nothing
Nothing
> joinMaybe (Just Nothing)
Nothing
> joinMaybe (Just (Just 4))
Just 4
```

The Maybe x returned by joinMaybe will contain an x only as long as both of
the Maybe wrappers are Just. Otherwise, we'll get Nothing. joinMaybe
effectively joins, or flattens, two Maybes into one.

Intuitively, we can think of Maybe x as being a closed box potentially
containing some x. A physical box. So, if we have a Maybe (Maybe x), then it
means we have a box potentially containing yet another box inside, in turn
potentially containing an x. In order to see if there is an x at all, first we

must open the outer box, and only then we can proceed to open the inner box, if any, to see what's in there. joinMaybe corresponds to the idea of opening the *outer* box.

Generally, boxy and containery analogies don't make sense for most functors, not even for lists. But for Maybe, this analogy is rather beautiful. It conveys the idea of *time*, by highlighting that we must open the outer box *before* the inner box, and the idea of *dependence*, in the sense that whether we can open the inner box at all *depends* on what we find inside the outer box.

And, yes, by applying mkAp to joinMaybe, we obtain a suitable implementation of <*> that we can use when defining Maybe's Applicative instance.

```
instance Applicative Maybe where
  pure = Just
  (<*>) = mkAp joinMaybe
```

Does it work? You bet it does.

```
> (+) <$> Just 1 <*> Just 4
Just 5
> (+) <$> Just 1 <*> Nothing
Nothing
> (+) <$> Nothing <*> Just 4
Nothing
> (+) <$> Nothing <*> Nothing
Nothing
```

And, not that we are explicitly paying attention to it, but laziness is preserved too. A second Maybe will never be scrutinized unless strictly necessary.

```
> (+) <$> Nothing <*> undefined
Nothing
```

Great. No surprises. So, what does all of this mean? Are mkAp and these *joining* functions special somehow? Take a guess...

One hundred ninety-one

Any Functor for which we can define pure and this *joining* function is an Applicative, yes, but it's also something else. We'll see.

We've joined lists, we've joined Maybes, and now it's time we joined Parsers.

```
joinParser :: Parser (Parser a) -> Parser a
joinParser ppa = Parser (\s0 ->
  case runParser ppa s0 of
    Left e -> Left e
    Right (!n0, s1, pa) ->
      case runParser pa s1 of
        Left (!n1, msg) -> Left (n0 + n1, msg)
        Right (!n1, s2, a) -> Right (n0 + n1, s2, a))
```

Despite the different type, this implementation of joinParser is pretty much the same as our most recent implementation of pair for Parser. The only thing that changed, really, is that whereas pair received the two Parsers to compose as separate input parameters, joinParser, as strange as it may seem, receives one of them *inside* the other one. Or rather, in functorial terms, it receives a Parser *producing* another one. Functors produce, that's key. But why would we ever have a Parser producing another one? Well, we'd better ask ourselves *why not*.

Consider pure, the function that allows us to put *any* value of our choosing into some Applicative functor. Into a Parser, say.

```
pure :: ∀ x. x -> Parser x
```

Any value. *For all* x. So what if we applied pure to False to construct a Parser Bool, say, and then we applied pure once again to that result? Could we do that? Sure, why not.

```
pure (pure False) :: Parser (Parser Bool)
```

There it is. The double Parser. Yes, it is a rather silly Parser, but it *is*, and we can *join* it, that's the point.

```
> runParser (joinParser (pure (pure False))) "hello"
Right (0, "hello", False)
```

joinParser (pure (pure False)) doesn't achieve anything that pure False wouldn't accomplish on its own, but this is mostly a consequence of using pure to construct our Parsers rather than involving Parsers capable of more interesting things.

```
> runParser (pure False) "hello"
Right (0, "hello", False)
```

Thinking about one Parser being *inside* another doesn't really work for the same reason that thinking about Functors as containers doesn't work. They are generally *not* containers, even though some, like Maybe, can be considered one. When we think about Functors, we must think of them being able to *produce* something. That's what a type like Parser (Parser x) should convey to us. A Parser will do its job and, if successful, will produce another Parser as result. It's not really about one of them being inside the other, but about one of them *depending* on the successful execution of a *previous* one.

And yes, just like we did it for lists and Maybe, we can implement `<*>` for Parser by applying `mkAp` to `joinParser`.

```
(<*>) :: Parser (a -> b) -> Parser a -> Parser b
(<*>) = mkAp joinParser
```

Does this mean that `<*>` and `joinParser` are equivalent? No, not at all. Sure, in our example, we could have accomplished the same as `joinParser` (pure (pure False)) did, say, using `pure id <*> pure False`. But this mostly has to do with pure's rather moot achievements. However, we can do unboring things too. You see, we've opened a door to somewhere fascinating. Come.

One hundred ninety-two

Let's implement a new Parser, our fanciest one yet. Let's implement parseCharTwice, a Parser that parses a *same* character twice, producing the Char in question as result.

```
> runParser parseCharTwice "aa"
Right (2, "", 'a')
> runParser parseCharTwice "b"
Left (1, "Not enough input")
> runParser parseCharTwice "cd"
Left (1, "Expected 'c' got 'd'")
> runParser parseCharTwice "xxxyz"
Right (2, "xyz", 'x')
```

We *could* implement this Parser using Applicative and Alternative vocabulary by trying to parse "aa" first, trying "bb" if that doesn't work, then trying "cc" if that doesn't work, etc.

```
parseCharTwice :: Parser Char
parseCharTwice =
  liftA2 const (parseChar 'a') (parseChar 'a') <|>
  liftA2 const (parseChar 'b') (parseChar 'b') <|>
  liftA2 const (parseChar 'c') (parseChar 'c') <|>
  ...
```

However, as you can probably imagine, trying these many possibilities one after the other is not very efficient. Particularly if we consider that Char's cardinality says there are 1114112 distinct Chars we'd have to take into account. Unfortunately, this is the best we can do if we are limited to the Applicative and Alternative vocabulary. But if we extend our vocabulary with joinParser, then we can do better.

```
parseCharTwice :: Parser Char
parseCharTwice =
  joinParser (fmap (\c -> fmap (const c) (expect1 c))
                   parseChar)
```

It's not the most elegant code, but it gets the work done just fine.

```
> runParser parseCharTwice "xxz"
Right (2, "z", 'x')
> runParser parseCharTwice "xyz"
Left (1, "Expected 'x' got 'y'")
```

parseChar, of type Parser Char, will successfully parse *any* Char leading the input being parsed. parseCharTwice achieves its goal by providing parseChar's output to expect1, a function of type Char -> Parser Unit where the resulting Parser *expects* the given Char to lead the input being parsed. This is easily achieved using fmap.

```
fmap expect1 parseChar :: Parser (Parser Unit)
```

Good. There are two problems with this expression, though. The first one is that we have a Parser (Parser x), rather than just a Parser x. Well, that's why we have joinParser, isn't it?

```
joinParser (fmap expect1 parseChar) :: Parser Unit
```

At this point, we have created a Parser that accomplishes what we wanted, at least regarding the successful consumption of input.

```
> runParser (joinParser (fmap expect1 parseChar)) "xxz"
Right (2, "z", Unit)
> runParser (joinParser (fmap expect1 parseChar)) "xyz"
Left (1, "Expected 'x' got 'y'")
```

However, there is an additional problem. We were really looking forward to getting that repeated Char as output from this new Parser, yet all we got was Unit. Worry not, this can be easily fixed by fmapping \c -> fmap (const c) (expect1 c) over parseChar, rather than just expect1. This replaces expect1's Unit output with c.

Thus concludes our implementation of parseCharTwice, where the existence of that second parser created with expect1 *depends* on the output from a *previous* successful execution of another Parser, namely parseChar. It was joinParser who made it possible.

One hundred ninety-three

Strictly speaking, as we said before, we could have implemented parseCharTwice in a *very* inefficient manner without using joinParser, by relying solely on Parser's Applicative and Alternative vocabulary. Generally, this is true for any Parser, because if we think about it, we can *always* try to guess what the entire input to our Parser will be, and occasionally we will be right. It's a bit like those folks playing the lottery time and time again just in case. However, seeing as there are infinitely many Strings we could provide as input to our Parsers, and *many* ways in which we could successfully parse them, hopefully we all agree that doing this is insane and often practically unfeasible.

So, in part, that's why we have joinParser. Many times, having a Parser whose whole existence *depends* on a *previous* Parser's output, like we saw in parseCharTwice, makes sense for practical reasons.

But more importantly, as we'll see later on for Functors other than Parser, a joining function may be necessary for *fundamental* reasons, and not just as a practical convenience. This is not the case with Parser because we can always try and guess what the input is, even if that takes virtually forever and we die before making the right guess. But hey, it's our life, our death, our choice.

And anyway, liftA2 and <*> are implemented in terms of joinParser, so we gain nothing by avoiding it.

One hundred ninety-four

Let's formalize things a bit. If for a particular Applicative functor, the idea of a functorial value *depending* on a *previous* functorial *output* makes sense, then we say this functor is a Monad, too.

```
class Applicative f => Monad f where
  join :: f (f a) -> f a
```

All Monads are Applicatives. This is why we see an Applicative superclass here. Not all Applicatives are Monads, though, and we'll see some examples of that soon. For now, let's write some Monad instances.

```
instance Monad Parser where
  join = joinParser

instance Monad Maybe where
  join = joinMaybe

instance Monad [] where
  join = concat
```

What? Were you expecting something new? Not this time. We know what join does, we implemented it multiple times before. All that remains, really, is to get comfortable with it. Let's start working towards that by looking at the Monad laws.

First, we have an *identity* law saying that fx, join (pure fx) and join (fmap pure fx) all mean the same. And indeed, we saw some of this when we toyed around with pure and joinParser before.

```
> runParser parseChar "hello"
Right (1, "ello", 'h')
> runParser (join (pure parseChar)) "hello"
Right (1, "ello", 'h')
> runParser (join (fmap pure parseChar)) "hello"
Right (1, "ello", 'h')
```

See? All the same. This law is often expressed in a nicer looking point-free manner:

```
join . fmap pure  ==  join . pure  ==  id
```

Second, we have an *associativity* law:

```
join . fmap join  ==  join . join
```

This law doesn't look like the associativity laws we are used to seeing because join is not a binary operator. However, if we look hard enough and assume a value named fffx of type f (f (f x)), then it should be easier to see. We are saying that there's no difference between first joining the two outer f layers and then the inner one, using join (join fffx), or first joining the two inner layers and then the outer one, using join (fmap join fffx). It's all the same.

Finally, there is a *naturality* law saying that there's no difference between applying a function a -> b to the a in f (f a) before or after joining those f layers into one.

```
join . fmap (fmap f)  ==  fmap f . join
```

And, by the way, did you notice how these laws mention pure, fmap and join, but not liftA2 nor <*>, even though Applicative is a superclass of Monad, suggesting that perhaps we should be worrying about it? Well done, you are paying attention. We'll come back to this later on.

One hundred ninety-five

Let's implement another fancy Parser, fancier than parseCharTwice, the fanciest one yet. This Parser shall be one able to obtain the Natural number 3910, say, from the input "43910", where that first digit, '4', states how many of the following characters should be considered part of the Natural number we are trying to parse. No less, no more. We'll implement it using join, and we'll call it parseSizedNatural.

```
> runParser parseSizedNatural "43910"
Right (5, "", 3910)
> runParser parseSizedNatural "3391"
Right (4, "", 391)
> runParser parseSizedNatural "0"
Right (1, "", 0)
> runParser parseSizedNatural "85"
Left (2, "Not enough input")
```

The most interesting examples, actually, are the ones where the leading digit asks for *less* digits than the ones available in the input:

```
> runParser parseSizedNatural "43910111111"
Right (5, "111111", 3910)
> runParser parseSizedNatural "13910111111"
Right (2, "910111111", 3)
```

Had we used parseNatural, rather than parseSizedNatural, all of the digits
fed to the Parser would have been consumed.

```
> runParser parseNatural "43910111111"
Right (11, "", 43910111111)
```

But parseSizedNatural is different, its behaviour really *depends* on what
that first digit is. This is particularly noticeable when the leading character
is '0', meaning that no parsing beyond said first character is required at all.

```
> runParser parseSizedNatural "03910111111"
Right (1, "3910111111", 0)
```

How do we implement parseSizedNatural? Step by step, as usual, correctly
solving the small problems first.

Small problem number one: Continuing with our "43910" input example,
we need to parse that leading character '4' into the Natural number 4.
Well, that's just fmap naturalFromDigit parseDigit, isn't it?

```
> :t fmap naturalFromDigit parseDigit
fmap naturalFromDigit parseDigit :: Parser Natural
> runParser (fmap naturalFromDigit parseDigit) "43910"
Right (1, "3910", 4)
```

Indeed. That was quick. Now, on to small problem number two. We'll
need a function that, given a Natural number, will parse *that* many Digits
from the input String, returning them listed in order of appearance. In
our "43910" example, this would be the function that given the Natural
number 4, obtained by parsing that first '4', will parse the following *four*
characters "3910" into the Digits [D3, D9, D1, D0].

```
parseSomeDigits :: Natural -> Parser [Digit]
parseSomeDigits n = sequence (replicate n parseDigit)
```

That was easy. replicate n parseDigit gives a value of type [Parser Digit],
and sequence transforms that into Parser [Digit].

```
> runParser (parseSomeDigits 3) "12345"
Right (3, "45", [D1, D2, D3])
> runParser (parseSomeDigits 0) "12345"
Right (0, "12345", [])
> runParser (parseSomeDigits 3) ""
Left (0, "Not enough input")
```

Actually, performing sequence after replicate is such a common thing to
do that there's a handy function called replicateM doing this for us.

```
replicateM :: Applicative f => Natural -> f a -> f [a]
replicateM n fa = sequence (replicate n fa)
```

Great. We now have parseSomeDigits, a *function* of type Natural -> Parser [Digit], and we have fmap naturalFromDigit parseDigit, a *functorial value* of type Parse Natural. At this point, all we can do with them is fmap one over the other, so that's what we do.

```
fmap parseSomeDigits (fmap naturalFromDigit parseDigit)
    :: Parser (Parser [Digit])
```

Look at that. A Parser producing yet another Parser producing a list of Digits. Quick, let's do what type-checks, let's apply join to this beast and see what happens.

```
join (fmap parseSomeDigits (fmap naturalFromDigit parseDigit))
    :: Parser [Digit]
```

Fascinating. But, actually, we wanted parseSizedNatural to be a Parser Natural, didn't we? That's easy to fix, all we have to do is apply naturalFromDigits, of type [Digit] -> Natural, to the [Digit] output of this entire thing using fmap.

```
parseSizedNatural :: Parser Natural
parseSizedNatural =
    fmap naturalFromDigits
        (join (fmap parseSomeDigits
                    (fmap naturalFromDigit parseDigit)))
```

This got a bit ugly. Don't worry, we'll clean it up soon enough. The important question to ask ourselves is whether this works. And indeed, it does.

```
> runParser parseSizedNatural ""
Left (0, "Not enough input")
> runParser parseSizedNatural "04"
Right (1, "4", 0)
> runParser parseSizedNatural "10"
Right (2, "", 0)
> runParser parseSizedNatural "15"
Right (2, "", 5)
> runParser parseSizedNatural "31"
Left (2, "Not enough input")
> runParser parseSizedNatural "43910"
Right (5, "", 3910)
> runParser parseSizedNatural "2528973"
Right (3, "8973", 52)
```

Having a Parser's whole existence *depend* on a *previous* Parser's output just can't be done with fmap, pure, <*> and liftA2 alone. We need join too. We saw that in parseCharTwice before, and we see it again here in parseSizedNatural where expect1 c *depends* on that c, the output of a *previous* Parser.

One hundred ninety-six

Are we comfortable with Monad and join already? Most certainly not. Perhaps cleaning up parseSizedNatural, using the bits of knowledge we acquired so far, will help.

```
parseSizedNatural :: Parser Natural
parseSizedNatural =
  fmap naturalFromDigits
      (join (fmap parseSomeDigits
                  (fmap naturalFromDigit parseDigit)))
```

Relying on the Monad laws, we'll rearrange our expressions so that they look better. We'll chase beauty. We'll *always* chase beauty, for beauty is often a symptom of something better. In programming, anyway.

First, there is a naturality law that says that whether we fmap before or after joining, it's all the same. We can use this to change the place where we use fmap naturalFromDigits.

```
parseSizedNatural :: Parser Natural
parseSizedNatural =
  join (fmap (fmap naturalFromDigits)
            (fmap parseSomeDigits
                  (fmap naturalFromDigit parseDigit)))
```

Second, according to the Functor laws, which are implied by the Monad laws, fmap g . fmap f is the same as fmap (g . f). Using this law, we can bring fmap naturalFromDigits and parseSomeDigits closer together.

```
parseSizedNatural :: Parser Natural
parseSizedNatural =
  join (fmap (fmap naturalFromDigits . parseSomeDigits)
            (fmap naturalFromDigit parseDigit))
```

Notice how these laws allow us to move terms around without worrying at all about what they are trying to accomplish. For all we care, these terms we are dealing with could be called anything else, be doing anything else, and our refactoring of this code would have been equally correct.

Our implementation of parseSizedNatural is still not ideal, but compared with the previous two versions, we can more easily make sense of what's going on. It certainly looks more neatly arranged.

That join (fmap ... situation is interesting. We also encountered it before when implementing parseCharTwice.

```
parseCharTwice :: Parser Char
parseCharTwice = join (fmap (\c -> fmap (const c) (expect1 c))
                            parseChar)
```

This seems to be a recurring theme. And it makes sense, because if we want a functorial value f y to *depend* on the output of a previous

functorial value f x, then we'll obviously need to fmap a function of type x
-> f y over f x. This is what the outermost fmap in both parseCharTwice
and parseSizedNatural is doing. However, doing this results in a value of
type f (f y), rather than just f y, which is why we must join afterwards.
Yes, this is something we'll need to do every time a functorial value
depends on the output of another one. First we fmap, then we join. This is
such a common thing to do, actually, that in Haskell we have a name for
it.

```
bind :: Monad f => f x -> (x -> f y) -> f y
bind fx g = join (fmap g fx)
```

Look, naming things is hard. And all things considered, once we develop
an intuition for it, we'll see that bind is quite a decent name. The idea is
that the x output from f x gets *bound* to the input of x -> f y. And within
said function, we can do whatever we need to do with x in order to return
an f y.

Let's use bind in parseSizedNatural. All we have to do is mechanically
replace all occurrences of join (fmap *this that*) with bind *that this*.

```
parseSizedNatural :: Parser Natural
parseSizedNatural =
  bind (fmap naturalFromDigit parseDigit)
       (fmap naturalFromDigits . parseSomeDigits)
```

That's better. Notice how the two input parameters of bind are in the
opposite direction than those of fmap before. This is a welcome change,
actually, seeing how we can read things from left to right and top to
bottom now, in the same order in which these Parsers are executed, the
same order in which the input is consumed. Yes, this is an improvement.
Would bind improve parseCharTwice, too?

```
parseCharTwice :: Parser Char
parseCharTwice = bind parseChar (\c -> fmap (const c) (expect1 c))
```

Yes, much nicer. parseChar runs first, and then its output Char is bound to
the name c in the accompanying function, which determines that expect1
c must run next. Clearly, expect1 c *depends* on c, the output of the
previously performed parseChar.

One hundred ninety-seven

Here's a more complex example. Let's write a Parser that, given an input
with the pattern *xxyzyxz*, where *x*, *y*, and *z* could be *any* character, will
return said characters *x*, *y* and *z*. That is, we want to implement a Parser
(Char, Char, Char), let's call it parseXYZ, that given "aabcbac" as input, say,
produces ('a', 'b', 'c'), that given "1123213" produces ('1', '2', '3'),
etc. This should be fun.

Dealing with the first two characters in the pattern *xxyzyxz* will be very easy for us, since it is essentially parseCharTwice, which we implemented only pages ago. Let's start with that.

```
parseXYZ :: Parser (Char, Char, Char)
parseXYZ = bind parseCharTwice (\x -> ...
```

Great. If parseCharTwice succeeds, its output will be bound to the name x. Next in our pattern *xxyzyxz* comes yet another unknown character. Let's uncover it using parseChar, binding it to the name y.

```
parseXYZ :: Parser (Char, Char, Char)
parseXYZ = bind parseCharTwice (\x ->
              bind parseChar (\y -> ...
```

Then comes yet another unknown character, which we'll call z.

```
parseXYZ :: Parser (Char, Char, Char)
parseXYZ = bind parseCharTwice (\x ->
              bind parseChar (\y ->
                bind parseChar (\z -> ...
```

Finally, we expect y to appear again, then x, then z. Simply applying expect to these three Chars, in that order, will do the job. However, there's one additional detail we need to take care of. expect [y, x, z], if successful, produces Unit as output, not (x, y, z) as we want, so we must modify its output using fmap too.

```
parseXYZ :: Parser (Char, Char, Char)
parseXYZ = bind parseCharTwice (\x ->
              bind parseChar (\y ->
                bind parseChar (\z ->
                  fmap (const (x, y, z))
                       (expect [y, x, z]))))
```

Thus concludes our implementation of parseXYZ, which, of course, works as expected.

```
> runParser parseXYZ "xxyzyxz"
Right (7, "", ('x', 'y', 'z'))
> runParser parseXYZ "xxyzyxa"
Left (6, "expected 'z' got 'a'")
> runParser parseXYZ "xxyzybz"
Left (5, "expected 'x' got 'b'")
> runParser parseXYZ "ooxgxog"
Right (7, "", ('o', 'x', 'g'))
```

Our implementation of parseXYZ is quite straightforward, but we can do even better. In Haskell, for ergonomic reasons that will become apparent soon, the bind function is not really called bind. Instead, intended to be used as an *infix* operator, it goes by the name of >>=.

```
(>>=) :: Monad f => f x -> (x -> f y) -> f y
(>>=) = bind
```

Nothing changed regarding the implementation of this function, only its name. Rather than saying bind uno dos, we must now say uno >>= dos. Is >>= better than bind? Let's see how parseXYZ changes with it.

```
parseXYZ :: Parser (Char, Char, Char)
parseXYZ = parseCharTwice >>= (\x ->
              parseChar >>= (\y ->
                parseChar >>= (\z ->
                  fmap (const (x, y, z))
                       (expect [y, x, z])))))
```

No, it's not immediately obvious that this is better. Perhaps we should shake things up a bit more. First, seeing as we have so many parentheses stating quite clearly how these expressions associate together, maybe we can avoid the ever increasing indentation altogether by letting those parentheses do their work.

```
parseXYZ :: Parser (Char, Char, Char)
parseXYZ = parseCharTwice >>= (\x ->
            parseChar >>= (\y ->
            parseChar >>= (\z ->
            fmap (const (x, y, z))
                 (expect [y, x, z])))))
```

We know, though, that the syntax rules of Haskell say that a lambda expression starts with a \ and extends *all the way* to the end of the expression, unless explicitly limited by a parentheses somehow. So, in fact, we can just drop those parentheses entirely.

```
parseXYZ :: Parser (Char, Char, Char)
parseXYZ = parseCharTwice >>= \x ->
            parseChar >>= \y ->
            parseChar >>= \z ->
            fmap (const (x, y, z))
                 (expect [y, x, z])
```

With this arrangement, we can see more clearly now, from top to bottom, the individual Parsers being composed in the same order in which they are executed. Well, except for that last expect, which shows up in a rather weird place due to our usage of fmap. Worry not, we can fix that.

A while ago we learned that given a functorial value fa of type f a and a function g of type a -> b, rather than using fmap for applying g over fa as fmap g fa, we could use pure g <*> fa and achieve the same. Well, we can pull a similar trick using just >>= and pure.

```
fmap g fa  =  fa >>= \a -> pure (g a)
```

Relying on this equality, let's replace our fmap in parseXYZ with uses of >>= and pure.

```
parseXYZ :: Parser (Char, Char, Char)
parseXYZ = parseCharTwice >>= \x ->
           parseChar >>= \y ->
           parseChar >>= \z ->
           expect [y, x, z] >>= \Unit ->
           pure (const (x, y, z) Unit)
```

const (x, y, z) Unit is the same as (x, y, z), though, isn't it? Right, so let's drop that const nonsense.

```
parseXYZ :: Parser (Char, Char, Char)
parseXYZ = parseCharTwice >>= \x ->
           parseChar >>= \y ->
           parseChar >>= \z ->
           expect [y, x, z] >>= \Unit ->
           pure (x, y, z)
```

And, actually, we are not really using that Unit input anymore, so let's ignore it altogether using _.

```
parseXYZ :: Parser (Char, Char, Char)
parseXYZ = parseCharTwice >>= \x ->
           parseChar >>= \y ->
           parseChar >>= \z ->
           expect [y, x, z] >>= \_ ->
           pure (x, y, z)
```

We have to admit that, from an ergonomic perspective, this is much better. Moreover, those pointy arrowy thingies really convey a sense of direction, of order. uno >>= dos >>= tres, lovely.

And finally, one last touch. Seeing as we are *not* using the Unit output from expect [y, x, z] anymore, we can use >> instead of >>= to avoid some of the noise.

```
(>>) :: Monad f => f a -> f b -> f b
fa >> fb = fa >>= \_ -> fb
```

That is, fa >> fb establishes an execution order between the functorial values fa and fb, but unlike in fx >>= g, the second functorial value does *not* depend on the functorial output from the first. We can see in our implementation that we are clearly ignoring the output produced by fa, returning fb right away.

Strictly speaking, the implementation of >> doesn't even need to use >>=, seeing how it's only relying on a functor's Applicative capabilities, rather than its Monadic ones. We could implement >> in terms of liftA2 instead, say.

```
(>>) :: Applicative f => f a -> f b -> f b
(>>) = liftA2 (\_ b -> b)
```

However, because of unfortunate historical reasons, >> actually imposes a Monad constraint on f, rather than an Applicative one. Believe it or not,

Applicative functors are a relatively recent discovery. Certainly more recent than Monads. So, for this reason, Haskell's oldest vocabulary sometimes talks about Monadic functors in situations where a mere Applicative functor would suffice. Anyway, not important. The important thing is that, seeing as we are not using expect's Unit output to determine what the next Parser should be, we can just replace >>= with >> in that one case.

```
parseXYZ :: Parser (Char, Char, Char)
parseXYZ = parseCharTwice >>= \x ->
           parseChar >>= \y ->
           parseChar >>= \z ->
           expect [y, x, z] >>
           pure (x, y, z)
```

The noise is gone. What remains is what we truly meant to say. First we run parseCharTwice, then we run parseChar, then another parseChar, then we expect the previously discovered Chars to appear once again, and finally we purely produce said three Chars in order of appearance. Isn't it lovely?

One hundred ninety-eight

Let's also modify parseSizedNatural so that it uses >>=.

```
parseSizedNatural :: Parser Natural
parseSizedNatural =
  fmap naturalFromDigit parseDigit >>=
  (fmap naturalFromDigits . parseSomeDigits)
```

Technically speaking, due to the operator precedence rules between . and >>=, which we refuse to take into account so as to preserve our sanity, we *could* drop the parentheses around fmap naturalFromDigits . parseSomeDigits too, and everything would still behave as if the parentheses had been there.

```
parseSizedNatural :: Parser Natural
parseSizedNatural =
  fmap naturalFromDigit parseDigit >>= fmap naturalFromDigits .
  parseSomeDigits
```

Or would it? It doesn't matter, because we'd never dare burden ourselves, nor others, with having to figure out where the implicit grouping of expressions happens. So we'll put the parentheses back in, and never again mix strange infix operators as part of a same literal expression without extra explicit parentheses, whether necessary or not.

```
parseSizedNatural :: Parser Natural
parseSizedNatural =
  fmap naturalFromDigit parseDigit >>=
  (fmap naturalFromDigits . parseSomeDigits)
```

Much better. Are we done? Is parseSizedNatural as handsome as it could be? Maybe. Maybe not.

Ever since we started our parseSizedNatural ramblings, we've been using fmap naturalFromDigits . parseSomeDigits, our function of type Natural -> Parser Natural, in a point-free manner. That is, rather than writing \x -> g (f x), say, giving x a lead role in our function, we've been writing g . f, avoiding any mention of x, and only focusing on how g and f compose. Technically, there's nothing wrong with that. After all, composition is what programming is all about. However, as the number of expressions we are dealing with as part of a same literal expression grows, we may benefit from switching from a point-free style to a more explicit style where we name our inputs. By doing this, we'll often discover interesting opportunities for reorganizing our code that, perhaps, we wouldn't have thought of otherwise. First step, eta-expansion.

```
parseSizedNatural :: Parser Natural
parseSizedNatural =
  fmap naturalFromDigit parseDigit >>= \n0 ->
  fmap naturalFromDigits (parseSomeDigits n0)
```

Good. Now, we seem to be fmapping a function over parseDigit in order to convert its Digital output to a Natural number. That's fine and necessary, but, as implied by the Monad laws, rather than fmaping a function over a Parser on the *left* side of the bind operator >>=, we can apply said function to the Parser's output on the *right* side of the bind operator, and the result will be exactly the same. Let's do this.

```
parseSizedNatural :: Parser Natural
parseSizedNatural =
  parseDigit >>= \dig0 ->
  fmap naturalFromDigits (parseSomeDigits (naturalFromDigit dig0))
```

Great. That last line is a bit contrived, though. Perhaps we could replace the use of fmap with a use of >>= and pure, as we learned before when implementing parseCharTwice.

```
parseSizedNatural :: Parser Natural
parseSizedNatural =
  parseDigit >>= \dig0 ->
  parseSomeDigits (naturalFromDigit dig0) >>= \digs ->
  pure (naturalFromDigits digs)
```

Is this better than before? Meh. For us, this was mostly an exercise showing that when we have types, laws and sensible compositions, we can move things around without paying *any* attention to the details which, they say, make our program special. They don't.

Something's changed, though. Something subtle. Previously, our attention went to worrying about how things compose, *why* things compose, to the *relationship* between things rather than things themselves. But now, when we look at parseSizedNatural, when we look at parseXYZ, it's *time*, it's *things*, it's *actions* who take the spotlight. *Do this, then do that, then take these and put them over there.* Composition is still there, we'll never get rid of it, we don't want to get rid of it. But to the untrained eye, composition is not the protagonist anymore.

If we were to describe this latest parseSizedNatural out loud, we'd say that first we parse a Digit, *then*, taking that Digit into account, we parse some more Digits, and *then* we produce a suitable Natural number. *Then.* We call this style of programming where the passage of time takes the lead role, where "then" always seems to be the most important word, *imperative programming.* Monads, *bind*, is how we encode this programming style in Haskell. We sometimes even read >>= out loud as "then".

One hundred ninety-nine

Tricked, once again. We said Monad was this typeclass with just one method, join, but as usual, that was not true. Conceptually, *yes*, a monad is an Applicative functor for which we can implement a joining function, but that's not the exact definition that Haskell uses in the Monad typeclass. And strictly speaking, a monad only cares about the pure part of an Applicative functor, which, in case you didn't notice, is not so different from join in what it achieves. pure and join are merely two different ways of constructing a functorial value f x.

```
pure ::      x  -> f x
join :: f (f x) -> f x
```

Mind you, join imposes a Monad constraint on f. And pure, belonging to the Applicative typeclass, an Applicative one. But we get the idea. For didactic purposes, this is great. We can either put an x inside an f, or join two f layers into one. In practice, however, rather than talking about join, the Monad typeclass talks about the more ergonomically aware infix operator >>=. Here's what the actual typeclass looks like in Haskell.

```
class Applicative f => Monad f where
  (>>=) :: f a -> (a -> f b) -> f b
```

That is, our Monad instances will have to define >>= directly, rather than join. There's no join method. There is a join *function*, though, defined for all Monads in terms of >>=, but that's beside the point.

```
join :: Monad f => f (f x) -> f x
join ffx = ffx >>= id
```

Anyway, without further ado, let's start by defining the Monad instance for Parser. We already know what it's supposed to do.

```
instance Monad Parser where
  (>>=) :: Parser a -> (a -> Parser b) -> Parser b
  pa >>= f = Parser (\s0 ->
    case runParser pa s0 of
      Left e -> Left e
      Right (!n0, s1, a) ->
        case runParser (f a) s1 of
          Left (!n1, msg) -> Left (n0 + n1, msg)
          Right (!n1, s2, b) -> Right (n0 + n1, s2, b))
```

If we compare the implementation of >>= with that of join before, or even with that of <*> or liftA2, we'll see that they are almost the same. We still run two Parsers. The only difference is in *how* that second Parser comes to be. In the case of <*> and liftA2, those two Parsers are provided as separate input parameters. In the case of join, which takes a value of type Parser (Parser x), the second Parser will be the result of a successful execution of the first one, conveying the idea of *order*. And in fa >>= g, the second Parser is obtained by applying g to the output produced by fa, which conveys the idea of order too, seeing as we wouldn't be able to apply g until *after* obtaining an a as a consequence of having executed fa.

Two hundred

We are repeating almost the same code time and time again, but that's mostly because we are learning. In practice, at most we'll need to write this once. If a Functor is a Monad, which implies it's an Applicative too, then the definitions of fmap, <*> and liftA2 come for free. All we ever need to manually define is >>= and pure. We knew this already, though.

A while ago we learned how liftA, defined exclusively in terms of Applicative vocabulary, served as a suitable implementation of fmap.

```
liftA :: Applicative f => (a -> b) -> f a -> f b
liftA g fa = pure g <*> fa
```

With this, the Functor instance for Parser, or any other Applicative functor, can be as straightforward as this:

```
instance Functor Parser where
  fmap = liftA
```

In fact, the Applicative laws require that this be the case. We are free to hand roll an fmap implementation if we want to, but ultimately, it *must* behave exactly like liftA.

Similarly, we saw how mkAp, given a joining function, always gives us a suitable implementation for <*> that we can use when defining an Applicative instance for some Functor.

```
mkAp :: Functor f
       => (f (f b) -> f b)
       -> (f (a -> b) -> f a -> f b)
mkAp join = \fg fa ->
  join (fmap (\g -> fmap g fa) fg)
```

Theoretically, this is perfect. In practice, mkAp has two engineering problems we need to solve. First, mkAp takes a joining function explicitly when it could be relying on the fact that f is a Monad for which join is already implicitly defined. We can fix this by simply imposing a Monad constraint on f, and using our recently defined join directly.

```
mkAp :: Monad f => f (a -> b) -> f a -> f b
mkAp fg fa = join (fmap (\g -> fmap g fa) fg)
```

This change solves a documentation problem, too. As authors of a function explicitly taking a joining function as input, we would have been forced to write some comments explaining what's the expected behaviour of said function. However, that behaviour is best described by the Monad laws, so by merely requiring a Monad constraint on f we've avoided the need for some redundant documentation.

But second, and more importantly for operational reasons, mkAp mentions fmap in its implementation. The same fmap that could be defined to be liftA, which in turn would refer back to mkAp had we used it to implement <*>. This would make our program diverge. This is bad. Unfortunately, we can't avoid using fmap if we use join to implement mkAp, but we can avoid it if we rely on >>= and pure instead.

```
mkAp :: Monad f => f (a -> b) -> f a -> f b
mkAp fg fa = fg >>= \g ->
             fa >>= \a ->
             pure (g a)
```

Very straightforward. We first bind fg's a -> b output, then we bind fa's a output, and then we merely apply one to the other in the only way possible, wrapping the result in f using pure. And, as we avoided any mention of fmap, for practical purposes, this implementation of mkAp is superior to the one that used join, even if both of them mean the same theoretically. This new implementation of mkAp comes with Haskell, but it's actually called just ap.

```
ap :: Monad f => f (a -> b) -> f a -> f b
ap fg fa = fg >>= \g -> fa >>= \a -> pure (g a)
```

ap is to <*> what liftA is to fmap. That is, ap is an Applicative feature defined in terms of a Monadic vocabulary, just like liftA is a Functorial feature defined in terms of Applicative vocabulary. With ap, here's how we'd define the Applicative instance for Parser.

```
instance Applicative Parser where
  pure = \x -> Parser (\s -> Right (0, s, x))
  (<*>) = ap
```

We can't avoid defining pure by hand, but we can most certainly avoid manually implementing <*>. In fact, we should, seeing how the Applicative laws require that if our Applicative functor is also a Monad, then <*> must have the same behaviour as ap. So we don't gain much by manually implementing <*> without ap anyway.

Alternatively, seeing how the Applicative typeclass only requires us to define *either* <*> *or* liftA2, rather than defining <*> in terms of ap, we could have defined liftA2 in terms of liftM2. Notice the M in that name.

```
liftM2 :: Monad f => (a -> b -> c) -> f a -> f b -> f c
liftM2 g fa fb = fa >>= \a -> fb >>= \b -> pure (g a b)
```

To sum up, when our Functor is an Applicative, by means of liftA we can avoid manually implementing fmap. And when our Functor is a Monad, we can avoid manually implementing <*> or liftA2 by means of ap or liftM2. So, no more repeating ourselves over and over again. I repeat, no more repeating ourselves

Two hundred one

Let's explore another monad, the Maybe monad.

```
instance Monad Maybe where
  (>>=) :: Maybe a -> (a -> Maybe b) -> Maybe b
  Nothing >>= _ = Nothing
  Just a  >>= f = f a
```

Previously, we reasoned about the Maybe monad through join and the cute analogy of putting a box inside another box. That worked, but let's try something else this time.

Let's imagine we have a String that perhaps could be parsed into an Integer number using a function called integerFromString of type String -> Maybe Integer. If the String is not the expected representation of an Integer number, we get Nothing.

```
integerFromString :: String -> Maybe Integer
```

It doesn't matter how integerFromString is implemented, we only care about its type.

Then, let's say we also have a function for converting an Integer number to a Natural number if possible. That is, if the given Integer number is non-negative. Otherwise, we get Nothing.

```
naturalFromInteger :: Integer -> Maybe Natural
```

And finally, let's say we have a function for converting a Natural number into a `Digit`, but only insofar as said number is smaller than *ten*.

```
digitFromNatural :: Natural -> Maybe Digit
```

What all these functions have in common is that they can *fail* somehow, and this potential failure is conveyed by `Maybe`.

Now, suppose that using only these three functions, we want to create a new function that tries to obtain a `Digit` from some input `String`. Let's call it `digitFromString`.

```
digitFromString :: String -> Maybe Digit
digitFromString str =
  case integerFromString str of
    Nothing -> Nothing
    Just int -> case naturalFromInteger int of
                  Nothing -> Nothing
                  Just nat -> digitFromNatural nat
```

This works just fine, but there are two problems with `digitFromString`. First, this is not the most efficient solution to this problem. But we'll just ignore that problem and focus on the next one, which is that the implementation is just ugly. Look at what's happening. If `integerFromString str` succeeds, and then `naturalFromInteger int` succeeds after it, and then `digitFromNatural nat` succeeds after both of them, then we'll finally have our `Digital` result. Each of these steps *depends* on the *previous* step succeeding. Well, isn't this *exactly* what >>= allows us to deal with, *without* having to explicitly pattern match on the `Just` and `Nothing` constructors?

```
digitFromString :: String -> Maybe Digit
digitFromString = \str ->
  integerFromString str >>= \int ->
  naturalFromInteger int >>= \nat ->
  digitFromNatural nat
```

In this new version of `digitFromString` we still see our initial input `str` and our intermediate results `int` and `nat`. However, the explicit pattern matching on `Nothing` and `Just` is gone, making our implementation significantly shorter. This is what >>= achieves in the `Maybe` monad. It allows us to *compose* the successes and failures of the intermediate steps, binding the successful output of each of them, if any, to a particular name that can be processed further.

```
> digitFromString "hello"
Nothing
```

In this example, `"hello"` made `integerFromString` fail, so we got `Nothing`. In the following one, `"-14"` successfully parses as the Integer number -14, but then this number fails to become a `Natural` by means of `naturalFromInteger`, so we get `Nothing` as output as well.

```
> digitFromString "-14'
Nothing
```

The next example provides "30" as input, which eventually becomes a Natural number, but then it fails to become a Digit by means of digitFromNatural. Again, we get Nothing as result.

```
> digitFromString "30"
Nothing
```

But, if we give "2" as input, all of the intermediate steps succeed, including the last digitFromNatural, and we get Just a Digit as result.

```
> digitFromString "2"
Just D2
```

In retrospect, what >>= does for Maybes is merely a subset of what it does for Parsers. That is, while >>= composes all of the failure, consumption length and leftover management features of Parsers, it only composes failure for Maybe, since that's all Maybe has to worry about. In other words, we haven't learned anything new.

Two hundred two

Unfortunately, even though our recent digitFromString can fail for many distinct reasons, at best we get Nothing if something goes wrong. Wouldn't it be nice to get a message saying *why* something went wrong, too? Yes it would. Worry not, for Either x has monadic features too, and as we saw many times before, we can use that x to convey the reason *why* something failed.

```
digitFromString :: String -> Either String Digit
digitFromString = \str ->
  note "Not an integer"
       (integerFromString str) >>= \int ->
  note "Not a natural number"
       (naturalFromInteger int) >>= \nat ->
  note "Not a digit"
       (digitFromNatural nat)
```

This new version of digitFromString is not so different from the one before. The only thing that changed is that rather than using >>= to compose Maybes, we are using it to compose Eithers whose Left payload is a String explaining *why* this function fails, if it does. >>= takes the following type in our example:

```
(>>=) :: Either String a
     -> (a -> Either String b)
     -> Either String b
```

Consequently, we use note, of type x -> Maybe a -> Either x a, to convert the Maybe values resulting from our uses of integerFromString and friends, into more descriptive values of type `Either String whatever` that we can bind using >>=. And what do we get in return? See for yourself.

```
> digitFromString "hello"
Left "Not an integer"
> digitFromString "-14"
Left "Not a natural number"
> digitFromString "30"
Left "Not a digit"
> digitFromString "2"
Right D2
```

Helpful, isn't it? Now, if `digitFromString` fails, at least we'll know why. The definition of the `Monad` instance for `Either x` is straightforward.

```
instance Monad (Either x) where
  (>>=) :: Either x a -> (a -> Either x b) -> Either x b
  Left x  >>= _ = Left x
  Right a >>= f = f a
```

In other words, if the given `Either x a` is `Left`, then there's just nothing we can do besides returning yet another `Left` with that same payload. But if the given `Either x a` is `Right`, then we'd have an `a` to which to apply `f` in order to obtain an `Either x b` we can return. Luckily for us, because of parametricity, this is the only implementation we could give to >>=, so there's no room for us to even consider a different implementation.

And, as promised before, if we wanted to define the `Functor` and `Applicative` instances for `Either x` in terms of this `Monadic` vocabulary, we could.

```
instance Applicative (Either x) where
  pure = Right
  liftA2 = liftM2

instance Functor (Either x) where
  fmap = liftA
```

Of course, we don't *need* to do any of this, for the `Functor`, `Applicative` and `Monad` instances for `Maybe` and `Either x`, among others, already come with Haskell out of the box. We are just learning, that's why we write about these things.

So, why do we have the `Applicative` typeclass if we can always define our `Applicative` instances using monadic vocabulary? Wouldn't we be better off if `pure` was a method of the `Monad` typeclass and we got rid of `Applicative` altogether? Well, *no*, because as we said before, just like not all `Functors` are `Applicative` functors, not all `Applicative` functors are `Monads`, meaning that >>= will sometimes be missing.

Intuitively, if we can envision some kind of *ordering* when composing two functorial values, in the sense that the existence of a functorial value *depends* on the output of a previous one, then *yes*, that Applicative functor is a Monad too. Otherwise, it is *not* a Monad.

And all that talk we had before about how in fg <*> fx, say, there was some kind of ordering forcing fg to happen before fx? That was not entirely true, I'm afraid. It was true for the Applicative functors we explored, yes, but it is not true for *all* Applicative functors in general. Nothing in the Applicative laws says anything about there being an ordering to fg and fx in fg <*> fx. It's just that all the Applicative functors we saw were Monads too, and this imposed the extra requirement that <*> behave just like ap, and liftA2 just like liftM2. We were bound to bind, this was the reason for the ordering of evaluation of our functorial values.

Generally speaking, unless an Applicative functor is a Monad too, then there is *no* requirement regarding whether fg or fx should be evaluated first in fg <*> fx. For all Applicative cares, they could even be evaluated at the same time.

You want to see an example, don't you? An example of an Applicative functor that is *not a Monad*? We'll have to wait, I'm afraid. It's not that easy. For now, let's continue exploring Monads.

Two hundred three

Lists are Monads too. We know that, of course, because we were able to define a joining function for them a while ago. Namely, concat. Let's write a proper Monad instance for lists, though, using >>= rather than join, as required by Haskell.

```
instance Monad [] where
    (>>=) :: [x] -> (x -> [y]) -> [y]
    []         >>= _ = []
    (x : rest) >>= f = mappend (f x) (rest >>= f)
```

If >>= is applied to an empty list, then there's nothing we could possibly do, so we return []. But if it is applied to a non-empty list, then we concatenate the result of applying f to the first element of the list, which results in an expression of type [y], with a *recursive* application of >>= to the same f and the rest of the list, another expression of type [y]. Here, mappend has type [y] -> [y] -> [y].

What do we get from this? Similar to how fmap does it, >>= will apply a given function to *all* the elements in the list. However, whereas fmap takes a function x -> y that will convert each element of the input list to *one* element in the output list, >>= takes a function x -> [y] that converts each element of the input list to *zero or more* elements in the output list. If we compare the types fmap and >>= —or rather, its flipped version— we'll see that they are not that different.

```
       fmap :: Functor f => (x ->     y) -> f x -> f y
  flip (>>=) :: Monad   f => (x -> f y) -> f x -> f y
```

By the way, flip (>>=) goes by the name =<<, too. You see, the arrows
point in the opposite direction, to the left, signaling that in g =<< fa it's fa
who gets evaluated first, and g afterwards.

Anyway, if we were to fmap a function of type x -> f y over an f x, we'd
end up with a value of type f (f y) rather than just f y. That is, whereas
fmap doesn't prevent an extra f layer from being introduced, >>= makes
sure we never have more than one of those.

What we are saying, in concrete terms, is that if we *bind* a value of type [x]
to a function x -> [y], then the result will have type [y] and not [[y]] as
fmap's would. And why is this useful? Can't we just concat that [[y]] into
[y]? Sure, go ahead, do that. But remember, concat is just another name
for join, itself implemented in terms of >>=, so we wouldn't really be
avoiding the bind, would we? We are bound to bind.

So, even without looking at any examples, we can see what fx >>= g does.
It applies a function g returning a list to each element in the list fx, which
results in a *list of lists* that then gets joined as a single list. That's all. From a
fundamental point of view, there's nothing new in >>=. But, from an
operational perspective, we acquired some new interesting vocabulary for
working with lists, a fascinating new way of composing them.

Let's try it. Take replicate 2 for example, the function of type a -> [a]
which, when applied to a value, creates a list containing that value twice.

```
> replicate 2 True
[True, True]
```

If we fmap replicate 2 over "xyz", a value of type [Char], also known as
String, then we get a value of type [[Char]].

```
> fmap (replicate 2) "xyz"
["xx", "yy", "zz"]
```

But if rather than fmapping, we *bind* replicate 2 using >>=, then we get yet
another [Char] as result.

```
> "xyz" >>= replicate 2
"xxyyzz"
```

"xxyyzz" is merely the result of applying join to ["xx", "yy", "zz"], which
is exactly what we should have been expecting.

```
> join (fmap (replicate 2) "xyz")
"xxyyzz"
```

Great. Now let's try something more interesting. Let's duplicate the
element only when it is an even Natural number, otherwise we'll replace it
with 9.

```
> [1, 2, 3, 4] >>= \x -> bool [9] [x, x] (even x)
[9, 2, 2, 9, 4, 4]
```

Fascinating. We couldn't have accomplished that with fmap alone. Nor with <*>, for that matter. We most certainly need to involve >>= or join somehow.

```
> join (fmap (\x -> bool [9] [x, x] (even x))
             [1, 2, 3, 4])
[9, 2, 2, 9, 4, 4]
```

Can we *drop* elements from this list, too? Say, can we drop those that are odd while still duplicating those that are even? Of course we can.

```
> [1, 2, 3, 4] >>= \x -> bool [] [x, x] (even x)
[2, 2, 4, 4]
```

The function we bind to a list [a] has type a -> [b], implying that *each* of the individual as in the input list has a say on what [b] should be. If this function chooses to say that for some a it should be [], there would be nothing wrong with that. For that a, this function would simply *not* contribute anything to the final [b], and that's fine. If we look at this step by step, using fmap and join, we can easily see what's really going on.

```
> fmap (\x -> bool [] [x, x] (even x)) [1, 2, 3, 4]
[[], [2, 2], [], [4, 4]]
> join [[], [2, 2], [], [4, 4]]
[2, 2, 4, 4]
```

What about using >>= more than once? That worked with other Monads before. Does it work with lists as well?

```
> [1, 2] >>= \a -> [3, 4] >>= \b -> [(a, b)]
[(1, 3), (1, 4), (2, 3), (2, 4)]
```

Do you recognize that result? Look closer. That's right, it's the *cartesian product* of the lists [1, 2] and [3, 4]. We know that the cartesian product is what we get from pair.

```
> pair [1, 2] [3, 4]
[(1, 3), (1, 4), (2, 3), (2, 4)]
```

But pair is just a different name for liftA2 (,).

```
> liftA2 (,) [1, 2] [3, 4]
[(1, 3), (1, 4), (2, 3), (2, 4)]
```

And we also know that liftA2 is just another name for liftM2, the function that achieves the same thing that liftA2 does, but using a Monadic vocabulary rather than an Applicative one.

```
> liftM2 (,) [1, 2] [3, 4]
[(1, 3), (1, 4), (2, 3), (2, 4)]
```

Here is the definition of liftM2.

```
liftM2 :: Monad f => (a -> b -> c) -> f a -> f b -> f c
liftM2 g fa fb = fa >>= \a -> fb >>= \b -> pure (g a b)
```

Now, let's beta-reduce liftM2 (,) by replacing all occurrences of g with (,).

```
pairM :: Monad f => f a -> f b -> f (a, b)
pairM fa fb = fa >>= \a -> fb >>= \b -> pure (a, b)
```

And let's use lists as our Monad. That is, let's replace f with lists in the type.

```
pairList :: [a] -> [b] -> [(a, b)]
pairList fa fb = fa >>= \a -> fb >>= \b -> pure (a, b)
```

We know that pure x, in the case of lists, means the same as [x]. That is, pure merely creates a list with *one* element in it. So let's replace pure (a, b) with [(a, b)] too.

```
pairList :: [a] -> [b] -> [(a, b)]
pairList fa fb = fa >>= \a -> fb >>= \b -> [(a, b)]
```

Now, let's apply this new function to [1, 2], beta-reducing it right away. In other words, let's replace fa with [1, 2] of type [Natural].

```
pairList12 :: [b] -> [(Natural, b)]
pairList12 fb = [1, 2] >>= \a -> fb >>= \b -> [(a, b)]
```

And finally, let's do the same with fb and [3, 4].

```
list1234 :: [(Natural, Natural)]
list1234 = [1, 2] >>= \a -> [3, 4] >>= \b -> [(a, b)]
```

Why, look at that. This is *exactly* the same thing we wrote in GHCi. No wonder we got the cartesian product of our two lists. In other words, we can always use >>= and pure, rather than <*> or liftA2, to do any kind of cartesian operation on our lists.

```
> [1, 2] >>= \x -> [3, 4] >>= \y -> [x * y]
[3, 4, 6, 8]
> [1, 2] >>= \x -> [3, 4] >>= \y -> [x + y]
[4, 5, 5, 6]
> [1, 2] >>= \x -> [3, 4] >>= \y -> [x < y]
[True, True, True, True]
```

But we can do *so much more* than just replicating what Applicative already achieved...

Two hundred four

Take filter, for example.

```
filter :: (a -> Bool) -> [a] -> [a]
```

This function, which we've already met before, takes a list of as and a predicate on a —that is, a function a -> Bool— and discards from said list the elements for which the predicate is False.

```
> filter odd [1, 40, 2, 5, 22]
[1, 5]
```

We can easily implement filter using >>=.

```
filter :: (a -> Bool) -> [a] -> [a]
filter p as = as >>= \a -> case p a of
                              True -> [a]
                              False -> []
```

Quite straightforward. Each of the as in the given list gets bound as an input to the function \a -> case p a ..., which decides whether to keep that a, by producing a non-empty list containing it, or to discard it by producing an empty list. This is fine, but we can improve its looks a bit using guard.

```
guard :: Alternative f => Bool -> f Unit
guard True  = pure Unit
guard False = empty
```

By applying guard to a Boolean value, we get a functorial value that either "succeeds" using pure Unit if the given Bool is True, or "fails" using empty from the Alternative typeclass if False. How is this useful? Well, if our f happens to be an Alternative, like lists are, and moreover a Monad, then we can use guard as some kind of checkpoint in our neatly arranged monadic code, preventing all code after an unsuccessful guard from being executed at all.

```
filter :: (a -> Bool) -> [a] -> [a]
filter p as = as >>= \a -> guard (p a) >> pure a
```

The Alternative instance for lists, which we haven't seen before, is as follows:

```
instance Alternative [] where
  empty = []
  (<|>) = (++)
```

Remember, ++ is the infix list concatenation operator. Perhaps surprisingly, this instance has exactly the same implementation as the Monoid instance for lists does, which should remind us of the *monoidal* origins of Alternative through our fantasy Monoidalt typeclass of yore.

```
instance Monoid [x] where
  mempty = []
  mappend = (++)
```

Anyway, not important. Knowing that empty means [], and knowing that pure x means [x], we can reason about how guard accomplishes what it does. If the Bool given to guard is False, then we end up with [], which when composed using >>=, or any of the operators implemented in terms of it like >>, <*> or even <$>, results in yet another empty list.

```
> guard False :: [Unit]
[]
> guard False >> [1, 4, 5]
[]
> guard False >> [1, 4, 5] >>= \a -> replicate a a
[]
```

But, if the Bool we give to guard is True, then this results in [Unit], a value that isn't particularly interesting in itself, but when composed using >>= or derivatives, doesn't prevent any subsequent functorial work from being performed.

```
> guard True :: [Unit]
[Unit]
> guard True >> [1, 4, 5]
[1, 4, 5]
> guard True >> [1, 4, 5] >>= \a -> replicate a a
[1, 4, 4, 4, 4, 5, 5, 5, 5, 5]
```

Great. It's useful to keep in mind that guard has nothing to do with lists, but with Alternative. Meaning that we can use this with *any* Alternative. For example, with Maybe.

```
> guard False :: Maybe Unit
Nothing
> guard False >> Just 4
Nothing
> guard True :: Maybe Unit
Just Unit
> guard True >> Just 4
Just 4
```

Or, why not, with Parser. Take expect1 x, for example, the Parser that expects x to be the character leading the input being parsed. We could implement it using guard.

```
expect1 :: Char -> Parser Unit
expect1 x = parseChar >>= \y -> guard (x == y)
```

guard already produces Unit if the given Bool is True, so there's nothing else to be done after checking that y, the Char we discover using parseChar, equals the expected x.

```
> runParser (expect1 'a') "abc"
Right (1, "bc", Unit)
> runParser (expect1 'x') "abc"
Left (1, "empty")
```

It works. Well, kinda. This isn't exactly an improvement over the expect1 we had before. The error message we get now when things go wrong is terribly uninformative, and moreover, it reports 1 rather than the correct 0 as the position where parsing failed. Can you see why?

Two hundred five

Here's a contrived example showcasing the elegance of acknowledging the monadic structure of lists. *Pythagorean triples.*

A Pythagorean triple consists of three positive integers a, b and c, where $a^2 + b^2 = c^2$. For example, 3, 4 and 5 are a Pythagorean triple because $3^2 + 4^2 = 5^2$. Let's write some Haskell code to enumerate Pythagorean triples. Let's build a list with infinitely many of them.

We'll be relying on Euclid's formula for generating these triples. The formula says that given any two natural numbers m and n, where $m > n$ and $n > 0$, a Pythagorean triple (a, b, c) can be obtained by using $m^2 - n^2$ as a, $2 \times m \times n$ as b, and $m^2 + n^2$ as c. Writing this in Haskell is easy.

```
pyths :: [(Natural, Natural, Natural)]
pyths = enumFrom 1 >>= \m ->
         enumFromTo 1 (m - 1) >>= \n ->
         pure (m*m - n*n, 2*m*n, m*m + n*n)
```

Before we try to understand this, let's see it work.

```
> take 9 pyths
[(3, 4, 5), (8, 6, 10), (5, 12, 13),
 (15, 8, 17), (12, 16, 20), (7, 24, 25),
 (24, 10, 26), (21, 20, 29), (16, 30, 34)]
```

Notice that pyths is an *infinite* list lazily coming to existence as we try to evaluate it, so in order to display it in GHCi, we must take only a handful of them. Otherwise, GHCi would never finish displaying it. Is, for example, (7, 24, 25) a Pythagorean triple? Yes it is, for $7^2 + 24^2 = 25^2$, that is, 625. Alright, let's look at the implementation of pyths now.

First, we have some new helper functions. enumFrom 1 constructs an *infinite* list *enum*erating the Natural numbers, starting *from* 1.

```
> take 8 (enumFrom 1)
[1, 2, 3, 4, 5, 6, 7, 8]
```

That is, in pyths, where we bind enumFrom 1 to \m -> ..., our m will first be 1, then 2, then 3, then 4, etc. This m corresponds to the previously mentioned m in Euclid's formula.

```
pyths :: [(Natural, Natural, Natural)]
pyths = enumFrom 1 >>= \m -> ...
```

On the next line we have enumFromTo 1 (m - 1), which does what its name suggests. It enumerates numbers *from* 1 *to* m - 1 inclusively. Euclid's formula requires that $n > 0$ and $n < m$, so, by enumerating all Natural numbers from 1 to m - 1, we are listing all the sensible values our n could take. Thus, we bind this enumeration, this list, to \n -> ....

```
pyths :: [(Natural, Natural, Natural)]
pyths = enumFrom 1 >>= \m ->
        enumFromTo 1 (m - 1) >>= \n -> ...
```

At this point we have n, a Natural number greater than 0, and m a Natural number greater than n. All we have to do now is use Euclid's formula to calculate the elements of our triple *a*, *b* and *c*, and produce that triple using pure.

```
pyths :: [(Natural, Natural, Natural)]
pyths = enumFrom 1 >>= \m ->
        enumFromTo 1 (m - 1) >>= \n ->
        pure (m*m - n*n, 2*m*n, m*m + n*n)
```

Slick, isn't it? Now let's spice things up a bit. Not all Pythagorean triples are equal, some are said to be *primitive*. The primitive Pythagorean triples *(a, b, c)* are those for which *a*, *b* and *c* are *coprime*. That, is, they have no common divisor other than *one*. Let's change pyths so that rather than generating all Pythagorean triples, it only generates the primitive ones. Why? Well, it doesn't matter, does it? It's just an example, and it's important that we can tell the example apart from what really matters. Namely, binding lists.

The most obvious way to accomplish this would be to filter the entire output of pyths in order to discard non-primitive Pythagorean primitives *after* they have been produced as above. Yes, that would work. However, let's try something else. Euclid's formula also says that if its *m* and *n* are coprime, and at least one of them is even, then the Pythagorean triple generated by the formula will be a primitive one. So, if we guard against any undesirable choices of m and n, we should be able to avoid producing non-primitive triples altogether.

```
pyths :: [(Natural, Natural, Natural)]
pyths = enumFrom 1 >>= \m ->
        enumFromTo 1 (m - 1) >>= \n ->
        guard (even m || even n) >>
        guard (gcd m n == 1) >>
        pure (m*m - n*n, 2*m*n, m*m + n*n)
```

The expression gcd m n calculates the greatest common divisor between m and n, which, if equal to 1, implies that m and n are coprime. Does this work?

```
> take 5 pyths
[(3, 4, 5), (5, 12, 13), (15, 8, 17),
 (7, 24, 25), (21, 20, 29)]
```

It works. For example, notice how (8, 6, 10) is not mentioned anymore, for it's not a primitive Pythagorean triple.

We are using *two* guards in our implementation of pyths. This is fine, but if desired, we could have written a single guard instead, enforcing that the *conjunction* of our two Bools be satisfied.

```
guard ((even m || even n) && (gcd m n == 1)) >> ...
```

We don't really have to pay much attention to gcd, the function that gives us the greatest common divisor between two numbers. It already comes with Haskell out of the box, doing exactly what it promises. However, its implementation is rather beautiful, so let's contemplate it in awe anyway.

```
gcd :: Natural -> Natural -> Natural
gcd 0 b = b
gcd a b = gcd (rem b a) a
```

Awe. Strictly speaking, in Haskell, gcd is defined to work with all Integer numbers, not just Natural. But for our purposes, this works just fine. And by the way, rem x y gives the *remainder* from dividing x by y into an integer number. For example, rem 14 5 is 4, for dividing 14 by 5 results in 2 plus a remainder of 4.

There is one last small detail we need to address in pyths. We see (3, 4, 5) in our list, but we don't see (4, 3, 5), which is the primitive Pythagorean triple obtained by merely swapping the first two elements of the other primitive triple. This absence is a consequence of how Euclid's formula comes up with these triples. However, if we wanted to include the swapped triples in our result, we could easily do so. All we'd have to do is produce *two* triples for each suitable combination of m and n, rather than just one. Each of them with their first two elements swapped, of course.

```
pyths :: [(Natural, Natural, Natural)]
pyths = enumFrom 1 >>= \n ->
        enumFromTo 1 (m - 1) >>= \n ->
          guard (even m || even n) >>
          guard (gcd m n == 1) >>
          case (m*m - n*n, 2*m*n, m*m + n*n) of
            (a, b, c) -> [(a, b, c),
                          (b, a, c)]
```

Yes, the case expression at the end, which we are using to give a, b and c names, is a bit noisy. We'll learn interesting ways to deal with this later on. Anyway, this does what we wanted it to do.

```
> take 6 pyths
[(3, 4, 5), (4, 3, 5),
 (5, 12, 13), (12, 5, 13),
 (15, 8, 17), (8, 15, 17)]
```

Are Pythagorean triples interesting to us? Not at all. You can forget all about them now. Please do. As we said at the beginning, this was merely an example intended to show the elegance of using the monadic features of a list.

Anyway, enough about lists. And enough about Monads and Applicatives and Functors. Once again, we've learned way too much. Let's put this new knowledge to good use, let's get comfortable with it. Let's build a programming language.

Two hundred six

What if there were no types? Worse, what if there were no expressions other than functions? Would we still want for something? Yes, of course, but way less than you'd think.

```
data Expr = Lam String Expr
          | App Expr Expr
          | Var String
```

We met Expr a long time ago, when we encountered the Lambda calculus for the first time. Back then, we said that Expr was all we needed in order to effectively compute anything that can be computed. For example, we learned that if we wanted to express a function as a value of type Expr, such as the identity function, we could express it as Lam "x" (Var "x"). This corresponds to Haskell's \x -> x, but without the nice syntax. Well then, let's build a programming language on top of Expr, with cute syntax and all.

Now, don't be fooled. Ours won't be a fully fledged programming language like Haskell, but the basics will be there. There's only so much we can fit into this book. In any case, you don't need to learn any of this by heart. This exercise in building a language is for you to get a general idea of how a programming language is built and what are some of the common challenges, while learning some day-to-day Haskell as we go. So tag along, take it easy, pick up the interesting pieces and quickly move on.

Our toy language will use the syntax x → x to convey Lam "x" (Var "x"), the same as \x -> x conveys in Haskell. Notice there is no leading backslash \ in said x → x. We just decided that. We can decide whatever we want, it's our language, our syntax, our rules. Relax.

Technically, all the functions in our language take just one input parameter, as witnessed by the single name field in the Lam constructor. However, by making the function arrow → associate to the *right* as in Haskell, we will enable functions like a → (b → (c → d)) to be written as a → b → c → d instead, without the redundant parentheses. While not strictly necessary, at times, this will make the syntax more convenient for us humans.

We'll also need some syntax to express the application of a function f to some value x, corresponding to a use of the App constructor for Expr. In Haskell, we write this as f x, merely juxtaposing f and x. In our toy language, we'll do exactly the same. Why not? Obviously, either f or x could be more complex expressions themselves, like a → a or g x. In that case, as in Haskell, we'll require extra parentheses around each of these expressions. For example, (a → a) (g x).

And finally, also like in Haskell, the expression f a b c shall mean exactly the same as ((f a) b) c. That is, we'll make sure function application associates to the *left*, and as a consequence of this, we'll be able to leave out those parentheses if desired, giving the impression that f is a function taking three parameters while still being able to reason about it as taking just one at a time, making the whole endeavour much more convenient. The currying of functions really is a handy thing.

Knowing all of this, we can start implementing a Parser for Expressions that have been written using this brand new syntax. And remember, there are no types in our language. Or rather, there are types, but we've chosen not to acknowledge them, so we won't see them anywhere, and our program will fail unpredictably if we get our Expr wrong. Fun, right? Not fun.

Ready? Let's go.

Two hundred seven

Let's start by writing a Parser, which is something we know how to do. Our Parser will take the source code for an expression as input, and produce a value of type Expr as output. This shouldn't be too hard, considering how comfortable we are with Parsers at this point, and how straightforward the syntax we chose is.

```
parseExpr :: Parser Expr
parseExpr = asum [ parseLam, parseApp, parseVar
                 , inParens parseExpr ]
```

parseExpr will try each of the four Parsers given to asum, in the order they have been specified, until one of them succeeds. The first three are supposed to tackle the individual constructors Lam, App and Var, as their names indicate. The fourth one is merely a recursive execution of parseExpr expecting parentheses to be surrounding the input. This is because we want things like a → b, (a → b) and ((a → b)) to mean the same. If people want to put redundant parentheses around expressions, so be it. Let's start with inParens.

```
inParens :: Parser x -> Parser x
inParens p = expect1 '(' >>
                 p >>= \x ->
                 expect1 ')' >>
                 pure x
```

This function takes a Parser x as input, which it names p, and returns yet another `Parser` x which first ensures that there is a left parenthesis `'('` leading the input, then runs the original p, then ensures that there is a right parenthesis `')'`, and finally, if all of this succeeds, it purely produces the x that came out of p.

There's nothing monadic about inParens, by the way. Nothing in it truly depends on the output of a previous `Parser`, so we could have expressed this using only `Applicative` vocabulary, rather than involving `>>=`. For example, here's an aesthetically disappointing but correct implementation using `liftA3`.

```
inParens :: Parser x -> Parser x
inParens p = liftA3 (\_ x _ -> x)
                    (expect1 '(')
                    p
                    (expect1 ')')
```

Often, we prefer to use `Monadic` vocabulary like `>>=`, even if not strictly necessary, simply because it makes things a bit easier on the eyes.

And as long as we are exploring different ways of writing things, let's introduce the `Applicative` bird operators, too.

```
(*>) :: Applicative f => f a -> f b -> f b
(*>) = liftA2 (\_ b -> b)

(<*) :: Applicative f => f a -> f b -> f a
(<*) = liftA2 (\a _ -> a)
```

Presumably called *bird* operators because, well, just look at them. The `<` and `>` parts look like beaks, and `*` like eyes.

These operators compose the effects of two functorial values f a and f b in the same order in which `liftA2` or `<*>` would compose them, but rather than somehow combining the outputs a and b into a new value c, they merely keep just one of either a or b, discarding the other. The beak points to the output they keep. With these operators, we could instead implement inParens as follows.

```
inParens :: Parser x -> Parser x
inParens p = expect1 '(' *> p <* expect1 ')'
```

Slick, isn't it? We are not really interested in the output from expect1 `'('`, only in the effect it has in ensuring that `'('` is present in the input being parsed. So, we compose its parsing effect with that of p, but keep just p's output. And similarly, we want to perform expect1 `')'` after all of expect1 `'('` and p, but we are not at all interested in its output, so we point the beak someplace else.

By the way, `*>` is conceptually the same as `>>`, which we already encountered before. So, in reality, there's nothing truly new for us here. The main difference between them is their fixity. Whereas `>>` has the same fixity as >>=, allowing us to write slick monadic code without having to add noisy parentheses, <* and *> have a fixity setup that allows them to get along well with each other in situations like the one in inParens. Remember what we said before, >> is implemented in terms of a Monadic vocabulary rather than an Applicative one, but that's mostly a historical curiosity. In principle, there are no significant reasons for *> and >> to have different names, and a new incarnation of Haskell would do just fine with only one of them.

Two hundred eight

Back to parseExpr. Let's look at the important stuff now. We need to implement all of parseLam, parseApp and parseVar. Let's start with the latter, with parseVar, the simplest of them all.

The purpose of the Var constructor is to identify a "name" being used as an expression so that its meaning can be properly determined later on. That's all. For example, consider the following expression:

```
f x
```

We are seeing an application of a function named f to an expression named x. But what are f and x, exactly? Well, it doesn't really matter yet. We can represent each of them as Var "f" and Var "x" respectively, App one to the other, and be on our merry way. Later on, someone will hopefully bind actual values to these names using Lam, and everything will work out just fine. But that doesn't concern Expr, which at this point is merely the representation of our language's syntax as a Haskell datatype.

What is a "name", anyway? As pretty much everything else, it's up to us to decide. To keep things simple, we hereby decide that a "name" shall be one or more consecutive lowercase characters between the *a* and the *z* of the English alphabet. Let's start small. Let's first parse *one* character between *a* and *z*.

```
parseAZ :: Parser Char
parseAZ = parse1 (\c ->
  case elem c "abcdefghijklmnopqrstuvwxyz" of
    True -> Right c
    False -> Left "Expected 'a'..'z'"))
```

elem, of type Eq x => x -> [x] -> Bool, tells us whether the x given as first parameter is contained in the list [x] given as second parameter. We are using it here to check that c, the Char we are trying to parse, is between *a* and *z*. This is not a very efficient approach to searching, mind you. elem will check the list Chars one by one, in the order they are listed, insofar as it's necessary in order to determine whether the Char in question is there or not. Yes, elem is inefficient. Picture yourself wondering what *zenzizenzizenzic* means and having to read all the words in the dictionary in order, without skipping any, until we can finally see what it means. Terribly inefficient. There are more efficient ways to implement this, but we don't care about them at the moment. This is good enough for us, for now.

Knowing how to parse *one* character between *a* and *z*, we can simply parse *one or more* of them using some in order to finally obtain a "name".

```
parseName :: Parser String
parseName = some parseAZ
```

Actually, let's allow hyphens to be part of a name too, so that we can write things such as player-two or tuna-sandwich.

```
parseName :: Parser String
parseName = some (parseAZ <|> fmap (const '-')
                                   (expect1 '-'))
```

Come to think of it, our expect1 function, currently of type Parser Unit, might as well produce as output the Char we give it as input, when successful. It's not strictly necessary, but it can make things a bit more convenient at times. Like now.

```
expect1 :: Char -> Parser Char
expect1 a = parse1 (\b ->
  case a == b of
    True -> Right a
    False -> Left ("Expected " <> show a <> " got " <> show b))
```

And, as long as we are worrying about this, we could also change expect so that it has type String -> Parser String and does a similar thing. We can imagine how its implementation would change, there's no need to write it down. Anyway, with this new expect1, the implementation of parseName becomes a bit easier on the eyes.

```
parseName :: Parser String
parseName = some (parseAZ <|> expect1 '-')
```

Now, a "name" is not just any String, but rather, it is a String that satisfies the constraints we decided it should satisfy, like being at least one character long and only containing characters between *a* and *z* or a hyphen. With this in mind, we should probably give a different type to these "names", lest we accidentally mix them up with other less specific Strings.

```
data Name = Name String
```

Name is merely a wrapper around String. Ideally, anytime we see a value of type Name, we should be able to safely assume that we are dealing with a well-formed name. Now, nothing prevents us from manually constructing a malformed Name by applying the Name constructor to an undesired String like "*/hs!heLLo!". Worry not, we'll deal with that in the future. For now, let's just assume our Names are always as well-formed as desired.

```
parseName :: Parser Name
parseName = Name <$> some (parseAZ <|> expect1 '-')
```

This new version of parseName merely wraps the output of some ..., of type String, as a value of type Name. It achieves this by using <$>, which, if we recall, is just an infix name for fmap. So, assuming we modify our Expr type to talk about Name rather than String everywhere *names* are involved, we can proceed to implement parseVar in a very straightforward manner.

```
parseVar :: Parser Expr
parseVar = Var <$> parseName
```

parseVar simply modifies the Name output from parseName by applying the Var constructor to it. There's nothing more to it. Here's the updated definition of Expr, talking about Names rather than Strings.

```
data Expr = Lam Name Expr
          | App Expr Expr
          | Var Name
```

This concludes our implementation of parseVar and friends.

```
> runParser parseVar "x"
Right (1, "", Var (Name 'x"))
> runParser parseName "tuna-sandwich"
Right (13, "", Name "tuna-sandwich")
> runParser parseVar "89'
Left (0, "Expected 'a'..'z'")
> runParser parseVar "t!!"
Right (1, "!!", Var (Name "t"))
```

Two hundred nine

Next, let's implement parseLam. Remember, in this case we are trying to parse things such as foo → bar.

```
parseLam :: Parser Expr
parseLam = parseName >>= \n ->
           some whitespace >>
           expect1 '→' >>
           some whitespace >>
           parseExpr >>= \e ->
           pure (Lam n e)
```

We parse the Name to which to bind the input of this lambda expression, some whitespace, an arrow symbol →, some more whitespace, the Expr that will be the body of this lambda expression, we put the Name and the Expr in a Lam constructor, and we are done.

Again, notice that there's nothing monadic about this code. We are using

>>= and >> merely as a matter of syntactic convenience, but we could as well have used <*>, the birds, and other `Applicative` friends.

What about `whitespace`? How is that implemented?

```
whitespace :: Parser Char
whitespace = parse1 (\c ->
  case elem c " \n\r\t" of
    True -> Right c
    False -> Left "Expected whitespace")
```

Similar to `parseAZ` before, we check whether the `Char` we are currently trying to parse is an `element` among `" \n\r\t"`, all different ways of saying "whitespace". For example, through the space-bar in our keyboard, our enter key, or our tabulation key.

Nobody really knows why `'\n'` means "new line", `'\t'` "tabulation", nor why all other *character escape sequences* mean what they do. Through a series of secret rituals, this lore has been passed down for generations, and that's why we are still able to interpret them. Now, it's your turn. Prepare to receive this knowledge.

Just kidding. You can refer to the Haskell documentation to discover what each of these characters mean. It's not very important.

In any case, `whitespace` succeeds if it encounters one of these. In `parseLam` we use `some whitespace` to ensure that we encounter at least *some* whitespace. That is, we expect `whitespace` to succeed once, but possibly more times, too.

And while we are worrying about whitespace, perhaps we should make sure we ignore any extra whitespace we find within our parentheses in `inParens`, too. Yes, let's do that.

```
inParens :: Parser x -> Parser x
inParens p = expect1 '(' >>
             many whitespace >>
             p >>= \x ->
             many whitespace >>
             expect1 ')' >>
             pure x
```

Notice that we are using `many` this time, rather than `some`.

```
many :: Applicative f => f x -> f [x]
many fx = some fx <|> pure []
```

Whereas `some` expects *one or more* executions of `f x` to succeeed, `many` expects *zero or more* of them. Both of these functions come out of the box with Haskell. Are you confused by the names `some` and `many`? You are doing great, then. So are we.

And for completeness, let's be sure we skip any whitespace leading parseExpr, too.

```
parseExpr :: Parser Expr
parseExpr = many whitespace >>
            asum [ parseLam, parseApp, parseVar
                 , inParens parseExpr ]
```

This will give us a lot of flexibility regarding how we physically arrange our code. For example, the following would be valid syntax now:

```
x →       y →
z → perhaps
undesirable
```

We are almost there.

Two hundred ten

Finally, parseApp. This one will be tricky, so pay close attention. This is a valid function application in our language:

```
f x
```

And so are these, which merely add some redundant parentheses to f x.

```
(f) x
f (x)
(f) (x)
```

Also, there's no need for f and x to be plain names. They could, for example, be lambda expressions.

```
(a → a) x
f (a → a)
(a → a) (a → a)
```

Or even function applications applied to function applications.

```
(((f a) b) c) (g (h d))
```

Generally speaking, in our language, in a function application f x, each of f and x can be either a plain name lookup like foo or z, or *any* other expression, as long as it's surrounded by parentheses. Let's make a first attempt at implementing parseApp.

```
parseApp :: Parser Expr
parseApp =
  asum [parseVar, inParens parseExpr] >>= \f ->
  some whitespace >>
  asum [parseVar, inParens parseExpr] >>= \x ->
  pure (App f x)
```

This implementation would cover all the examples we just described. But consider (f a) b and f a b. We want these two expressions to mean exactly the same. The only difference between them is the presence or not of some redundant parentheses. However, as far as our current Parsers are concerned, those parentheses are mandatory.

```
> runParser parseApp "(f a) b"
Right (7, "", App (App (Var (Name "f"))
                       (Var (Name "a")))
                  (Var (Name "b")))
```

(f a) b parses successfully as the result of the application of f to a, whose result is further applied to b. This is good, this is what we want. But look at what happens when we try to parse f a b instead.

```
> runParser parseApp "f a b"
Right (3, " b", App (Var (Name "f"))
                    (Var (Name "a")))
```

Not good. The application of f to a parses successfully, but then we get " b" as leftovers. The problem is that parseApp didn't attempt to parse anything beyond the first function parameter a. We need to fix that.

When we look at the expression f a b c d, either in Haskell or in our own language, we must acknowledge that we are merely looking at syntactic sugar for (((f a) b) c) d. Now, do you see all of those parentheses piling up to the *left*? What does it remind you of? What was that thing that piled up parentheses on the left? That's right, it was the *left fold* of a list.

```
foldl' :: (y -> x -> y) -> y -> [x] -> y
```

When we left-fold a list, say [b, c, d], using some operation like + and an initial accumulator a, what we are doing is turning said list into an expression ((a + b) + c) + d. The parentheses pile up to the left, with the initial accumulator at the leftmost position. Now, imagine if all of a, b, c and d were values of type Expr, and our operation was App. What would happen? App (App (App a b) c) d of course, which is exactly what we want. Let's implement this.

As before, the initial value for our fold, the leftmost expression in our source code, will be the Expr resulting from either parseVar or yet another parseExpr wrapped in parentheses.

```
parseApp :: Parser Expr
parseApp =
  asum [parseVar, inParens parseExpr] >>= \x -> ...
```

This corresponds to parsing the a in a b c d, say. What comes next is some whitespace and b. And then some more whitespace and c, and some more whitespace and d. Generally speaking, after parsing the leftmost expression, we'll always have a sequence of one or more expressions afterwards, each of them preceded by whitespace. And just like our a, these expressions will be either a simple name lookup, or a more complex expression necessarily wrapped in parentheses. Let's get some of that, then.

```
parseApp :: Parser Expr
parseApp =
  asum [parseVar, inParens parseExpr] >>= \f ->
  some (some whitespace >>
        asum [parseVar, inParens parseExpr]) >>= \xs ->
    ...
```

Now we have f of type Expr, and xs of type [Expr], containing at least one Expr in it. All that remains is to foldl' this using App as our combining operation.

```
parseApp :: Parser Expr
parseApp =
  asum [parseVar, inParens parseExpr] >>= \f ->
  some (some whitespace >>
        asum [parseVar, inParens parseExpr]) >>= \xs ->
  pure (foldl' App f xs)
```

This implementation may be a bit hard on the eyes, but it works just fine.

```
> runParser parseApp "f a b"
Right (5, "", App (App (Var (Name "f"))
                      (Var (Name "a")))
                (Var (Name "b")))
```

Can we clean it up? Sure, here's an initial attempt at improving it by using an applicative style, rather than a monadic one.

```
parseApp :: Parser Expr
parseApp = foldl' App
  <$> asum [parseVar, inParens parseExpr]
  <*> some (some whitespace >>
            asum [parseVar, inParens parseExpr])
```

In this implementation, we applicatively apply foldl' App to functorial values of type Parser Expr and Parser [Expr], leading to a Parser Expr result. From an aesthetical point of view, this is perhaps a bit better than before. However, there's still an issue with repetition. We are saying asum [parseVar, inParens parseExpr] twice. What about avoiding that, too? Sure, we can do that, but maybe later.

For now, surprisingly perhaps, we are done. Really, that's all it takes for us to write a Parser for expressions in our toy language. The tally says we wrote about 45 lines of code in total, of which 30% were types. And mind you, that includes the entire definition of our language, too.

Two hundred eleven

Let's write some programs in our language, then. Granted, we haven't yet implemented anything that *computes* Exprs as something meaningful, but that shouldn't prevent us from writing some code that parses successfully and, eventually, once everything else is in place, will achieve the particular goal we programmers had in mind. We already saw how to implement the identity function:

```
x → x
```

Remember, just like in Haskell, names don't really matter to the language. We could have used a name other than x and the meaning of our program would be the same.

```
what-is-a-name → what-is-a-name
```

Great. The identity function, that's our first program. Now let's try something more interesting. Let's try to implement the constant function, Haskell's const.

```
x → y → x
```

This function takes two parameters as input, and returns the first one, ignoring the second. Easy enough.

And by the way, in case it wasn't made sufficiently clear, let's remind ourselves that we are writing *expressions*, not types, even if they look very much like Haskell types. It's very easy to get lost in the beauty of this syntax and forget what's going on.

Let's try something more complicated, let's implement Haskell's flip, the higher-order function that converts a function x -> y -> z to a function y -> x -> z.

```
f → (y → x → f x y)
```

Things are getting trickier. Notice that the parentheses are there only to emphasize that this function takes f as input, a function, and returns yet another function taking y and x as input. But, strictly speaking, those parentheses are redundant.

```
f → y → x → f x y
```

Our language doesn't understand types, so we can't really specify what the type of f is. Nor that of x, nor that of y. We can see that f is applied to x and y in a different order than that in which they are provided to y → x → ..., and from this fascinating event we deduce that f is expected to be a function taking two parameters as input in a particular order. But insofar as our language's "type-system" is concerned, f might as well have been a horse. Nothing prevents us from writing an "ill-typed" implementation of flip such as f → (y → x → f y x). Or even f → (y → x → x x x x x x), why not? Let's go wild, right? Who cares? There are no types anyway.

That's silly. Types are there, whether we acknowledge them or not. Only a function can be applied, only two numbers can be multiplied, only two booleans can be conjoined. If we get this wrong, as we most likely will without types, our expressions will be gibberish. We are far from civilization, we better be careful.

In any case, does our correct implementation of flip parse? Sure it does.

```
> runParser parseExpr "f → (y → x → f x y)"
Right (19, "", Lam (Name "f")
                  (Lam (Name "y")
                    (Lam (Name "x")
                      (App (App (Var (Name "f"))
                                (Var (Name "x")))
                        (Var (Name "y"))))))
```

It barely fits on this page, but if we look hard enough, we'll see everything is there. Luckily we'll never have to read things like this beyond these toy examples.

Two hundred twelve

We know how to define functions, we know how to apply them, and, well, that's all we know how to do. Actually, that's all we *have*. What about other things, like, say, telling things apart as we do with sum types in Haskell, or putting them together as pairs? Surely that's something we'll want to do. Indeed. Let's start with sum types. Here's Haskell's Either:

```
data Either a b = Left a | Right b
```

A Haskell value of type Either a b contains *either* a value of type a *or* one of type b, as discriminated by the choice of constructor. Easy enough. Except in our language we don't have types nor constructors, so we can't *exactly* replicate this. Moreover, we don't have pattern-matching.

```
foo :: Either Natural String -> Bool
foo = \e -> case e of
              Left n -> even n
              Right s -> isEmpty s
```

Here, foo will transform a given Either to a Bool by pattern-matching on its distinct constructors Left and Right, and depending on which of them it encounters, it will perform a computation suitable for their respective Natural or String payloads.

```
> foo (Left 6)
True
> foo (Right "Hello")
False
```

Pattern-matching is handy. But, it turns out, it's not strictly necessary for us to discriminate between different constructors.

```
foo :: Either Natural String -> Bool
foo = \e -> either even isEmpty e
```

This new implementation of foo doesn't use pattern-matching at all.

Instead, it relies on the higher-order *function* named either to apply even to the Natural number in e if e was constructed with the Left constructor, or isEmpty to the String value in e if it was constructed with Right.

```
> foo (Left 6)
True
> foo (Right "Hello")
False
```

This new foo achieves exactly the same as the one before *without* explicitly pattern matching on the Left and Right constructors. Well, almost. In truth, we are cheating, because the reason why this works at all is because either *itself* is doing the pattern-matching.

```
either :: (a -> c) -> (b -> c) -> Either a b -> c
either = \f g e -> case e of
                     Left a -> f a
                     Right b -> g b
```

That is, we haven't *really* avoided pattern-matching on the Either constructors, have we? In foo, we merely delegated the hard work of pattern-matching on the Left and Right constructors to the either function. Could we avoid that? Could we prevent either *itself* from pattern-matching on the Either constructors? Not exactly, but kind of. Pay close attention.

First, simply as a matter of convenience for what comes next, let's rearrange the arguments to the either function so that the value of type Either a b is its *first* parameter, rather than its third. This is not strictly necessary, but it'll make our exploration more convenient.

```
either :: Either a b -> (a -> c) -> (b -> c) -> c
```

Here's what the latest foo function would look like using this new either which merely rearranged the order of its input parameters.

```
foo :: Either Natural String -> Bool
foo = \e -> either e even isEmpty
```

Great. We still get an Either named e as input, and apply either to it and other things in order to eventually obtain a Bool. Now, imagine for a minute that rather than taking an Either Natural String as input and applying either to it, we were given an either already partially applied to said Either Natural String as input. Why? For reasons that will become apparent soon. Hang in there. We want to be able to say this:

```
> foo (either (Left 6))
True
> foo (either (Right "Hello"))
False
```

That is, rather than applying foo to a value of type Either Natural String directly, we want to apply foo to the either function *itself partially applied* to a value of type Either Natural String. What would change? Well, for one, the type of fco.

```
foo :: ((Natural -> Bool) -> (String -> Bool) -> Bool) -> Bool
foo = \f -> f even isEmpty
```

The f that foo now receives as input, which we expect to be an either partially applied to some value of type Either Natural String, will be applied to even, a function of type Natural -> Bool, and to isEmpty, a function of type String -> Bool, in order to obtain a value of type Bool that foo can return. The type of f merely follows from this usage. Haskell would fully infer it as such, were we to ask for its logical guidance.

Why does this matter, anyway? What have we accomplished? Well, look again at what we've written. Really, look. Do you see *any* mention of Either, Left, Right, or of the either function? Me neither. Puff. Gone. All that remains are functions.

Sure, in our example we used either and Left and Right to construct the function to which we applied foo.

```
> foo (either (Left 6))
True
> foo (either (Right "Hello"))
False
```

However, nothing prevents us from implementing the function we give to foo without using either, Left and Right *at all*. All foo expects is a function of type (Natural -> Bool) -> (String -> Bool) -> Bool, and as long as we are able to provide one, it would be fine.

```
> foo (\g _ -> g 6)
True
> foo (\_ h -> h "Hello")
False
```

Said function receives *two* other functions as input, here named g and h. If we are pretending to be the Left constructor, then all we have to do is apply the first function g to what would have been the payload of the Left constructor. In our case, as in all previous examples, 6. Otherwise, if we want to pretend to be the Right constructor, we can apply the second function h to what would have been the payload of the Right constructor, namely "Hello" in our case.

In other words, \g _ -> g 6 achieves *exactly* the same as either (Left 6) would, and _ h -> h "Hello" achieves *exactly* the same as either (Right "Hello") would. They do so for or *our* version of either anyway, the one whose order of arguments has been rearranged, but a different arrangement can be arranged, too.

Alright. Let's give names to \g _ -> g a and _ h -> h b, for any choice of a and b, not just 6 and "Hello".

```
left :: a -> ((a -> c) -> (b -> c) -> c)
left a = \g _ -> g a

right :: b -> ((a -> c) -> (b -> c) -> c)
right b = \_ h -> h b
```

Look at that. The left *function*, applied to what is supposed to be the payload of the Left constructor, gives us a function of type (a -> b) -> (b -> c) -> c, *exactly* like the one obtained by partially applying either to a value Left a, *exactly* like the one foo expects. And similarly, we have right taking care of the Right side of things. Does it work? Of course, why wouldn't it?

```
> foo (left 6)
True
> foo (right 8)
False
```

Functions. All we used were functions. Sure, things are noisier this way, but in principle, they work just fine. (a -> c) -> (b -> c) -> c is what we call the *Church-encoding* of the datatype Either a b. These two are *isomorphic* representations of a same idea, and as such, at any time we can replace one with the other without any loss of information.

Here is the proof that we can convert *from* the Church-encoded version of Either a b to an actual Either a b:

```
fromChurch :: ((a -> c) -> (b -> c) -> c) -> Either a b
fromChurch = \f -> f Left Right
```

Well, almost. Technically, we can't really write down the type of fromChurch as such. If we stare at this type long enough, we may realize why. Stare at it.

The problem is that we are saying the type that the function f ultimately returns is *some* c, while in fact, our usage of f forces c to be Either a b. That is, it's not *some* c what f returns, but Either a b exactly. Haskell notices our very specific choice and complains that the type we wrote for f is less precise than its usage otherwise indicates.

- Couldn't match type 'c' with 'Either a b'
  'c' is a rigid type variable bound by
  the type signature for:
    fromChurch
      :: forall a c b
      . ((a -> c) -> (b -> c) -> c) -> Either a b

A very precise error message from our friendly compiler, GHC. So, in truth, the type of fromChurch is a bit more precise than we said it was.

```
fromChurch :: ((a -> Either a b) ->
               (b -> Either a b) ->
               Either a b)
            -> Either a b
```

And of course, we can go in the opposite direction too. Here's the proof that given a value of type Either a b, we can convert it *to* its Church-encoded representation.

```
toChurch :: Either a b -> ((a -> c) -> (b -> c) -> c)
toChurch e = \f g -> case e of
                       Left a -> f a
                       Right b -> g b
```

Contrary to the fromChurch case, c can truly be anything here. This makes sense, for toChurch doesn't have any expectations regarding what c should be, so it's really up to f, g, and the eventual use of their output to decide what c it should be. As long as they agree, as parametricity mandates, toChurch will gladly accommodate their needs.

Do you recognize toChurch? Maybe if we remove those redundant parentheses...

```
toChurch :: Either a b -> (a -> c) -> (b -> c) -> c
```

That's right. toChurch is simply our rearranged either function. And *why* does the Church-enconding of a datatype matters to us? Because this encoding, named as such in honour of Alonzo Church, who first devised the Lambda calculus, only uses *functions* to accomplish what it does. Functions, the *only* means of abstraction in our language so far.

Two hundred thirteen

Without further ado, let's introduce our first ever datatype other than functions. Pretend datatype, anyway.

$$a \to (f \to g \to f\ a)$$

Left? Right? *Left*, that's right. Such is the definition of what we just called left in Haskell. That is, $a \to (f \to g \to f\ a)$ is the Church-encoded version of Haskell's Left constructor for Either a b. Similarly, here is its Righteous counterpart:

$$b \to (f \to g \to g\ b)$$

What can we do with that? Why, anything we would dare do with an Either a b in Haskell. For example, say we wanted to apply a function named ding to the a in a Church-encoded Either a b, if there happens to be an a at all. Otherwise, we'll apply a function dong to the b.

$$e \to e\ ding\ dong$$

This minuscule function takes as input an expression e expected to have been obtained by using left on some a or right on some b, and will apply ding to said a *or* dong to said b respectively. Obviously, seeing as in our language we have no types, e could be something else and things would fail. But, let's cross our fingers and hope for a second that e is what we expect it to be. In that case, the following equalities would be true:

```
(e → e ding dong) (left a)   =  ding a

(e → e ding dong) (right b)  =  dong b
```

But wait, what are left and right? We said a → (f → g → f a) was a suitable implementation of what *in Haskell* we called left, and similarly b → (f → g → g b) was a suitable implementation for what *in Haskell* we called right, but we never gave actual names to these things in *our* language, did we? Come to think of it, we never named the counterparts of Haskell's id, const nor flip either. Help. We need names for these things. Or, do we?

Two hundred fourteen

In Haskell, names like id, const, not or flip are defined at the *top-level,* thus they are available for us to use anywhere in our programs. We can mention const in an expression, and unless this name has been shadowed locally to mean something else —which by the way, will cause the compiler to warn us that we are doing something fishy— Haskell will resolve it to its top-level definition just fine.

```
> const 3 5
3
```

We don't have this in our toy language yet. We don't have a way to introduce top-level names. We only know how to introduce names by binding them in a lambda expression. Have we lost? Not really.

Imagine for a second that Haskell didn't allow us to refer to top-level names as it does now. Imagine we wanted to use const, but the name const didn't automatically resolve to the expression named const defined as a top-level name. How sad. Could we use const in an expression anyway? Kind of. We'd have to take const as an extra input to our expression, and expect that whoever plans to use our expression provides us with a suitable const as input at that time.

```
> (\const -> const 3 5) (\x _ -> x)
3
```

See? We got exactly the same result as before, but no top-level names were involved. In other words, we are suggesting that we don't really *need* to support top-level names in our language just yet. However, as you can hopefully see, this approach can become cumbersome really quickly. Just imagine having to explicitly take *all* of const, flip, fmap and left, say, as extra inputs to an expression that plans to use these now not-so-top-level names:

```
const → flip → fmap → left → ...
```

Rather noisy, isn't it? Yes. But for now, it'll have to be enough.

Two hundred fifteen

So, what does it mean to *run* a program, an Expr? For our current intents and purposes, it mostly means repeatedly performing *beta-reductions* on it until we can't do so any more.

Is (a → f → g → f a) z an expression that we should be able to *run*? Yes, yes it is, because we can beta-reduce by having the expression z take the place of the expression a inside the body of a → f → g → f a. That is, after one beta-reduction step, we end up with f → g → f z. Are we done? Can we perform yet another beta-reduction? No, we can't. f → g → f z is a function, but we have nothing to which to apply it, so this is it. This is the result of running our program. Quite disappointing, isn't it?

What does it mean, though? It's not obvious. If we look hard enough, we might recognize f → g → f z as the Church-encoded version of Left z, but maybe that's just a coincidence. Usually, when running our programs, we expect something more practical out of of them. We expect them to interact with the environment where they run somehow, and ultimately, produce some meaningful output in return. We'll work towards that later on, but for now, let's just focus on beta-reducing our expressions until we can't do it any more, a process we call *evaluating* our program. Let's start small.

```
eval :: Expr -> Expr
eval = \case
  App fun arg -> case eval fun of
    Lam name body -> eval (sub name arg body)
    expr1 -> App expr1 arg
  expr0 -> expr0
```

This Haskell function, eval, does what we've been doing so far each time we've talked about beta-reduction.

The first thing to notice is that this function will only do interesting things if it encounters an App constructor. If it doesn't, then it returns the same expression it was given as input, expr0. For now, this is fine. Later on we'll be smarter about these non-App scenarios.

If the expression we are dealing with is the Application of some expression fun to another expression arg then first we proceed to recursively evaluate fun to check if it is in fact a Lambda expression what we are trying to apply, and if it is, then we substitute with arg, in the body of the lambda expression, every reference to the name being *bound*. Finally, we recursively evaluate the resulting expression. We'll see the implementation of sub shortly.

On the other hand, if recursively evaluating fun resulted in an Expression other than Lam —which, because of how our recursion happens, can only mean Var— then we simply name it expr1 and return App expr1 arg, hoping that somebody else figures out what to do with that.

We can't evaluate this App expr1 arg any further yet, and perhaps nobody ever will, but, we may at least have evaluated fun to something more meaningful. For example, we may have gone from ((x → x) f) a, where fun originally corresponded to (x → x) f and arg to a, to the simpler f a. But without knowing what f and a are we can't really do anything beyond this, which is what the expr1 -> App expr1 arg pattern is saying. However, in reducing (x -> x) f to f, we made any future reduction work to be carried by somebody else, *if* any, much simpler.

Notice how the evaluation order we are giving to our language is *lazy*, as is Haskell's. That is, in our language, in a function application foo bar, only foo is necessarily evaluated right away. bar, instead, will only be evaluated *if* foo so requires. Changing our evaluator so that ours is a strict language, rather than a lazy one, would be quite straightforward. All we'd have to do is make sure we force the evaluation of eval arg and make use of that result, rather than arg, in the application of sub.

That's it. Those are the very basics of what it means to evaluate a lambda expression. Simple, isn't it?

Notice, however, that while ours is a lazy language, we will not be implementing the idea of thunks. In particular, we will not be doing anything to prevent a same expression from being evaluated more than once. Remember, in Haskell, when we say a + a, said a will be evaluated at most once. The equivalent expression in our toy language will not have such an optimization. It could, but due to time and space constraints of this book, it won't.

Implementing a strict language is much, much easier than a lazy one. Much easier. A lot. So much so, that one can't help but wonder how crucial this factor has been in the widespread development of hundreds of strict languages, compared to only a handful of lazy ones. Their designers, perhaps, were just being lazy.

Two hundred sixteen

The trickiest part of evaluating our small Lambda calculus, Expr, actually happens in sub, the variable substitution step.

```
sub :: Name -> Expr -> Expr -> Expr
sub n0 new = \case App fun arg -> ...
                   Var n1 -> ...
                   Lam n1 body -> ...
```

sub n0 new x is supposed to replace with the expression new, all references to the name n0 in the expression x. That is, everywhere Var n0 appears inside x, we replace it with new. Well, almost everywhere. Variable substitution is a tricky business.

Let's start with the case of pattern matching on the App constructor, the most straightforward case.

```
App fun arg -> App (sub n0 new fun) (sub n0 new arg)
```

That is, in case of App fun arg, we recursively apply sub n0 new to both fun and arg, and construct an App again. The expectation is that n0, to the extent that it appears inside either fun or arg, will be replaced with new. Simple enough. The next case deals with the Var constructor.

```
Var n1 -> bool (Var n1) new (n1 == n0)
```

In this scenario, we check whether n1, the name that appears as a payload of the Var we are dealing with, is equal to n0, the name we are trying to substitute with new. If it is, then we found what we were looking for, so we simply substitute it with new. Otherwise, if the Var mentions any other name, we just leave it as it is.

Two hundred seventeen

Incidentally, let's take a look at bool for a moment, the function we just used to handle the Var case in sub:

```
bool :: x -> x -> Bool -> x
bool x _ False = x
bool _ x True  = x
```

Let's rearrange the order of the arguments to bool so that the Bool input comes first.

```
bool :: Bool -> x -> x -> x
bool False x _ = x
bool True  _ x = x
```

Do you recognize that? Doesn't it remind you a bit of what we did with the either function? That's right, x -> x -> x is the Church-encoded form of boolean values. For example, if we partially apply our bool function to a True value, then we end-up with a function of type x -> x -> x that when applied to two values of type x, returns the second one. Yes. This is how we Church-encode booleans.

The choice of whether we return the first or second x depending on what we want to mean *true* or *false* was rather arbitrary. We could have done it the other way around. The important thing is to choose something and stick with it.

We can imagine having a function called false as follows:

```
f → t → f
```

And another one called true with this similar, but different, implementation:

```
f → t → t
```

If we compare them with left, implemented in our toy language as a → (f → g → f a), and right, implemented as b → (f → g → g b), we should be able to see the resemblance. Contrary to the Either constructors, the Bool ones don't carry a payload, so things are different in that sense. Rather than taking f and g, functions to be applied to the respective payloads, the Church-encoded representation of Bool merely returns the f or t that was provided to it, as necessary.

Does this work? Can we do something more interesting with it? Of course we can. For example, here is the implementation of not, the negation of booleans.

```
b → (f → t → b t f)
```

That is, given a Church-encoded boolean b, this function returns yet another Church-encoded boolean f → t → b t f that merely switches the order in which f and t are provided to b, so that f becomes the outcome of b being *true*, and t otherwise. So, for example, whereas true x y results in y, (not true) x y results in x. Extra parentheses just for emphasis. Here's the equational reasoning that proves this:

```
not true x y
  == (b → (f → t → b t f)) true x y
  == (f → t → true t f) x y
  == (t → true t x) y
  == true y x
  == (f → t → t) y x
  == (t → t) x
  == x
```

And once we have true, false and not, coming up with other boolean combinators is easy. For example, here is and, the conjunction of two booleans a and b:

```
a → b → (f → t → a f (b f t))
```

Notice how b is evaluated only if a is true. Generally speaking, seeing how ours is a lazy language, a Church-encoded boolean f → t → ... will only ever evaluate *either* f or t, not both. There's nothing special about this, it's just a consequence of the evaluation of Exprs being lazy. Similarly, here is or, the disjunction of two booleans a and b:

```
a → b → (f → t → a (b f t) t)
```

It's quite beautiful to see and and or side by side, to appreciate their resemblance.

Two hundred eighteen

A funny thing we should know about Haskell's bool is that nobody uses it. Well, I suppose some do.

Most languages out there use a special syntax for making decisions depending on a boolean value, rather than a normal function application or explicit pattern matching. Haskell, accommodating as it is, allows us to use this syntax too, if so desired.

```
if x then y else z  ==  bool z y x
```

That's it. Those two expressions mean and accomplish exactly the same.

```
> if 3 == 5 then "red" else "blue"
"blue"
> bool "blue" "red" (3 == 5)
"blue"
```

Hopefully we can see that this *if expression* must be an ad hoc feature of Haskell's syntax. If it weren't, Haskell would have interpreted this as the application of a function named if to a handful of arguments. But as we just saw, it didn't.

As we learned some time ago, we always want the actions we take based on a boolean value to be executed lazily. That is, for example, only if it's true that we've run out of fruit shall we go to the market to buy some more. However, as we said some time ago, laziness is *not* a feature most languages out there have. So, languages come up with a special *if expression* construct to make sure the expressions in the *then* and *else* branches are never evaluated unless necessary, depending on what the boolean value being scrutinized says.

In Haskell, this *if expression* is merely a convenience to accommodate newcomers looking for a familiar construction, but it is not truly necessary. We have bool, we have pattern matching, we don't *need* anything else. In most other languages, however, this syntax is indeed truly necessary. Imagine for a second that Haskell was a strict language. That is, a language where the only possible implementation we could give to the bool function was this:

```
fool :: x -> x -> Bool -> x
fool !f !t = \case False -> f
                   True  -> t
```

This implementation of bool, here named fool to highlight what we hope needs no further clarification, *forces* the evaluation of both f and t *before* deciding which one of them will be the ultimate value of this expression, depending on whether the Bool it gets is True or False.

```
> bool undefined "monkey" True
"monkey"
> fool undefined "monkey" True
"*** Exception: Prelude.undefined
```

This foolish behavior is exactly the one we would get in every strict language. Generally speaking, the bool we know and love is a function that, except in Haskell and a handful other lazy languages, cannot be implemented at all. Not with such grace, anyway. But it gets worse. Most languages don't even support the beautiful idea of pattern-matching on different constructors to tell apart one thing from the other. For example, our toy language doesn't. So, that's why they have this ad hoc syntax to treat booleans specially instead. You can think of *if expressions* as pattern matching for booleans and booleans only.

```
> if 3 == 5 then undefined else "blue"
"blue"
> if 3 /= 5 then "red" else undefined
"red"
```

How do languages allow this syntax to lead to a different behavior than the one a function would have? They make it part of their core language, and give it extraordinary evaluation rules. That's what they do. For example, were we interested in adding support for *if expressions* to our toy language, we could extend Expr this way:

```
data Expr = Lam Name Expr
          | App Expr Expr
          | Var Name
          | If Expr Expr Expr
```

This new constructor, If, would take three other Exprs as payload, corresponding to the x, y and z expressions in Haskell's if x then y else z. There would be a Parser converting whatever weird syntax we come up with to this If constructor, of course, and finally, eval would do what we expect it to do. That is, return y or z depending on what x says it should happen. Fortunately, our toy language is lazy, so we won't needing any of this. You can forget all about If now, other than the fact that you'll *need* it if you ever construct a strict language of your own.

Two hundred nineteen

Anyway, back to sub. This is what we have so far:

```
sub :: Name -> Expr -> Expr -> Expr
sub n0 new = \case
   App fun arg -> App (sub n0 new fun) (sub n0 new arg)
   Var n1 -> bool (Var n1) new (n1 == n0)
   Lam n1 body -> ...
```

Remember, sub n0 new x is trying to replace with the expression new all the references to the name n0 inside x.

Let's deal with the Lam case, by far the trickiest one. Here's what needs to happen: First, if we encounter a situation where the name we are trying to substitute, n0, equals the name being bound in this Lambda expression, n1, then there's nothing we need to do, and we can just return Lam n1 body unmodified.

```
Lam n1 body
   | n1 == n0 -> Lam n1 body
   | ...
```

Why? Consider the following expression written in our toy language:

$(x \rightarrow (x \rightarrow x))$ a b

This reduces to b, of course, *not* to a, as easily proved by some equational reasoning:

```
(x → (x → x)) a b
  == (x → x) b
  == b
```

What happens is that the innermost x —the rightmost, if you will— refers to the x that gets bound the *closest* to its usage site. That is, to the x that gets bound to b, not to the one that was bound to a before. So, it would be a terrible mistake for us to try and substitute said innermost x with a, and this is why we stop any further substitution of this name as soon as we encounter a Lambda expression binding *the same name* we are trying to substitute.

Great, we are being careful. Let's continue. Let's see what happens when n0 is different from n1, then.

```
Lam n1 body
   | n1 == n0 -> Lam n1 body
   | n1 /= n0 -> ...
```

The straightforward solution would be to return Lam n1 (sub n0 new body). That is, we keep the name being bound by Lam as it is, n1, and we replace all occurrences of n0 in the body of the Lambda expression with new. Unfortunately, this would be wrong.

```
Lam n1 body
   | n1 == n0 -> Lam n1 body
   | n1 /= n0 -> Lam n1 (sub n0 new body)
```

Sure, this would work just fine in situations like the following:

```
(x → y → x) a b
  == (y → a) b
  == a
```

That is, when we apply x → y → x to a, a will take the place of x in y → x to become y → a, which, later on, when applied to b becomes a. This is fine, nothing surprising there. However, let's look at what happens when the name we are trying to substitute also happens to be present in the expression to which we are applying our function. Let's change our example so that rather than applying our function to something named a, we apply it to something named y.

```
(x → y → x) y b
```

In principle, nothing should change. If we were to try this in Haskell using True and False as our choices of y and b, say, the result would be True.

```
> (\x -> \y -> x) True False
True
```

However, our broken evaluator behaves differently this time:

```
(x → y → x) y b
  == (y → y) b
  == b
```

That is, we get the b as result, the second argument to which we applied x → y → x, rather than y, the first one. This is wrong. This is the opposite of what we saw before, both in our language's (x → y → x) a b, which resulted in a, and in our Haskell example. The problem is in our substitution program.

```
sub :: Name -> Expr -> Expr -> Expr
sub n0 new = \case
  Lam n1 body
    | n1 /= n0 -> Lam n1 (sub n0 new body)
    | ...
  ...
```

Looking at the concrete case of evaluating our failing example ((x → (y → x)) y) b, here with extra parentheses for explicitness, the first Application we'll have to reduce is that of x → (y → x) to y. In doing that, we'll have assigned Name "x" to n0, Name "y" to n1, the expression Var (Name "y") will be our new, and the expression Var (Name "x") will be our body. So, when we finally say Lam n1 (sub n0 new body), what we are saying is that in body —that is, in Var (Name "x")— we should replace all occurrences of Var (Name "x") with Var (Name "y"), and use all of that as the body of a new Lambda expression that binds its input to Name "y". That is, Lam n1 (sub n0 new body) means Lam (Name "y") (Var (Name "y")), the identity function. Applying this to Var (Name "b") later on, will of course result in Var (Name "b") again. We broke something.

The problem lies in the fact that n1, the name being bound by the Lambda expression we are scrutinizing, appears *free* in new, by which we mean it's not being *bound* within that expression. We can't have that. Look, what can we say about f, g, and ∗ in the following expression?

    g → x → f (g x)

Both g and x are being *bound* as part of this expression. f, however, isn't. Where does f come from? Who knows. f is what we call a *free* variable, for it is *not* bound within this expression. Presumably, an outer lambda will bind the name f at some point:

    f → ... → g → x → f (g x)

From the point of this new *entire* expression, f is a *bound* variable. But as far as the g → x → f (g x) part is concerned, unaware of f's binding, f is *free*. Confusing? Perhaps. Here's a simpler example:

    foo

In the expression above, foo is a *free* variable. In the following example, however, foo is a *bound* variable:

    foo → foo

And we are not referring to one of the two appearances of foo in particular, but rather, to the name foo anywhere in this expression. foo is a *bound* variable everywhere in this expression.

With this new knowledge, let's go back to our substitution problem. We said that we could *not* replace references to a particular name being bound by a lambda expression *if* said name appears as a *free* variable in the new expression we are trying to replace it with. This is unfortunate. So, what do we do? We cheat, of course. If we leave the new expression as it is, free variables and all, and instead focus on coming up with a different and *unique* name to bind in the lambda expression we are scrutinizing, we'd avoid any overlap between said name and the free variables in new. No overlap, no problem.

That is, if we are trying to substitute the name g with the expression f x in x → g x, say, and failing to do that because x, the name being bound in x → g x also appears *free* in f x, even if these two xs potentially refer to different expressions, then perhaps what we can do is *not* use x as the name we bind in x → g x, and use for example y instead, leaving us with y → g y rather than x → g x. And this works just fine, because other than the different names we are giving to these variables, there's nothing semantically different between these two functions. We say that these two functions are *alpha-equivalent*, and we call the process of converting one to the other *alpha-conversion*. We briefly mentioned this a long time ago, actually, when we were Haskell toddlers.

Here is more or less what our implementation will look like, once we add an alpha-conversion step.

```
sub n0 new = \case
  Lam n1 body
    | n1 == n0 -> ...
    | isFree n1 new ->
        case freshName new of
          n2 -> Lam n2 (sub n0 new
                          (sub n1 (Var n2) body))
    | ...
  ...
```

That is, when we are trying to substitute references to the name n0 with the expression new inside a Lambda expression that binds the name n1, but n1 also happens to appear as a *free* variable inside new, then we need to come up with a fresh name that is *not* among the free variables of new, which we'll call n2, set n2 as the name that the returned Lambda binds, substitute all references to the name n1 in the body of the lambda with references to the name n2, and only then substitute references to the name n0 with the new expression in this modified body.

We'll look at freshName and isFree soon. Don't worry about them just yet.

Alpha-conversion is a very tricky matter. Many have perished here. We could have more interesting types to help us. For example, we could change the type Expr so that it mentions the names of the variables that appear free in it, preventing substitution of any of those names elsewhere. We could, yes, Haskell can do that. However, we won't, and in any case, there are approaches to solving the problem of overlapping names that are less error-prone than alpha-conversion. Were we being serious about building a language, and not too concerned about fitting all the content in a handful of pages, we'd take a look at them. Unfortunately —fortunately, actually— this book is neither about alpha-conversion, Lambda calculus nor evaluators. We are just passing by.

Two hundred twenty

Let's take a quick diversion to clean up what we just wrote.

```
case freshName new of
  n2 -> Lam n2 (sub n0 new (sub n1 (Var n2) body))
```

Look at what we are doing. We are using a case expression, a mechanism more often used to discriminate between *multiple* patterns or constructors, to give the name n2 to the expression freshName new so that we can use it more than once without having to repeat it nor evaluate it again. This is fine, but there is also a different and at times more convenient way of writing this: A let expression. Brace yourself for some new Haskell syntax.

```
let n2 = freshName new
in Lam n2 (sub n0 new (sub n1 (Var n2) body))
```

A let expression has the shape let n = *foo* in *bar*, and what it means is that within *bar*, we can use the newly minted name n to refer to the expression *foo*. In our example, we *let* n2 be a new name for the expression freshName new *in* Lam n2 .... Handy. And just like when defining top-level names in our Haskell code, we can give an explicit type to the names we bind. Often we don't *need* to, we can leave the type out and Haskell will infer it for us if possible, as we did just now. But types are there to help us more easily understand what's going on, so we'll often want to add them anyway.

```
let n2 :: Name
    n2 = freshName new
in Lam n2 (sub n0 new (sub n1 (Var n2) body)
```

Great. We can even use let to introduce *multiple* names at a time. For example, that last line is a bit noisy. Maybe we'd like to give a name to that inner application of sub to clean things up a bit.

```
let n2 :: Name
    n2 = freshName new
    body1 :: Expr
    body1 = sub n1 (Var n2) body
in Lam n2 (sub n0 new body1)
```

Whether this is better than before is up for debate. It's mostly a matter of style.

By the way, notice how the definition of body1 refers to n2, a name introduced as part of the *same* let expression. Fascinating. Generally speaking, the names being introduced in a let expression can be used within the body of the let expression —that is, within the expression mentioned in the in part— or within the expressions that are being named themselves. And perhaps even recursively, why not? Here, for example, we define the famous length function, unnecessarily using a let expression inside.

```
length :: [x] -> Natural
length xs =
  let f :: Natural -> [t] -> Natural
      f !n = \case [] -> n
                   _ : rest -> f (n + 1) rest
  in f 0 xs
```

In let expressions we can use equations, bangs, \case expressions, recursion, and everything else we can use in top-level names. There's no difference in that regard. It's mostly a matter of keeping the names we introduce local, rather than making them available for everybody else at the top-level. In this example, nobody outside the guts of length can access f.

And by the way, notice how in our example, xs, the name that is bound to the input of the length function, is a name to which we could refer within the definition of f if we wanted to. However, since we are not doing that, seeing as f's behavior is completely independent from xs, we could rewrite things a bit so that we only take xs as input when truly necessary and not before.

```
length :: [x] -> Natural
length =
  let f :: Natural -> [t] -> Natural
      f !n = \case [] -> n
                   _ : rest -> f (n + 1) rest
  in \xs -> f 0 xs
```

While previously our entire let expression had type Natural, as determined by the expression written in its in part, now it has type [x] -> Natural. In other words, this time we used a let expression to construct a function \xs -> s 0 xs, which may be a bit surprising, but is fine. And while we are shifting code around, we could also do without that eta-expansion \xs -> ... xs.

```
length :: [x] -> Natural
length =
  let f :: Natural -> [t] -> Natural
      f !n = \case [] -> n
                   _ : rest -> f (n + 1) rest
  in f 0
```

Also, notice how while in length we talk about a type-variable x, in f we mention a different type variable t. This makes sense, because despite the fact length's *usage* of f will specialize t to be x, f itself could work with any t. For example, here's a very silly implementation of length where in order to come up with our answer, we add the length of an empty list of Bools, which is obviously *zero*, to the actual length of the list we are concerned with.

```
length :: [x] -> Natural
length =
  let f :: Natural -> [t] -> Natural
      f !n = \case [] -> n
                   _ : rest -> f (n + 1) rest
  in \xs -> f 0 ([] :: [Bool]) + f 0 xs
```

But please don't do this. We only did it here to show that it can be done, to demonstrate that names introduced with let expressions can be as polymorphic as they need to be, just like top-level names can.

Two hundred twenty-one

Finally, here's the full implementation of sub. Tricky, but short.

```
sub :: Name -> Expr -> Expr -> Expr
sub n0 new = \case
  App fun arg -> App (sub n0 new fun) (sub n0 new arg)
  Var n1 -> bool (Var n1) new (n1 == n0)
  Lam n1 body
    | n1 == n0 -> Lam n1 body
    | isFree n1 new ->
        let n2 = freshName new :: Name
            body1 = sub n1 (Var n2) body :: Expr
        in Lam n2 (sub n0 new body1)
    | otherwise -> Lam n1 (sub n0 new body)
```

We haven't seen otherwise before. You'd think it is some kind of special Haskell syntax, but it's not.

```
otherwise :: Bool
otherwise = True
```

That is, writing otherwise in a guard, or in any other place where a Bool value is expected, is the same as writing True. Meaning, in this case, that the guard will *always* succeed, making it suitable to use as some sort of "catch all" of last resort. In our case, if n1 == n0 or isFree n1 new are True, we do something, *otherwise* we do something else.

otherwise doesn't add much to our understanding of Haskell, but it's very widely used, so we better get comfortable with it, rather than True, being used in guards. Let's take a look at isFree, now.

```
isFree :: Name -> Expr -> Bool
isFree n e = elem n (freeVars e)
```

Straightforward. We merely use elem, of type Name -> [Name] -> Bool here, to find out whether n is in freeVars e. That is, among the *free* variables of e.

Know that repeatedly looking for things in lists is a terrible idea from a performance perspective. And, given sub's implementation, we'll be doing it a lot. But for our current *didactic* needs, looking for things in lists is an excellent thing to do. Don't worry, though. There are many data structures more efficient than liked lists for this and other purposes, it's just that we won't be exploring them in this book.

Here's the implementation of freeVars, the function that will list all of the *free* variables in an expression. Remember, a free variable is a variable that is mentioned in an expression without also having been *bound* in it.

```
freeVars :: Expr -> [Name]
freeVars = \case
  Var n -> [n]
  App f x -> freeVars f <> freeVars x
  Lam n b -> filter (/= n) (freeVars b)
```

In Var n, n is obviously free since it's being mentioned there, yet it hasn't been bound. App f x merely concatenates together the free variables in f with the ones in x. Lam n b is the only case that deserves any special attention. There, we find all the free variables in b, the body of the lambda expression, and exclude from them any occurrence of n, the name being bound, using filter. Easy enough. And by the way, (/= n) is a partial application of the infix operator /= to one of its arguments. We could have writen (\a -> a /= n) instead, and it would have been the same.

Are we done? Not really. We are still lacking one function which we mentioned but never implemented: freshName.

```
freshName :: Expr -> Name
freshName e =
  let f :: String -> String
      f s | isFree (Name s) e = f ('x' : s)
          | otherwise = s
  in Name (f "x")
```

The purpose of freshName e is to return a Name that does *not* appear among the free variables of e. Our implementation achieves this in a crude and inefficient way. Essentially, if the name "x" is *not* among the free variables of e, we return that. Otherwise we try "xx", otherwise "xxx", otherwise "xxxx", otherwise "xxxxx", and we keep trying until we find the first Name made out of 'x's that does *not* appear among e's free variables. We are not particularly interested in what Name we come up with, but rather on its being *unique*. With that in mind, a bunch of 'x's are as good as any other Name. These are not for human consumption, but for the happiness of the machine.

And we are done. There's nothing left to do. We implemented sub, isFree, freeVars and freshName, and with that we have a fully functional eval that we can use to evaluate our programs, our Exprs.

For example, imagine we had an Expr corresponding to (x → f → f x) z (a → a):

```
example :: Expr
example = App (App (Lam (Name "x")
                        (Lam (Name "f")
                             (App (Var (Name "f"))
                                  (Var (Name "x")))))
                   (Var (Name "z")))
              (Lam (Name "a")
                   (Var (Name "a")))
```

Applying eval to example, would result in z, just as we expect it would:

```
> eval example
Var (Name "z")
```

Isn't that nice? We can parse our tiny language, and we can execute it. It's not much, but the fact that programming works at all doesn't seem so magical now, does it? We knew nothing, we had nothing, and look at where we are now, building languages and all.

Two hundred twenty-two

It's important to highlight that what we are building is not a compiler, but an *interpreter*. We haven't said much about interpreters yet, but it's high time we did.

A compiler takes the source code of our program as input and produces an executable version of that program as output, which can *then* be run to achieve the goal the programmer had in mind. An interpreter, on the other hand, takes the source code of our program as input and *executes it right away*, without having to convert it to an executable version first.

Interpreters sound better, right? Wrong. It's complicated. There are trade-offs between these two choices. For us, it turns out that implementing an interpreter is easier, which is why we are doing that. But we are abandoning a couple of things as a consequence of this choice.

It all comes down to compilers having the opportunity to observe the source code of our program in full without having to execute it. Compilers can use this valuable time to identify errors in names, they can type-check expressions, they can rearrange poorly performing code so that it becomes more efficient, and most importantly, they will translate our source code to a completely different language that can actually be executed.

Usually, this target language will be the primitive language understood by the machine where the program will ultimately run, but not necessarily. Sometimes, a compiler will output source code to be interpreted in a different language. That's acceptable too. We could have, say, a compiler that took Haskell source code as input, and produced source code in our own toy programming language as output to be interpreted later on. In the eyes of the compiler, and in ours as users finally able to run that Haskell program, the generated source code would be an acceptable executable form for a compiler to produce as output. Unless we were hoping for something else, that is, but a compiler wouldn't know anything about that.

Generally speaking, a compiler will always translate source code in one language, the *source* language, to source code in another language, the *target* language, to be interpreted right away without further analysis.

Sometimes this target source code is expected to be executed through a language interpreter, and sometimes it is to be interpreted directly by the computer, the machine, the hardware, without any aid from an intermediate language. In order to build such a compiler, however, we need to speak the language of the machine, and I'm afraid we won't be learning how to do that in this book.

Interpreters, on the other hand, optimistically try to run our program as they encounter new expressions to evaluate. Reckless. Wild. They don't care that among the next expressions our program may attempt to multiply a frying pan by a moose, likely rendering the previous computations moot, tainting all following computations too. Bad, bad boys. Not all is lost, though. It's still possible for interpreters to type-check expressions as they encounter them, right before executing them, hopefully aborting the execution of the program if said on-the-fly type-checking fails. Sure, a compiler could have prevented the ill-typed program from even starting to execute at all, but some safety is better than no safety at all. Seatbelts. When the alternative is to blindly perpetuate a mistake, aborting the execution of our program, *crashing*, as counter-intuitive as it may seem, becomes a desirable feature. *Crash program, crash,* for while crashing is undesirable and perhaps even catastrophic, perpetuating what's known to be wrong, or worse yet, what's believed to be right while facts prove otherwise, is just wicked, nefarious, a sign of neglect. *Abort interpreter, abort,* keep us from actively wreaking havoc.

Interpreters are excellent didactic vehicles. For everything else, compilers give us much more.

Two hundred twenty-three

A program runs. It's interpreted. Great. Now what? What does it mean to run the example from before, (x → f → f x) z (a → a), and get z as output? Not much. In fact, it probably means something is broken, since this is a reference to a thing named z which will never be resolved to something more meaningful, for nobody has said what z is.

It would be better if our program returned a value that made sense in a standalone manner, a value with no references to unresolved names. For example, if our program is supposed to tell us about the battery load in our computer, maybe getting a number representing such load as a percentage would be acceptable. To keep things simple, let's say we'll output a natural number between 0 and 100.

How do we represent a natural number in our toy language? Tricky, isn't it? We don't really have a way of saying 3 or 12 literally. Perhaps the obvious approach, considering what we have discovered so far, would be to Church-encode these numbers. Can we? Should we? Let's see.

In order to Church-encode Either, we created separate functions corresponding to each of our constructors. We had a function a → f → g → f a corresponding to the Left constructor, and a function b → f → g → g b corresponding to the Right constructor. When partially-applied to their respective first arguments, both of these functions result in yet another function f → g → z, the Church-encoded version of Either a b, where z will be either f a or g b, depending on whether we are dealing with a Left a value or a Right b.

In other words, generalizing things a bit, the Church-encoded version of a datatype usually takes as input as many functions as constructors the datatype has, and applies just one of them depending on which constructor was used to obtain this Church-encoded representation. In our language we don't have constructors in the Haskell sense yet, but conceptually, we can think of functions producing these Church-encoded representations as constructors.

How do we use this knowledge to model natural numbers, then? First we must identify which are the constructors for a natural number. In Haskell, any of the infinitely many natural numbers can be written down literally, like 3 or 891045, these are acceptable constructors for a value of type Natural. *Infinitely many?* that ain't good. We can't deal with infinitely many constructors in a Church-encoded representation expected to take as input one function per constructor. We need a finite number of constructors to Church-encode naturals. Luckily, we explored this idea before. We know that natural numbers have an alternative representation as an *inductive* datatype consisting of just two constructors.

```
data Nat = Zero | Succ Nat
```

That is, a natural number, here having the type Nat, is either Zero or the Successor to another Natural number. With this, for example, we represent 0 as Zero, 1 as Succ Zero, 2 as Succ (Succ Zero), etc. Great, we have a finite amount of constructors. Let's pick our Church-encoded representation now.

This is the function that we will call zero, corresponding to the Zero constructor of the Nat datatype:

```
s → z → z
```

And this is the function that we will call succ, corresponding to the Succ constructor of the Nat datatype, with redundant parentheses for emphasis.

```
n → (s → z → s (n s z))
```

zero would mean *zero*, of course, succ zero would mean *one*, succ (succ zero) would mean *two*, etc. How does this work? Both zero and succ n, for any choice of n, evaluate to a function s → z → x, where said x will be s recursively applied to z as many times as the natural number we want to represent indicates. In other words, if a Church-encoded natural was supposed to mean *zero*, then x will be z. If it was supposed to mean *one*, then x will be s z. If it was supposed to mean *five*, then x will be s (s (s (s (s z)))).

And surprisingly perhaps, with only this, we can implement the addition of natural numbers:

```
x → y → (s → z → x s (y s z))
```

Scary, I know. Let's assume that whole function goes by the name of plus, as in the namesake of the + operator. plus x y adds two Church-encoded natural numbers, resulting in yet another Church-encoded natural number s → z → x s (y s z). Is this true? Let's try and add together *one* and *two*, and verify that indeed we end up with *three*.

```
plus (succ zero) (succ (succ zero))
  == succ (succ (succ zero)))
```

succ zero is *one*, and succ (succ zero) is *two*. The expectation here is that adding these together, we end up with succ (succ (succ zero)), by which we obviously mean *three*.

Equational reasoning time. We'll first inline the definition of plus, and subsequently beta-reduce its two input parameters.

```
plus (succ zero) (succ (succ zero))
  == (x → y → s → z → x s (y s z))
     (succ zero) (succ (succ zero))
  == (y → s → z → succ zero s (y s z))
     (succ (succ zero))
  == s → z → succ zero s (succ (succ zero) s z)
```

Now, if it was up to our *lazy* evaluator, nothing else would need doing, seeing as the outermost expression is an unapplied lambda expression, which would be left as is until somebody forced its evaluation. But since we are learning, let's ignore that for a bit and keep evaluating things *inside* the lambda expression anyway, so that we can prove a thing or two.

```
  == s → z → (n → (s → z → s (n s z)))
             zero s (succ (succ zero) s z)
  == s → z → (s → z → s (zero s z))
             s (succ (succ zero) s z)
  == s → z → (z → s (zero s z))
             (succ (succ zero) s z)
  == s → z → s (zero s (succ (succ zero) s z))
```

Exhausting, isn't it? Feel free to skip to the end. I'll continue reducing this until it's not possible to do so anymore.

```
  == s → z → s ((s → z → z) s (succ (succ zero) s z))
  == s → z → s ((z → z) (succ (succ zero) s z))
  == s → z → s (succ (succ zero) s z)
```

See? This is why we have robots. It makes no sense to do these things by hand.

```
== s → z → s ((n → (s → z → s (n s z)))
              (succ zero) s z)
== s → z → s ((s → z → s (succ zero s z)) s z)
== s → z → s ((z → s (succ zero s z)) z)
== s → z → s (s (succ zero s z))
== s → z → s (s ((n → (s → z → s (n s z))) zero s z))
== s → z → s (s ((s → z → s (zero s z)) s z))
== s → z → s (s ((z → s (zero s z)) z))
```

We are almost there, I promise. Maybe we should have added together smaller numbers...

```
== s → z → s (s (s (zero s z)))
== s → z → s (s (s ((s → z → z) s z)))
== s → z → s (s (s ((z → z) z)))
== s → z → s (s (s z))
```

And we are done. No further reductions can be performed. This is as far as we can go. Did we add *one* and *two* up to *three*? Indeed. We didn't get succ (succ (succ zero) exactly, but we got s → z → s (s (s z)), which is the Church-encoded version of the natural number *three* to what succ (succ (succ zero)) eventually reduces:

```
succ (succ (succ zero))
   == (n → (s → z → s (n s z))) (succ (succ zero))
   == s → z → s (succ (succ zero) s z)
   == ... grunt work better left to the machine ...
   == s → z → s (s (s z))
```

In other words, we just proved that plus does what it says it does.

And just like that, if we are patient enough, we can do a myriad of other interesting things with these Church-encoded natural numbers. Like, say, multiply two numbers a and b:

```
a → b → (s → z → a (b s) z)
```

We won't do any further equational reasoning about this because we just can't stand the boredom, but you are welcome to try.

So, yes, we can represent numbers with nothing but functions. Is this something we want to do? No, of course not. We were just making a point. Nobody says "Oh no, the battery charge is at s → z → s (s (s (s (s (s (s (s z)))))))) percent, we are running out of time". That's silly.

Two hundred twenty-four

Let's add some *literal* natural numbers to our language. These are not strictly necessary, as we just demonstrated, but they'll be a great convenience for us. We would like a literal expression 3 to mean *three*, we want this number to be represented internally as the Haskell Natural number 3, and we want to forget about Church-encoding our numbers.

Now, this will be tricky, because all we have in our language, in Expr, are functions. It's Haskell who knows about Naturals. Our toy language knows nothing about them. So, what can we do?

Language implementors often encounter this situation. They would like their language to support a feature provided by the platform where programs written in said language will eventually run, but they understand that said feature is foreign to the means of abstractions provided by their own language. So, they need to bridge the gap somehow.

In our case, we want our toy language to reuse Haskell's representation of natural numbers as the Natural datatype. For that, we'll need to modify Expr so that it understands that it needs to treat these Natural numbers specially, different than the rest of the expressions in the language. Naturals in our language will become *primitive* expressions. That is, expressions that cannot be expressed in terms of other expressions in the language. Not exactly, anyway. Sure, we can use functions to encode something that *means* the same as a Haskell Natural number, for example, using Church-encoding, but it wouldn't be *exactly* the same as a Haskell Natural number in terms of how it is represented internally.

This is a bit disappointing, because our language will necessarily get more complex, but it is understandable.

```
data Expr = Lam Name Expr
          | App Expr Expr
          | Var Name
          | Nat Natural
```

We added a new constructor to Expr named Nat carrying a Haskell Natural number as payload. Simple enough, right? Wrong. The problem with extending our core language with new primitive expressions is that we'll need to add special support for them everywhere we've dealt with Exprs before. For example, now we need to be able to evaluate Nats:

```
eval :: Expr -> Expr
eval = \case
  Nat !x -> Nat x
  ... -> ...
```

Sure, evaluating a Nat x doesn't do anything besides forcing the evaluation of that x, but we still need to worry about it. Here, and everywhere else we deal with Exprs. So we need to decide whether this is something we'd like to support or not going forward, because if we do, then *everything* will become more complex for non-fundamental reasons.

We are going to accept this added complexity and cost if it means we'll have literal and more performant numbers in our language. We really want those. So, after modifying eval —and sub too, although we don't show it here— we proceed to adding a Parser for our Nat constructor.

```
parseNat :: Parser Nat
parseNat = fmap Nat parseNatural
```

We wrote parseNatural, of type Parser Natural, many chapters ago. We don't need to repeat it here. All we have to do is fmap the Nat constructor over parseNatural, name this new contraption parseNat, and add it to our most recent parseExpr so that natural numbers written as part of an expression in our toy language will be correctly picked up.

```
parseExpr :: Parser Exp¬
parseExpr = many whitespace >>
            asum [ parseLam, parseApp, parseNat
                 , parseVar, inParens parseExpr ]
```

Does it work? Sure.

```
> runParser parseExpr "f 198 9"
Right (7, "", App (App (Var (Name "f"))
                       (Nat 198))
                  (Nat 9))
```

Great. We can write down natural numbers. Unfortunately, that's pretty much all we can do with them. How can we, say, add them? The plus function we implemented before is not good anymore, seeing as we've stopped dealing with Church-encoded naturals. So, what now? Let's take a diversion.

Two hundred twenty-five

Consider what would happen in Haskell, were we to refer to an undefined name.

```
> foo
<interactive>:1:1: error: Variable not in scope: foo
```

Not in scope indeed. Here, the Haskell REPL, as it reads, evaluates, prints and loops, will try to resolve names like foo to their actual value, and complain out loud if some of those names haven't been defined. Our evaluator, however, whines not.

```
> eval (Var (Name "foo"))
Var (Name "foo")
```

It should, though. We've written our evaluator in such a way that if there's nothing useful that can be done to evaluate an expression, it simply returns it unmodified. Let's change this behaviour, let's teach our evaluator to whine about missing names.

The first thing we'll need to modify is the result type of eval. Rather than having it return Expr, let's have it return `Either EvalError Expr`.

```
data EvalError = NameError Name
```

We'll likely be adding more constructors to EvalError in the future, but for now, being able to whine about a Name not being defined shall be enough. Defined *where*, though? As eval stands today, Var references to Names defined inside the Expr we are evaluating as a consequence of having been *bound* by a Lam constructor, are always substituted in place right away. That is, we *never* end up with a Var reference to a Name that could have been substituted with an actual value but wasn't. In other words, whatever Var remains, however improbable, must be undefined. With this understanding in mind, let's simply change eval so that it reports a NameError when it encounters a Var.

```
Var n -> Left (NameError n)
```

Now, evaluating an expression like id 3, say, leads to the expected NameError, for there's no such thing named id in our language.

```
> eval (App (Var (Name "id")) (Nat 3))
Left (NameError (Name "id"))
```

That's a bit disappointing, though, isn't it? I'm sure we'd appreciate having names such as id, const, true and friends successfully refer to the things we expect them to. That is, to the *identity function*, the *constant function*, etc. Can we do that?

Two hundred twenty-six

Consider what would happen in Haskell, were we to refer to an undefined name.

```
> foo
<interactive>:1:1: error: Variable not in scope: foo
```

Not in scope indeed. This suggests that there exists a *scope* where variable names mean something, where variables *are*. Maybe we need to tap into that magical place somehow, put id, const and friends in there, in that *scope*, hoping they can be found when searched for. Yes, let's do that.

The *scope* of our expressions comprehends the variable names that have been bound by the many Lambda constructors that wrap a particular expression. However, we haven't really had to think about this explicitly because, rather than *looking up* the meaning of these names when we encounter references to them, we've been using a substitution mechanism that *replaces all references* to a particular name as soon as an expression gets bound to said name. There's nothing wrong with that, it's fine, but we'd like to enhance this with some variable name look-up, too.

Let's change our evaluator so that when dealing with Var (Name "id"), say, rather than complaining right away that the name id is not defined, it will try to look up id in this "scope" thing first. Maybe we are lucky and we find id there.

There where? Let's add a new input parameter to eval where we can look up names. Let's call it Env, short for *environment*, suggesting that the evaluation of an Expression doesn't take place in the void, but in an Environment where, among other things, we'll find the names *in scope*.

```
eval :: Env -> Expr -> Either EvalError Expr
eval env = \case
  Var name ->
    case envGet name env of
      Nothing -> Left (NameError name)
      Just ... -> ...
  ... → ...
```

That's the gist of it. If we are trying to evaluate a Var, rather than outright failing with NameError, we first try to *get* the expression associated with that name from the environment. Only if that fails, we fail.

```
envGet :: Name -> Env -> Maybe Expr
```

Given a Name and an Env, whatever that Env might be, envGet will find the expression associated with it, if any. For example, if we assume our Env has a definition of id in it, then looking up its Name should result in something interesting.

```
> envGet (Name "id") myEnv
Just (Lam (Name "x") (Var (Name "x")))
```

Is this enough? Maybe. Maybe not. First attempts at a poorly understood problem rarely are, and we are still trying to understand what we are doing. Let's see how far we can take this.

Two hundred twenty-seven

We have a way of finding whether a particular Name is in scope, and if it is, also the Expression associated with it. That's what envGet is for. But, how do those Names get in that scope in the first place? We'll probably need a corresponding envPut function too, won't we? A function that *puts* in what envGet later *gets* out. Sure, why not. Remember, it's up to us to make these things up, we don't need to ask for anybody's permission. If envPut is what we want, envPut we shall have. It might be wrong, of course, but that shan't constrain our imagination so soon. Later, perhaps.

```
envPut :: Name -> Expr -> Env -> Env
```

Presumably, envPut will take a Name, the Expression it refers to, the Env to which we want to add this Name, and it'll return a new, modified Env. Yes, this makes sense. And we'll probably need some sort of empty Environment too. In the beginnning, we'll need to envPut things somewhere, and an empty Env seems like a good starting point.

```
envEmpty :: Env
```

How do we implement Env itself? It doesn't matter, really. We can treat it as an abstract, opaque entity that responds as expected to uses of envGet and envPut, and that should suffice.

But *when* will we use any of this? Ah, that question... Think about Haskell for a second. How do we introduce new names in Haskell? We have a couple of ways, but most recently we learned about let. Can we let it be our inspiration? let's see.

```
let a = 1 in a + 2
```

The value of this whole expression is 3. First we introduce the name a as a new way of saying 1, and then, in the in part of this let expression, we add this a to 2 to finally obtain 3. Yes, let is a fine way of introducing names. Presumably, saying let a = ... made Haskell perform an operation not unlike envPut, and referring to this a later on triggered something like an envGet call.

Do we need a let in our language? Actually, no. Just like we don't need actual numbers and we can get by with Church-encoded numbers, we don't *need* a let. But, oh, how we *want* one.

```
(\a -> a + 2) 1
```

We can use normal lambda expressions to bind names, sure, we don't need let. But the ergonomics of doing that are not great. Consider this scenario:

```
let even :: Natural -> Bool
    even 0 = True
    even n = odd (n - 1)

    odd :: Natural -> Bool
    odd 0 = False
    odd n = even (n - 1)
in ...
```

This let introduces two names, even and odd, but this time, these names *recursively refer to each other.* How would we translate that into a mere lambda expression? It's not obvious, is it? We can't do the \even odd -> ... trick just as easy as we did before, for odd needs to know about even and even about odd. It can be done, but it's awkward. It's also very beautiful. But still, awkward. I encourage you to grab a pen and piece of paper and think about it for a while. But only for a while, lest you get a headache. Chances are you will get a headache. So, rather than trying to encode these name introductions as lambda bindings, we'll just support let in all its recursive beauty as a core expression in our language.

```
data Expr = Lam Name Expr
          | App Expr Expr
          | Var Name
          | Nat Natural
          | Let Env Expr
```

Handwaving things a bit, a Haskell expression like let a = *foo*; b = *bar* in *qux* —which by the way is valid Haskell syntax, with that ; and all— would translate to an Expr more or less like this:

```
Let (envPut (Name "a") foo
       (envPut (Name "b") bar
           envEmpty))
     qux
```

Do we need some new syntax for this Let? We do, we do. And, can it be a very beautiful syntax too? Sure, it can be as beautiful as our imagination permits. However, said syntax shall exist only in our imagination, for we won't be implementing it, nor the evaluation of Let, in this book. It's a tricky thing to do, and we only have so many pages we can write. We'll profit more from actually using let in Haskell, than from implementing it in our toy language. Let's be wise about it.

The important thing to notice here is that if our core language supported *let* expressions —as it magically does, by the power of our imagination— we could use this mechanism to easily bring names into scope by means other than lambda expression bindings. We could have names like id and const being brought into scope using Let, and afterwards, eval looking up the definitions for these names as necessary when evaluating expressions that refer to them.

Two hundred twenty-eight

Even if we are not going to actually implement *let* expressions in our language beyond mere wishful thinking, there's something we should know about them. Consider the following Haskell example:

```
let x = y * 3
    y = 2
in let y = 10
   in  x
```

What do you expect the value of this expression to be? Think about it. Think about it a bit more. Done thinking? Alright, the correct answer is 6. What happens is that when the definition of the name x refers to the name y, it means the y that is in scope at the time when the expression for x is defined. That is, at a time when y meant 2. That later on we introduce a different y with a different value 10 before evaluating x doesn't affect the fact that, by y, our definition of x meant 2. The expression y * 3 committed to the y it knew back then and no other.

This idea of tying names to the values they had in the scope of the definition that refers to them, rather than the scope where they are eventually used, is called the *static* or *lexical* scoping of variables. The name *lexical*, from the Ancient Greek λέξις meaning *word*, evokes the idea that by merely looking at the words where a name is mentioned, we should be able to figure out what was meant by that particular name at that time.

Alternatively, had the example above resulted in 30 rather than 6 —implying y meant 10, not 2— then we would have been dealing with a *dynamic* scoping of variables, where the value a name referenced by an expression takes is looked up in the *current* evaluation scope, rather than in the scope where that name was defined. But don't worry too much about it, dynamic scope is a terrible idea anyway, so you can forget all about it. Or better yet, actively avoid it. Forever. Haskell, and pretty much every other decent programming language out there, uses lexical scoping. And our imaginary implementation of name resolution will do so too.

For those who build languages, implementing lexical scoping is harder than implementing dynamic scoping. Way harder, because the evaluator needs to keep track of the names that *were* in scope at the time when an expression was defined. Merely using the names that are in scope where a definition is looked up is not enough anymore. Luckily for us, all of this is being implemented in our minds, so we don't really have to sweat and struggle with the many intricacies concerning this task.

It's called a *closure*, by the way. An expression, together with the environment wherein it's supposed to be evaluated, is called a *closure*. With lexical scoping rules, a closure is created when a name is introduced, containing both the expression associated with that name and the enviroment to which said expression had access when defined. The language evaluator is responsible for keeping track of those closures so that they can be retrieved and used later on as necessary.

Two hundred twenty-nine

Our language adopted Haskell's Natural numbers as a *primitive datatype*. But what about, say, adding two Natural numbers together? We can't really do that with our current Expr vocabulary. The only means of computation we have there are Lambda expressions and their Application, but the addition of Natural values using Haskell's Natural number addition function + doesn't quite fit there: + is a Haskell function, not a function in our toy language. We need to bridge that gap. We need to, once again, extend our core language so that it supports this feature, the addition of Natural numbers, as a *primitive operation*.

There are a couple of different ways to accomplish this, but to continue with Lam's tradition of taking just one input parameter at a time, even though in our case Haskell's + most certainly takes two, we'll add a new constructor Op to the Expr datatype, processing one parameter at a time. Op stands for *operation*, of course. *Primitive operation.*

```
data Expr = Lam Name Expr
          | App Expr Expr
          | Var Name
          | Nat Natural
          | Let Env Expr
          | Op (Expr -> Either String Expr)
```

Expr is growing, isn't it? Op carries a function Expr -> Either String Expr inside, which, as its type indicates, will transform the given Expr to some other Expr if possible. Using Op, this is how we'd encode Haskell's +:

```
opPlus :: Expr
opPlus = Op (\case Nat a -> Right (Op (\case
                       Nat b -> Right (Nat (a + b)))
                       _ -> Left "Not a natural number")
                   _ -> Left "Not a natural number")
```

That is, if the function inside the outermost Op is applied to an Expr constructed with Nat, then we name its Natural payload a and return another Op which, again, if applied to a Nat, will allow us to obtain another Natural number b to which it'll finally add the previous a using Haskell's +. Then, we pack the result inside yet another Nat constructor, and we are done.

Obviously, it could be that we are accidentally trying to apply this Op -wrapped function to something other than Nats, in which case we return Lefts with some meaningful description of the unfortunate situation. Which, by the way, in this case conveys the idea of a *type error*. This is how an unplanned type-checker is born. You see? Types are always there, whether we acknowledge them or not. If we try to add two Expressions other than Nats, then the evaluator shall convey this error right away and prevent any further execution of the program.

Now, we've simplified things a bit here. Depending on how our evaluator performs substitutions and variable lookups, it could be that rather than always applying Ops to Nats directly, sometimes we get Variables refering to Nats instead. In that case, there will need to be some deeper collaboration between Ops and our evaluator so that those Variables can be resolved to their primitive Natural values before they can actually be added using +.

Two hundred thirty

Op is a bit different from all the previous Expr constructors in the sense that it's only ever intended to be constructed internally, as part of our language interpreter, and not written by the users of said language using the language's own syntax. We write Ops in Haskell itself, not in *our* language.

In fact, at this point, as language implementors, we could, and probably should, split Expr into two different datatypes: one used exclusively for evaluation purposes, where all the Expr constructors we've seen so far are present, including Op, and another datatype listing only the expressions that have a corresponding representation as syntax, used exclusively for parsing and other purely syntactic purposes. This latter form is often called the *AST*, or *Abstract Syntax Tree* of the language.

```
data AST = AST_Lam Name AST
         | AST_App AST AST
         | AST_Var Name
         | AST_Nat Natural
         | AST_Let [(Name, AST)] AST
```

The name suggests that this is a suitable *abstract* representation for our *syntax* as a *tree*. That is, as the branches in a tree from nature are themselves made out of other branches, syntax is made out of more syntax. Yes, we are dealing with a *syntax tree* somehow. Yet, as far as our datatype is concerned, it doesn't really matter what said syntax looks like visually. It is *abstract*, thus the name.

Obviously, Op is excluded from this AST datatype because it is not part of our syntax, let alone its branchful abstract counterpart.

Having this datatype, for example, our language Parsers would produce values of type AST rather than Expr, to be converted to Exprs later on when it's finally time to evaluate them and not before. Until then, we may use values of this type AST for purposes of a more syntactic nature than those supported by Expr.

Why, though? What's so interesting about having our AST and expressions represented using different datatypes? For one, there's that priceless, young sense of accomplishment that comes from having all the constructors in our datatype serve a purpose. Op was excluded from AST because it served no purpose there. But now, all the AST constructors really do have a role as syntax in our language, and similarly, all the Expr constructors do too, in the sense that they all have some unique evaluation semantics. But more importantly, the reason why we want the abstract syntax tree and the expressions to be accurately described using *different* datatypes is so that they can evolve separately. Haskell, for example, is notorious for parading an extensive syntax, yet having only a small set of core expressions it must ever evaluate. This, in turn, tames the complexity of the related evaluation machinery by having it be as complex as it needs to be, but not more.

Two hundred thirty-one

Taking this AST idea to the extreme, some languages, Haskell included, expose vocabulary to talk about the AST of the language to programs written in the language itself.

What? Why? Imagine Haskell didn't have a special syntax for *if expressions*, yet we, *users* of Haskell —that is, people programming *in* Haskell, rather than programming the Haskell language itself— would like to add support for it. Something like when x then y else z. Yes, we'll use the word when rather than if in order to avoid making things more confusing. If we had tools to manipulate the language AST from within the language, then we could do this without having to ask for anybody's permission. Let's see how.

We are going to simplify things *a lot* in this following example.

```
when :: AST -> AST
when [x, "then", y, "else", z] = ["bool", z, y, x]
when other = other
```

Obviously, our AST is not a list of Strings as our patterns seem to suggest, but please ignore that and focus on what matters. As we said, we are simplifying things a lot, so let's adjust our expectations.

Normally, when x then y else z would be interpreted as the application of a function named when to five other expressions. But, what if we could use when not as a *function* taking expressions as input and producing other expressions as output, but rather, as some sort of syntactic transformation taking *syntax* as input and producing more *syntax* as output? What if writing when x then y else z in Haskell was transformed to bool z y x at a syntactic level, before any attempt to evaluate when x then y else z as a function application? We'd have a different expression to evaluate, but this expression would give us the exact behavior we were looking for. Imagine the possibilities.

This is part of what's commonly referred to as *meta-programming*, where we use programming to program the programs we program in our programming language. Quite often, in scenarios like these, things such as when are not called functions but *macros*. This is a fuzzy term, tough. Some languages call them something else, some use the word *macro* to mean something different, and others like Haskell still call them functions. Functions with a special type that only makes sense as a meta-programming step, though, so all is well in the world.

With some imagination we could, for example, use this mechanism to implement *if expressions* in languages with strict evaluation semantics, even if the core language didn't have explicit support for them. In fact, that's how *if expressions* are often implemented in languages where, syntactically, meta-programming blends in nicely with the rest of the language. Like Lisp, which is not a single programming language but infinitely many of them. You should learn about Lisp some time, it's interesting. To be fair, to a great extent, our toy language is one of these Lisps too, so you already know a thing or two about it.

Two hundred thirty-two

So, how do we use Ops if we can't use our language syntax for working with them? Well, we can't use the language syntax to *define* new primitive Operations, but we can most certainly refer to them by name.

The idea is that our interpreter will give a name to each of these Op expressions, likely using the Let mechanism we introduced before, and users of the language will be able to refer to said primitive Operations by Name and Apply them just as they would any other Expression. For example, plus 3 4 could be trying to say something along the lines of opPlus (Nat 3) (Nat 4). Of course, this implies that we modified our evaluator so that it knows how to Apply not only Lambda expressions, but also primitive Operations. Yet another interesting exercise we'll leave to our imagination.

For example, before running an Expression provided by the user, the interpreter could wrap it in a Let expression that assigns the name plus to opPlus, the primitive Operation we just defined, allowing a later reference to this name using Var to find it.

```
Let (envPut (Name "plus") opPlus ...)
    (...the actual user program to run goes here...)
```

Obviously, this happens behind the scenes, and users of our language need not care about the details of how it happens. Just like when we start GHCi and magically there are things called True, const or fmap there, in our toy language there will be something called plus.

With this, and our imaginary evaluator which now knows how to Apply an Op, if users of our programming language want to add two primitive natural numbers together, by applying the name plus to said numbers, as we did in plus 3 4, they'd be effectively adding primitive natural numbers using primitive operations without abandoning the comfortable surface syntax of our beautiful, unnamed toy language. But, you see, nowhere does the user of the language construct this plus function. From the point of view of the programmer that writes a program in our language, plus exists, it's part of the environment where the program executes, so it can be used. But there's no way to look at what's inside plus, nor would the programmer be able to recreate it without having control over the interpreter itself. It's primitive, it is part of the language, we are thankful for it, we embrace it, we use it, and we move on.

Two hundred thirty-three

Generally speaking, *primitives* are one of the ways in which a programming language integrates with the underlying platform where a program eventually runs.

Just like Haskell offers Natural to our toy language, the computer itself offers things that not even Haskell itself has. For example Natural itself, which in Haskell we use as if it were any other Haskell datatype, is often implemented using further primitive numeric datatypes offered by the machine where the program runs. This is why we have machines at all. They are supposed to give us the most primitive pieces on top of which we build everything else. Ultimately, we are limited by what these machines can or cannot do.

Mostly, they can do numbers, but perhaps more importantly, depending on how you look at it, they allow us to interact with the environment where our programs run. And by enviroment we don't mean that Env thing from before, but the actual *physical* environment where our program exists.

Computers interact with the internet to which they are connected, with the display screen on which we watch a movie, with the keyboard we use to type our programs, and with many other things like that. And all these interactions must be programmed, too. The machine offers the vocabulary we can use to interact with the outside world, but it's our job as programmers to use that vocabulary. So let's interact with that environment now.

Two hundred thirty-four

Interaction number one. Let's interact with our computer's keyboard, or any similar text input device you may have in that future of yours. Some kind of stick, maybe? A holo-something? It doesn't matter, we'll be ready for it anyway.

So, how do we do this? Actually, first, what do we *mean* when we talk about interacting with the keyboard, with the environment?

Say we have a function named uppercase of type String -> String that transforms any textual input it receives to uppercase.

```
> uppercase "chocolate"
"CHOCOLATE"
```

We typed all of those words with our keyboard. Is *that* what we mean by interacting with the keyboard, with the environment? No, not at all.

uppercase "chocolate" is part of our program definition. We already know about the "chocolate" at compilation time. There's no interaction with the environment happening in this program. We know, by the time we write our program, what the input to the uppercase function will be. There are no *unknowns* that must be discovered from the environment where and when the program eventually runs. None.

In fact, we knew so much at compilation time already, that we could have discarded the uppercase application altogether, directly replacing it with the expected output.

```
> "CHOCOLATE"
"CHOCOLATE"
```

Why not? Even before running uppercase "chocolate" we knew what we'd be getting back, so we weren't gaining much by that uppercase application, anyway. Who needs computers, right?

When we talk about interacting with the environment we are saying that it won't be until *runtime*, until the moment that our program is executed, that we'll discover *from* that environment a previously unknown value that our program can use as *input* for some of its calculations. Imagine for example that the purpose of our program was to tell us where we are in the world. We run it today and it says "Asia", we run it next week and it says "America". That is, assuming we travel a lot. Clearly, this program needs to access the environment where it runs to somehow figure out where in the world it is running. *That's* what we mean by "interacting with the environment".

But that's only half of the story. It turns out that just like our program will be able to discover things from the environment where it runs, the environment itself could discover things from the program, too. For example, it could be that after taking some inputs from the environment, the program processes them, and then, somehow, conveys the processed result back to the environment. Think, for example, of a videogame. We press a key, the videogame notices this, makes our character jump, and finally communicates this change in our character's state back to us by having the display screen render a new, different image about our videogame hero with his feet up in the air. We notice that. The environment changed.

In other words, these interactions with the environment are bidirectional. There is always some kind of *input* or *output*, or both, going in or out of our program as it runs. Unsurprisingly, in Haskell, where interesting ideas have types of their own, these environment interactions have a distinct marker in their types. Standing for *input, output*, we have IO.

Two hundred thirty-five

Here's one of these interactions that gets some input from the environment, getLine. This interaction expects us to type an entire line of text with our keyboard, and as soon as we are done, as signaled by the pressing of the *enter* key, we'll obtain that line of text as a String.

```
getLine :: IO String
```

The type of getLine says that it is an IO interaction that, as a consequence of having interacted with the environment somehow, will provide us with a String.

It shouldn't matter to us how getLine is implemented. To a good extent, we can consider it a primitive of the language. Strictly speaking, it is not a primitive, but that doesn't matter much to us at this moment. The documentation for getLine says what it does, its type says what this is, and that's pretty much all we need to know about it. All that remains is to perform this interaction with the environment.

Now, GHCi has some special magic for this. We'll profit from it. We know that if we submit to the REPL an expression of type, say, Bool or String, then GHCi will, behind the scenes, render it using show and print it back to us.

```
> True
True
> "ban" <> "ana"
"banana"
```

But if we submit to the REPL a value of type IO x, for some choice of x, then rather than just showing this IO x value —which, anyway, being the description of an *interaction* with the environment, has no sensible visual representation— GHCi will *perform* the IO interaction in order to obtain that x, and only then, after the interaction succeeds, will it show the obtained x.

```
> :t getLine
getLine :: IO String
> getLine
```

Nothing happens? Clack, clack, clack, my friend. getLine *interacts* with the environment, and what we are witnessing here is that interaction. getLine is *waiting* for us to clack that line away. We won't be getting that String in return until we type it, followed by a touch of the enter key.

```
> getLine
Where's my hat?
"Where's my hat?"
```

As promised, getLine got us a line of text, as a String, as a consequence of having interacted with the environment where it ran. Afterwards, GHCi showed it just as it would any other String. And in case it wasn't clear, that first line immediately after > getLine is supposed to have been typed by us, manually, using the keyboard.

Congratulations, we have interacted with the environment. Explicitly, anyway. In truth, we've already done it hundreds of times before.

Two hundred thirty-six

It'd be a shame if all we could do with that String produced by getLine was watch it be printed back to us in the REPL. Fortunately, that's not the case. We can do with that String anything we could do with any other String.

```
getLine :: IO String
```

How, though? getLine is a value of type IO String, not just String, so if we were to apply to it a function like, say, uppercase, of type String -> String, the type-checker would rightly complain.

```
> uppercase getLine
<interactive>:4:11: error:
  • Couldn't match type 'IO String' with 'String'
    Expected type: String
      Actual type: IO String
```

This is not unlike what would happen if we tried to apply the same uppercase function to a Maybe String, say.

```
> uppercase (Just "fantasy")
<interactive>:5:11: error:
  • Couldn't match type 'Maybe String' with 'String'
    Expected type: String
      Actual type: Maybe String
```

We've dealt with this scenario before, however, so we know that not all is lost. In this Maybe String scenario, we could easily pattern-match against the Maybe constructors and apply uppercase to the String in there, if any. Or even better, we could profit from the fact that Maybe is a Functor and fmap the uppercase function over our Maybe String, achieving exactly the same result.

```
> fmap uppercase (Just "fantasy")
Just "FANTASY"
```

Well, just like Maybe, [], Parser, Either x and so many other type constructors, IO is a Functor too, which means that we can use the same trick to uppercase the String *produced* by getLine, our value of type IO String.

```
> fmap uppercase getLine
Where's my hat?
"WHERE'S MY HAT?"
```

Easy there, no need to yell. It's probably there somewhere. The type of fmap, as we know, is `Functor f => (a -> b) -> f a -> f b`. So, seeing as in this example IO is our f, and String is both our a and b, the type of `fmap uppercase getLine` must necessarily be `IO String`. We can even give this loud IO interaction a name, too.

```
getLoudLine :: IO String
getLoudLine = fmap uppercase getLine
```

This is how we profit from these IO interactions. We can't really look inside IO as we can inside Maybes and so many other Functors. These IO things, once built, are a black box. We can only interact with their output through their functorial features.

```
> getLoudLine
I found it!
"I FOUND IT!"
```

Two hundred thirty-seven

What if we were really, really curious about the environment, and rather than getting *one* line from it, we tried getting *two*. Would that be possible? Would we even have computers if it wasn't?

The IO type constructor is an Applicative functor too, and, as we've probably forgotten by know, Applicative functors are there to allow us to combine two or more functorial values somehow. To accomplish this, among other tools, we had liftA2.

```
liftA2 :: Applicative f => (a -> b -> c) -> f a -> f b -> f c
```

That is, given a way to combine two values of type a and b into a single value of type c, and two *functorial* values f a and f b, we'd get yet another functorial value f c producing the desired combination c as output. In our case, we'll be pairing up the two Strings obtained from getLine and getLoudLine, both functorial values of type `IO String`, in a tuple, by using `(,)` of type `String -> String -> (String, String)` as the combining function given to liftA2.

```
> :t liftA2 (,) getLine getLoudLine
IO (String, String)
```

That is, we have created yet another IO interaction that produces not one but two Strings as a consequence of performing *two* separate interactions: getLine first, and getLoudLine afterwards.

```
> liftA2 (,) getLine getLoudLine
tomato
CHEESE
("tomato", "CHEESE")
```

We can confirm that getLine happened *before* getLoudLine by considering the order in which we provided the input lines, and noticing that the second line we wrote, "cheese", is the one that got yelled back to us. That is, the second line was obtained by means of getLoudLine, after getLine had already finished interacting with the environment.

So, yes, IO is an Applicative functor too. This allows us to use liftA2, sure, but also more interesting tools such as that sequence we know and love.

```
sequence :: Applicative f => [f a] -> f [a]
```

We'll learn later on that the type of sequence is even more general than what we've seen so far, but for now, this suffices. With sequence, we can turn a list of IO interactions producing outputs of a same type into a single IO interaction that runs all of those IO interactions in the order they were listed, collecting and re-producing their respective results in the same order they were originally obtained. For example, imagine we had a list of IO interactions producing a String each.

```
foo :: [IO String]
foo = [ getLine, pure "hello", fmap reverse getLine
      ] <> fmap (fmap uppercase) (take 2 foo)
```

Fancy list. Applying sequence to it would create a new IO interaction that when run, would run each of the listed interactions and collect their outputs, as promised.

```
> :t sequence foo
sequence foo :: IO [String]
> sequence foo
abc
def
ghi
["abc", "hello", "fed", "GHI" "HELLO"]
```

As soon as the IO interaction is done, GHCi prints back to us the resulting [String].

Beautiful, isn't it? Not only can we perform these IO interactions, but also we can name them, compose them, and manipulate them as we would any other Haskell value.

Two hundred thirty-eight

The predictable ordering of our composed IO interactions, however, where one interaction always happens *after* another one, is not really a feature of Applicative functors, but of Monads. We know this. It's *bind*, >>=, who says *this* happens before *that*. But of course, as we learned before, if an Applicative functor happens to be a Monad too, then we'll see tools like liftA2, sequence or <*> enforce the same ordering as >>= would. That's how things are supposed to be according to the Applicative laws. But still, the ordering really comes from the fact that our type constructor, IO, just like

Maybe, Parser and many others, is a Monad.

In practical terms, IO being a Monad means that we can use >>= to say *this* happens before *that*, but more importantly, it means we can have past interactions influence the behavior of future interactions, if any. Let's look at an example to better understand this.

Imagine we enjoyed poetry, and as such, we are always eager to read what other poets have to say. So, being the fine programmers we are, we go ahead and write a program that allows poets to type poems in our computer, one verse at a time. So far, unsurprisingly, poets haven't been able to agree on what a verse is. But for our purposes, we'll say a verse is one or more lines of nonsense grouped together as if they were a paragraph. A poem is usually made of one or more of these verses. Now, verses can have different number of lines, different rhymes, etc. So, how do we know when a verse ends? When will our verse-getting interaction with the environment stop? We'll just assume that a verse ends as soon as a poet writes an empty line. Finally.

Let's write the program that does that. It'll get from the environment every line the poet writes, until it encounters an empty one.

```
getVerse :: IO [String]
getVerse = getLine >>= \case
             "" -> pure []
             x  -> fmap (x :) getVerse
```

getVerse is not particularly different from the monadic code we wrote back when implementing Parsers. First we perform getLine, *and then* —that's more or less how we read >>=, remember— *if* the line that we got is empty "", we stop the interaction by purely returning []. We know that pure, of type Applicative f => a -> f a, creates a functorial value that does nothing interesting other than producing the given a. In our case, we are creating a "fake" IO interaction that doesn't really interact with the environment yet produces the given [] as output. Otherwise, if the line we got from getLine is not empty, we name said line x and prepend it to the list of Strings produced by a recursive call to getVerse. Of course, since getVerse is a functorial value, we must fmap this prepending function (x :) over it.

You see, a decision is being made about whether we should perform getVerse again or not *depending* on a *previous* IO interaction output. This is the essence of what it means to be monadic. It's not really about the ordering, but about the *dependence* of functorial values on previous outputs.

```
> getVerse
Roses are red.
Violets are blue.
We just wasted two lines.
And now we waste the fourth too.

[ "Roses are red.", "Violets are blue.",
  "We just wasted two lines.",
  "And now we waste the fourth too." ]
```

Notice the promised blank line marking the end of our verse, right before
GHCi shows us the result of our IO interaction with the environment.
Our implementation of getVerse noticed this empty line too, and aborted
any further interaction by purely returning [].

Two hundred thirty-nine

Using IO interactions we have obtained now, many times, inputs *from* the
environment. But we haven't really provided anything *to* it. Or, have we?
When Haskell evaluates an expression that we type into the REPL and
then prints it back to us, isn't that a form of interaction *to* the
environment, too? Indeed. Let's see if we can do something like that.

```
putStrLn :: String -> IO Unit
```

putStrLn, presumably standing for "put out a String line", is an IO
interaction that will put out —that is, print, display on the
environment— the given String as a standalone line of text. And as a result
of this interaction with the environment, we'll get a moot Unit back as
result. That is, we won't be getting anything useful from the environment
this time, but we will be providing something *to* it. We can try this in
GHCi as before.

```
> putStrLn "ding"
ding
```

GHCi, internally, has been performing putStrLn automatically for us all
along. Here we are doing it manually. The result, however, is a bit
different from just typing "ding" into the REPL:

```
> "ding"
"ding"
```

If we do that, what we see in return is the textual representation of "ding"
rendered using the Show instance for Strings. This is *not* the same as
displaying "ding" itself. Do you see the difference? There's "ding" the
String, and then there's the String that contains the representation of
"ding" the String. The latter, which we obtain using show, is the one that
GHCi shows when trying to render an expression.

```
> putStrLn (show "ding")
"ding"
```

This approach makes sense, considering that we expect the REPL to be able to display things other than Strings too, such as Bools, Natural numbers, etc. If we were to apply putStrLn of type String -> IO Unit to True, say, of type Bool, the type-checker would complain. showing that Bool first, though, would help.

```
> True
True
> putStrLn (show True)
True
```

So, yes, the REPL interacts with the environment, and it does it in two directions. It *reads* some input from the environment using an IO interaction like getLine, and after evaluating the input it was given as a Haskell expression, it *prints* out a representation of the result using putStrLn before starting this same loop again.

Two hundred forty

Handwaving things a bit, we could implement the Haskell REPL ourselves.

```
repl :: IO x
repl = putStr "> " >>
       getLine >>= \input ->
       let expr = parseExpression input
           output = evaluateExpression expr
       in putStrLn (show output) >>
          repl
```

Handwaving, we said. putStr, which we hadn't seen before, is like putStrLn but it doesn't automatically add a newline marker after putting out the given String. For us, this is useful, because it leaves the text input cursor to the right of the newly printed >, rather than below it. That way, as soon as we start typing, words appear *beside* >, as we've come to expect from our REPL, rather than on the line below.

> █

We call this thing, this moment, a "prompt". Any time we have a REPL-like interface waiting for input from its environment, *prompting* the user for it, that's a *prompt*. It's useful for said user to be able to identify this moment by noticing an accompanying distinct marker, which, confusingly, we also call *prompt*. So, REPL authors tend to include those too. In our case, in repl, we picked > as our prompt.

After prompting the user to input something, which we obtain with getLine, we proceed to parse that as a Haskell expression using a hypothetical function parseExpression, whose result we then evaluate using yet another hypothetical function evaluateExpression, whose result we finally display using putStrLn and show as we learned before. Obviously, parseExpression and evaluateExpression being as hypothetical as they are, means that our code doesn't quite work. But in theory, it should. To be fair, we did ignore the fact that parseExpression and evaluateExpression could fail. In a real-world scenario, we would have dealt with that too. Finally, we loop back into repl.

The type of our repl interaction is interesting. It says IO x. That is, ∀ x. IO x. In other words, repl is an IO interaction that can produce *any* chosen x. Wait, what? How? That can't be true. Well, as you can probably imagine, it isn't. repl is an interaction that runs forever, time and time again. It never delivers said x. It goes on and on reading, evaluating, printing and looping. But as the traffic lights we mentioned a while ago, repl does this while being useful, productive. It's infinitely productive, we could say. And by "infinite" we mean until the user forcefully closes the program somehow, just like when we close GHCi by any mechanism offered by our operating system. We could, of course, easily alter our interaction so that it doesn't run forever, so that it terminates when it encounters some special input. In GHCi, for example, entering :quit at the prompt causes GHCi to stop looping and quit. We did something like this in getVerses, which stopped as soon as it encountered an empty line.

So, this is how a REPL is built. We could implement one for our toy language, too, by relying on parseExpr and the evaluator we wrote before, properly dealing with any failure scenarios from any unsuccessful parsing or evaluation. We could, yes, but we won't.

Two hundred forty-one

Tradition has it that a book about programming must start by greeting the world by saying "hello, world". Ours started a long ago, but we want to say hi, too.

```
> putStrLn "hello, world"
hello, world
```

But not like that. There's a whole ceremony to perform. It might seem like a joke, but this exercise serves at least three purposes other than greeting us. First, to get the newcomer familiar with the language syntax at a high-level. Second, to show the basic structure of a full-blown program, one that we can *execute*. And third, to ensure that we have all the tools in place to compile and execute our programs, implying that this *hello, world* example is always accompanied with instructions about how to run it. So far, we haven't exactly done right by these goals.

Let's tackle the syntactic aspects first. Here's what the source code for a full blown *hello, world* Haskell program looks like.

```
module Main where

main :: IO Unit
main = putStrLn "hello, world"
```

Great. New things. In Haskell, code is written in files, each corresponding to a *module*. We haven't really mentioned modules yet, but that's where top-level Haskell code lives. We'll look into them in a minute. For now, suffice it to say that here we are defining a module named Main, and in it, an IO interaction named main that performs some work and returns Unit, that is, nothing of use.

So far we've been trying all our examples in GHCi, and for exploratory purposes that's fine. However, as programmers building a product, at some point we'll probably want to run that product in a standalone manner, without us or our users having to interact with it through the REPL. For this, we'll instruct GHC, our Haskell compiler, to convert our source code into a standalone executable form. And this executable form, when eventually executed, will perform the IO interaction named main found in the Haskell module named Main. The names Main and main are merely a convention, and can be easily overridden with some configuration parameters to GHC if we are bored.

So, how do we do this execution? Moreover, how do we compile our program? As a first step, we should make sure we put our code in a file named Main.hs. Then, assuming we have GHC installed, we open a command-line and run the following in the directory where the Main.hs file is:

```
% ghc Main.hs -o hello
```

Here we are asking GHC to compile Haskell source code found in the file Main.hs, and if successful, generate an executable named hello as output. In this example, % is our command-line prompt, much like > has been our GHCi prompt so far. So, ignore that part. After running this, GHC will start compiling our Haskell program:

```
[1 of 1] Compiling Main        ( Main.hs, Main.o )
Linking hello ...
```

And once that's done, we'll have a standalone program called hello that we can execute right away by typing ./hello in our command-line.

```
% ./hello
hello, world
```

Why, hello to you too!

Two hundred forty-two

Let's get something out of the way. The Unit type and its Unit constructor? They don't exist.

```
data Unit = Unit
```

Not literally, anyway. Of course there's a well-known unit type with its one constructor, but it's not called Unit, it's called (). That's right, ().

Obviously, () is some ad hoc syntax specially supported by Haskell. We wouldn't be able to define the literal name () ourselves. For example, the following is invalid Haskell syntax:

```
data () = ()
```

Nevertheless, the Haskell language will successfully deal with *mentions* of () as if it had been defined as above. This is not unlike the special syntax Haskell has for tuples and lists: We can't *define* the tuple or list types and constructors ourselves using their funny syntax, but we can use them just fine. Other than its literal name, everything we've learned about Unit is true about ().

```
> ()
()
> :type ()
() :: ()
```

Hopefully all of this nonsense clarifies why we opted to talk about Unit so far and not (). From now on, though, we'll do (), since that's what most Haskell code uses. A very widespread convention.

For example, the type of putStrLn, which we previously suggested was String -> IO Unit, is really String -> IO (). Also main, the entry point to our program, was not of type IO Unit but IO (). And so on.

Out loud, we still pronounce () as *unit*.

Two hundred forty-three

Programmers willing to use some of the top-level code defined in a particular module must *import* that code first.

By default, Haskell *imports* —that is, brings into scope— all the top-level names that are *exported* by a module named Prelude. This is where things such as Bool, fmap, length and Maybe are defined. This automatic import of the Prelude module happens in GHCi, too, which is why we can readily refer to these names right away in the REPL without having to import them manually. However, some of the names we've been using are exported by modules other than Prelude, and those do need to be imported explicitly. We haven't been doing this import ceremony ourselves yet because it's not particularly fundamental to our learning, but strictly speaking, we should have.

Consider `join`, for example, a function exported by a module named `Control.Monad`. If we try to refer to the name `join` as soon as we start GHCi, we get an error saying the name is not in scope.

```
> :t join
<interactive>:1:1: error: Variable not in scope: join
```

To fix this, we must bring `join` into scope by importing it from the `Control.Monad` module.

```
> import Control.Monad (join)
> :t join
join :: Monad m => m (m a) -> m a
```

Great. Other interesting names we know and love are exported from `Control.Monad`, too. For example, `guard`. However, when we wrote `import Control.Monad (join)` we said we only wanted to import the name `join` from that module, so `guard` is still not in scope.

```
> :t guard
<interactive>:1:1: error: Variable not in scope: guard
```

We can import more than one name from a same module by enumerating them in the import statement. For example, saying `import Control.Monad (join, guard)` will bring into scope both `join` and `guard` from the same `Control.Monad` module.

```
> import Control.Monad (join, guard)
> :t join
join :: Monad m => m (m a) -> m a
> :t guard
guard :: Alternative f => Bool -> f ()
```

Alternatively, we can only mention the module from which we want to import things, leave out the list of names to import, and *all* the names will be imported. For example, by saying `import Control.Applicative` we would bring to scope all of the names exported by that module, which include things like `liftA2` and `some`.

```
> import Control.Applicative
> :t liftA2
liftA2 :: Applicative f
       => (a -> b -> c) -> f a -> f b -> f c
> :t some
some :: Alternative f => f a -> f [a]
```

Of course, the need for this import ceremony applies not only to GHCi, but to Haskell modules too. For example, let's go back to our most recent `Main` module example.

```
module Main where

main :: IO ()
main = putStrLn "hello, world"
```

We were able to mention all of IO, () and putStrLn because these names are exported from Prelude, the module magically imported by Haskell unless we ask otherwise. But if we wanted to refer to a name other than those exported by Prelude, then we would need something else. For example, say we want to use liftA2 to write some code that will read two lines from the environment. Since liftA2 is exported by a module other than Predule, we must import it explicitly. We do so right below the line where we define what the name of our module will be. Namely, below module Main where.

```
module Main where

import Control.Applicative (liftA2)

getTwoLines :: IO (String, String)
getTwoLines = liftA2 (,) getLine getLine

main :: IO ()
main = getTwoLines >>= \(a, b) ->
       putStrLn b >> putStrLn a
```

Hadn't we imported liftA2, the code wouldn't have compiled.

Two hundred forty-four

Haskell modules import names from other modules that export things. Any module can import and export names. By default, if we don't explicitly mention what names the module we are defining shall export, it will export every new name being introduced in it. For example, our recent Main module exported all of getTwoLines and main, the two names newly defined in Main. But, had we preferred to export only main, say, and leave getTwoLines unexported, we could have done so using a syntax similar to the one we use when explicitly importing names.

```
module Main (main) where
```

That is, a module can specify the names it wants to export by listing them separated by commas in between those parentheses, preventing any name not mentioned there from being made available for other modules to import.

```
module Foo (bar, baz) where
```

The privacy awarded by being able to restrict our exports is quite useful. On the one hand, by not exporting the less interesting names, we are encouraging those importing this module to only pay attention to the names that matter, leading to a better user experience overall, seeing as users will have less things to worry about and understand. But more importantly, this privacy creates safety.

Two hundred forty-five

Say we want to represent email addresses in Haskell. We can easily do it using values of type `String`.

```
myEmail :: String
myEmail = "bob@example.com"
```

But that's very disappointing, so please, never do that. There are two main problems with this. We are familiar with the first one already: The meaning of this `String` is completely lost in the big picture. For example, imagine we had a function for constructing an IO interaction that would send an email to a given email address, having a given subject and content.

```
sendEmail :: String -> String -> String -> IO ()
```

String, String, String? Judging by the scenario we just described, we can assume that these `String`s refer to the recipient's email address, the subject, and the email content. However, we can't tell from the type of `sendEmail` which is which. What if we provide them in a different order? What if we put the content where the subject should be? That would be unfortunate.

Unconvinced? Do the same exercise, but rather than getting your inspiration from an IO interaction that sends emails, picture an IO interaction taking as input the geographical latitude and longitude where a missile should land and explode on enemy territory. We better don't mix those up!

```
launchMissile :: Integer -> Integer -> IO ()
```

We write programs being aware that people, including ourselves, are likely to make mistakes. It's our responsibility as programmers to prevent these mistakes from ever happening. Hoping for the best doesn't work, mistakes *will* happen. Thankfully, we have types to keep us from harm.

```
sendEmail :: Address -> Subject -> Content -> IO ()
```

```
launchMissile :: Latitude -> Longitude -> IO ()
```

Much better. It's now impossible to mix these things up, be it accidentally or due to incompetence. As soon as we try to make a mistake, the type-checker will yell at us and refuse to compile our program. Obviously, this won't prevent us from making the silly decision of initiating warfare. Or worse, sending an email when we shouldn't have. Those problems are outside the scope of Haskell.

So, yes, we want email addresses to have a type of their own. That's how we'll be able to tell them apart from every other String-like value such as

Subject, Content, and obviously String itself. Generally speaking, we want every value that belongs to a distinct enough set of things to have a type of their own, even if ultimately they may share the same underlying representation as other types. In this case, they are all fundamentally Strings. In the case of Latitude and Longitude, they are both fundamentally numbers of some sort. But from the point of view of the type-checker they are different, and that's good.

So, how do we do this? How do we introduce this Address type? We know how, there's nothing new for us here.

```
data Address = Address String
```

Here, we are defining Address to be a new type that carries a String as a payload. That's all. We did exactly the same thing a while ago when we were building our toy programming language and decided we wanted Names to have their own type, lest we accidentally mixed them up with other Strings.

```
data Name = Name String
```

That's it, really. Email addresses are Addresses now, they are not Strings nor Subjects nor Names. They can't be confused with each other, accidentally or not.

Having Address, we can proceed to define an email address of this type.

```
myEmail :: Address
myEmail = Address "1234"
```

Wait, what? That ain't no email address. Clearly, that is a password.

Two hundred forty-six

Having a type, Address, is only part of the story. It helps, but it doesn't really take us all the way where we want to go. What we are missing is a way to ensure that we never have a value of type Address wrapping a String other than a valid email address. However, as long as the Address constructor is readily available for us to use, we can't really prevent this unfortunate situation from happening. To address this, we'll restrict the usage of the Address constructor. Rather than Address the constructor, we'll force users to use addressFromString the *smart constructor*.

```
addressFromString :: String -> Maybe Address
addressFromString s | isValidEmailString s = Just (Address s)
                    | otherwise            = Nothing
```

We'll leave isValidEmailString, of type String -> Bool, up to our imagination. We can picture it checking whether the String contains an @ sign, is non-empty, etc. The important thing is not isValidEmailString, but addressFromString itself. This function is *smart* about which Strings it wraps with the Address constructor. Just those it deems worthy of the name will carry the banner, Nothing else will make it through, Strings containing something other than an email address will be rejected, they will not become values of type Address.

Smart constructor is merely a fancy name for a function that serves this particular purpose of checking that the guarantees supposed to be conveyed by a particular type —here Address, conveying the idea of a valid email address— are indeed respected by all the values carrying this type. Just like a normal constructor, a smart constructor constructs values of a particular type. But, in addition, it'll reject invalid inputs by means of Maybe, Either and the like. Even though technically, the String inside Address could have been anything, addressFromString prevented that, granting us the ability to *assume* that every value of type Address represents a well-formed, valid email addresses.

Smart constructors, in combination with "dumb" type wrappers such as Address or Name, are some of the cheapest, simplest, yet most effective tools we have at our disposal to guarantee that the software we build is correct. Use the dumb types and smart constructors, make them part of your basic vocabulary.

Two hundred forty-seven

Now we have Address the type and addressFromString the smart constructor, which we want people to use, but we also have Address the constructor which we *don't* want people to use, lest they wrap with it a String containing something other than a valid email address, breaking our assumptions, our reasoning. Can we prevent people from using the Address constructor at all? Why, yes, all we have to do is *not* export it from the module where it's defined.

```
module Email (Address, addressFromString) where

data Address = Address String

addressFromString :: String -> Maybe Address
addressFromString = ...
```

In this example, we defined a module named Email that exports a type named Address and a function named addressFromString. According the first line of code, this module *does not* export the *constructor* named Address, only the type named Address. If we wanted to export the constructor named Address too, then we would have written something a bit different.

```
module Email
  ( Address(Address)
  , addressFromString
  ) where
```

The export syntax Foo(Bar, Qux) means "export the type Foo together with its constructors Bar and Qux". If we omit the constructors from our export list and write just Foo instead, then only the Foo type will be exported. This is exactly what we did in order to avoid exporting the Address constructor. Whether the names of a type and its constructors are different like Foo, Bar and Qux, or equal like Address and Address, is irrelevant. It may be confusing to us humans, but Haskell will do just fine.

So, outside this module, the Address constructor will simply not exist. It won't be available for people to use, wrongly or otherwise. If they want a value of type Address, they'll have to go through addressFromString. Within this module, however, we can go wild. Address, the constructor, is here for us to use and potentially misuse. But this is fine. This is a restricted access area where only we, the grand creators of this type, will have to be careful at least once. We *need* access to the Address constructor in this module anyway, otherwise we wouldn't be able to implement addressFromString, the smart constructor which we *do* want to export.

Two hundred forty-eight

When we pattern-match against values of a particular type in Haskell, we do so against the constructors of that type. So, what happens if said constructors are not available because they haven't been exported? Like Address, the constructor for the Address type, which we chose not to export from the module we named Email. Well, not pattern-matching, that's for sure.

Just kidding. We have at least two alternatives to deal with this. But first, let's clarify what it is that we expect to get out of pattern-matching against the Address constructor, if we could.

```
addressToString :: Address -> String
addressToString (Address x) = x
```

A String, of course. That's what we expect to get out of this pattern-maching. What else could it be? That's what's inside the Address constructor anyway.

And this is exactly one of the solutions to the pattern-matching problem. Rather than exporting a constructor to pattern-match on, we export a function that outputs what a successfully matched pattern would have given us. In this case, a String.

```
module Email
  ( Address
  , addressFromString
  , addressToString
  ) where
```

Now, not only can users of this Email module mention the Address type in their programs and use addressFromString to construct values of that type, but also they can extract the String inside it by means of addressToString. And of course, said returned String being a String, could be pattern-matched against if necessary as any other String could.

```
isBob's :: Address -> Bool
isBob's x = case addressToString x of
              "bob@example.com" -> True
              _ -> False
```

Two hundred forty-nine

A second alternative allowing users to pattern-match on values of type Address without forcing us creators to export the corresponding constructor is to define a new pattern. Yes, that's right, a new pattern. Just like we can create types, functions and typeclasses out of the blue, we can create patterns to be matched against, too. Haskell can do that.

When we define a constructor for a datatype, we get a pattern for it. But really, constructors and patterns are different things, and we can have one without the other.

Let's make a small change. Let's rename our Address constructor to something else. By doing that, we'll be able to repurpose the word Address for the pattern we'll create.

```
data Address = UnsafeAddress String
```

As we know, we can name constructors anything we want, and at this time UnsafeAddress seems like an excellent choice. The name somewhat warns us that using this constructor for building values of type Address is not safe, in the sense that this constructor can't guarantee that the accompanying String is a well-formed email address. That's why we created the addressFromString smart constructor at all. Now, having freed up the Address name, we proceed to create a new standalone pattern named Address, with this new code we'll write at the top-level, inside our Email module.

```
pattern Address x <- UnsafeAddress x
```

We are defining a *pattern* named Address, carrying a payload of type String here named x. This x, as shown to the right of the leftward <- arrow, in turn comes from pattern matching on UnsafeAddress x. In other words, we are saying that rather than mentioning the UnsafeAddress constructor when pattern matching on values of type Address, we can mention this newly minted pattern called Address, and the result shall be the same. All that remains now is for our Email module to export this pattern, too.

```
module Email
  ( Address
  , pattern Address
  , addressFromString
  , addressToString
  ) where
```

Notice the word pattern there in the export list. Now, users of the Email module get to enjoy the meaning conveyed by the Address type, the safety granted by the smart constructor addressFromString, and the convenience of still being able to pattern-match against values of type Address, even if the constructor is not exposed to the world. We pattern-match against this new pattern, Address, as we would against any constructor.

```
isBob's :: Address -> Bool
isBob's (Address "bob@example.com") = True
isBob's _ = False
```

At this point, we may want to avoid exporting addressToString, as it's been made redundant by the existence of the new Address pattern. However, we'll likely find ourselves reaching out for this function way too often, each time we want to include an email address as part of a bigger String, so we'll keep it. Actually, one might argue that this whole pattern business was redundant, seeing as in practice addressFromString is sufficient and likely to be used more than the Address pattern. And one would be right. Nevertheless, here we are, with new knowledge of what a pattern can be, knowledge a chapter ago we didn't have. Throw away the Address pattern now if you will, for us it was mostly the learning vehicle.

We've barely touched the surface of what we can accomplish with patterns and their pattern syntax, but that's as far as we'll go. You can check out the GHC reference manual for the details if you are interested.

Two hundred fifty

What's wrong with our implementation of addressFromString?

```
addressFromString :: String -> Maybe Address
addressFromString s
  | isValidEmailString s = Just (Address s)
  | otherwise            = Nothing
```

Nothing wrong? Nothing wrong. However, this implementation implies that there exists a function isValidEmailString with type String -> Bool, and *that* is wrong. Sort of.

We talked about the disappointing nature of Bools a while ago. Bools don't convey much meaning, so we are pretty much always better off avoiding Bools in favour of richer types. isValidEmailString is a function expected to tell us whether a given String contains a valid email address or not, but in conveying this knowledge as a boolean value, the proof that this input String contained an email address was lost. Wouldn't it have been better to preserve a proof of this fact? Indeed, but that's what addressFromString already does. If the String that addressFromString receives as input is a valid email address, then we'll get a value of type Address as output, a *proof*. Otherwise, we won't.

```
addressFromString :: String -> Maybe Address
```

But we *did* implement addressFromString in terms of isValidEmailString. So, what are we saying exactly? Are we going in circles? In circles we are going.

isValidEmailString shouldn't exist. If for some reason we needed a predicate to filter out some Strings, say, we could use something along the lines of isJust . addressFromString without having to give this composition a name. Not all things deserve a name. While each name we introduce could certainly serve a useful purpose, it'll also burden users of our code with more things to pay attention to, to worry about, and it's rarely useful to worry about things that are not fundamental to the problem we are trying to solve. Whether a String is an email address or not is fundamental to our problem, but an answer to this question in boolean form, when we already have the answer that we need in a richer Maybe Address form, makes isValidEmailString obsolete. We can always *discard* information from said Maybe to obtain a Bool if necessary, that's what isJust is for.

```
isJust :: Maybe x -> Bool
isJust (Just _) = True
isJust Nothing  = False
```

There's isNothing, too, doing the obvious thing.

So we agree that addressFromString is good, and that isValidEmailString, while useful at times, is not worthy of a name. It could easily be derived from addressFromString.

Two hundred fifty-one

How do we implement addressFromString, then, without relying on isValidEmailString which shouldn't exist? We parse. We don't just check that things are what we expect them to be. We parse.

```
addressFromString :: String -> Maybe Address
addressFromString s = case runParser parseAddress s of
                        Right (_, "", address) -> Just address
                        _ -> Nothing

parseAddress :: Parser Address
parseAddress = ...parse and produce an Address...
```

We can now accept parseAddress as the source of truth for what an email
address is or isn't, and addressFromString as merely a convenient shortcut
for running the Parser. The work parseAddress will have to do in order to
verify that a String is indeed a valid email address is as complex as it would
have been in isValidEmailString. However, now, as a consequence of that
effort, we get a richer output, a *proof*, in the form of a value of type
Address. This is better than a Bool.

As a bonus, seeing as we have nothing to lose and how our users could
surely benefit from it, our module will export parseAddress too, making it
possible for Parsers outside this module to build bigger Parsers where
email addresses are just a small piece.

```
module Email
  ( Address
  , pattern Address
  , addressFromString
  , addressToString
  , parseAddress
  ) where
```

Now *this* is a fine module. Add some documentation, examples, and a
clear description of what a well-formed email address is supposed to look
like, and we are good to go. Ship it.

Two hundred fifty-two

Our module being called Email suggests that we may want to export things
other than Address and related functions from it. For example, we may
want to export types like Subject, Content and that sendEmail function we
discussed before. It's up to us, really, to decide what to export from a
particular module. While doing so, however, it's important we keep our
users' experience in mind.

And by "users" we mean ourselves and the other programmers who will
be using the Haskell code we export. Programmers are users too. They use
source code, functions, types and tools made by other programmers, and
expect them to effectively and efficiently deliver what they promise in a
neatly organized, well communicated and polished manner, as any
consumer does of any product. Users choose, and if they can't find what
they are looking for, they choose something else. So we better make our

Haskell modules pristine, lest users suffer and leave.

Generally, it's a good idea to export from a same module things that are somewhat related. This has nothing to do with what Haskell needs, but with what users expect when discovering our module for the first time. They'll look at the names the module exports, read their types and documentation, and hopefully figure out how to put things together in order to achieve what they need. Or even *discover* what they need. Remember, types are there to help us reason about solutions, but also problems. By exporting related names from a same module, and by avoiding a prominent appearance of less interesting names, we'll likely make this task easier for our users.

A second thing to consider is how users are expected to use names exported from this module. There are three possible answers to this question. The first one is that users will explicitly list the names they want to import from this module.

```
import Email (Address, addressFromString, addressToString)
```

This is fine. However, if users are likely to require many names from this module at a time, forcing them to list each one individually will make their experience less than ideal, since these listings will need to be written and maintained in every module that requires our names. An alternative, then, is to encourage users to import names implicitly.

```
import Email
```

However, this practice creates a few other problems that we, as authors, must be aware of. First, it's now unclear which names have been brought to scope by this import statement. This is not particularly worrisome in itself, because there's nothing wrong with having imported a name but not having used it. However, as soon as we start importing more modules, it becomes unclear which name comes from which module, which makes maintaining and understanding the code at hand much harder. And we will be importing more modules. Many of them. It's not uncommon for mid-sized Haskell modules to import twenty or more modules.

Another problem with importing names from a particular module implicitly —which to a lesser extent affects explicitly imported names, too— is the fact that names that perhaps made sense within the context of the module that defined them, might not make sense anymore in a different one. For example, consider Address. Obviously, within the context of emails, an *address* has a unique and well-understood meaning. However, if we are bringing the name Address to scope in a module where we deal with physical product purchases, say, then something called *address* may refer to a shipping address or billing address, rather than an email address. This will be confusing for us. And for the compiler, too, if for example there's more than one thing named Address. To tackle this issue, and in general, to improve the internal organization of our modules, we have *qualified* imports.

```
import qualified Email
import qualified SuperProject.Data.Shipping
```

By adding the world `qualified` to our import statement, we've changed things so that if we want to use a name exported by any of these modules, we *must* prefix it with the name of the module. For example, rather than saying addressFromString, we'd have to say Email.addressFromString.

```
> import qualified Email
> :t addressFromString
<interactive>:3:1: error:
   • Variable not in scope: addressFromString
   • Perhaps you meant 'Email.addressFromString'
     (imported from Email)
> :t Email.addressFromString
Email.addressFromString
   :: String -> Maybe Email.Address
```

If we are lucky enough, like above, GHC's fine error messages will point us in the right direction. Now, if we want to talk about an email address within the module that imports things this way, we have to say Email.Address, and if we want to talk to talk about a shipping address as understood by the module called SuperProject.Data.Shipping, we need to say SuperProject.Data.Shipping.Address, which, clearly, is very long and disappointing. Fortunately, Haskell allows us to rename modules when importing them.

```
import qualified Email
import qualified SuperProject.Data.Shipping as S
```

Notice the as S part. Rather than SuperProject.Data.Shipping.Address, we must now say S.Address. We can rename modules to anything we want, provided it looks like a module name. Saying as Shipping or as Foo, for example, would have been fine too.

And why does any of this matter? Because being aware of how our modules and names will be used, we can make choices that will lead to a better programming experience. For example, had we designed our Email module for qualified usage, we could have exported a function named send rather than sendEmail, to be used as Email.send. Programming is not only about correctly understanding and solving problems, but about ergonomics too.

Two hundred fifty-three

Modules. Modules everywhere. Names. Names everywhere. But IO interactions in one place and one place only. The main function in the Main module is the only place where IO interactions can be performed. It sounds restrictive, I know, but don't be scared. As we'll see later on, this is *exactly* what we want.

```
echo :: IO ()
echo = getLine >>= putStrLn
```

Consider echo here. This IO interaction obtains a String from the environment using getLine and echoes it back to the user using putStrLn, an IO interaction taking a String as input and producing a moot () as output. We use putStrLn for how it pokes the outside world, not for its output value, so getting () out from it is fine. We achieve all of this by using >>=, *bind*, which allows us to compose the two functorial values into a single new one by establishing ordering and dependency between them. In this case, our functorial values are IO interactions.

But merely defining echo doesn't *perform* what it says it will. This is not unlike what happens when we define any other name. We are only *naming* something, we are not using it yet. For echo to happen, for echo to be performed, it must be part of the IO interaction called main. If it isn't, then it won't.

```
main :: IO ()
main = echo
```

Now, yes, executing this program will *perform* the steps described by echo.

```
% ghc Main.hs -o example1
[1 of 1] Compiling Main      ( Main.hs, Main.o )
Linking hello ...
% ./example1
foo
foo
```

What about this similar scenario?

```
main :: IO ()
main = void (sequence (take 1 [echo, echo, echo]))
```

Does it perform echo three times?

```
% ghc Main.hs -o example2
[1 of 1] Compiling Main      ( Main.hs, Main.o )
Linking hello ...
% ./example2
foo
foo
```

Nope. Just the one time, as requested by our use of sequence and take 1. For our IO interactions to be performed, they must be *bound*, in the monadic sense of the word, into the IO interaction called main. Merely mentioning a particular IO interaction in main won't magically perform it. It must become one with it, they must bind >>= or be composed in some other functorial way.

void, by the way, has nothing to do with the Void datatype we saw before, which conveyed the idea of zero, of a type for which a value can never exist. The poorly named but widely used void function, traditionally exported from the Data.Functor module, is something completely different.

```
void :: Functor f => f x -> f ()
void = fmap (const ())
```

That is, it'll discard the output of a functorial value and replace it with (). It will *not* discard the actual IO interaction, as this would be a blatant violation of the functor laws which say that fmaping over a functorial value can't ever change the *functorial* part of it, only the output it produces. The functor police would come for void if it did that. In our recent example we used void to replace the output produced by our application of sequence with (), seeing as IO () is the type main is supposed to have, not IO [()] as it would have had, had we not used void. Remember, the type of sequence is —approximately, more on this later— Applicative f => [f x] -> f [x], and knowing echo's type we can easily infer the rest.

Back to main. The fact that only main can perform IO interactions doesn't mean that we must literally write all our IO interactions in there. It just means that directly or indirectly, any IO interaction we want to execute as part of our program, as big or small as our program might be, must be composed in this value called main. In our most recent example, for example, getLine and putStrLn were part of the IO interaction described by main. Yet, they weren't mentioned in main literally. Instead, they were bound by echo, who was in turn bound in main. Actually, echo *was* main. And for all we care, for all Haskell cares, echo could have been defined in a different module, far away from Main, and it would have been fine too. Actually, getLine and putStrLn are defined in their own modules already, and nonetheless everything worked just fine. In other words, the fact that only main can *perform* IO interactions doesn't prevent us from writing IO interactions anywhere we want, monadically, applicatively, functorially, however we want.

Why is it, though, that only main can IO?

Two hundred fifty-four

One, two, three. A, B, C. Order. Our programs will likely interact with the environment in more than one way, and we want those interactions to happen in a particular order. First this, then that. If the goal of our program was to play a song, *first* it would need to access the song contents somehow, and only *then* can it start sounding the song through the speakers. Doing these interactions in a different order wouldn't make sense, so we want to prevent that.

The monadic features of IO enforce the order of each of the small IO interactions that make up the main interaction, we know that. However, if it was possible to initiate IO interactions from other places, at overlapping times, then even if the steps within a single IO interaction happened in a known order, nothing would prevent these steps from unpredictably overlapping with steps being carried out by other IO interactions initiated elsewhere.

Picture yourself, Mr. Main, cooking. You have a recipe listing ingredients, and a carefully prepared list of steps to follow in a particular order. As long as you have the necessary resources and follow the laid out steps, the two pizzas will come out fine. So you check the fridge and pantry, confirm that you have enough utensils and ingredients, and set out to cook. You start by taking a bowl, and mixing in it 300g of plain flour, one teaspoon of dry yeast, and 200ml of lukewarm water. You add a pinch of salt and half a teaspoon of olive oil, and then you knead this mixture for about five minutes until you have a smooth and elastic dough. After 40 minutes of it resting at room temperature, you'll split the dough in half, roll out the two pieces very thinly, put them in trays, add tomato sauce, cheese, and cook them in the preheated oven for about 10 minutes at 250°C. Then, you eat. It's an easy recipe.

However, it turns out that after those 40 minutes of resting you won't be able to do any of that, because you didn't take into account Mrs. Main, who had cookies in progress, too. She's taken over the oven and the trays. Luckily, your pizza dough can sit in the fridge for up to a day, so there's no need to panic.

But you see the problem, don't you? It's not Mrs. Main, she's fine. And it's not you either. It's both of you *together*. While your respective interactions with the environment were in order within the context of each of your recipes, you assumed an undisturbed timeline and full control over the kitchen resources. But you should have known better. You and Mrs. Main should have come up with an agreement about it. "I'll cook tonight, honey, you cook tomorrow", or something along those lines. In Haskell, this whole problem is avoided by having a single entry point to our programs. Haskell gives us the undisturbed timeline, the full control over our resources insofar as they are available to Haskell. There's no need to coordinate with other parts of the program about who goes first, because there's only ever one timeline to worry about, the one that starts and finishes in main.

Now, this is not to say that a single program can't do more than one thing at a time. Indeed it can. For example, our hypothetical music player could be playing a song and at the same time downloading from the internet the song it will play next. This ability to do or worry about more than one thing at the same time is called *concurrency*. Inevitably, a successful concurrent program will require the concurrent interactions to coordinate with each other so as to avoid what you and Mrs. Main couldn't. However, this is not something we need to deal with unless we explicitly ask for it. And even if we do, we'll still have a predictable ordering of execution while *setting up* these concurrent affairs. Haskell has some fascinating vocabulary to deal with concurrency. For now, however, we'll go alone, straight ahead, through the one door in front of us, main.

Two hundred fifty-five

The fact that as soon as it starts, our program goes into performing this IO interaction called main, doesn't imply that all of our program will be written with this IO interaction in mind. For example, consider this program:

```
main :: IO ()
main = getLine >>= \a ->
       let b = foo a
       in putStrLn b
```

We get a line of input from the environment using getLine, we process it somehow through foo, and finally we print the result back to the environment using putStrLn. This is a very common thing to do, and a large number of programs are shaped this way. A compiler, even, is not too different from this. It reads source code as input, transforms it into its executable form, and finally outputs that executable form back to the environment. Sure, a compiler will probably read more than one line at a time, and perhaps it'll finally output its result to a file rather than printing it out, but the general idea is the same. The real essence of what the program does is not in our IO interactions but in foo, the function that actually processes the input.

```
foo :: String -> String
```

We can see in its type that foo is completely unaware of the IO interactions taking place. foo, the bulk of our program, is a function that takes raw input to be processed and returns processed raw output. It could be a huge function that worries about a million different things, but it'll never have to worry about where the input came from nor where it will go afterwards.

While it's true that ultimately we want programs to interact with the environment somehow, the large majority of the functions that make up our programs will remain oblivious to this fact. IO interactions happen at the boundaries between our program and the outside world, and even in highly interactive programs such as games, these interactions are seldom the program's essence.

Two hundred fifty-six

Functions in Haskell are different from functions in most other languages. And by "different" we mean *better*.

```
qux :: Bool -> Integer
```

We don't need to look inside qux to understand that, at most, it can only ever produce two different Integers as output. One, when qux is applied to True, and the other one when qux is applied to False. We don't know *which* Integers will be produced, but it doesn't matter. What truly matters is that we can *reason* about the inputs and outputs of a function with confidence. The input to this function could only ever be what its type says it could be, and likewise, its output. No inputs other than the given Bool will affect the outcome of this function, and the consequences of having applied qux to said Bool can only be observed through its formal output parameter, the Integer.

If all of this sounds redundant, it's because it is. If you struggle to understand why any of this matters, congratulations, you are doing just fine. We call this property *purity*. In Haskell, all functions are *pure*. Meaning, among other things, that the possibilities of a Haskell function are fully determined by its formal input and output parameters. To better appreciate why this matters, let's imagine what an *impure* function would be like. We can't *write* one because there's no such thing as an impure function in Haskell, but nobody can take our imagination away from us.

Imagine a function that took an Integer as input and returned yet another Integer as output, a function that multiplied its input by ten, say. So far, nothing special. We can even give this function a type and a name.

```
zoink :: Integer -> Integer
```

Except when it's raining. On rainy days, this "function" multiplies its input by two, not by ten. Then it broadcasts the result to all of our family members by email, and finally returns the number seven as its formal output. For example, if we applied zoink 3 on a sunny day, the result would be 30, but if we applied it on a rainy day, the result would be 7 and our family members asking themselves why they each got a number 6 in their inbox. This nonsense is one of the many things an "impure function" could do. Good luck reasoning about this function and how it relates to the rest of our program.

Generally speaking, we don't want functions to be impure. Why would we? We cherish our ability to reason about functions by looking at their input and output parameters, and we'll fight to defend that. However, at times, we may want to do things depending on external events such as rain, and we may or may not want to send an email. Can't Haskell do that? Have we lost? We haven't, it can.

Two hundred fifty-seven

One way to tackle the rain problem is to *not* tackle it at all. That is, rather than checking *within* our function whether it rains or not, we'll make the interaction with the environment somebody else's problem, and just take rain as a formal input to our function instead.

```
zoink :: Weather -> Integer -> Integer
zoink Sunny n = n * 10
zoink Rainy n = ?
```

And similarly with that family email. We won't send any emails from within zoink. Instead, we'll output instructions for somebody else to send them.

```
zoink :: Weather -> Integer -> Either Integer (Integer, FamilyEmail)
zoink Sunny n = Left (n * 10)
zoink Rainy n = Right (7, FamilyEmail (n * 2))
```

If we are told that it's Sunny, we multiply the given number by ten. If Rainy, we return instructions for a FamilyEmail containing n * 2, alongside the number 7, which we said we also wanted to return for some reason.

This continues to be a silly function, nothing could take that away from it. But, it is a *pure* silly function at least. We can reason about the possibilities merely by looking at its type, as we've been doing all along.

Checking the weather and sending emails? Not zoink's problem anymore.

Two hundred fifty-eight

Still, at some point, we'll need to see about that rain. We can keep doing what zoink did all the way up the call stack, eventually reaching main, and let main deal with all the IO interactions. That way, the bulk of our program will remain pure, with only a handful interactions happening in main.

```
main :: IO ()
main = getWeather >>= \w ->
        case zoink w 37 of
           Left n -> putStrLn (show n)
           Right (n, fe) ->
              sendFamilyEmailSomehow fe >>
              putStrLn (show n)
```

Still a silly program, but a familiar silly one, not all too different from the main example we saw before where we read one line from the environment, processed it somehow, and then printed back the results. The only things that changed since then regarding our interactions is that rather than getting our input from the environment using getLine of type IO String, we get it using a hypothetical getWeather of type IO Weather, and that besides using putStrLn to print that number, we may or may not send an email somewhere. IO, pure processing, and then some more IO. As said before, this is a very common thing to do, we'll see it very often.

This approach however, for practical purposes, is often insufficient. Imagine a program that must do many IO interactions. If all functions everywhere deferred their need to interact with the real world to main, then main would be huge. What we'll do instead, is have main delegate some of its responsibilities to other parts of the program. Indirectly, main will still be the only one *initiating* the performance of all the IO interactions, we know that, but these interactions won't be written down literally in main.

Two hundred fifty-nine

Here's a different example for us to study, a bit more straightforward than the one before.

```
evenish :: Integer -> IO Bool
evenish x = getWeather >>= \case
                Sunny -> pure (even x)
                Rainy -> pure (odd x)
```

Still silly, though. evenish is a function that on Sunny days tells us whether a given number is even, and on Rainy days whether it's not.

We may be tempted to say that evenish is an "impure function", seeing how its output doesn't depend solely on its formal input parameters but also on the weather. But we know that there's no such thing as an impure function in Haskell, so this must be wrong.

One way to reason about this is to acknowledge that evenish is *not* a function that takes an Integer as input and impurely produces a value of type Bool as output, but rather, it is a function that given an Integer, *purely* produces a value of type IO Bool. A subtle, but profound difference.

evenish x evaluates to an IO interaction that, when performed later on, perhaps years from today, depending on what the weather is then, will tell us whether the x we chose back then, today, was even or not. Coming up with this IO interaction, however, doesn't require us to check what the whether is like today. Rain or shine, evenish 6 will always produce the same IO interaction, the one that on Sunny days says True and on Rainy days says False. evenish is a pure function. Its output, the IO interaction that produces a Bool, is fully determined by its Integer input.

Haskell remains pure.

Two hundred sixty

All Haskell functions are pure, including those that lead to interactions with the environment. In our imagination, however, as a matter of convenience, we may choose to accept them as impure. That's fine, Haskell doesn't care what we believe in. In fact, it does an excellent job at *safely* allowing us to blur the line, in our minds, between what's pure and what's not.

```
evenish :: Integer -> IO Bool
evenish x = getWeather >>= \case
                Sunny -> pure (even x)
                Rainy -> pure (odd x)
```

Can you see the difference between this code and the previous definition of evenish? Well, you shouldn't, nothing's changed. However, our perspective will. We'll say now what we didn't say before, that evenish is a function which takes an Integer and *impurely* produces a Bool. The impurity is IO. The impurity is what forces our function to have type Integer -> IO Bool rather than just Integer -> Bool. Every so-called "impure function" will have their output wrapped in the IO functor, the type-checker won't let us have it otherwise.

There are many ways to write evenish, but all of them will at some point use >>= to access the Weather output produced by getWeather, and this is what taints our output with that IO marker.

```
(>>=) :: IO a -> (a -> IO b) -> IO b
```

In our example, getWeather is our IO a. We can use >>= to observe that a, sure, but in return, we must produce an IO b. And then we are stuck, because as we know, there's no way to remove that IO wrapper except through its execution in main.

And the same is true for other functorial tools implemented in terms of >>=, such as fmap or <*>. All of them eventually result in a functorial IO value. We can't avoid it.

```
fmap  ::    (a -> b) -> IO a -> IO b
(<*>) :: IO (a -> b) -> IO a -> IO b
```

So, yes, we are forever trapped in IO. Once we commit the sin of binding an IO interaction, we are forever bound to IO. We can't leave. We are stuck. IO forever taints everything it touches. And, *thankfully* it does. This way, the execution of this now impure and sinful function will necessarily have to be coordinated with every other IO interaction in our program, directly or indirectly, from main. This will allow us to guarantee, for example, that we never try to get a reading from our weather sensors without having successfully set up the sensors themselves first, seeing how the type checker will force us to order things, and how we can leverage that. *First this, then that.*

So, what's wrong with every function being "impure"? With tainting our every function with IO ? Well, we know what evenish does because we wrote it, so we might be tempted to say "it's not too bad". But what about harm here?

```
harm :: Integer -> IO Bool
```

Scary, isn't it? From its type, we can't tell what harm does, but more importantly, we can't tell what it *doesn't*. We can only hope its name was a mistake.

Mastery expects us to find the right balance between purity and IO. In practice, this mostly means avoiding IO interactions as much as possible, as we've successfully been doing so far.

Two hundred sixty-one

Let's see how we can accomplish some of the things for which we'd traditionally want IO interactions, without doing any IO interactions. Let's do some magic, otherwise known as card tricks.

```
data Deck = ... not important ...
```

A magician will require a Deck of cards and some vocabulary for interacting with it in order to successfully perform some tricks. For example, there needs to be a way of picking a card out of the Deck.

```
pickTopCard :: Deck -> (Deck, Maybe Card)
```

Picking a Card, here from the top of the Deck, results in that one Card, if any, together with the rest of the Deck. We must necessarily return a new Deck alongside our Card because if we were to reuse the original Deck later on, we'd find our Card still in there. Obviously, we aren't supposed to reuse the original Deck that way, that's not the kind of magic we are talking about. That's why we return a new Deck, one Card shorter.

Picking from places other than the top of the Deck will surely be trickier, but the *type* of our picking interaction will remain more or less the same. We'll always get a Card together with the remainder of the Deck.

We'll also need a way of putting that Card back onto the Deck of cards later on.

```
addTopCard :: Card -> Deck -> Deck
```

The output Deck contains the Card. The input Deck, hopefully, doesn't. *Hopefully*, because nothing is keeping us from accidentally reusing the original Deck from before, the one that did have the Card in it.

What else will a magician need? Let's see... we'll need a way to split a Deck into two smaller Decks, and a way of putting those smaller Decks back together.

```
splitDeck :: Deck -> (Deck, Deck)

joinDecks :: Deck -> Deck -> Deck
```

And of course, we'll probably need to shuffle our Deck at some point.

```
shuffle :: Deck -> Deck
```

We'll need more things, too. But for now, this should suffice to appreciate the common theme among these functions, which is that they all take at least one Deck of input, transform it somehow into a new Deck, and expect us to use this new Deck, not the original one, in the next Deckful interaction.

What's important to notice here is that while we are dealing with multiple values of type Deck, conceptually, we consider all of them to be different observations, states, of *the same* Deck over time, of which we will only ever care about the most recent one. We never want to reuse an old observation of a Deck.

Optionally, some of these interactions return an additional value too, such as that second half of the Deck in splitDeck, or that Card in pickTopCard.

```
Deck -> (Deck, something)
```

This is a very common theme in programming. We get an input, we modify it somehow, and we return it alongside something else. For example, we saw this back when working in our Parser type, whose last incarnation was something along the lines of:

```
data Parser x = Parser (String -> (String, x))
```

That is, our Parser was essentially a function that given a String containing the raw input to be parsed, would parse a value of type x from that input and return it alongside a modified String containing leftovers from that original input String. Yes, our actual Parser type was a bit more complicated, but this part alone is sufficient to highlight the common pattern that interests us now.

We only ever want to use these leftovers as input for a subsequent Parser. Otherwise, were we to reuse the same original input, we'd be consuming already consumed input once again. The fact that we *could* reuse that input is great, though. This is what allowed our Parsers to backtrack. However, backtracking implies that a later Parser has failed to parse what it expected to parse, so we backtrack to try something different. That is, we can't really consider backtracking as part of the idealized scenario in which parsing succeeds, where our most recent leftovers must necessarily be the input for the Parser being executed immediately afterwards.

What about allegedly more complex things, such as a database? Well, it's not too different. We can add things to a database to obtain a modified database.

```
add :: Thing -> Database -> Database
```

We can delete things from it to obtain a new database that excludes a particular thing.

```
delete :: Thing -> Database -> Database
```

Or we can lookup things in the database, perhaps by name.

```
lookup :: Name -> Database -> Maybe Thing
```

All of these functions take a Database as input and return a Database as output alongside something else. What? Isn't it obvious that they all do that? Perhaps it'll be clear if we make their types more redundant. Let's start with add and delete.

```
add :: Thing -> Database -> (Database, ())
```

```
delete :: Thing -> Database -> (Database, ())
```

There's nothing interesting about returning a superfluous (), that's why we opted to have add and delete return only the modified Database. In principle, however, there's nothing wrong with saying that add and delete take a Database and return a modified version of it alongside a moot (). It's silly, but it is not wrong. If we do this, add and delete will fit the pattern we first identified when dealing with the magician's Deck. That is, Database -> (Database, something).

Then there's lookup. Contrary to add and delete, this function doesn't modify the given Database at all, which is why it doesn't need to return a new one. lookup only cares about the Thing it may find. However, to convince ourselves that lookup could fit the Database -> (Database, something) arrangement too, we can change it so that it returns the same Database it was given as input, unmodified, alongside the desired Maybe Thing output.

```
lookup :: Name -> Database -> (Database, Maybe Thing)
```

Yes, the output Database is superfluous, just like () was for add and delete. But also, it is not wrong.

So, while add and delete only care about the Database they modify, and lookup only cares about the value it can extract from this Database, both can pretend to care about what their peers do to find a common ground, a unifying pattern.

We can, of course, have functions that truly care about both things. Imagine, for example, a function that deleted duplicate Things from the

Database, reporting back to us which were those Things that were duplicate.

```
deleteDuplicates :: Database -> (Database, [Thing])
```

The returned Database is potentially different from the input Database, and the returned list of Things is valuable information. There's nothing superfluous about the type of deleteDuplicates, just like there wasn't anything superfluous about pickTopCard or splitDeck.

Once again, we find that the order in which we interact with our Database, just like we interacted with our Deck before, is terribly important. We may find a Thing in a Database at some point in time, but we'll definitely not find it anymore *after* we decide to delete it. If we expect a Database to honor its name, we must be careful to always use its most recent version as time goes by.

We've found a pattern. And a pattern, once properly identified, can often be abstracted away.

Two hundred sixty-two

The state. The root of all evil. The enemy. The one who creates and spreads misery. That's who we've identified. That's who we'll limit to its minimal expression lest it ever grows and, inevitably, makes us pay and suffer for its crimes.

A thing is. But then it changes, and it is not anymore. Now, there's a new thing in its stead. The old thing is now forbidden. It's not gone, but we cannot rely on it anymore. As time goes by, this happens time and time again. Some of these changes may even seem to be happening at the same time, making the whole affair a lot more confusing. We refer to this ever-changing thing as the *state* of our program, of our Database, of our Parser, of our Deck. This is the pattern, this is the problem we are trying to solve.

A Database changes. The raw String input available to a Parser changes. A Deck, too.

```
Database -> Database
```

```
String -> String
```

```
Deck -> Deck
```

And while changing, sometimes we have some extra information we can offer, some new information we can *produce* as part of this change. Like the Things that we've found to be duplicate when deleting them from the Database, or the value a Parser was able to parse while preparing the leftovers for a subsequent Parser, or the Card at the top of our Deck.

```
Database -> (Database, [Thing])

String -> (String, x)

Deck -> (Deck, Maybe Card)
```

So, we always have a thing that changes, and we always have something extra that we can produce alongside this change. Sometimes, that something extra is a boring (), as it was in Database's add, but there's nothing wrong with that.

```
add :: Thing -> Database -> (Database, ())
```

In all these cases, what we have is a State transformation.

```
data State s x = State (s -> (s, x))
```

We are introducing a new type, State, as a wrapper around this pattern of modifying a thing, here of type s, and producing some extra output of type x while doing so.

Keep in mind at all times that State s x is *not* the state value itself, but rather a function-like thing, a way to *transform* a state value of type s.

We can now use State in some of our state transformations. For example, here's the function that that creates a transformation that adds a Thing to a Database.

```
add :: Thing -> State Database ()
```

Here's the one that picks a Card from the top of our Deck, if any.

```
pickTopCard :: State Deck (Maybe Card)
```

And here's the one that deletes and reports duplicate Things from the Database.

```
deleteDuplicates :: State Database [Thing]
```

All of these transform, or at least *could* transform, their respective Deck or Database values. And as a part of this transformation they *produce* some extra *output* such as those Things or that Card.

The implementation of these State transformations is straightforward. All we need to do is use the functions we had before as the payload to the State constructor. For example, if we pretend for a second that Deck is just a type synonym for [Card], then the full implementation for pickTopCard would be something like this:

```
pickTopCard :: State Deck (Maybe Card)
pickTopCard = State (\case [] -> ([], Nothing)
                           x : rest -> (rest, Just x))
```

That is, we wrap in the State constructor a function that given a Deck, returns the top Card if any, together with the updated Deck that excludes it.

Two hundred sixty-three

Produce, output. Those words alone should immediately prompt us to wonder whether a State transformation is a Functor. That is, whether we can implement fmap in such a way that the functor laws hold. Here's a first attempt.

```
instance Functor (State s) where
   fmap :: (a -> b) -> State s a -> State s b
   fmap g (State f) = State (\s0 -> let (s1, a) = f s0
                                     in (s1, g a))
```

Our fmap takes a function a -> b, a State s a, and must return a State s b whose a has become a b by means of the given function. This new State transformation is one which *receives* a value of type s to transform, applies to it the original state transformation function, the one inside the passed in State s a, and finally converts the obtained (s, a) to an (s, b) by applying the given function a -> b to that a.

In our implementation, we used the name s0 to refer to the original value of type s, and s1 to refer to the updated one. Notice how the original state s0 is not used anymore after we've applied f to it. In fact, s0 is not even mentioned in the value returned by this function at all. This is good, we want the original state value to be forever lost, lest we accidentally refer to it in the future when we are not supposed to. This is exactly the problem State solves. And notice how the type of fmap doesn't tell us whether our implementation should return the updated state value rather than the original one. Our *goals* do, however, so we better do just that.

Yes, State, or rather State s, is a Functor. This means that we can fmap over a particular State transformation to modify the output it produces. For example, imagine we wanted to modify the output of deleteDuplicates so that rather than reporting the Things it deletes, it merely reports how many duplicate Things were there.

```
deleteDuplicatesCount :: State Database Natural
deleteDuplicatesCount = fmap length deleteDuplicates
```

fmapping the length function, of type [a] -> Natural, over the functorial value deleteDuplicates of type State Database [Thing], will result in a functorial value State Database Natural where that Natural tells us how many Things were there in the list.

Two hundred sixty-four

We mentioned *time*, too. *Time* is deeply ingrained in State transformations, obviously, considering that the state values *before* and *after* a State transformation happens are potentially different. And *time*, in Haskell, means Monad. Let's see if we can implement >>=, *bind*, for the State functor. This one is a bit trickier than fmap, so pay attention.

```
instance Monad (State s) where
  (>>=) :: State s a -> (a -> State s b) -> State s b
  State f >>= k = State (\s0 -> let (s1, a) = f s0
                                    State g = k a
                                in g s1)
```

Given a State transformation producing an a, and a function to obtain a new State transformation producing a b from that a, we return said latter State transformation.

Notice that when we say let State g = k a we are *pattern-matching* on the value k a, of type State s b to extract the function of type s -> (s, b) that's in it, which we name g. We do this because while our function k already gives us the State transformation we are ultimately supposed to return from this application of >>=, at the time when we have access to both k and a we are *not* supposed to be constructing a State s b, but rather a value of type (s, b). So, by pattern-matching on k a we obtain the function that, when applied, will give us exactly what we want. This function, g, takes a value of type s of which we have two. We have s0 representing the original state, and s1 representing the newer, updated state. Guess which one we want to use.

What does it mean for a State transformation to be a Monad? Well, it means exactly what we want. That is, it means that we can completely avoid worrying about accidentally reusing an old state when a new one is available. All that state handling is now embodied in what it means for State to be a monad.

```
pickTwoCardsShuffleAndPickOneMore :: State Deck [Card]
pickTwoCardsShuffleAndPickOneMore =
  pickTopCard >>= \yc1 ->
  pickTopCard >>= \yc2 ->
  shuffle     >>
  pickTopCard >>= \yc3 ->
  pure (catMaybes [yc1, yc2, yc3])
```

As pickTwoCardsShuffleAndPickOneMore shows, there's no explicit manipulation of the Deck state value anymore, which completely eradicates the possibility of us using the wrong one. Compare this with an implementation that explicitly threads the ever-changing Deck through the code.

```
dangerous :: Deck -> (Deck, [Card])
dangerous d0 =
  let (d1, yc1) = pickTopCard d0
      (d2, yc2) = pickTopCard d0
      d3        = shuffle d2
      (d4, yc3) = pickTopCard d3
  in (d4, catMaybes [c1, c2, c3])
```

See? Noisy. Much harder to get right. Much easier to get wrong.
Something that should be impossible to get wrong, *the passage of time*, is
suddenly the easiest thing to lose track of amidst so much noise. In fact,
we deliberately got things wrong just now. Did you notice? It's fine if you
didn't. This is exactly why we embrace Monads, why we embrace tools.

Here's another State transformation moving the Card at the top of the Deck
to the bottom.

```
moveTopToBottom :: State Deck ()
moveTopToBottom = pickTopCard >>= \case
                    Nothing -> pure ()
                    Just c  -> addBottomCard c
```

So, is this whole state problem only about making sure we always use the
most recent state value? It doesn't seem so bad now, we could probably
live with this. Wrong. The state is a complex evil, and as we continue our
journey we'll discover challenges significantly trickier than this, such as
having multiple parties interacting with a same state value at once,
concurrently. This? This is nothing.

Two hundred sixty-five

Before we continue demystifying the State, let's take a small diversion.

```
shuffle :: Deck -> Deck
```

If you've ever played a game of cards, I'm sure you will agree that shuffling
the cards thoroughly and unpredictably is necessary for a fair match. It
should be impossible for players and observers to predict what card will
pop next from the deck, after we've shuffled it. Anything less would be
wrong.

So, how do we do this? Maybe shuffle could interleave cards somehow.
Say, first card to the bottom, second card to the front. Third to the
bottom, fourth to the front, and so on. Good idea? No. If one knew what
was in the original deck, and one knew about the shuffling procedure, one
could easily calculate which card lands where on the new and supposedly
shuffled deck. And it wouldn't be too hard for players to know the
position of some of the cards in the original deck. Imagine, for example,
that they just played a hand and are now returning the cards to the deck. If
they pay attention, they'll learn a lot. Maybe not everything, but enough.

And we can't expect the shuffling procedure to be secret either. It's written there, in Haskell code, anybody can read it. So, no, shuffling this way is a terrible idea.

What about doing a similar interleaving, but rather than starting from the first card, we skip as many leading cards as the topmost card indicates, and only then we start working? Say, if the topmost card is the queen of diamonds, we skip queen many cards before starting to move cards around. And maybe we do this twice for extra safety. Clever, right? Left. Yes, it will be a bit harder for an observer to predict the position of the cards in the shuffled deck. Harder, but not impossible. The unpredictability of the shuffled deck relies on not knowing which card was originally on top, something that, at least, won't be a secret for the player that returns the last card before the deck is shuffled. Pass.

What we need is a way to shuffle a deck where the unpredictability of the shuffled deck doesn't rely on keeping neither the shuffling process nor the original deck secret. What we need is cryptography.

Cryptography is the branch of knowledge that worries about the secrecy, integrity and authority of data. How can we be certain that a deck is properly shuffled? Was it shuffled by somebody who had the authority to do so? Has somebody tampered with the deck? Can we confirm that the shuffled deck contains all the cards it's supposed to contain, without revealing their position in the deck? These are some of the questions that cryptography asks and tries to answer.

To shuffle in a cryptographically secure manner, we can't rely on the original deck nor on the shuffle function being secret. We assume that everybody knows about them. Instead, an additional input, fully independent from the deck and the shuffling code, a secret, will need to be taken into account.

```
shuffle :: Secret -> Deck -> Deck
```

Secret will be some sort of random number. And when we say *random*, we mean a truly unpredictably random number, such as a number obtained by repeatedly rolling dice. Nobody but the card dealer must know about this Secret, because knowledge of it, together with knowledge about the original deck and about how shuffle operates, potentially allows one to determine the final position of the cards in the shuffled deck.

The shuffle function must make use of this Secret while determining its card moving routines, this will prevent outside observers from predicting what the final deck will look like, for they know not the Secret. How shall shuffle use the Secret? Well, this is the part of cryptography we are not supposed to talk about. Pretty much all of the ideas that we may come up with, unless we are well versed on this topic and can mathematically prove that our reasoning is correct, will be wrong. Cryptography is very subtle.

One wrong movement and *boom*, all the secrets could be revealed. But, to get a general idea, we could imagine that Secret represents how many times to repeat a card movement operation or how many leading cards to skip. This is similar to how in the previous example we allowed the topmost card in the original deck to influence our shuffling, except this time it's not the topmost card, but the Secret.

Beauty lies in the eye of the beholder, they say. Well, I don't think so. Cryptography is always beautiful, regardless of any beholding. Unfortunately, because this isn't a book about cryptography, we won't explore its beauty in here. Wouldn't that be nice, though? Beautiful.

Two hundred sixty-six

Our State is still missing a couple of pieces. Isn't it always? Anyway, a State transition is a Functor and a Monad, but it is also an Applicative functor. We don't really have to ask ourselves whether it is, because that's the way things are. Something that's a Monad, must necessarily be an Applicative functor too. So, without further ado, we go and write the corresponding instance. Conceptually, we should have written this instance before having written the Monad one. In practical terms, however, doing it the other way around is usually easier.

```
instance Applicative (State s) where
  pure :: a -> State s a
  pure = \a -> State (\s -> (s, a))
  (<*>) :: State s (a -> b) -> State s a -> State s b
  (<*>) = ap
```

Let's start with <*>, the easy one. Well, actually, implementing <*> by hand is often a bit tricky. However, as we learned before, if our chosen Applicative functor is a Monad too, then we can use ap from the Control.Monad module as a default correct implementation of <*> in terms of >>= and pure. We'll use that. This is why we say writing the Monad instance first is easier.

Actually, if we recall, the same trick applies to the Functor instance, whose fmap implementation can be just liftA or liftM for Functors that are also Applicatives or Monad respectively. Easy, isn't it?

```
instance Functor (State s) where
  fmap = liftA
```

Anyway, back to pure. Given an a, we are trying to create for it a functorial value of type State s a. That is, a state transformation that produces that same a. We don't really have a transformation to do to that state value s, so we make up a function s -> (s, a) which merely returns the state value without modifying it, put it inside the State constructor, and we are done.

So, yes, State transformations are Applicative functors. Great, what can we do with this? We can, for example, use <*> to combine two State transformations.

```
lookupBreadAndButter :: State Database (Maybe Thing, Maybe Thing)
lookupBreadAndButter = (,) <$> lookup "bread" <*> lookup "butter"
```

Or maybe sequence ten Stateful transformations. Why not?

```
moveTopToBottomTenTimes :: State Deck [Card]
moveTopToBottomTenTimes = sequence (replicate 10 moveTopToBottom)
```

Two hundred sixty-seven

It turns out we often don't really *need* to use the State constructor directly, as we did in our most recent pickTopCard.

```
pickTopCard :: State Deck (Maybe Card)
pickTopCard = State (\case [] -> ([], Nothing)
                          x : rest -> (rest, Just x))
```

State transformations come with some very useful vocabulary, get and put, that allows us to avoid explicitly dealing with the State constructor.

```
get :: State s s
get = State (\s -> (s, s))
```

get is a functorial value that produces as its output the current state value. The current Deck of cards, say.

And in the opposite direction, we have put, a function that will set the most recent state value to whatever we want it to be. Provided it is of the same type s, that is.

```
put :: s -> State s ()
put s = State (\_ -> (s, ()))
```

That is, put will create a new State transition that merely discards the state value it gets as input, replacing it with something else. put produces a boring () as output, seeing as there's nothing else interesting that it could produce. Well, I suppose it could output the former state value, but we can achieve that by other means nonetheless.

Anyway, having get and put, we can rewrite pickTopCard in such a way that, by using them, we can avoid dealing with the State constructor explicitly, which may at times feel awkward. The idea is to use get, rather than the State constructor, to get access to the Deck in pickTopCard, and *after* we are done doing whatever it is that we want to do with that Deck, we put it back as the current state value.

```
pickTopCard :: State Deck (Maybe Card)
pickTopCard = get >>= \deck ->
              case deck of
                []      -> pure Nothing
                x : rest -> put rest >> pure (Just x)
```

Notice how in this example, we only put a new state value if it's at all different from the one before. If we find out that our Deck is already empty, then we don't really need to put a new Deck at all, for this state is equal to the one would be putting anyway.

This new implementation is not necessarily an improvement over the one before. Still, hopefully we can imagine how get and put may come in handy in longer State transformations.

In any case, we'll always be able to mix together uses of get, put, the State constructor, and any other way of obtaining these State transformations such as pure, fmap or sequence. So, at any time, we are free to pick whichever is most convenient. However we build them, we can always *bind* State transformations together using >>=.

```
shuffleTwiceFlipDeckPickTop :: State Deck (Maybe Card)
shuffleTwiceFlipDeckPickTop =
  replicateM 2 shuffle >>
  State (\deck -> (reverse deck, ())) >>
  pickTopCard
```

Two hundred sixty-eight

Alright. We have a State transformation, say, of type State s x. Now what? We can continue adding more steps to that transformation as we've been doing so far, or we can *run* it to actually transform an original state value of type s into an updated s and accompanying brand new x.

```
runState :: State s x -> s -> (s, x)
```

Say we have a Deck of cards consisting of cards a, b and c. Transforming it using pickTopCard will give us a new Deck consisting of the topmost card, say a, alongside the rest of the Deck.

```
runState pickTopCard [a, b, c]  ==  ([b, c], Just a)
```

The implementation of runState is trivial:

```
runState :: State s x -> s -> (s, x)
runState (State f) s = f s
```

The State constructor already carries a function of type s -> (s, x) which we can apply to an initial state value s, so there's not much that needs to be done. Essentially, runState is just getting rid of that State wrapper. This becomes more apparent if we add some redundant parentheses in runState's type.

```
runState :: State s x -> (s -> (s, x))
runState (State f) = f
```

Two hundred sixty-nine

Sometimes we are not really interested in the updated state value resulting
from our State transformation, but only in the accompanying functorial
output x. For those cases, we have evalState.

```
evalState :: State s x -> s -> x
evalState m = snd . runState m
```

We haven't seen snd so far, surprisingly. In Haskell, we can use the tightly
but appropriately named fst and snd functions to take the *first* and *second*
elements out of a pair, respectively.

```
fst :: (a, b) -> a
fst (a, _) = a

snd :: (a, b) -> b
snd (_, b) = b
```

In evalState, we could have pattern-matched on the output of runState,
but using snd is just as fine.

Similarly, we may be interested only on the updated state value s and not
in the functorial output x, in which case we may reach out for execState.

```
execState :: State s x -> s -> s
execState st = fst . runState st
```

Probably because of their unevocative names, these two functions
evalState and execState are a bit elusive and easy to forget. Worry not,
we'll just remember runState and it'll be fine.

We wn't frget fst nd snd, that's fr sure.

Two hundred seventy

The state value we transform, over and over again, is reminiscent of the
value we accumulate while we foldl, say, a list. And if we recall, there was
something particular about the accumulator in a fold. It had to do with
the strictness, or laziness, of its evaluation.

As we go about writing our State transformations, we must sometimes
pay attention and ensure that the state we transform gets fully evaluated,
lest we end up creating many thunks that we could have avoided. For
example, consider this state transformation.

```
addOne :: State Integer ()
addOne = get >>= \x -> put (x + 1)
```

Each time we perform addOne we are deferring the evaluation of x a bit more. x + 1 is a thunk that contains in it yet another thunk, the one for x. In small scenarios, this is not a problem, but if we were to do things such as replicateM 12345 addOne, say, we'd be creating at least 12345 unnecessary thunks, each of which consumes computery resources such as memory and time.

Ideally, we should be getting rid that inner thunk at least, the one for x. We can accomplish this by forcing the evaluation of x by means of seq or a *bang* !, as we've learned before.

```
addOne :: State Integer ()
addOne = get >>= \ !x ->
         put (x + 1)
```

This new code forces the evaluation of x before evaluating put (x + 1), keeping our laziness strict enough so that we avoid unnecessary thunks. Notice we left some whitespace in between \ and !x. This is because otherwise the Haskell parser gets confused about them being immediately next to each other. Alternatively, using seq, we can write:

```
addOne :: State Integer ()
addOne = get >>= \x ->
         seq x (put (x + 1))
```

Remember, seq a b forces the evaluation of a to *weak head normal form* before proceeding to return b. For some reason, it's very common to see seq used in *infix* form, rather than prefix form.

```
addOne :: State Integer ()
addOne = get >>= \x ->
         x `seq` put (x + 1)
```

Yes. In Haskell, every prefix function can be used in infix form by surrounding its name with backticks `.

```
seq     a b  ==  a `seq`     b
fmap    a b  ==  a `fmap`    b
mappend a b  ==  a `mappend` b
```

This is like using infix functions in prefix form by surrounding them with parentheses like in (+) or (:), except the other way around. Obviously, this only works for functions taking two arguments. In some situations this can come in handy. For example, in Haskell we have a function called isPrefixOf that tells us whether a list is at the very front of another list.

```
isPrefixOf :: Eq x => [x] -> [x] -> Bool
isPrefixOf [] _ = True
isPrefixOf _ [] = False
isPrefixOf (a:as) (b:bs) | a == b -> isPrefixOf as bs
                         | otherwise -> False
```

Because of this function's name, if we were to apply it as isPrefixOf foo bar, it wouldn't be clear whether foo is expected to be a prefix of bar or vice-versa. On the other hand, if we write foo `isPrefixOf` bar, it becomes apparent that we are wondering if foo is a prefix of bar, and not the other way around.

Generally speaking, however, prefix functions in infix position, or infix functions in infix position even, suffer from being in infix position. Or rather, *we* suffer because of their being in infix position and their order of precedence not being clear when more than one infix application shows up in a same expression. It's the *infix operator soup* problem we mentioned a while ago. So, be careful, don't take these backticks lightly.

Two hundred seventy-one

While we are on the topic of syntax, let's introduce some more of it. Consider this monadic value and its many binds >>=.

```
ding :: IO Natural
ding = getLine >>= \a ->
       getLine >>= \b ->
       putStrLn b >>= \_ ->
       putStrLn a >>= \_ ->
       pure (length a + length b)
```

We understand this program, there's nothing wrong with it. However, it is true that those uses of bind >>= get a bit noisy after a while. We can clean up things a bit using >> rather than >>= _ -> in some of those lines, but Haskell offers something even better: The "do notation", arguably one of the most notable pieces of syntax in the Haskell language.

```
dong :: IO Natural
dong = do
  a <- getLine
  b <- getLine
  putStrLn b
  putStrLn a
  pure (length a + length b)
```

Hopefully you'll find this is a bit easier on the eyes. Haskell's "do notation", in reference to the leading do keyword, is another way for us to write our monadic values without having to write down the bind operator >>= literally. That's pretty much it. While they *look* different, ding and dong mean exactly the same. As part of the compilation process, Haskell will translate the code in dong to be the one that we see in ding.

The leading do is not the name of a function, but a keyword. It's something special in the syntax of the Haskell language, like when we write class, let or instance. After the do keyword, we write a series of *statements*, one after the other on separate lines. Some of these statements will involve a name, followed by a leftwards arrow <-, further followed by a monadic value. The output produced by that monadic value will be bound to the name on the left. For example, a <- getLine, will get translated to getLine >>= \a -> ....

Some statements won't include the leftwards arrow <-, nor anything to the left of it. That is, they will only mention the monadic value part, as in putStrLn a. In this case, the output from the monadic action will simply be discarded.

The last statement of all must be a monadic expression on its own, without the leftwards arrow <-. This is the monadic expression that will produce the final output of the whole do block. For example, in dong above, the Natural output of the whole IO interaction is length a + length b.

It's a rather mechanical translation. What may not be immediately obvious, however, is that we can combine this do notation with uses of the many monadic and functorial operators we already know and love.

```
dung :: IO Natural
dung = do
  (a, b) <- (,) <$> getLine <*> getLine
  putStrLn b >> putStrLn a
  pure (length a + length b)
```

In this new code we are combining do notation with uses of <$>, <*> and >> to achieve the same thing we achieved before in ding and dong. It's handy, this do notation. We'll use it a lot.

One thing we should be aware of, though, is that while similar, do notation is very different from let expressions. Consider these two examples side by side.

```
let a = foo        do a <- foo
    b = bar           b <- bar
```

Yes, the do keyword can be on the same line as the first statement, that's fine. In the let example, we are saying that a equals foo and that b equals bar. Or perhaps we are saying it the other way around, b first and a second. There isn't really an ordering between these two equalities, none of them happen before or after the other. In the do notation example, however, things are very different. First we are performing foo and binding its *output* to the name a, and *then* we are performing bar, binding its *output* to the name b. There's a clear ordering here, and the things we are naming are not the same things a let expression would. There's nothing new about this, this is exactly what we already had with >>=. It's just a matter of getting comfortable with this new, handy syntax.

And yes, this works with *every* monad. For example, we can write a State transformation this way.

```
moveTopToBottom :: State Deck ()
moveTopToBottom = do yc <- pickTopCard
                     case yc of
                         Nothing -> pure ()
                         Just c  -> addBottomCard c
```

We can write IO interactions as we just did, or we can write Parsers.

```
inParens :: Parser x -> Parser x
inParens p = do expect1 '('
                many whitespace
                x <- p
                many whitespace
                expect1 ')'
                pure x
```

We can use this do notation for anything, as long as what we are using it
for is a monadic value. That is, a functorial value where said functor
happens to be an instance of the Monad typeclass, like IO, Parser or State s
are. Isn't it pretty?

Two hundred seventy-two

Actually, that was a lie. Sorry. It's such a fascinating teaching vehicle, the
lie. No wonder that's how most education is done. You do understand
what the do notation is about now, don't you? See? Fascinating.
Unfortunately, many a student never learns any better, and then goes out
to preach what's wrong. It's dangerous. Let's fix this.

The do notation doesn't just translate into Monadic values using binds >>=,
but also, at times, when possible, it will translate to Applicative functorial
values using but <*>. If you recall, Applicative functors are more readily
available than Monadic ones, so that's a good thing. It means we can do more
things than we thought we could. Let's see why, let's see how.

```
foo :: Monad f => f ()
foo = do a <- pure ()
         b <- pure a
         pure b
```

A very silly example, foo, but it serves our purpose. If we ask GHC to *infer*
the type of this expression foo, it says Monad f => f ' () as we see above.
However, if we ask about this other expression bar, it says something
different.

```
bar :: Applicative f => f ()
bar = do a <- pure ()
         b <- pure ()
         pure b
```

What's the difference? Can you tell? What was it that Monads add to
Applicative functors? The difference is that when we have Monadic

vocabulary available, we can have functorial values depending on previous functorial outputs by means of *bind* >>=. But when we are restricted to Applicative vocabulary only, we can at most combine different functorial outputs somehow. We definitely can't have functorial values depend on past outputs without bind >>=.

This is exactly the difference between foo and bar above. In foo, we see a <- pure () and b <- pure a. That is, we see the the functorial value pure a *depending* on that a, an output from a previous functorial value. Whereas in bar, when we say a <- pure () and b <- pure (), none of these functorial values depend on a previous functorial output. Haskell notices this too, and infers a simpler Applicative constraint on f rather than a Monad one. This is good. And yes, we can still use bar with choices of f that are Monads, for being a Monad implies being an Applicative functor too, which is what bar cares about.

As a proof that bar can be implemented when only Applicative vocabulary is available, here's an implementation of bar that avoids do notation altogether, using <*> instead.

```
baz :: Applicative f => f ()
baz = (\a b -> b) <$> pure () <*> pure ()
```

See? No bind >>=. When compiling the code using do notation in bar, GHC will generate something that looks just like baz. On the contrary, we just cannot express foo without relying on >>=. Try it, it won't work. And no, you can't cheat and skip applying pure to that output a from a previous functorial value. If it helps, imagine you are not applying pure but a function that does perform meaningful functorial effects. You wouldn't want to skip applying that one, would you?

Here's another example using do that only exploits the Applicative features of the chosen functor.

```
inParens :: Parser x -> Parser x
inParens p = do expect1 '('
                many whitespace
                x <- p
                many whitespace
                expect1 ')'
                pure x
```

Obviously, a Parser supports the Monadic bind >>= as well, but in all truth, we are not exploiting that here, so this code would have worked just as fine back when our Parser was just an Applicative. Remember that time? We were so young back then. As a proof, here's how we'd translate inParens to rely only on <*>, similarly to how GHC will do it behind the scenes.

```
inParens :: Parser x -> Parser x
inParens p = (\_ _ x _ _ -> x) <$> expect1 '('
                               <*> many whitespace
                               <*> p
                               <*> many whitespace
                               <*> expect1 ')'
```

It looks ugly, doesn't it? Good thing we have the do notation to clean up this mess.

In any case, it's easier to *think* of do notation as translating to uses of >>= rather than uses of <*>, so that's what we do. Also, do's translation to <*> is a relatively recent addition to the Haskell language, so it's common for people to forget that this exists. You are the next generation, though, so you get to know these things.

do does one more useful thing too. However, we'll leave that one for later.

Two hundred seventy-three

While we are on the topic of lying, lets reveal the ugly truth about something else.

```
data State s x = State (s -> (s, x))
```

It turns State is not defined like that, but this other way.

```
data State s x = State (s -> (x, s))
```

Actually, funny story, State it's not defined *exactly* that way either, but we'll see about that later on. One step at a time.

See the difference? It's not much, we just swapped the x and s in that tuple. Why? Your guess is as good as mine. For practical purposes, not much changes, both definitions work just fine. However, there's something beautiful about s -> (s, x) that's missing from s -> (x, s). It has to do with the functorial nature of all the pieces involved.

Before, we defined the Functor instance for *our* State, the one with s -> (s, x) inside, using liftA. That's fine. However, we could as well have defined it manually, without relying on that magical tool.

```
instance Functor (State s) where
   fmap f = State . fmap (fmap f) . runState
```

Beautiful, isn't it? We define fmap for State s relying only on the functorial nature of its constituent parts. We fmap once to modify the output of the function s -> (s, x), and then we fmap a second time to modify the second element in the tuple (s, x). Remember, two-element tuples are functorial values with their second element as functorial output. We also remove the State wrapper and add it back using runState and State, but that's unimportant here.

The aesthetics of implementing `fmap` for `State (s -> (x, s))`, on the other hand, are somewhat disappointing.

```
instance Functor (State s) where
    fmap f = State . fmap (\(x, s) -> (f x, s)) . runState
```

Something is gone. *It's just beauty*, they say, *not important*. But see, beauty was there before, and now it's gone. *Why?*

This is irrelevant to users of `State` who just deal with `State` from outside, but those manually using the `State` constructor will always be reminded of this, as they wonder why their code is often more awkward than it needs to be for no fundamental reason.

Don't neglect aesthetics, they matter. It's unlikely an accident that so many of our foundational pieces are as straightforward, simple and beautiful as they are.

Two hundred seventy-four

Monads for this, monads for that, monads for everything. Many monads, each doing one interesting thing, but unfortunately, not yet doing many things together. We can have `State` transformations *or* we can have `IO` interactions *or* `Maybe` we can succeed, but we can't yet have `State` transformations *and* `IO` interactions *and* `Maybe` succeed at the same time. Or can we? After all, `Parser`, for example, is a `Monad` that must deal with some of these concerns at the same time. It deals with failure, with consuming input, with leftovers. Let's see if we can properly identify those concerns and do something about them.

```
data Parser x = Parser (String -> Either (Natural, String)
                                          (Natural, String, x))
```

If we recall, our `Parser` is essentially a function that takes a `String` to be parsed into an x and, if it's possible to obtain said x, returns it alongside the number of consumed `Character`s as well as any leftover `String`. Otherwise, if parsing is not successful, it returns the position of the `Character` that first failed to parse, alongside an error message as a `String`. Composing multiple `Parser`s together using *bind* `>>=` makes sure these leftovers and errors are carried over as expected.

So, *errors*. Obviously our `Parser` deals with errors. We can see, however, that it's not the `Parser` itself who ultimately supports the idea of errors, but rather, it is the `Either` that's inside our `Parser` who does. And `Either` is a `Monad`.

And *consuming* the input, too. A `Parser` consumes input, sure. But, again, it's not the `Parser` itself, but the fact that a `Parser` is modeled as a *function*, which allows for this consumption. And moreover, this input is not only consumed, but it is eventually *transformed* into leftover input. And didn't we have State transformations for modeling scenarios like this? Indeed. And State is a Monad.

What about the number of consumed Characters? There are different ways to approach that, we'll see, but for the time being, we can consider it a State transformation too. One that takes as input the number of Characters consumed so far, which initially will obviously be 0, and after parsing, whether successfully or not, adds to this amount the number of newly consumed Characters.

So, as we can see, a Parser's concerns are made up of smaller concerns, each of which is a Monad on its own. We are, in a sense, composing all of these smaller Monads into a bigger one. However, we are doing it in a rather ad-hoc way, manually dealing with our state transformations, for example, rather than reusing State directly. We are mashing things together, rather than composing them. Let's see if we can clean this up and truly compose these smaller monads instead.

Two hundred seventy-five

To our disappointment, Monad composition is not really a thing that exists. Not in the sense we are used to, anyway. For example, if we take two functions f and g, we can easily compose them into a new function f . g using the function composition operator (.). Or if we take two Monoids, we can compose them into a new one using mappend. But there's no such thing for Monads. There's nothing that can take two arbitrary monadic values, or rather, two type constructors for two particular Monads, and return a new one that is also a Monad combining the effects of them both. We'll see why, shortly, but first let's learn about Functor composition. Yes, Functors compose just fine.

```
data Compose g f a = Compose (g (f a))
```

If both g and f are Functors, then Compose g f is also a Functor. What does it mean? Well, whatever the Functor instance for Compose g f, which has only one possible implementation, says it means.

```
instance (Functor g, Functor f) => Functor (Compose g f) where
  fmap :: (a -> b) -> Compose g f a -> Compose g f b
  fmap h (Compose gfa) = Compose (fmap (fmap h) gfa)
```

In other words, fmapping over a functorial value of type Compose g f a allows us to modify the a being produced by a functorial value f a, which is in turn being produced by another functorial value g (f a). This allows us to treat these two nested functorial values as if they were just one functorial value.

```
foo :: [Maybe Integer]
foo = [Just 1, Nothing, Just 3]
```

Normally, if we wanted to add one to each of the Integers in foo, we'd need to fmap once over foo to target the outermost Maybe Integer output, and once again to target the innermost Integer output.

```
> fmap (fmap (+ 1)) foo
[Just 2, Nothing, Just 4]
```

But, if we wrap foo in Compose, we can get the same result by fmaping just once over it.

```
bar :: Compose [] Maybe Integer
bar = Compose foo

> fmap (+ 1) bar
Compose [Just 2, Nothing, Just 4]
```

We have that extra Compose wrapper now, of course, but we can easily get rid of it using pattern-matching when necessary, or better yet, we can introduce a helper function for removing that wrapper.

```
unCompose :: Compose g f a -> g (f a)
unCompose (Compose gfa) = gfa
```

For practical purposes, what we gain from Functor composition is similar to what we gain from function composition. For example, if we have a function that adds 1 and another function that multiplies by 7, composing these functions will result in a new function that first adds 1 and then multiplies by 7. Or the other way around, depending on the order of our composition. Composing functors is similar. If one of the functors is Maybe, say, which allows us to convey failure or absence, and the other is [], which allows us to list zero or more elements, then the composition of Maybe and [] would be a functor where, depending on the order of the composition, is a list where some of its elements are optional, or a list of non-optional elements that is itself optional.

Alternatively, forgetting our interpretation of this composition and focusing on the types, if we have two functions f of type a -> b and g of type b -> c, we can compose these functions as g . f even before having access to their input a, and work further with this composition before it's applied to produce an output, if ever. That is, we can work with f and g independently from a. Similarly, Compose g f a, or rather the partially applied Compose g f, lets us deal with the functorial aspects of g and f without paying too much attention to what a will be later on. And the other way around, too. If all we care about is a, then we can focus on it and ignore how many functorial values we must peel until we get to it, and what they convey.

```
> fmap (+ 1) [3]
[4]
> fmap (+ 1) (Compose [Just 3])
Compose [Just 4]
> fmap (+ 1) (Compose [Compose [Right 3, Left 9]])
Compose [Compose [Right 4, Left 9]]
```

So, yes, Functors can be composed in a predictable manner using Compose, and what we end up with is yet another Functor that preserves the structure of the two original Functors. This doesn't mean that using Compose is a particularly common thing to do explicitly, but that doesn't make the composition of Functors any less true.

Two hundred seventy-six

All of the Functors we composed are Monads too. However, in our examples we didn't make use of any of their monadic features, so they might as well not have been Monads and nothing would have changed. When we say that Monads don't generally compose, what we mean in practical terms is that if we try to implement a Monad instance for Compose g f, with g and f themselves being arbitrary Monads, we will fail.

```
instance (Monad g, Monad f) => Monad (Compose g f) where
  Compose gfa >>= k = Compose (gfa >>= \fa ->
                              let fgfb = fa >>= \a ->
                                          pure (unCompose (k a))
                              in magic? fgfb)
```

Try it. Or try a different approach. You'll get stuck some way or another, but it's nonetheless an interesting exercise. So, *no*, in general, Monads do not Compose. And what we mean by this, exactly, is that it's not possible to implement a Monad instance for Compose g f where g and f are *arbitrary* Monads, similarly to how we did with Functors.

If we know what g and f are, though, we can give a Monad instance for that particular composition. For example, here's how we'd compose State and Maybe.

```
instance Monad (Compose (State s) Maybe) where
  Compose sya >>= k =
    Compose (State (\s0 ->
      case runState sya s0 of
        (Nothing, s1) -> (Nothing, s1)
        (Just a, s1) -> runState (unCompose (k a)) s1))
```

Very mechanical. We are just following the types, going wherever they take us. Well, mostly. We are dealing with State transformations here, so we must ensure we always use the most recent state value and such.

Sometimes we can get away with a Monad instance for Compose g f where only one of g or f are polymorphic.

```
instance Monad g => Monad (Compose g Maybe) where
  Compose gfa >>= k = Compose (gfa >>= \case
                        Nothing -> pure Nothing
                        Just a -> unCompose (k a))
```

One way or another, when defining a Monad instance for Compose g f, we'll need to know things about g, f, or both. We can't implement a Monad instance for a fully polymorphic Compose g f, we must do so for a particular choice of g and f. In the worst case scenario, we'd be writing one instance per unique combination of g and f.

But suppose for a second that we decide to go through the hurdle of implementing bespoke Monad instances for specific combinations of g and f in Compose. It turns out that, in practice, *using* Compose explicitly is rather annoying anyway, since we'd have to deal with that Compose wrapper each time we *bind* with >>= or do notation.

```
foo :: Compose (State Integer) Maybe String
foo = do x <- Compose (State (\s -> (Just (show s), s + 2)))
         n <- Compose (State (\s -> (Just s, s * 10)))
         _ <- Compose (State (\s -> (Just (), s + 1)))
         pure (concat (replicate n x))
```

Annoying, isn't it? Anyway, it does what we want.

```
> runState (unCompose foo) 5
(Just "5555555", 71)
```

Namely, it composes the State transformations with the possibilities of failure according to Maybe. Here's an example exploiting the latter.

```
bar :: Compose (State Integer) Maybe String
bar = do x <- Compose (State (\s -> (Just (show s), s + 2)))
         n <- Compose (State (\s -> (Nothing, s * 10)))
         _ <- Compose (State (\s -> (Just (), s + 1)))
         pure (concat (replicate n x))
```

Compared to the previous foo, in bar the functorial value supposed to produce the output n fails with Nothing, causing bar to bail out early without producing that promised String.

```
> runState (unCompose bar) 5
(Nothing, 70)
```

Notice how even if we don't get the final String, we do get to the final Integer value of our State transformation. This time, however, it's 70 and not 71 as it was before, demonstrating how the effects from State and Maybe have indeed been composed. Past functorial values, be them from the Maybe or the State half, do influence any and all subsequent functorial values in their composition.

Why did we get that Integer? Because the type of bar says so. Remember that Compose g f a is just a wrapper around g (f a), so Compose (State

Integer) Maybe String is essentially State Integer (Maybe String). And
State s x, in turn, is just a wrapper around s -> (x, s), so State Integer
(Maybe String) is essentially Integer -> (Maybe String, Integer). See?
Given an Integer, we get a Maybe String and an Integer back.

Notice that if we compose our Monads in the opposite order Compose Maybe
(State Integer) String rather than Compose (State Integer) Maybe String,
the result will be different. Compose Maybe (State Integer) String is a
wrapper around Maybe (State Integer String), and it says right there that
Maybe, just maybe, we'll have a State transformation to perform.

In other words, the order in which we Compose our Monads is meaningful.
This is fine. This is expected. Monads are Functors, and generally speaking,
Functor composition is not commutative except for a few select Functors.
That is, g (f a) and f (g a) don't necessarily accomplish the same.

Two hundred seventy-seven

Applicative functors do Compose just fine, by the way.

```
instance (Applicative g, Applicative f)
  => Applicative (Compose g f) where
  pure :: a -> Compose g f a
  pure = Compose . pure . pure
  liftA2 :: (a -> b -> c)
         -> Compose g f a
         -> Compose g f b
         -> Compose g f c
  liftA2 k (Compose gfa) (Compose gfb) =
    Compose (liftA2 (liftA2 k) gfa gfb)
```

pure does the obvious thing, wrapping a in f first, then wrapping f a in g,
and finally wrapping all of g (f a) in Compose. The implementation of
liftA2, as it's often the case, is very similar to that of fmap. That is, it
performs liftA2 twice, once per each Applicative functor being composed,
and finally wraps the result in our Compose wrapper. For comparison, here's
fmap again, also known as liftA.

```
fmap :: (Functor g, Functor f)
     => (a -> b) -> Compose g f a -> Compose g f b
fmap k (Compose gfa) = Compose (fmap (fmap k) gfa)
```

Why do Applicative functors Compose this way and Monads don't? Monads
allow for functorial values to depend on the output from previous
functorial values, and this forces the composition of two Monads f and g to
become deeply intertwined, since both f and g affect subsequent
functorial values involving the composition of f and g. That's not the case
with Applicative functors, though, where a functorial value can't possibly
depend on a previous functorial output, so their composition doesn't

need to know nearly as much about them.

So, what can we do in Haskell regarding Monads if we can't Compose them? Do we give up? No, of course not. We fight harder, we tear down this wall.

Two hundred seventy-eight

Haskell composes Monads, yes. Kinda. It's different, invasive, but it works, and it does so in a relatively ergonomic manner. Rather than using Compose g f for arbitrary or specific Monads g and f, we use something else, specific to *one* of the effects being composed. A *monad transformer*.

A monad transformer is a type constructor that *transforms* a particular Monad, extending it with a particular set of new effects. For example, if we had IO interactions on the one hand, and wanted to extend those with some State-like state transformation support, we'd use a "state monad transformer" to extend the IO monad. What does this monad transformer look like? Behold.

```
data StateT s m a = StateT (s -> m (a, s))
```

StateT is one of the *many* monad transformers. Strictly speaking, StateT s is the monad transformer, not just StateT. This is akin to how our old State s was a Monad, and not just State. This monad transformer is intended to add to an arbitrary Monad of our choice, m, state transformation effects similar to those State had.

The T in StateT stands for *transformer*, this is how monad transformers are traditionally named. The s and a are the same as they were on State. That old State and this new StateT are similar, but they are not the same. Look.

```
data State  s  a = State  (s ->   (a, s))
```

```
data StateT s m a = StateT (s -> m (a, s))
```

Remarkable similarity. So similar, that a wise choice of m, as we'll see later on, allows StateT to fully replace State. There is no State, really, it's always been StateT.

Pursuing our motivational example, if we wanted to compose state transformations of some value of type s with IO interactions as conveyed by the IO Monad, then we would find ourselves dealing with functorial values of type StateT s IO a.

```
funnyGetLine :: StateT Natural IO String
funnyGetLine = StateT (\s0 ->
  if s0 > 100
    then pure ('", s0)
    else do
      x <- getLine
      pure (x, s0 + length x))
```

funnyGetLine is a single functorial value in StateT Natural IO that *depending* on the current state value, may or may not perform the IO interaction getLine. *If* it does, it will then modify the state value by adding to it the length of the obtained String. Finally, it will produce said String as functorial output.

```
> runStateT funnyGetLine 5
Book
("Book", 9)
> runStateT funnyGetLine 200
("", 200)
```

In the first example, because the initial state value is 5, the getLine IO interaction is performed and we see "Book" as the String output, and 9 as the final state value. That is, 4, the length of "Book", plus the initial state value of 5. In the second example, since the initial state value is 200, getLine is never performed and we get what we get.

runStateT is to runState what StateT is to State.

```
runStateT :: StateT m s a -> s -> m (a, s)
```

```
runState  :: State    s a -> s ->   (a, s)
```

Other than their different types, their implementations are the same. They just remove the State or StateT wrapper, allowing the underlying function to be applied to an initial state value.

```
runStateT (StateT f) = f
```

```
runState  (State  f) = f
```

So, from the type of runStateT follows that runStateT funnyGetLine 100 must be an expression of type IO (String, Natural), which GHCi will gladly perform for us.

Did funnyGetLine compose the effects of a state transformation with those of an IO interaction? Let's see. There were IO interactions depending on the initial state value, and there were new state values depending on those IO interactions. So, yes, StateT s IO did compose state transformation effects with IO interaction effects. However, that's not enough. The whole point of us coming up with this weird idea of monad transformers is about the *composition* of two Monads being itself a Monad, and we haven't seen that yet. That is, we haven't composed two functorial values in StateT s m using *bind* >>= yet. Let's do that. First, we must implement the relevant Monad instance.

```
instance Monad m => Monad (StateT s m) where
  (>>=) :: StateT s m a -> (a -> StateT s m b) -> StateT s m b
  sma >>= k = StateT (\s0 -> do
              (a, s1) <- runStateT sma s0
              runStateT (k a) s1)
```

Straightforward, isn't it? It's not too different from the Monad instance for State s. The first thing to notice is the extra Monad m constraint on this instance. This is saying that for StateT s m to be a Monad, m itself must be a Monad too. This shouldn't be surprising at all, seeing how we are trying to compose *monads* after all.

Our goal is to return a StateT s m b, so we start by using the StateT constructor, which takes as argument a function of type s -> m (b, s). This is a state transformation, so we know that the s we receive, which we name s0, is the initial state we'll be transforming first through sma, and then through the StateT s m b obtained from k. Having named this initial state value s0, we now try to construct a functorial value of type m (b, s). We provide s0 as initial value to sma using runStateT, here of type StateT s m a -> s -> m (a, s), and as soon as we have that functorial value m (a, s), we *bind* it to get a hold of the a and the new s. We chose to use do notation for this, but we could have used the iconic bind operator >>= instead. Same thing. What's important here is that by *binding* this functorial value we are forcing m to perform its Monadic effects in order to produce the a and the s. This event is where the Monad constraint on m comes from. Had m been Maybe, say, this would be a moment when depending on whether we were dealing with Nothing or Just, the rest of the computation would be affected. Maybe there's no (a, s) output at all. Of course, we don't *manually* check what this functorial value is. We can't, because we know nothing about m other than the fact it is some Monad. All we can do is *bind* it and let things happen however they need to happen.

So, if and when we get that a, we apply k to it and use runStateT again, this time using the newest s1 as initial state value. And then we are done, for runStateT (k a) s1 is an expression of type m (s, b), which is exactly what we were looking for.

Does StateT s m compose the effects of a state transformation of s with those of the Monad m? Yes, yes it does. The state transformation effects are obviously happening because we just found ourselves passing around s0 first, then s1, and finally returning the state value resulting from runStateT (k a) s1. As for m, well, we saw ourselves binding functorial values in m, and we even considered what would happen had m been Maybe. So, yes, StateT s m composes state transformation effects with those of m.

```
superfunny :: StateT Natural IO (String, String)
superfunny = funnyGetLine >>= \a ->
             funnyGetLine >>= \b ->
             pure (a, b)
```

What do you think will happen when we run superfunny? Trick question. That's the thing about monadic values. *It depends. We don't know.* It depends on what the initial state values for the respective computations are, and on what the outcomes of the IO interactions funnyGetLine *may* perform are. You see? Functorial values representing different effects depending on each other.

```
> runStateT superfunny 90
these are 23 characters
(("these are 23 characters", ""), 113)
> runStateT superfunny 202
(("", ""), 200)
> runStateT superfunny 90
four
more
(("four", "more"), 98)
```

Beautiful, this is what we wanted all along. Obviously, as every other Monad, StateT s m is a Functor and an Applicative too, but there's no need to dwell on such small details.

```
instance Monad m => Functor (StateT s m) where
   fmap = liftM

instance Monad m => Applicative (StateT s m) where
   pure a = StateT (\s -> pure (a, s))
   liftA2 = liftM2
```

See? Small details. As usual, we can use liftM and liftM2 to get our fmap and liftA2 implementations through the Monad instance for StateT that we just implemented. That's where the Monad constraint on these instances comes from. We could have defined <*> to be ap, instead of defining liftA2, but the final result would have been the same. Whatever you prefer, really.

Two hundred seventy-nine

StateT s m a and our original State s a are similar, but they are not the same. As currently defined, they are not compatible with each other. Luckily, that's not how they are defined in Haskell. Well, StateT is defined that way, yes.

```
data StateT s m a = StateT (s -> m (a, s))
```

But State isn't. Behold.

```
type State s a = StateT s Identity a
```

Look at that. State was a lie. State is not a full blown type on its own, but merely a type synonym for a StateT monad transformer on top of the monad called Identity. We've seen Identity before, but only as a Functor. Well, surprise, it's a Monad too.

```
data Identity x = Identity x
```

The Identity functor, if you recall, was the functor that did nothing functorially interesting. That is, Identity x is essentially the same as x. They are isomorphic, we can always convert an x to an Identity x and back without any problem. What's special about Identity is that it's a Functor.

```
instance Functor Identity where
   fmap f (Identity a) = Identity (f b)
```

And an Applicative.

```
instance Applicative Identity where
   pure = Identity
   Identity f <*> Identity a = Identity (f a)
```

And a Monad.

```
instance Monad Identity where
   Identity a >>= k = k a
```

How is this of any use? Well, if we pick the Monad below StateT to be Identity, then the StateT constructor will carry a value of type s -> Identity (a, s) inside. But we just said that Identity x is essentially the same as x, so we could remove that Identity wrapper without affecting the meaning of this function. That is, StateT s Identity a is essentially the same as saying s -> (a, s). Which, if we recall, is exactly how we defined our original non-transformer State.

```
data State s a = State (s -> (a, s))
```

In other words, we don't need a non-transformer State implementation because StateT on top of an Identity monad accomplishes the same. Except for some extra wrapping and unwrapping on this Identity constructor, that is. And yes, we do talk of a monad transformer as being *on top of* another Monad. Much of the vocabulary we'll soon learn builds on top of this metaphor.

What do we accomplish by combining StateT with Identity? Nothing that our old non-transformer State wouldn't have accomplished on its own. Identity doesn't add any effect to our computation. It doesn't introduce the possibility of failure, it doesn't interact with the environment, it doesn't transform a state value. It's there, it occupies space, it fits an expected shape, but that's about it.

State s a is defined to be a *type synonym* for StateT s Identity a for two reasons. First, because already having a more capable StateT, there's not much motivation for having to maintain a less capable implementation alongside. And second, because this allows us to reuse StateT vocabulary in State. For example, remember get and put from before?

```
get :: State s s
get = State (\s -> (s, s))

put :: s -> State s ()
put s = State (\_ -> ((), s))
```

They are actually defined in term of StateT, and not State.

```
get :: Monad m => StateT s m s
get = StateT (\s -> pure (s, s))

put :: Monad m => s -> StateT s m ()
put s = StateT (\_ -> pure ((), s))
```

That is, exactly the same get and put are available for any choice of m, including Identity. That's good. For example, here's a function called swap that will set a given value as the new state value, returning the old one, implemented using get and put.

```
swap :: Monad m => s -> StateT s m s
swap s1 = do s0 <- get
             put s1
             pure s0
```

We run this state transformation on top of Identity, and it works.

```
> runStateT (swap 2 :: State Integer Integer) 1
Identity (1, 2)
> runStateT (swap 2 :: StateT Integer Identity Integer) 1
Identity (1, 2)
```

And we run it on top of other Monads like lists or IO interactions, and it works too.

```
> runStateT (swap 2 :: StateT Integer [] Integer) 1
[(1, 2)]
> runStateT (swap 2 :: StateT Integer IO Integer) 1
(1, 2)
> runStateT (swap 2 :: StateT Integer Maybe Integer) 1
Just (1, 2)
```

Hadn't State been merely a type synonym for StateT, we couldn't have used get and put as currently defined in terms of StateT.

Oh, and one last thing. Since State is not a true datatype anymore, but just a type synonym, we'll benefit from having a handy function to construct State s a values. Let's call it state.

```
state :: (s -> (a, s)) -> State s a
state f = StateT (Identity . f)
```

Two hundred eighty

StateT s m a is a functorial value in StateT s m, producing an output of
type a. StateT s m is a Monad, which implies it's an Applicative and a Functor
too. And finally, StateT s is a monad transformer, which means that m is
expected to be a Monad too. All of these things are described by a new
typeclass, MonadTrans, for which monad transformers like StateT s
implement instances.

```
class (forall m. Monad m => Monad (t m)) => MonadTrans t where
    lift :: Monad m => m a -> t m a
```

The (forall m. Monad m => Monad (t m)) constraint is something new that
we haven't seen before, even if to the untrained eye it might not seem so.
It says that if t is known to be a MonadTransformer, then *for all* ms that are
known to be Monads, t m is known to be a Monad too. This implication
requires that a Monad instance for t m be available, of course. For example,
in the case where t is StateT s, a Monad m => Monad (StateT s m) instance is
expected to be defined alongside the MonadTrans (StateT s) instance.

What's so new about all of this? Look at the m. In the definition of the
Monad class, this m appears on the class head. That is, to the right of the fat
arrow =>.

```
class Applicative m => Monad m where ...
```

But in the definition of the MonadTrans class, it doesn't.

```
class (forall m. Monad m => Monad (t m)) => MonadTrans t where ...
```

Instead, the m is universally quantified with a forall inside those
parentheses, meaning that m is only accessible within them. Those
parentheses, thus, are *not* optional. The Monad m => Monad (t m) constraint
mentions this universally quantified m, which means that there must be a
Monad (t m) instance compatible with *all* ms that are known to be Monads.

These constraints universally quantifying a type within them are called
quantified constraints for obvious reasons. What may not be obvious is
why we *want* this. The practical consequence of this is that if the type-
checker knows that some specific t is a MonadTransformer, and it knows
that some specific m is a Monad, then it also knows that t m is a Monad, and this
fact doesn't need to be explicitly mentioned anywhere in our source code
but here.

This is the first time we encounter a quantified constraint, but mostly
because we deliberately avoided talking about them before. Some of the
classes we've seen before, like Bifunctor, benefit from a quantified
constraint superclass too.

```
class (forall x. Functor (f x)) => Bifunctor f where
    bimap :: (a -> c) -> (b -> d) -> f a b -> f c d
```

The idea is the same as before. If f is a `Bifunctor`, then f applied to something, whatever that something is, must be a `Functor` too.

Two hundred eighty-one

Let's focus on `lift`.

```
lift :: Monad m => m a -> t m a
```

`lift` is a method available to every `MonadTransformer` that allows us to take a functorial value in `m`, the `Monad` that this monad transformer `t` is extending somehow, and convert it to a functorial value in `t m`. Why is this necessary? We'll see, we'll see. It might be tricky to appreciate `t m a` as a functorial value at first, but adding a few redundant parentheses might help.

```
lift :: Monad m => m a -> (t m) a
```

As for the name "lift", we did say that we talk about monad transformers being *on top* of `Monad`s. And in English, "lift" means "to bring up something from below". Strong imagery. Here, `t` is the `MonadTransformer` on top of `m`, so a function that brings a functorial value in `m` to a functorial value in the realm of `t m` must necessarily be called `lift`. There's no question about it. Given the metaphors we've chosen to talk about monad transformers, `lift` is the correct name. Except, of course, for the confusion it may cause with `liftA`, `liftM`, `liftA2`, `liftM2`, etc.

Anyway, here's an alternative implementation of that `funnyGetLine` from before, this time using `lift` to bring functorial values from the *lower* `Monad`, here `IO`, to the `MonadTransformer` sitting on top of it, and embracing `get` and `put` as means to avoid the explicit usage of the `StateT` constructor.

```
funnyGetLine :: StateT Natural IO String
funnyGetLine = do s <- get
                  if s > 100
                    then pure ""
                    else do
                      x <- lift getLine
                      put (s + length x)
                      pure x
```

Way more pleasant than before, isn't it? It says `lift getLine`, don't miss that part. Here we are combining functorial values in `StateT Natural IO` such as `get` and `put`, with functorial values in `IO`, which by grace of `lift` are *lifted* into the `MonadTransformer` where `get` and `put` live. Can we *not* write `lift getLine`, and instead write `getLine` alone? No. See what GHC's fine type-checker has to say about it.

- Couldn't match type 'IO' with 'StateT Natural IO'
  Expected type: StateT Natural IO String
    Actual type: IO String
- In a stmt of a 'do' block: x <- getLine
  ...

Just like we can't bind, say, a functorial value in IO with a functorial value in Maybe, we can't bind a functorial value in a MonadTransformer with a functorial value in the Monad that underlies that transformer. lift is necessary. Yes, sprinkling lift here and there brings in some noise, but don't worry too much about it, as we'll replace it with something else later on. Meanwhile, here's the MonadTrans instance for StateT s.

```
instance MonadTrans (StateT s) where
  lift :: Monad m => m a -> StateT s m a
  lift ma = StateT (\s -> ma >>= \a -> pure (a, s))
```

Very mechanical. lift adds a StateT wrapper that keeps the state value as is. That's all. What else could it do without violating the law?

Ah, yes. As every other interesting typeclass worthy of a name, MonadTrans has laws that instances must abide by. First, as usual, there's an *identity law*.

```
lift . pure  ==  pure
```

This *identity law* says that whether we purely produce a functorial output in t m, or we do it in m and then we lift it to t m, the resulting functorial t m a is the same.

Second, there's a *naturality law*.

```
lift (ma >>= k)  ==  lift ma >>= (lift . k)
```

This one is a bit trickier to see. The type of ma here is m a, and the type of k is a -> m b, for some ms, as and bs of our choosing. The *naturality law* says that lifting the functorial value ma >>= k, is the same as lifting ma first, and binding its output to a lifted version of k afterwards.

It's very common to see an *identity* law each time we talk about laws, we know this. As for the *naturality* law, this one often appears in one way or another when functorial values in a particular Functor are converted to functorial values in a different Functor. In our case, we are converting functorial values in m to functorial values in t m. Suffice it to say that this law is the one that allows us to either compose things first and then lift them, or lift things first and then compose them, without having to worry about these alternative approaches giving different results.

Two hundred eighty-two

A less ugly but not necessarily less confusing way to write the

MonadTransformer *naturality law* would be in partially applied form, using the leftward fish operator <=<.

```
lift . (g <=< f)  ==  (lift . g) <=< (lift . f)
```

Or maybe not. Maybe it's equally ugly. Or equally beautiful. One can never tell with these things. More beautiful, probably. What about the fish?

```
(<=<) :: Monad m => (b -> m c) -> (a -> m b) -> (a -> m c)
g <=< f = \a -> g =<< f a
```

This operator is the point-free version of =<<, which perhaps should be nicknamed the leftward fish tail operator. That is, <=< composes two functions returning a functorial value into a third one. The resemblance between <=< and the normal function composition dot operator . is unquestionable.

```
(<=<) :: Monad m => (b -> m c) -> (a -> m b) -> (a -> m c)
(.)   ::             (b ->   c) -> (a ->   b) -> (a ->   c)
```

While the dot . just composes functions so that one happens after the other, <=< does the same but for functions that return functorial values. If we imagine f to be a function of type a -> m b and g one of type b -> m c, then g <=< f of type a -> m c would perform f and then g, that's all. On the other hand, g . f wouldn't even type-check.

Functions in this a -> m b shape, where m is a Monad, are special. They mean a lot, since they are one of the basic pieces we'll need when establishing ordering and dependencies between functorial values, as witnessed by their appearance here in <=<, as well as in =<< or >>=. And, special things, we know, have names. These a -> m b functions are called *Kleisli arrows*. In other words, <=< is the operator for composing Kleisli arrows.

Of course, the fish could swim to the right too.

```
(>=>) :: Monad m => (a -> m b) -> (b -> m c) -> (a -> m c)
```

These are the fish, these are the laws.

Two hundred eighty-three

While we are on the topic of leaning left or right, and knowing that composing functions direction . this . in is a bit weird: Could we compose it in the other direction? Sure, but what would that operator look like? Good luck drawing a right-leaning dot.

Out of the box, Haskell comes with two very directional-looking operators that can be used for normal function composition. The leftwards one, <<<, works exactly like the dot . does.

```
(<<<) :: (b -> c) -> (a -> b) -> (a -> c)
(.)   :: (b -> c) -> (a -> b) -> (a -> c)
```

And the rightwards one, >>>, does the same but it takes the input parameters in the opposite order.

```
(>>>) :: (a -> b) -> (b -> c) -> (a -> c)
```

With it, we could write function compositions in >>> this >>> direction, rather than direction <<< this <<< in.

Two hundred eighty-four

Why do we have two operators in Haskell, the dot . and <<<, doing the same things? Good question. The answer is that we don't.

```
(<<<) :: Category x => x b c -> x a b -> x a c
```

This operator is a bit more general than the dot ., that's why we have it. The dot . is an operator for composing *functions*. The arrow looking operator, <<<, composes something related to that Category class. And there is a Category instance for functions, which is why everything type-checks when we pick (->) as our Category.

```
(<<<) :: (->) b c -> (->) a b -> (->) a c
```

Unreadable. But, if we write those (->) in infix manner, we'll see that this is just the function composition dot.

```
(<<<) :: (b -> c) -> (a -> b) -> (a -> c)
```

Two hundred eighty-five

What is this Category class? What does it have to do with composing functions? Beautiful questions.

Categories are the unifying theme throughout this book, throughout Haskell, throughout mathematics. Categories are why we are able to reason so soundly about every topic in our adventure, why seemingly unrelated things turn out to be related after all. Categories are things and the relationships between those things.

In mathematics, a *category* is made out of *objects* and things called *morphisms* which relate an object to another object in a particular direction. These morphisms can be *composed* such that if there is a morphism from an object a to an object b, and also a morphism from an object b to an object c, then there is a composition of these morphisms that goes from a to c. These a, b and c could be different objects or not. Additionally, among these morphisms, there is a special one called the *identity morphism* relating an object to itself, whose composition with any other morphism behaves exactly the same as that other morphism does on its own. In Haskell, this idea is represented by the `Category` typeclass.

```
class Category x where
  idm :: x a a
  (<<<) :: x b c -> x a b -> x a c
```

Instances of the `Category` class must implement the identity morphism `idm`, as well as <<<, the morphism composition operator. The x, thus, represents the idea of a morphism. For example, x a b is a morphism from an object a to an object b.

We know that normal Haskell functions (->) implement a `Category` instance because we said so before. Let's use (->) in place of this x, and see what happens. Due to parametricity, the implementation of this instance will be fully determined by the types.

```
instance Category (->) where
  idm :: a -> a
  idm = id
  (<<<) :: (b -> c) -> (a -> b) -> (a -> c)
  (<<<) = (.)
```

The mathematical definition of a *category* doesn't talk about functions, only about *objects* and *morphisms*. So, where are those? Let's pretend we haven't figured it out yet and deduct this by logical reasoning. If idm is supposed to be the *identity morphism* that relates an object to itself, and in this `Category` instance we are saying that the identity morphism is a *function* taking a Haskell *type* to the same type, then we can conclude that Haskell types like Int and Bool are the objects in this category, and that Haskell functions like id, length and not are the morphisms between those objects.

In other words, the implementation of the `Category` instance for Haskell functions and Haskell types just says that the *identity function* id is the *identity morphism* idm, and that *function composition* (.) is the *morphism composition* operation (<<<).

By the way, instead of repeating "the category of Haskell functions and Haskell types" time and time again, we can say "the *Hask* category". *Hask* is just a friendly name for this category, that's all. Some things are given names, some aren't.

If this sounds redundant and boring, it's because this is the only category we know so far. Things get more interesting beyond *Hask*.

Two hundred eighty-six

Functions being *one* instance of the `Category` class suggests that perhaps there are more. And indeed, we know of a few others. The most straightforward is Op, which if we recall, is just a backwards version of (->).

That is, Op a b is conceptually the same as b -> a.

```
data Op a b = Op (b -> a)
```

In the Category instance for Op, the objects are still Haskell types, but the morphisms are in the opposite direction compared to the category of normal Haskell functions. That is, while in the Hask category a morphism from object a to object b conveyed the idea of a Haskell function from type a to type b, a morphism from object a to object b in this Op category conveys the idea of a Haskell function from type b to type a. See? Backwards. Opposite direction.

```
instance Category Op where
    idm :: Op a a
    idm = Op id
    (<<<) :: Op b c -> Op a b -> Op a c
    Op f <<< Op g = Op (g . f)
```

The identity morphism for Op is what it is, there's not much to say about it. The composition of Ops, our morphisms, says that if we know how to go from object a to b by means of a Haskell function from type b to type a wrapped in Op, and we know how to go from object b to object c by means of a Haskell function from type c to type b wrapped in Op, then we also know how to compose those two morphisms into one that goes from said object a to said object c by means of a Haskell function from type c to type a wrapped in Op. And don't let all this Op-wrapping distract you. We know that Op a b means b -> a, so focus on that.

If we play with these two instances in GHCi, we'll see that while <<< seems to be composing the functions (+ 1) and (*10) in the same order in the following two examples, it isn't. Only the *morphisms* are composed in the same order, not the underlying functions. The rightmost morphism first, the leftmost morphism second. But the morphisms in these two categories just look similar. In reality, they mean different things, so naturally we get different results.

```
>       (   (+ 1) <<<     (*10)) 5
51
> getOp (Op (+ 1) <<< Op (*10)) 5
60
```

The correspondence between morphisms and functions in the Hask category feels very natural to us, so it's tempting to assume that morphisms and functions are the same in every category. But no, they aren't. Again, <<< composes *morphisms*, not functions. The fact that our Op morphism means "backwards function" is as irrelevant to <<< as the fact that (->) means "normal function".

Oh, and getOp is the function of type Op a b -> b -> a that does the obvious thing. If we were feeling mathematical, we could also say that getOp is a *morphism* from the *object* Op a b to the *object* b -> a in the Hask category.

Perhaps this difference between functions and morphisms will be easier to see if we replace Op with a fantasy infix backwards function arrow symbol.

```
instance Category (<-) where
  idm :: a <- a
  (<<<) :: (b <- c) -> (a <- b) -> (a <- c)
```

The as, bs and cs, the *objects* related by *morphisms*, appear in the same order they appeared in <<<'s type signature in the Hask category, the category of normal Haskell functions and types. Yet, the function arrows *in the morphisms themselves* are backwards, thus, any morphism in this category will behave differently than those in Hask, whether it results from a composition or not.

This Op category, where we are essentially just flipping the arrows in our morphisms, should remind us of the concept of *duality*, which we once described as something having opposite relationships compared to those the original thing has. And that's exactly what we are seeing here. The relationships in the Op category are all in the opposite direction compared to those in the Hask category. It's the Opposite category, you see. We say that the Op category is the *dual* category to Hask. This Op category is called *Hask*^{OP}. Generally speaking, the dual of a category named *Whatever* will be referred to as *Whatever*^{OP}

Categories allow us to disconnect the ideas of types and functions from the ideas of things and relationships. That is, objects and morphisms. We can have relationships that aren't normal Haskell functions, like Op, and we can have things that aren't normal Haskell types, we'll see.

Another way to look at all of this is that in a category *we* decide what it means to a be an object and what it means for an object to be related to another object. In the Hask category *we* decided objects are Haskell types related by means of normal Haskell functions, and in the Hask^{OP} category *we* decided they are related by means of backwards Haskell functions. But whatever we choose, *morphism composition* and the *identity morphism* will always behave predictably in every category out there. All a category needs is an identity element and a suitable associative morphism composition operation. Everything else is just noise.

Two hundred eighty-seven

We couldn't give a Category instance for a literal backwards function arrow <- because such arrow is not valid Haskell syntax. Well, it is, but it's used for do notation, a completely unrelated purpose. Many chapters ago we also weren't able to give this literal backwards function arrow a

Contravariant instance for the same reason. We just can't have a <- b mean "function from b to a", so we reach out for this wrapper, Op.

```
data Op a b = Op (b -> a)
```

Our next Category, too, needs a similar wrapper, for it can't be written literally.

```
data Kleisli m a b = Kleisli (a -> m b)
```

We learned about Kleisli arrows two or three chapters ago. They are functions that happen to have this a -> m b shape where m is some Monad. They convey the idea that given a value of type a, we can obtain a *functorial output* b produced by some m. To make things easier, whenever we see Kleisli m, we can just think of "m the Monad and all its monadic vocabulary" and it will be fine. Well, Kleisli arrows form a Category too. Once again, the implementation of the relevant instance is fully determined by parametricity.

```
instance Monad m => Category (Kleisli m) where
  idm :: Kleisli m a a
  idm = Kleisli pure
  (<<<) :: Kleisli m b c -> Kleisli m a b -> Kleisli m a c
  Kleisli g <<< Kleisli f = Kleisli (g <=< f)
```

There's nothing here we haven't seen before. This is the Category of Kleisli arrows in Haskell, the *Kleisli* category, wherein objects are Haskell types as in our previous instances, and morphisms are Kleisli arrows wrapped in this Kleisli type because just like we can't write an instance for a backwards arrow <-, we can't write an instance for the -> m part in a -> m b. It's just not valid Haskell syntax. We need this Kleisli wrapper mostly just to please Haskell's syntax rules.

Ignoring the Kleisli wrapper for a minute, the identity morphism in this category is pure from the Applicative vocabulary, of type a -> m a, which we know to be a functorial version of id. No surprises there. Morphism composition is <=<, which we learned behaves like the function composition dot ., except for functions returning functorial values. No surprises there either.

Notice this Category instance is not given to the Kleisli type constructor directly, but rather to a partially applied Kleisli m. On the practical side of things, this is done to satisfy the kind requirements of the Category type class, but conceptually, this highlights that neither idm nor <<< care about this m nor the fact that it is a Monad. Let's recall what the Category class looked like.

```
class Category x where
  idm :: x a a
  (<<<) :: x b c -> x a b -> x a c
```

In our Kleisli category this x is being replaced by Kleisli m, but the Category class doesn't care about what this x is, rather, it cares about whether it behaves as expected according to the category laws. Namely, that there be a identity morphism and an associative morphism composition operation.

With this instance in place we can use <<< to compose Kleisli m a b morphisms instead of using <=< to compose functions of this a -> m b shape. Why would we do that? No idea. Honestly, unless we *need* to use this extra Kleisli wrapper for some obscure reason, dealing with a -> m b and <=< directly is more ergonomic. And the same inconvenience applies to Op, too. Were we mathematicians working with pen and paper, and not Haskell programmers, perhaps we would feel more comfortable with all of this. Perhaps.

```
> (naturalFromString <=< lookup 2) [(1, "1010"), (2, "2020")]
Just 2020
> runKleisli (Kleisli naturalFromString <<< Kleisli (lookup 2))
            [(1, "1010"), (2, "2020")]
Just 2020
```

See? Noisier. runKleisli, of type Kleisli m a b -> a -> m b, essentially just removes the Kleisli wrapper so that we can apply the actual Kleisli arrow a -> m b to our a.

What have we achieved from all this exploration of categories? For our immediate practical purposes, nothing. But we have confirmation now that seemingly disparate things such as normal functions, backwards functions and monads, can all be reasoned about using the same principles and vocabulary. Essentially, the fact that we can identify something as a category is telling us that everything is fine, that we are going in the right direction. Many of the fascinating ideas we've talked about such as functions, functors, monads and monoids are related to categories, that's why we can reason so predictably about them, with them. We've been going in the right direction all along. We already knew that, sure, but now we know why.

Two hundred eighty-eight

Let's discover a new toy category, just for fun. They are everywhere, we just need to learn to recognize the patterns. This will be the category of language translations. That is, spoken languages like Portuguese or French, not programming languages.

Texts in a particular language will be the objects in this category, and translations will be the morphisms taking these texts from one language to another. Are our morphisms correct? All we have to do is check that they satisfy the category requirements.

First, we need an identity morphism. Say you give me a document in German for me to translate into German and I just give it back to you. There, I "translated it from German into German". Happy? Decent identity morphism. You are getting the same object, unmodified.

Second, we need to compose morphisms. That is, compose translations. To keep things simple, we will assume a fantasy scenario where these translations are so perfect that all meaning is preserved. In this scenario, if I knew how to translate Arabic into Spanish, and you knew how to translate Spanish into Greek, then we could compose our skills so that together we could translate from Arabic into Greek via an intermediate Spanish step.

Third, our morphism composition operation must be associative. Let's say I know how to translate Arabic into Spanish, you know how to translate French into Italian, and both of us know how to translate Spanish into French. What the associativity law is saying is that whether you or I do the translation from Spanish into French, the result from using our combined skills to translate a document from Arabic into Italian must be the same. And indeed, it will be, because we said we'd assume this fairy tale scenario where all translations are so perfect that no meaning is lost, no matter who does what, meaning we will always end up with the same final document in Italian whichever path we take.

Real life translations are not nearly as perfect, though, so this example won't be a category out there. Composing a translation from language a to b with one from language b to c probably won't have the same result as translating from language a to c directly, and no two translators translate everything in exactly the same way, so translation composition is unlikely to be associative. Mostly, categories concern things that can be modeled perfectly in mathematics, and not fluffy real life endeavors. Still, hopefully we can see how categories can help us reason even beyond Haskell, and how we can benefit from knowing that we can confidently compose morphisms.

Two hundred eighty-nine

Monoids are beautiful, so they are categories too. Obviously. The corresponding instance doesn't fit nicely in Haskell's Category class, however, because even while Haskell is very good at allowing us to express many things, it doesn't allow us to express *all* the things. Or maybe it does, but sometimes it's just too awkward, so we pretend it is not possible. Yes, Category can be used as is for describing a Monoid as a category. It's just that it's ugly. So, pen and paper it will be. Mostly. The important takeaway is that, somehow, a monoid is a category too.

The monoid category has just one object, one thing, and from our Haskell point of view that one object will be a Haskell type for which there is a Monoid instance. The values inhabiting this Haskell type are not the objects, just like they weren't in the Hask category. Instead, each value is a *morphism* in our category. For example, if our monoid was String, then the String type itself would be the one object in the category, and "butter", "phone", "cabbage" and any other possible value of type String would be the morphisms. The category of the String monoid, thus, has one object and infinitely many morphisms.

This is a good example of morphisms which don't resemble functions at all. Let's see if we can wrap our head around this. The important thing to keep in mind is that a category doesn't really care about what the objects or the morphisms are, what they mean or what they look like. We have to let go. This is hard, I know, but we must try. A category cares about there being objects and morphisms, sure, but only insofar as it can make sense of them through the lens of composition. As long as we are able to *compose* said morphisms through an associative operation, and identify one such morphism as the *identity* morphism, we have a proper category. What do the objects look like? What do the morphisms mean? A category doesn't care, and we shouldn't either.

Let's focus on composition. That is, turning two values of some Monoid type into a third one. Isn't that exactly what mappend does, one of the core operations of the Monoid typeclass?

```
mappend :: Monoid m => m -> m -> m
```

Could mappend be the associative morphism composition operation? Let's see. mappend takes two values of a type m for which there is a Monoid instance and returns a third. We know that mappend is an associative composition operation because the Monoid laws say so. For the same reason, we also know that composing any value of type m through mappend with mempty, the Monoid's identity element, doesn't change the original m value at all. With all this knowledge, we can confidently say that mappend is our associative morphism composition operation with mempty being the identity morphism. We have identified all the pieces we need to a describe a category. Namely, the identity morphism and the associative operation for composing morphisms.

These ideas don't fit nicely in the Category typeclass, though. It's not obvious how Monoid's mempty and mappend can become Category's idm and <<<. However, if we stare long enough at these two typeclasses side by side, we will find that they look somewhat similar if we know what to ignore.

```
class Monoid x where
  mempty :: x
  mappend :: x -> x -> x

class Category x where
  idm :: x a a
  (<<<) :: x b c -> x a b -> x a c
```

The main difference seems to be that Category mentions those a, b and c
type parameters, the *objects* that the morphisms relate, while Monoid
doesn't. This makes sense because there is only *one* object in the monoid
category, so Monoid doesn't need to keep repeating the same thing over and
over again. Thus, it does without those extra type parameters. Monoid is,
essentially, a one-object Category where mempty means the same as idm and
mappend the same as <<<.

```
class Category x where
  idm :: x _ _
  (<<<) :: x _ _ -> x _ _ -> x _ _
```

So, can we give a Category instance to Monoid? Not exactly, but yes. Just like
we couldn't give an instance for <- or -> m for superficial reasons, but we
could still do it for Op and Kleisli m, here we will need some kind of
wrapper around Monoid too.

```
data Wonoid m a b = Wonoid m
```

We'll call it Wonoid to emphasize that this is a *W*rapper around a m*onoid*.
Clever, right? We should write a programming book with clever jokes like
this here and there, it would be fun. The m will be the type for which there
is a Monoid instance. We know this because at the value level, on the Wonoid
value constructor, we only see m, so m must be the type of our Monoid. The
curious thing about Wonoid m a b is that it doesn't make use of its last two
type parameters a and b, which is exactly what we did manually before
when we choose to ignore as, bs and cs while trying to find similarities
between the Category and Monoid classes. Of course, the type-checker will
still complain if we use a value of type, say, Wonoid m a b where a value of
type Wonoid m p q was expected, but why would it be expected? We know
our category has only one object, so there's no reason for any of these type
variables to be different. Internally, Wonoid m a b and Wonoid m p q are just
wrappers around m, and all of a, b, p and q are ignored. With this Wonoid
contraption, we can proceed to implement a Category instance.

```
instance Monoid m => Category (Wonoid m) where
  idm :: Wonoid m a a
  idm = Wonoid mempty
  (<<<) :: Wonoid m b c -> Wonoid m a b -> Wonoid m a c
  Wonoid l <<< Wonoid r = Wonoid (mappend l r)
```

We still see the as, bs and cs there, but, again, we can pretend they are all

the same category object, the same Haskell type, and move on. That's
what Wonoid will be doing internally, anyway.

```
instance Monoid m => Category (Wonoid m) where
  idm :: Wonoid m _ _
  (<<<) :: Wonoid m _ _ -> Wonoid m _ _ -> Wonoid m _ _
```

Beautiful. I mean, ugly. With this, we can proceed to check in GHCi that
Category (Wonoid m) and Monoid m are indeed isomorphic.

```
> mempty :: String
""
> idm :: Wonoid String a a
Wonoid ""
> mappend "for" "est"
"forest"
> Wonoid "for" <<< Wonoid "est"
Wonoid "forest"
```

Notice how in the mempty and idm examples we had to explicitly give a type
to our expression. This is because, otherwise, Haskell doesn't know which
Monoid instance to use. By setting m to String, we are giving enough
information for mempty and idm to find the right Monoid instance. The last
two type parameters to Wonoid can be anything, including free type
variables like in this example, as long as both are the same.

Anyway, other than the Wonoid wrapper, in this example we can see that
idm behaves as mempty and <<< as mappend. Notice that we made a choice
when deciding that <<< would correspond to mappend, rather than flip
mappend. We could have chosen the other way, though.

```
> Wonoid "for" <<< Wonoid "est" :: Wonoid String
Wonoid "estfor"
```

Which one is right? Which one is wrong? None of these implementations
violate the category laws when they say that this should be the associative
morphism composition operator having a particular relationship with the
identity morphism, and that's all that matters.

```
> Wonoid "x" <<< (Wonoid "y" <<< Wonoid "z")
Wonoid "xyz"
> (Wonoid "x" <<< Wonoid "y") <<< Wonoid "z"
Wonoid "xyz"
> Wonoid "xyz" <<< Wonoid ""
Wonoid "xyz"
> Wonoid "" <<< Wonoid "xyz"
Wonoid "xyz"
```

If we use mappend in the implementation, no matter in which order we
compose our Stringy Wonoids, we end up with Wonoid "xyz". An alternative
<<< implementation using flip mappend leads to a different result, of
course, but the composition still behaves as expected, returning the same
Wonoid "zyx" in all cases.

```
> Wonoid "x" <<< (Wonoid "y" <<< Wonoid "z")
Wonoid "zyx"
> (Wonoid "x" <<< Wonoid "y") <<< Wonoid "z"
Wonoid "zyx"
> Wonoid "zyx" <<< Wonoid ""
Wonoid "zyx"
> Wonoid "" <<< Wonoid "zyx"
Wonoid "zyx"
```

See? Both alternatives are fine. The important thing is to pick one and stick with it. The one we don't pick will be the dual category to the one we do pick.

Two hundred ninety

We know that in Wonoid m a b, the a and b are just for decor. As far as the type system is concerned, they could be anything because they are fully polymorphic. Conceptually, however, we know that these type variables can only *mean* one thing because there is only one object in our category. Namely, m. Can we use the type system to convey that? Sure. Is that something we want to do? Probably not, but mostly because this particular example is silly. We just don't care that much about the Category instance for Wonoid m. Working with mempty and mappend is more ergonomic than working with idm and <<<, just like working with <=< was in the case of Kleisli arrows, so we won't be using these Category instances a lot in practice. Knowing that monoids form a category is all we care about, we don't need any new Haskell code to be happy. Yet, let's consider how we could approach this matter if we cared enough, so that we can be ready for similar scenarios in the future. First, let's understand the *kinds* involved in the Category class.

```
class Category (x :: k -> k -> Type) where ...
```

The kind of x, the type parameter to the Category class, is k -> k -> Type. This is new. All the kinds we had seen so far were some combination of Type, conveying the idea of a Haskell type for which there could exist values at the term level, and the arrow ->, conveying the idea of a type constructor. But here, besides that, we have k. Just like there are variables at the term level such as the a in \a -> ..., and variables at the type level like the b in Maybe b, there are also variables at the kind-level like this k. This k, in lowercase as every other variable name in Haskell, is a placeholder for *any* kind we chose. Notice that because k appears twice, whichever kind we pick as k must be the same in both cases. So, k -> k -> Type is the kind of a type constructor that, when applied to two types of some arbitrary kind k, constructs a type of kind Type. And, by the way, k is just a name. We could have used z, boat or kangaroo as name instead of k.

So, what kinds can we use as k besides Type and some combination of ->

and Type? Any. It's up to us, really. Just like we can create types to solve problems, we can create kinds too, and herein lies one possible solution to this chapter's non-problem.

So far, when considering the values that the a and b in Wonoid m a b could take, we implicitly assumed that a and b would be of kind Type. That is, things like Int, String or Maybe Bool. But the problem with that assumption is that there are infinitely many types inhabiting this Type kind, not just *one* as expected of a monoid category known to have but one object. Can we limit this somehow? Sure. All we have to do is force the kind of the a and b type parameters to be of a particular kind inhabited by just one type. First, let's recall the definition of our Wonoid type.

```
data Wonoid (m :: Type) (a :: x) (b :: y) = Wonoid m
```

We didn't explicitly mention the kinds of m, a and b before because Haskell could infer them just fine, but implicitly, they were always there. m must have kind Type because a value of this type shows up as value level payload to the Wonoid value constructor, and only types of kind Type can have inhabitants at the value level. The kinds of a and b, on the other hand, here named x and y, remain fully polymorphic and independent from each other. Notice that because of said independence, the kind of Wonoid is even more polymorphic in a and b than the Category class requires. That's alright, we can always use more polymorphic things in places where less polymorphic compatible things are expected. Let's go ahead and redefine Wonoid to force a and b to both have the same kind, inhabited by only one type.

```
data Wonoid (m :: Type) (a :: One) (b :: One) = Wonoid m
```

What is One? We just made it up.

```
data One = One
```

In Haskell, saying data One = One defines a *type* called One with kind Type having a single value level inhabitant called One, but at the same time it also defines a *kind* named One with a single type level inhabitant also called One and no value level inhabitants. Yes, this is confusing. Very. We would very much prefer being able to write kind One = One to define this kind, but unfortunately that's not how Haskell works today. All we can do is hope that in a near future it does. So, yes, whenever we define a type and its values in Haskell, we are also defining a kind and its types having the same names. And, as if this wasn't awkward enough, we also have a special syntax to tell apart the type named One of kind Type, from the type named One of kind One.

```
> :kind! One
One :: Type
```

We learned about the :kind! command in GHCi a while ago. As the name suggests, it tells us the kind of the specified type. In this first example we ask about the kind of One and we get Type as an answer. That's not surprising, every other type we've seen so far has this same kind. See Bool here, for example.

```
> :kind! Bool
Bool :: Type
```

But look at this other example, where we ask for the kind of 'One, having that very important leading tick ' symbol.

```
> :kind! 'One
'One :: One
```

This is the syntax. 'One is of kind One. If we want to talk about One, the type of kind Type, then we write One. If, on the other hand, we want to talk about One, the type of kind One, then we write 'One instead, with that barely noticeable leading tick '. Of course, there are no values having type 'One because only types of kind Type have value level inhabitants, and 'One is no such type. The tick matters. This tick business is true for all Haskell datatype declarations we've seen so far, even those that come with Haskell out of the box. Consider how the Bool datatype is defined in Haskell.

```
data Bool = False | True
```

This defines the type Bool, of kind Type, with two value constructors False and True. We know that. But, additionally, it defines the types False and True of kind Bool, referred to as 'False and 'True.

```
> :kind! Bool
Bool :: Type
> :kind! 'False
'False :: Bool
> :kind! 'True
'True :: Bool
```

So, while our new Wonoid m a b is still fully type-polymorphic in m, a and b, these a and b must now be of kind One, which restricts its kind polymorphism and implicitly tells us that only 'One can ever take the place of a and b. That is, a and b can remain polymorphic when they appear in places like the type of idm or <<<, but anytime a concrete type takes the place of a or b in Wonoid m a b, then that concrete type must be 'One. Nothing else type-checks. See what happens if we try to type-check Wonoid String Integer Bool, for example.

```
> :kind! Wonoid String Integer Bool
<interactive>:1:15: error:
    • Expected kind 'One', but 'Integer' has kind 'Type'
    • In the second argument of 'Wonoid', namely 'Integer'
      In the type 'Wonoid String Integer Bool'
<interactive>:1:19: error:
    • Expected kind 'One', but 'Bool' has kind 'Type'
    • In the third argument of 'Wonoid', namely 'Bool'
      In the type 'Wonoid String Integer Bool'
```

Great. We achieved something. Do we need to change anything in our previous Category instance for Wonoid m in order to accomodate this new Wonoid m of kind One -> One -> Type?

```
instance Monoid m => Category (Wonoid m) where
  idm :: Wonoid m a a
  idm = Wonoid mempty
  (<<<) :: Wonoid m b c -> Wonoid m a b -> Wonoid m a c
  Wonoid l <<< Wonoid r = Wonoid (mappend l r)
```

No, the Category instance stays the same. One -> One -> Type, the new kind of Wonoid m, is compatible with the more kind polymorphic k -> k -> Type required by the Category class. By restricting the kinds of the second and third type-parameters to Wonoid we made it impossible to get a *concrete* type like Wonoid String Integer Bool to type-check, but situations where these type-parameters are polymorphic remain the same. In summary, all of this effort was mostly about scratching an itch, and not really about solving a real problem. We did learn about custom kinds though, didn't we? Those will be really handy in our Haskell adventures.

Our solution makes 'One the only possible type for a and b in Wonoid m a b. However, considering a and b are supposed to talk about the *objects* of our category, seeing 'One there is a bit confusing, for there is no such 'One object in our category. There is *one* object in our category, sure, but that one object is m, not 'One. For example, if our category is that of the String monoid, then all the values in our category will be of type String. That is, String would be the one object in our category, not 'One. Our 'One solution just *conveys* the idea of there being one object, but it doesn't mention m explicitly in a and b, where objects are supposed to be mentioned. We just interpret 'One to mean m, that's it. All in all, this solution is still a bit disappointing. Perhaps a bit less so than our original solution, but disappointing nonetheless. Can we do better? Sure, we just use the Monoid vocabulary instead of the Category vocabulary when dealing with Monoids in Haskell, as we've been saying all along.

Two hundred ninety-one

Let's look at One again.

```
data One = One
```

Doesn't it remind you of that Unit datatype we introduced a million years ago?

```
data Unit = Unit
```

But didn't Unit actually go by the name () in Haskell? Indeed. In other words, in our previous chapter, we could have defined Wonoid in terms of this kind, and we would have achieved the same.

```
data Wonoid (m :: Type) (a :: ()) (b :: ()) = Wonoid m
```

And just like we had to precede One with a tick ' to mean "the type named One of kind One", we have to write '(), also with leading tick, if we want to mean "the type () of kind ()", rather than "the type () of kind Type". It's ugly, but it works. For example, Wonoid String '() '() type-checks just fine, but Wonoid String () () doesn't.

Two hundred ninety-two

Remember when we said that the Category class looked like this?

```
class Category x where
  idm :: x a a
  (<<<) :: x b c -> x a b -> x a c
```

We lied. Once again, we lied. In reality, idm is not called idm, and <<< is not called that either.

```
class Category x where
  id :: x a a
  (.) :: x b c -> x a b -> x a c
```

The identity morphism and the morphism composition operation actually go by the names of id and dot ., names which conflict with the id function and the function composition dot exported by the Prelude module which we already know and love. Hate. Love. What prevents these duplicate names from truly conflicting is that they are exported from a different module Control.Category. So, if we want to refer to the identity *function* we write Prelude.id, and if we want to refer to the identity *morphism* we write Control.Category.id. That is, we just prefix the names with the name of the module where they come from. Just kidding, we can do better than that. Quite often, we will find lines like the following among the import statements in a Haskell module.

```
import Prelude hiding (id, (.))
import Control.Category (id, (.))
```

The first line says to import everything from the Prelude module except id and the function composition operator (.). That's what the hiding (a, b, c) syntax is for. The second line says to import id and (.) from

Control.Category. With these statements we are saying that we want to have access to the names id and (.) without having to fully type the name of their module of origin as prefix, yes, but we want them to be the ones from Control.Category, and not the ones from Prelude. This doesn't harm users of the function versions of id and (.) too much because, we know, there is a Category instance wherein morphisms are normal Haskell functions, so the Category versions id and (.) will type-check and behave exactly the same as the function versions when necessary. However, by importing the Category versions of these operators, those trying to compose Kleisli arrows or Ops will have easier access to the necessary tools. Of course, we'll often find ourselves importing a few more tools from Control.Category besides these two, such as the Category name.

Why doesn't Prelude simply re-export the Control.Category versions of these operators instead of exporting versions restricted to functions? Perhaps there is no right answer to this good question, but it may have something to do with the fact that categories are harder to understand than functions, and that type-inference works much better for the function versions because Haskell doesn't have to go guessing what type of morphisms we are dealing with. Functions, they are always functions. So maybe the current situation of Prelude not exporting the Category versions of these names by default is not too bad. Maybe.

By the way, we didn't lie when we said there is an operator named <<<. It's just that it's not part of the Category typeclass, but only a standalone name defined in the Control.Category module.

```
(<<<) :: Category x => x o c -> x a b -> x a c
(<<<) = (.)
```

And there's the rightwards version, too.

```
(>>>) :: Category x => x a b -> x b c -> x a c
(>>>) = flip (.)
```

There's nothing special about these operators compared to the normal morphism composition operator dot. They just look cool.

Why are we talking about categories and fish anyway? Weren't we learning about monad transformers before? Oh my, look at the time. We better get back on track.

Two hundred ninety-three

A MonadTransformer allow us to add the effects of a particular Monad to any other Monad. It isn't really composition of Monads, but something closer to piling Monads on top of each other. We use MonadTransformers to "build Monad towers", they say. We saw how StateT s m a, for example, added the state transformation effects of State s to m. So, what other monad transformers are there, besides StateT? To begin with, there's IdentityT, the transformer version of the Identity monad.

```
data IdentityT m a = IdentityT (m a)
```

What does it do? Nothing. I mean, nothing different than what the underlying Monad, m, can do on its own.

```
instance Monad m => Monad (IdentityT m) where
  ma >>= k = IdentityT (do a <- ma
                           runIdentityT (k a))
```

IdentityT exists, is a monad transformer, and that's pretty much it. Everything of essence that there is to say about IdentityT m a has already been said by m and a.

The runIdentityT function seen in our implementation merely removes the IdentityT wrapper.

```
runIdentityT :: IdentityT m a -> m a
runIdentityT (IdentityT x) = x
```

The kind of IdentityT, as that of every other monad transformer, is (Type -> Type) -> Type -> Type. Although it may be easier to appreciate the name "monad transformer" if we add some redundant parentheses to this kind.

```
(Type -> Type) -> (Type -> Type)
```

Every Monad has kind Type -> Type. We know this. For example, all of Maybe, State s, Either e, [] and IO have this kind. A monad transformer like IdentityT or StateT s takes one of these Monad of kind Type -> Type and *transforms* it into a different monad of the same kind Type -> Type.

So, what does IdentityT mean? When we use Identity *below* a monad transformer, as in StateT s Identity x, we are adding the effects of a state transformation to the moot effects of the Identity monad. On the other hand, when we use IdentityT *on top* of a monad, we are adding the moot effects of IdentityT to those of the underlying monad m. Same thing, but different. In other words, composing monads with IdentityT is not particularly interesting. It works, sure, but that's all.

Can we define Identity in terms of IdentityT like we defined State in terms of StateT? Not really. State s x is a type synonym for StateT s Identity x, but we can't have Identity be a synonym for IdentityT Identity, as that would result in an infinitely recursive type synonym.

```
type Identity = IdentityT Identity
```

If we try to type-check that definition, GHC complains that there is a cycle. We can have recursive data type definitions, but not recursive type synonyms.

Anyway, it's unlikely that we'll need to worry much about `IdentityT` in our Haskell life. The more interesting monad transformers lie elsewhere.

Two hundred ninety-four

Behold, the transformer variant of Maybe.

```
data MaybeT m a = MaybeT (m (Maybe a))
```

What do we have here? In `MaybeT m a`, `MaybeT` is the monad transformer that adds to the underlying `Monad`, m, the possibility of aborting the computation, preventing any further work and thus failing to produce an output value of type a. That is, it extends m with a short-circuiting effect, often used to convey an error situation.

```
runMaybeT :: MaybeT m a -> m (Maybe a)
runMaybeT (MaybeT x) = x
```

We may have figured this out by now, but just to be clear, as a matter of convention most monad transformers are called `WhateverT`, for some choice of `Whatever`, and have an accompanying `runWhateverT` function that gets rid of the `WhateverT` wrapper. We saw it with `StateT` and `runStateT` already, as well as with `IdentityT` and `runIdentityT`.

Unsurprisingly, `runMaybeT` removes the `MaybeT` wrapper and leaves us with m (Maybe a). The "run" name doesn't make much sense yet, I know, but it doesn't make sense in the same way that `getLine` of type `IO String` doesn't make sense. That is, `getLine` doesn't really "get a line", but rather, it represents a computation that when finally performed by `main` will "get a line". `runMaybeT` is similar. `runMaybeT` doesn't really *run* things, but rather, it is a functorial value in m that when actually performed by whomever can perform functorial values in m, will take the short-circuiting effects of `MaybeT` into account, which by then will have already been baked into the functorial value in m as described by the `Monad` instance for `MaybeT`. From that perspective, the "run" prefix makes more sense.

```
instance Monad m => Monad (MaybeT m) where
  tma >>= k = MaybeT (runMaybeT tma >>= \case
                        Nothing -> pure Nothing
                        Just a  -> runMaybeT (k a))
```

And in case you forgot, here's the type of bind `>>=` for `MaybeT`.

```
(>>=) :: MaybeT m a -> (a -> MaybeT m b) -> MaybeT m b
```

The bind function must return a value of type `MaybeT m b`, so it uses the `MaybeT` constructor on a value of type m (Maybe b). It constructs this functorial value in m by binding `runMaybeT tma`, an expression of type m (Maybe a), and scrutinizing its output using `\case` expression. If it's `Nothing`, then it purely returns `Nothing` right away, avoiding any further work and achieving our desired short-circuiting behavior. On the other hand, if Just a, then it applies k to that a, leaving us with a value of type MaybeT m b whose outer MaybeT wrapper we remove using runMaybeT, leaving us with a value of type m (Maybe b) just as we desired.

Of course, MaybeT m being a Monad implies that it is a Functor and an Applicative too.

```
instance Monad m => Functor (MaybeT m) where
    fmap = liftM

instance Monad m => Applicative (MaybeT m) where
    pure a = MaybeT (pure (Just a))
    liftA2 = liftM2
```

And there is a MonadTrans instance too, as expected of every monad transformer.

```
instance MonadTrans MaybeT where
    lift :: Monad m => m a -> MaybeT m a
    lift ma = MaybeT (fmap Just ma)
```

lift wraps the output of ma in Just, and then the entire functorial value in the MaybeT constructor. Let's use MaybeT, for example, on top of IO.

```
longers :: MaybeT IO (String, String, String)
longers = do
  a <- lift getLine
  b <- lift getLine
  if length b <= length a
     then MaybeT (pure Nothing)
     else do c <- lift getLine
             if length c <= length b
                then MaybeT (pure Nothing)
                else pure (a, b, c)
```

Very ugly. Don't worry, we'll clean it up later. I mean, *do worry*, but rest assured, things will improve. We are in the business of understanding things, remember. In time, beauty will come too.

longers reads up to three manually input lines a, b and c using the IO interaction getLine and returns them.

```
> runMaybeT longers
short
not so short
a very long line
Just ("short", "not so short", "a very long line")
```

However, it only does so if b is longer than a and c is longer than b. Moreover, if b is not longer than a, then c is not even read from the input at all. That is, longers has a short-circuiting behavior. As soon as a pair of Strings fails to satisfy our requirements, the whole process is aborted and further getLine IO interactions are avoided.

```
> runMaybeT longers
this is long
this ain't
Nothing
```

getLine, being the IO interaction that it is, needs to be lifted to be bound
in MaybeT IO. It reads two lines first, a and b, and only if b is longer than a
does it proceed to read a third line, c. Otherwise, it short-circuits the
computation, it aborts early, by purely producing Nothing as a functorial
value in IO, and then wrapping it in MaybeT. Remember, according to our
Monad instance for MaybeT m, if that m ever produces a Nothing, then the
computation is aborted.

And no, saying MaybeT (pure Nothing) is not the same as lift (pure
Nothing). Whereas a use of lift taking a value of type m (Maybe a)
transforms it into MaybeT m (Maybe a), the MaybeT value constructor takes a
value of type m (Maybe a) and transforms it into MaybeT m a. In other
words, lift (pure Nothing) doesn't cause MaybeT to abort its computation,
but MaybeT (pure Nothing) does. One functorial value always succeeds,
always produces that Maybe a, the other one always fails.

Success and failure. There was a name in Haskell for Functors that could
represent this idea, wasn't there? Oh hi there, Alternative old friend.

```
instance Monad m => Alternative (MaybeT m) where
  empty :: MaybeT m a
  empty = MaybeT (pure Nothing)
  (<|>) :: MaybeT m a -> MaybeT m a -> MaybeT m a
  tmx <|> tmy = MaybeT (runMaybeT tmx >>= \case
                          Just x  -> pure (Just x)
                          Nothing -> runMaybeT tmy)
```

To be empty is to fail. MaybeT (pure Nothing), this is exactly what we used in
our longers example just now. tmx <|> tmy, where both tmx and tmy have
type MaybeT m a, runs tmx in the hopes of producing an a. If it isn't possible
to obtain one such a, however, <|> runs tmy as fallback. Whether tmy
succeeds or fails does not concern <|>. Here's an Alternative
implementation of longers using empty.

```
longers :: MaybeT IO (String, String, String)
longers = do
  a <- lift getLine
  b <- lift getLine
  if length b <= length a
    then empty
    else do c <- lift getLine
            if length c <= length b
              then empty
              else pure (a, b, c)
```

Better. We can clean it up a bit more, though. If you recall, Alternative enables us to use the very handy guard function.

```
guard :: Alternative f => Bool -> f ()
guard True  = pure ()
guard False = empty
```

With it, our longers implementation can become significantly cleaner. Superficially, at least. Fundamentally, nothing changes.

```
longers :: MaybeT IO (String, String, String)
longers = do
  a <- lift getLine
  b <- lift getLine
  guard (length b > length a)
  c <- lift getLine
  guard (length c > length b)
  pure (a, b, c)
```

Isn't it beautiful now? It does read in a very straightforward sequential manner. *Do this, check that it's alright, then do that.* It reminds us of how we used to write code in our Parser monad, doesn't it?

Two hundred ninety-five

A lie. It's not really Alternative who allows us to use guard, but MonadPlus.

```
class Monad m => MonadPlus m where
  mzero :: m a
  mplus :: m a -> m a -> m a
```

Other than the name and the Monad superclass instead of an Applicative one, MonadPlus is pretty much the same as Alternative.

```
class Applicative m => Alternative m where
  empty :: m a
  (<|>) :: m a -> m a -> m a
```

In other words, the true definition of guard is not the one we saw before in terms of Alternative, but this one in terms of MonadPlus.

```
guard :: MonadPlus m => Bool -> m ()
guard True  = pure ()
guard False = mzero
```

Why? Surely this redundancy is pointless, isn't it? Mostly, it is. You must keep in mind that Haskell is old, and back when man discovered Haskell, there was no Alternative, there was only MonadPlus. Strictly speaking, MonadPlus has an additional law compared to Alternative, one that defines what to expect when mzero and the monadic bind >>= interact.

```
mzero >>= f  ==  mzero
m >> mzero   ==  mzero
```

For practical purposes, for Applicatives that are Monads too, mzero and mplus will be defined to be equal to empty and <|> respectively, so there's no need

to dwell too much on this. For example, here's the MonadPlus instance for
MaybeT m.

```
instance Monad m => MonadPlus (MaybeT m) where
  mzero = empty
  mplus = (<|>)
```

And so on, for every other Monad that is also an Alternative.

Two hundred ninety-six

Maybe is a Monad, and it has a MonadTransformer counterpart, MaybeT. This is
true for Maybe, State s, Identity and most other Monads out there. For
example, Either e has its transformer version, too.

```
data EitherT e m a = EitherT (m (Either e a))
```

Nothing unexpected about EitherT. If we compare it with MaybeT, we'll see
it's doing essentially the same, but with Either rather than Maybe.

```
data MaybeT m a = MaybeT (m (Maybe a))
```

What does EitherT e m a accomplish? The same as MaybeT does, except its
short-circuiting behavior can be accompanied with a message of type e,
just like binding functorial values in Either e allows us to fail with a value
of type e. Remember that?

```
> Right 123 >> Left "oops" >> Right True
    :: Either String Bool
Left "oops"
```

As soon as the Left gets involved, past accomplishments are lost, and
potential future achievements are prevented from ever seeing the light of
day. This is true in Either, and it is true in EitherT too, as conveyed by its
Monad instance.

```
instance Monad m => Monad (EitherT e m) where
  tma >>= k = EitherT (runEitherT tma >>= \case
                        Left e -> pure (Left e)
                        Right a -> runEitherT (k a))
```

runEitherT, akin to runMaybeT, merely removes the EitherT wrapper.

```
runEitherT :: EitherT e m a -> m (Either e a)
runEitherT (EitherT x) = x
```

EitherT e m has Functor and Applicative instances in the trivial way, as any
other Monad. And being a monad transformer, EitherT e has a MonadTrans
instance too.

```
instance MonadTrans (EitherT e) where
  lift :: Monad m => m a -> EitherT e m a
  lift ma = EitherT (Right <$> ma)
```

So, for example, if we wanted to enrich our previous longers example with error messages, we could build it using EitherT String, say, rather than MaybeT.

```
longers :: EitherT String IO (String, String, String)
longers = do
  a <- lift getLine
  b <- lift getLine
  guard (length b > length a)
  c <- lift getLine
  guard (length c > length b)
  pure (a, b, c)
```

Feeding longers with lines of the expected lengths succeeds as expected, this time resulting in a Right value, not a Just one.

```
> runEitherT longers
short
not so short
a very long line
Right ("short", "not so short", "a very long line")
```

Otherwise, if we feed the wrong lines to longers, we get a Lefty.

```
> runEitherT longers
this is long
this ain't
Left ""
```

Wait. It fails as expected, sure, but where does that "" payload on the Left constructor come from? There's a little detail we skipped.

Our longers code is using guard, and as you know, guard is defined in terms of MonadPlus. And if we look at how the MonadPlus instance for EitherT e m is defined, we'll see something interesting.

```
instance (Monad m, Monoid e) => MonadPlus (EitherT e m) where
  mzero = EitherT (pure (Left mempty))
  mplus = ...more about this later...
```

If you recall, saying guard False is exactly the same as saying mzero. And as we can see, in the case of EitherT e m, where that e is a Monoid, mzero means "make this functorial value avoid any further work, and use mempty as the final value to be conveyed by the Left constructor". So, if we pick String to be our e, say, as we did in our longers example, then mzero means "abort with "" as payload", for "" is the meaning of mempty, the Monoid identity element, for String. In other words, things worked in longers, but almost by accident because String happened to be a Monoid. Had we chosen a different type on the Left side of EitherT, one that was not a Monoid, type-checking would have failed. For example, if we change the type of longers to be EitherT Integer IO ..., then GHC complains loud and clear.

- No instance for (Monoid Integer)
  arising from a use of 'guard'
- In a stmt of a 'do' block:
  guard (length b > length a)

That is, for us to exploit the MonadPlus instance of EitherT e m, the one that enables us to use guard as we do in longers, our e must be a Monoid, as clearly conveyed by the constraints on the MonadPlus instance for EitherT e n. And Integer is no such Moncid.

Anyway, while longers works if using String as error message type in EitherT, getting an empty String as error message is disappointing. Let's fix that. Let's make a function named lowercase left that creates a failing EitherT e with a value of that type e as payload.

```
left :: Monad m => e -> EitherT e m a
left = EitherT . pure . Left
```

The resulting functorial output, a, can be left fully polymorphic because our functorial value will not produce it nor interact with it in any way, so it can pretend to be whatever we want it to be. Having left, we can rewrite longers to make use of it and convey useful error messages when error messages are due.

```
longers :: EitherT String IO (String, String, String)
longers = do
  a <- lift getLine
  b <- lift getLine
  if length b <= length a
     then left "The second line is too short"
     else do
       c <- lift getLine
       if length c <= length b
          then left "The third line is too short"
          else pure (a, b, c)
```

Feeding the wrong input once again, we get something a bit more useful this time.

```
> runEitherT longers
this is long
this ain't
Left "The second line is too short"
```

Much better, isn't it?

Two hundred ninety-seven

We skipped the definition of mplus for EitherT before. It's interesting, so let's take a look at it. Remember, mplus is one of the methods of the MonadPlus typeclass. It is essentially the same as <|> from the Alternative typeclass, representing the idea of "either this or that functorial value".

```
mplus :: (Monad m, Monoid e)
      => EitherT e m a -> EitherT e m a -> EitherT e m a
mplus tma tmb =
  EitherT (runEitherT tma >>= \case
              Right a -> pure (Right a)
              Left e0 -> runEitherT tmb >>= \case
                Right b -> pure (Right b)
                Left e1 -> pure (Left (mappend e0 e1)))
```

What's curious about this implementation is that if both tma and tmb result
in failing Left values, each with their particular payload of type e, these
two es are mappended together into a single e. For example, consider the
following functorial values.

```
foo :: EitherT [Integer] Indentity ()
foo = left [1, 2, 3]

bar :: EitherT [Integer] Identity ()
bar = pure "what" >> left [4, 5]
```

What do you think will happen if we apply mplus to foo and bar?

```
> runEitherT (mplus foo bar)
Left [1, 2, 3, 4, 5]
```

Careful, though. If we pick the wrong e, we'll end up mappending the
wrong things.

```
> runExceptT (mplus (left "bad") (left "terrible"))
Left "badterrible"
> runExceptT (mplus (left ["much"]) (left ["better"]))
Left ["much", "better"]
```

Anyway, it's nice to see both mzero and mplus exploiting that Monoid
constraint. And the Alternative folks too, of course.

```
instance (Monad m, Monoid e) => Alternative (EitherT e m) where
  empty = mzero
  (<|>) = mplus
```

Whether Alternative is defined in terms of MonadPlus or the other way
around is irrelevant. In fact, we'd probably be better off if one of these
typeclasses disappeared. But here we are, in sickness and in health.

Two hundred ninety-eight

EitherT is not called EitherT out there in the wild, but ExceptT. Why?
Nobody knows, it's a mystery. The letters in these words are very close to
each other in keyboards, so maybe it was a typing accident.

```
data ExceptT e m a = ExceptT (m (Either e a))
```

Accordingly, runEitherT is not called runEitherT but runExceptT.

```
runExceptT :: ExceptT e m a -> m (Either e a)
runExceptT (ExceptT x) = x
```

Other than that, everything else we learned about this monad transformer
was true. It's a disappointing name, given the close relationship between
Either and EitherT. I mean, between Either and ExceptT. It is what it is.

Two hundred ninety-nine

And then there were more monad transformers. Many more. Thousands
of them. But we're done poking them this way for now. Let's poke them a
different way instead. Let's go back to our Parser.

```
data Parser x = Parser (String -> Either (Natural, String)
                                         (Natural, String, x))
```

A Parser has multiple monadic concerns such as taking care of leftovers,
keeping track of how many characters it has consumed so far, and dealing
with errors. However, these different monadic features or effects are not
currently being *composed*, but rather, they are being mashed together into
a big function. Ultimately, that's what always happens, things *will* get
mashed up into a big function so that the computer can execute it.
However, while thinking about these things, we could benefit from a
higher-level vocabulary, for example, by building our Parser out of monad
transformers.

```
data Parser x = Parser (ExceptT String (State (Natural, String)) x)
```

Is this better than before? Not necessarily. It's just different. Is this the
only possible representation using monad transformers? Not at all, just
one of them. The idea of this one representation is that we can use State to
keep track of the Natural number of characters processed so far, as well as
the raw String input available including leftovers, and we can add an
ExceptT layer on top that allows parsing to fail with a particular String
error message. As an example, here's parse1 written using this new internal
representation of Parser. This is a Parser we've seen multiple times before,
which tries to parse just one Char.

```
parse1 :: (Char -> Either String a) -> Parser a
parse1 f = Parser (ExceptT (state (\(n0, i0) ->
  case i0 of
    [] -> (Left "Not enough input", (n0, i0))
    c : i1 -> case f c of
      Left e -> (Left e, (n0, i0))
      Right a -> let !n1 = n0 + 1
                 in (Right a, (n1, i1))))))
```

Is this better than before? No, nor is it worse. It's just different. Other
functorial values using parse1 don't need to change, unless they were
already mentioning Parser's internals explicitly.

Despite changing Parser's guts, we can still continue to provide a runParser function with exactly the same type as before.

```
runParser :: Parser x
          -> String
          -> Either (Natural, String) (Natural, String, x)
runParser (Parser et) i0 =
  case runState (runExceptT et) (0, i0) of
    (Left e,  (n1, _ )) -> Left  (n1, e)
    (Right x, (n1, i1)) -> Right (n1, i1, x)
```

Notice how we now provide 0 as part of the initial state to indicate that no characters have been processed so far, and how we ignore any leftovers in case of errors. In our previous representation of Parser, these extra inputs and outputs were not there so we didn't have to worry about them, but they are part of the state value now, so we must pay attention to them. Let's take a look at the Monad instance for this new Parser representation.

```
instance Monad Parser where
  Parser et0 >>= k =
    Parser (ExceptT (state (\s0 ->
      case runState (runExceptT et0) s0 of
        (Left  e, s1) -> (Left e, s1)
        (Right a, s1) ->
          let Parser et1 = k a
          in  runState (runExceptT et1) s1)))
```

Since we are dealing with state transformations, we must make sure we always use the most recent state value, but other than that, the implementation mostly writes itself if we just follow the types. Is this better than our previous implementation of >>=, back from when Parser was not made out of monad transformers? Again, *no*, nor is it worse. Is there any benefit, then, to building monads out of smaller monad transformers? Sometimes. For example, look at the Applicative instance.

```
instance Applicative Parser where
  pure = Parser . pure
  liftA2 = liftM2
```

As usual, we get the implementation of liftA2 for free. But this time, we also get the implementation of pure mostly for free, by relying on the one from ExceptT. And similarly with other typeclasses which we won't implement here because it's getting boring and we already got the idea.

Three hundred

We made Parser out of monad transformers, but we didn't make Parser a monad transformer itself. Without this, folks won't be able to add parsing effects on top of other monads. Let's fix this, it's easier than we might expect. Let's recall what our current Parser looks like.

```
data Parser x = Parser (ExceptT String (State (Natural, String)) x)
```

State, remember, is just a type synonym for StateT Identity.

```
data Parser x = Parser (ExceptT String
                       (StateT (Natural, String) Identity) x)
```

Now, all we have to do is replace the Identity we have at the bottom of our monad tower with a polymorphic one to be chosen by someone else.

```
data ParserT m x = ParserT (ExceptT String
                           (StateT (Natural, String) m) x)
```

We added a trailing ᵀ to this monad transformer's name because that's what tradition dictates. Having this ParserT, we could imagine ourselves wanting to write and use a parser that interleaves the parsing of two Characters with some IO interactions.

```
example :: ParserT IO (Char, Maybe Char)
example = do
  a <- parseChar
  lift isItRaining >>= \case
    True  -> do lift (putStrLn "It rains.")
                b <- parseChar
                pure (a, Just b)
    False -> do lift (putStrLn "It doesn't rain.")
                pure (a, Nothing)
```

It's a silly example, but it successfully illustrates the interleaving of effects from ParserT and some underlying monad. IO, in this case. Anyway, for example to type-check at all, we must implement a MonadTrans instance for ParserT so that we have access to lift.

```
instance MonadTrans ParserT where
  lift = ParserT . lift . lift
```

And we must update the other ParserT instances and primitives as well in order to account for the new internals. Very straightforward endeavors, all we have to do is fix the type-checker complaints one by one, adding binds >>= here and there. Here, for example, is Monad.

```
instance Monad m => Monad (ParserT m) where
  ParserT et0 >>= k =
    ParserT (ExceptT (StateT (\s0 ->
      runStateT (runExceptT et0) s0 >>= \case
        (Left e, s1) -> pure (Left e, s1)
        (Right a, s1) -> do
          let ParserT et1 = k a
          runStateT (runExceptT et1) s1)))
```

And here is a new version of parse1, the thing parseChar and many other core parsers are made of.

```
parse1 :: Monad m => (Char -> Either String a) -> ParserT m a
parse1 f = ParserT (do
  (n0, i0) <- lift get
  case i0 of
    [] -> left "Not enough input"
    c : i1 -> case f c of
      Left e -> left e
      Right a -> do
        let !n1 = n0 + 1
        lift (put (n1, i1))
        pure a)
```

Just for show, in this new implementation of parse1 we chose to use get, put, left and pure to interact with the ExceptT (StateT (Natural, String)) m a inside ParserT, instead of using the ExceptT and StateT constructors directly as we did last time. It's mostly a cosmetic decision, it all means the same.

To execute these new ParserT things, we'll need a runParserT that understands its new internals. Our old runParser without the trailing T won't do.

```
runParserT :: Monad m
           => ParserT m x
           -> String
           -> m (Either (Natural, String) (Natural, String, x))
runParserT (ParserT et) i0 =
  runStateT (runExceptT et) (0, i0) >>= \case
    (Left e,  (n1, _ )) -> pure (Left  (n1, e))
    (Right x, (n1, i1)) -> pure (Right (n1, i1, x))
```

runParserT is very similar to runParser from before. What changed is that we are now binding the result of the runStateT application instead of just pattern-matching on the runState application as before. This bind step forces m to do some work in order produce that output we'll be pattern-matching against, effectively composing our parsing effects with those of m.

```
> runParserT example "abcd"
It rains.
Right (2, "cd", ('a', Just 'b'))
```

Great, it works. At least now that it rains. We tried last week while the sun was shinning and the result was a bit different.

```
> runParserT example "abcd"
It doesn't rain.
Right (1, "bcd", ('a', Nothing))
```

It's important to keep in mind that the fact that we can interleave parsing with IO interactions, doesn't mean that we should do it. We most definitely shouldn't. That's crazy. We merely used IO as the underlying monad as an example. By interleaving IO interactions with parsing, we are potentially making otherwise pure and fully deterministic parsing decisions dependent on IO phenomenons such as the current weather. A more realistic choice of underlying monad would be State, for example, which could keep track of some extra state value as we parse our input. The important thing to understand is that this is possible. The way in which we choose to exploit this is up to us.

Three hundred one

Even if ParserT and runParserT are our core vocabulary now, we can still provide a Parser type synonym and a runParser function, for convenience, to be used in those situations where users don't need the monad transformer feature. We can achieve this in the same way State does, by hardcoding Identity as the underlying Monad.

```
type Parser a = ParserT Identity a

runParser :: Parser x
          -> String
          -> Either (Natural, String) (Natural, String, x)
runParser p i = runIdentity (runParserT p i)
```

The motivation for introducing type synonyms and functions with less free type variables in them, like these, is so that their users don't get distracted with irrelevant details whenever those details are indeed irrelevant. Sometimes it's worth doing this, sometimes it isn't. We must carefully analyze the trade-offs, for even while our users can now choose to ignore this m type parameter when desired, they do have two new names Parser and runParser to worry about. Moreover, type-checking errors from misuses of the Parser type synonym may mention ParserT Identity at times, instead of Parser, which can confuse users even more. In any case, even if we didn't export Parser and runParser, we wouldn't be taking anything fundamental away from our users. Remember, these names are just a convenience now. It's a delicate matter, designing code with usability in mind.

Three hundred two

When using the MonadTransformer approach to compose Monads on top of each other, we'll quickly realize that lifting things all the time becomes annoying. Moreover, when using lift we are only making reference to the Monad directly underneath the current MonadTransformer, which means that if we want to perform an action in a Monad two layers down, say, we have to lift twice.

```
example :: ParserT (StateT Natural IO) String
example = do
  s <- lift (lift getLine)       -- IO
  n <- lift get                  -- StateT
  expect (mconcat (replicate n s)) -- ParserT
  pure s
```

Get ready for yet another silly `example`.

```
> runStateT (runParserT example "NaNaNaNaNaNaNaNaBatman!") 8
Na
(Right (16, "Batman!", "Na"), 8)
```

Sorry about that. Our `example` reads a `String` from our keyboard input using `getLine`, and then tries to parse as many consecutive occurrences of that `String` as indicated by the current `Natural` value in `StateT`, before finally producing said `String` as output. In order to figure out the current state value, we use `get` from the `StateT` layer, which we reach by using `lift` once, because it's immediately below the outermost `ParserT` layer. To get the keyboard input we use `getLine` in the `IO` layer, which is twice removed from `ParserT`, so we do twice the `lift`ing.

A problem with this `example`, apart from its being silly, are all those `lift`s. Not only are they distracting, but also they force a particular order of our `MonadTransformer` layers. If we are one `lift` short or have one `lift` too many, the type-checker complains. That is, each time we need to use a monadic effect, we'll have to count the distance to the relevant monad layer, with our brain, and write that many `lift`s. Shouldn't we be doing more important things? Let's see if we can fix this. Ideally, we'd like to write something like the following, without any `lift`ing.

```
example :: ParserT (StateT Natural IO) String
example = do
  s <- getLine                   -- IO
  n <- get                       -- StateT
  expect (mconcat (replicate n s)) -- ParserT
  pure s
```

There are approximately seven million competing ways to approach this problem, each of them with their virtues and flaws. Different `Monads` just don't compose nicely, and attempts to workaround this often lead to usability or performance problems. However, that doesn't keep people from trying, and to different extents, different proposals make dealing with this a bit less annoying. We'll explore the status quo, the most widely known and used approach to deal with this issue.

Three hundred three

This is the type of `get` we know so far.

```
get :: Monad m => StateT s m s
```

But we don't want that. StateT s m is saying that the StateT layer must be on top of m, forcing a particular ordering of our layers, which is exactly what we want to avoid. Our task will be to use polymorphism to say something like "whatever my monad tower looks like, when I say get, use the first StateT layer you can find".

The get we learned about exists. We can import it from the Control.Monad.Trans.State module. However, in a different module named Control.Monad.State, without the Trans part, we find another version of get.

```
get :: MonadState s m => m s
```

This version of get is what we are looking for. MonadState s m is a constraint that says that somewhere in m there is a state transformation for a value of type s, and we can interact with it through get and put as we have been doing so far, but without having to explicitly lift. Crucially, the type of this get doesn't say anything about the ordering of MonadTransformer layers in m, nor about whether there are layers at all, which is exactly what we want. get is not a standalone function, but rather, it is part of the MonadState class itself.

```
class Monad m => MonadState s m | m -> s where
  get :: m s
  put :: s -> m ()
```

Ignore the | m -> s part for now, we'll explore that later. We can find the MonadState typeclass in the mtl Haskell library. A Haskell library is just a collection of Haskell modules distributed together, exporting related types and functions intended to solve a particular problem. Most of the code we've seen so far in this book comes from Haskell's base library, which, as the name suggest, exports the base on top of which we can start building more interesting stuff. Things like Bool, String or Monad are there. The base library comes out of the box with every Haskell installation, so we don't have to go and install it by hand. But, for example, StateT and its related vocabulary belong to yet another library called transformers, meaning that for any of our StateT examples to work, we have to install that transformers library somehow. How do we do that? In this book, we use wishful thinking. In real life, you'll need to figure it out yourself.

The idea of mtl's MonadState s m is that for each m we create, we should implement an accompanying MonadState s m instance too. When the m is an actual state-transforming monad, like StateT, then get and put will act directly on it.

```
instance Monad m => MonadState s (StateT s m) where
    get :: StateT s m s
    get = Control.Monad.Trans.State.get
    put :: s -> StateT s m ()
    put = Control.Monad.Trans.State.put
```

The implementation of MonadState for StateT just forwards the work to the get and put we had been using until now, the ones from the Control.Monad.Trans.State module. We don't need to write this particular instance for StateT, though, because it has been written by the mtl authors already. For monads other than StateT, where the state-transforming monadic features are on a MonadTransformer layer below the topmost, the MonadState instance implementation is expected to just lift recursive uses of get and put. For example, our ParserT is not intended to be used as a state-transforming monad, so, if there is a MonadState layer below ParserT, then the implementation of MonadState for ParserT will just lift all uses of get and put from the layer below. The s type parameter is always left polymorphic, meaning the instance will work regardless of the type of payload being transformed by the state-transforming layer.

```
instance MonadState s m => MonadState s (ParserT m) where
    get :: ParserT m s
    get = lift get
    put :: s -> ParserT m ()
    put = lift . put
```

In this implementation of get we use lift only once to target m, the Monad immediately below ParserT. However, the effect we use on m is itself get, which is expected to continue lifting deeper layers, one by one, until it reaches a StateT layer. This is an interesting example of recursive method calls, where each recursive call is relying on a different instance of the same class. For example, imagine our monad tower was ParserT (MaybeT (StateT String IO)). Using get on ParserT would perform get in the underlying monad layer, MaybeT. And, assuming there is a suitable MonadState instance for MaybeT, its implementation of get will perform get on its underlying monad layer, StateT. And this MonadState instance for StateT is the one that actually does some work without delegating to a deeper monad layer, so the recursion ends there. In other words, performing get in ParserT (MaybeT (StateT String IO)) becomes lift (lift get) as expected, but we don't write the lifts ourselves as users of get, only those implementing the MonadState class have to worry about doing that, once. We, as users, just write get and it works.

For completeness, here is the MonadState instance for MaybeT, which looks exactly like the one for ParserT.

```
instance MonadState s m => MonadState s (MaybeT m) where
    get = lift get
    put = lift . put
```

And the one for IdentityT too. These instances look the same because all they are supposed to do is lift the work up from the MonadState below.

```
instance MonadState s m => MonadState s (IdentityT m) where
  get = lift get
  put = lift . put
```

With this new knowledge, let's revisit our annoying example from before.

```
example :: ParserT (StateT Natural IO) String
example = do
  s <- lift (lift getLine)      -- IO
  n <- get                      -- MonadState
  expect (mconcat (replicate n s)) -- ParserT
  pure s
```

See? By using the get method from the MonadState class, which relies on the instance we just wrote for ParserT, we managed to drop the lift corresponding to the get effect in StateT. Let's see if we can get rid of the rest.

Three hundred four

The situation with getLine of type IO String is a bit different, but also a bit similar. Just like we have MonadState for abstracting over the idea of a Monad having state transformation features somewhere in it, we have MonadIO for pointing out that somewhere in our Monad we can perform IO effects.

```
class Monad m => MonadIO m where
  liftIO :: IO a -> m a
```

And there is no such thing as an "IO monad transformer". If there is a Monad made out of multiple MonadTransformer layers having access to IO effects, then the IO layer will always be at the bottom of that tower. All that liftIO does is bring that IO effect from the very bottom layer up to m. When m is IO itself, the implementation is very straightforward.

```
instance MonadIO IO where
  liftIO = id
```

In all other cases, the implementation is straightforward too, but in a different manner. All we need to do is lift the performance of liftIO from the layer immediately below, in exactly the same way get and put did it in MonadState.

```
instance MonadIO m => MonadIO (ParserT m) where
  liftIO = lift . liftIO
```

The implementations of the MonadIO instances for StateT, IdentityT, ParserT, ExceptT or any other Monad providing direct access to IO, look the same. With MonadIO, we can improve our example just a bit more.

```
example :: ParserT (StateT Natural IO) String
example = do
  s <- liftIO getLine          -- MonadIO
  n <- get                     -- MonadState
  expect (mconcat (replicate n s)) -- ParserT
  pure s
```

We changed that first line of code from lift (lift getLine) to liftIO
getLine. It's not the ideal we were expecting, of there being no lifting at all,
but it is an improvement. At least we don't have to explicitly count how
many lifts to perform anymore. We can say liftIO something, and
something will be done at the IO layer, however far the IO layer is from
where we currently are.

Can we get rid of that final liftIO? Not exactly. The thing is, getLine
comes out of the box with Haskell having type IO String. If it came with
type MonadIO m => m String instead, we could use it without explicitly
mentioning liftIO. But it doesn't. So, our alternatives are to either use
liftIO here, or define a new function called getLine' with trailing tick ' or
similar, hiding the liftIO under the rug.

```
getLine' :: MonadIO m => m String
getLine' = liftIO getLine
```

If we use this getLine' with trailing tick ' instead of the normal one
without it, we can "get rid of liftIO". Wink, wink.

```
example :: ParserT (StateT Natural IO) String
example = do
  s <- getLine'                -- MonadIO
  n <- get                     -- MonadState
  expect (mconcat (replicate n s)) -- ParserT
  pure s
```

The lifts are there, but we don't see them anymore.

Three hundred five

Can we have a MonadParser, too? Do we even need that? Perhaps
something that somehow allows us to talk about there being parsing
effects in a Monad, without explicitly mentioning ParserT? Sure. We just
need to figure out what this MonadParser class would look like.

If we search for inspiration in MonadState, we'll see that its methods talk
about somehow being able to perform state transformations via get and
put, which are the fundamental operations any state-transforming Monad
should support.

```
class Monad m => MonadState s m | m -> s where
  get :: m s
  put :: s -> m ()
```

And similarly, `MonadIO` straight up asks us to provide a proof that we are able to perform IO actions in our Monad.

```
class Monad m => MonadIO m where
  liftIO :: IO a -> m a
```

So, it looks like a `MonadParser` class should require its instances to implement the most primitive, fundamental, core operations that need to be supported in order to claim that a Monad supports parsing as we understand it. And we know the answer to that.

```
class Monad m => MonadParser m where
  parse1 :: (Char -> Either String a) -> m a
```

All our parsers can be built on top of parse1, we know this. So, as long as a chosen Monad can provide an implementation for it, we can give it a proper `MonadParser` instance. Or at least that's what we hope. The thing is, we are handwaving things a bit here Ideally, we'd accompany this typeclass definition with beautiful laws that instances should abide by, but we won't be doing that today, so let's just keep our fingers crossed.

For instance, here's the `MonadParser` instance for ParserT, using the exact implementation of parse1 from two or three chapters ago.

```
instance Monad m => MonadParser (ParserT m) where
  parse1 f = ParserT (do
    (n0, i0) <- lift get
    case i0 of
      [] -> left 'Not enough input"
      c : i1 -> case f c of
        Left e -> left e
        Right a -> do
          let !n1 = n0 + 1
          lift (put (n1, i1))
          pure a)
```

The type of this parse1 method is a bit more polymorphic than our previous parse1 function. Look.

```
parse1 :: MonadParser m => (Char -> Either String a) -> m a
```

And this is great, because nothing built off of parse1 pieces needs to explicitly mention ParserT anymore. Nothing.

```
parseChar :: MonadParser m => m Char
parseChar = parse1 Right

expect1 :: MonadParser m => Char -> m Char
expect1 a = parse1 (\case
  b | a == b -> Right a
    | otherwise -> Left ('Expected " <> show a <>
                          ' got " <> show b))
```

See? No mention of ParserT. In fact, ParserT is now just one possible implementation of MonadParser. If in the future we are able to come up with a different implementation, perhaps one that is more performant, we could easily switch between them without having to change any of our parsing code, for it will never mention ParserT explicitly. We'll see how to do that soon.

So, what about giving MonadParser instances to Monads that aren't themselves ParserT or similar parsing beasts, but rather, have parsing features provided by an underlying Monad in our tower? Easy. Just like we implemented MonadState for ParserT by lifting all state transformations from below, we could implement MonadParser for StateT, say, by lifting all parsing actions too.

```
instance MonadParser m => MonadParser (StateT s m) where
  parse1 = lift . parse1
```

And similarly for MaybeT, IdentityT and any other MonadTransformer that merely delegates the parsing work to a deeper MonadParser layer.

With all this new machinery, we can improve example once again.

```
example :: (MonadIO m, MonadState Natural m, MonadParser m)
        => m String
example = do
  s <- getLine'              -- MonadIO
  n <- get                   -- MonadState
  expect (mconcat (replicate n s)) -- MonadParser
  pure s
```

The term level implementation of our example remains the same, but its type has become as polymorphic as it can be in m. The lifts need not agree with the depth and ordering of the MonadTransformer layers anymore. All that's asked of m is that it support the necessary MonadIO, MonadState and MonadParser capabilities.

Three hundred six

But how can we run example now, if it doesn't mention what m is? Before, when its type was ParserT (StateT Natural IO), we knew we had to remove the ParserT layer first, then the StateT layer, and finally perform an IO action. But now, what? Now, it's up to us to decide what m is. If we decide that m shall be ParserT (StateT Natural IO), then so it shall.

```
> runStateT (runParserT example "NaNaNaNaNaNaNaNaBatman!") 8
Na
(Right (16, "Batman!", "Na"), 8)
```

And if we decide that it shall be StateT Natural (ParserT IO) instead, then it shall be that instead.

```
> runParserT (runStateT example 8) "NaNaNaNaNaNaNaNaBatman!"
Na
Right (16, "Batman!", ("Na", 8))
```

We are choosing the type of m implicitly when we choose in which order to peel our monad tower layers. For example, if we runParserT first and runStateT second, then m is inferred to be of type ParserT (StateT Natural IO). On the other hand, if we runStateT first and runParserT second, then m becomes StateT Natural (ParserT IO).

We can even sprinkle some extra useless MonadTransformers there, just for fun, and it will still work.

```
> runIdentityT (runParserT (runMaybeT (runStateT example 8))
                           "NaNaNaNaNaNaNaNaBatman!")
Na
Right (16, "Batman!", Just ("Na", 8))
```

In this case, m became StateT s (MaybeT (ParserT (IdentityT f))). Of course, different choices of m, and different choices of how to run or tear down that m, can lead to different results and result types as well. We saw this too. Generally, the composition of Monads is not commutative. That is, from the MonadTransformer point of view, putting BatT on top of ManT doesn't necessarily achieve the same as ManT on top of BatT does.

So, if we ever implement a MonadParser that is supposed to be more performant than ParserT, and we want to try it, all we have to do is use runMyFastWhateverT instead of runParserT in this one place. No need to change anything else. This exemplifies the main motivation for mtl and the rest of the competing bunch: They offer a way to abstract over the capabilities offered by different Monads. As users of monadic vocabulary, we just don't care about Monad implementations when thinking about the meaning of our code, we only want to talk about the capabilities of our Monads. That's what MonadState, MonadParser, MonadIO and the like allow us to do. We write our code assuming *some* implementation will be used, and that's all. lift? It was never about lift. It was about this.

We can imagine this decoupling being useful, for example, if we are building some missile-launching program. When is time to actually launch the missile because some politicians are not getting along, then we use the real life implementation. Boom. On the other hand, while testing this code during peacetime, we can use an implementation that just turns on a light bulb or something whenever the missile-launching effect is supposed to happen. No missile-launching code changed. No harm done. Just a choice between runBoomT and runBulbT somewhere.

Three hundred seven

There ain't no IO monad transformer, we claimed. But why is that? Let's fail to build one ourselves, and we'll understand.

```
data IOT m a = IOT (m ?)
```

The first issue is figuring out what to write there, where the ? placeholder is. Let's see if we can get some inspiration from other MonadTransformers.

```
data MaybeT m a = MaybeT (m (Maybe a))
```

```
data ExceptT e m a = ExceptT (m (Either e a))
```

It looks like what MonadTransformers often do is have m produce a value of the non-transformer version Monad type. MaybeT m a has m produce a Maybe a, ExceptT e m a has m produce an Either e a, etc. Let's try that, then. Let's have m produce an IO a.

```
data IOT m a = IOT (m (IO a))
```

We'll surely need a function for removing that IOT wrapper.

```
runIOT :: IOT m a -> m (IO a)
runIOT (IOT x) = x
```

And we'll definitely need to prove that IOT m is a Monad, before we can attempt to claim that IOT is a MonadTransformer. Alas, we can't.

```
instance ∀ m. Monad m => Monad (IOT m) where
  (>>=) :: ∀ a b. IOT m a -> (a -> IOT m b) -> IOT m b
  IOT w >>= f = IOT (do
    x :: IO a <- w
    let y :: IO (IOT m b) = fmap f x
        z :: IO (m (IO b)) = fmap runIOT y
    ? :: m (IO b))
```

In that last line, where once again we wrote a ? placeholder, we are supposed to come up with a value of type m (IO b), but the closest thing we have is z, of type IO (m (IO b)). If only we were able to remove that outermost IO layer we'd have our solution. However, removing that IO layer while in m corresponds to the idea of evaluating an *impure* interaction with the outside environment, like checking the current weather, in a potentially *pure* environment m, such as Maybe or Identity. So, no, IO can't be a MonadTransformer, as this would require violating everything IO stands for. That is, preventing outside interference to our programs unless we explicitly request otherwise by using IO.

Three hundred eight

Let's take a look at MonadState again. We skipped something.

```
class Monad m => MonadState s m | m -> s where ...
```

The | m -> s stuff, which we chose to ignore before, describes a *functional dependency* of this class. It says that if we know what m is, then from that knowledge alone we must be able to determine what s is too. Why is this useful? What does it enable? What does it prevent? Let's see.

On the more practical side of things, this helps with type-inference. For example, if we are dealing with a Monad type State Bool, then we expect get to produce a value of type Bool. This expectation is what's encoded in the functional dependency. Just from knowing that State Bool is our Monad, we know that state-transformation operations get and put must deal with values of type Bool. GHCi readily confirms this.

```
> :t put _ :: State Bool ()
<interactive>:1:5: error:
  • Found hole: _ :: Bool
    ...
```

We left a hole _ there so that the type-checker can tell us what the type of an expression taking its place should be, and it says Bool. The type-checker knows this because of the functional dependency tying State Bool to Bool. Or, more generally, State s to s, for all choices of s. If we try to put anything other than what the functional dependency mandates, we get a type-checking error saying exactly that.

```
> :t put 'x' :: State Bool ()
<interactive>:1:1: error:
  • Couldn't match type 'Bool' with 'Char'
      arising from a functional dependency between:
        constraint 'MonadState Char (StateT Bool Identity)'
          arising from a use of 'put'
        instance 'MonadState s (StateT s m)' at
    ...
```

So, seeing as Haskell always knows what the s in MonadState s m is supposed to be, just from knowing what m is, we never need to explicitly mention the s type in our code. In practice, inferring the type of m is often much easier than inferring the type of a standalone s every time we use put or get, so the functional dependency is there mostly to improve type-inference for s by tying it to m.

By the way, did you notice how even if we typed State Bool, we got a type-checker error saying something about StateT Bool Identity? This is one of the downsides of offering type synonyms as a convenience we mentioned before. When we get a type-checking error related to them, more often than not, the error will mention the actual type and not the type synonym. It can get confusing.

An additional benefit of MonadState's functional dependency is that it restricts us to having one instance per Monad at most. For example, if we attempt to define these two instances side by side, the type-checker will complain.

```
instance Monad m => MonadState Natural (StateT s m) ...

instance Monad m => MonadState Integer (StateT s m) ...
```

The problem is that, according to the functional dependency on the
MonadState s m typeclass, m must *uniquely* determine s. But here we have
one instance saying StateT s m determines a Natural type of state, and
another instance saying something else. We can't have that. We could have
this other thing, though.

```
instance Monad m =>
  MonadState Natural (StateT Natural m) ...

instance Monad m =>
  MonadState Integer (StateT Integer m) ...
```

There, StateT Natural m determines a Natural state, and StateT Integer m
determines an Integer state. That's fine, because StateT Natural m and
StateT Integer m are different Monads. However, why have two different
instances with probably exactly the same internal implementation, when
we could have a single polymorphic one that works for any choice of state
type?

```
instance Monad m => MonadState s (StateT s m) where ...
```

Moreover, having a single instance gives us the assurance that for all
choices of s, this implementation, internally, performs exactly the same
amount and kind of work. Perhaps this isn't the most important thing
from a conceptual point of view, though, seeing as MonadState comes with
its own set of laws that require its instances to be correct. Well, actually, as
it was the case with Alternative, people are still discussing what these laws
should be. But, more or less, they say that getting the state value returns
the most recently put state value. Anyway, it's nice to know that there's
only one implementation to worry about per Monad. Less stuff to maintain,
less performance matters to take into account.

We may not *need* functional dependencies, but programming without
them surely is painful, for we have to keep telling the type-checker about
the type of our state over and over again.

Three hundred nine

The syntax for establishing functional dependencies is relatively
straightforward.

```
class Foo a b c d | a -> b, c -> b d, a d -> c where ...
```

The functional dependencies are specified to the right of the vertical pipe
|. In this case, we are establishing three functional dependencies, separated

by commas. Or perhaps four, depending on how you count them. `a -> b` says that if we know what the type a is, then we can uniquely determine what b is. This is exactly what the `m -> s` functional dependency on `MonadState` looked like. Then, we have `c -> b d`. This says that if we know what c is, then we will also know what b and d are. We could have written `c -> b, c -> d` instead, and it would have achieved the same. And finally, a `d -> c` says that if we know what *both* a and d are, from that composite knowledge we can determine what c is.

Three hundred ten

What about other `MonadWhatever` typeclasses? What else do we have? Is there a `MonadIdentity` generalizing the idea of a `Monad` able to *purely* produce a functorial output of our choice? No. That's just `Applicative`'s pure half. A `MonadIdentity` typeclass wouldn't add anything new compared to what any `Monad` already offers, so there's no such thing as `MonadIdentity`. There could be a superclass to `Applicative` that provided only pure, but there isn't. Probably because a typeclass like that would suffer from lack of decent laws, considering most meaningful laws need to talk about the relationship between pure and `<*>`, and not about pure on its own. It's the relationship between things what matters the most, always remember that.

Maybe `MonadMaybe`? Yes, but also no. Let's think about a hypothetical `MonadMaybe`. What would be the core methods of a class intended to convey the possibility of a `Monad` sometimes failing to produce a functorial output, as exemplified by `Nothing`, and sometimes succeeding to do so, as exemplified by `Just`?

```
class Monad m => MonadMaybe m where
    nothing :: m a
    just :: a -> m a
```

Decent first attempt. Except neither base nor mtl export have a `MonadMaybe` like that. First, because just is exactly the same as `Applicative`'s pure. It constructs, out of a non-functorial value, a functorial value that does nothing except succeed. And second, because we already have `MonadPlus` and its mzero method achieving the same as nothing would. So, this `MonadMaybe` would be a bit redundant. Only a bit, though, for `MonadPlus` has that extra mplus method which our hypothetical `MonadMaybe` typeclass wouldn't need.

Know that older versions of Haskell had different homes for mzero and mplus, so we are not that crazy in wanting to explore this.

```
class Monad m => MonadZero m where
  mzero :: m a

class MonadZero m => MonadPlus m where
  mplus :: m a -> m a -> m a
```

The method names weren't exactly those, but it doesn't matter. What matters is that said MonadZero is exactly what we were expecting MonadMaybe to be. Why don't we have MonadZero in Haskell anymore? It's a contentious topic, but it might be because we can't have meaningful laws for mzero without describing it as mplus's identity element, and we don't like having lawless typeclasses for we can't truly make sense of them. So, mzero lives in MonadPlus. Good decision? Bad decision? We may never know.

There is a MonadFail class, too, that more or less is what we are expecting MonadMaybe to be, and then some.

```
class Monad m => MonadFail m where
  fail :: String -> m a
```

MonadFail is special. Yes, it is intended to convey the idea of the possibility of a Monad failing to produce an output, just like mzero does, but there's also an unexpected String there. Why? Because MonadFail is about Haskell's do syntax more than it is about failing.

```
foo :: MonadFail m
    => m (Natural, Natural)
    -> m Natural
foo mab = do (7, b) <- mab
             pure b
```

When foo does (7, b) <- mab, not only is it binding the functorial output from mab, but also it's pattern-matching on the tuple constructor expecting its leftmost payload to be 7. But what if it's 4 or 25 instead? Does our program crash because of a partial pattern match? No, it fails. This is what foo looks like when Haskell desugars the do notation.

```
foo :: MonadFail m => m (Natural, Natural) -> m Natural
foo mab = mab >>= \ab ->
            case ab of
              (7, b) -> pure b
              _ -> fail "Pattern match failed in do \
                         \expression at <some location>"
```

In other words, if mab's functorial output matches the (7, _) pattern, all is well and we proceed to execute the rest of the computation. Otherwise, if it's anything else, Haskell's desugaring of the do notation performs fail with some String describing the location where the pattern matching failed.

We can use fail without the do notation, of course, just how we can use >>= without it too.

By the way, unrelated, did you notice how we spread that literal String over multiple lines in the desugared version of foo?

```
"Pattern match failed in do \
\expression at <some location>"
```

A single literal String can be written across multiple lines if we end a line with a backslash \ and start the next line with a backslash \ too.

```
"And this works across \
\more than just tw\
\o lines, too."
```

Anyway, let's get back on track. The MonadFail instance for Maybe ignores the informative String altogether, but there's no harm in that.

```
instance MonadFail Maybe where
  fail :: String -> Maybe a
  fail _ = mzero
```

The documentation for MonadFail says that fail is expected to behave as mzero unless we have room for putting that informative String somewhere, in which case it should behave as mzero too, but also mention that String somewhere. So, mzero is what we have here. Maybe's mzero is Nothing, in case you forgot. MaybeT's instance is essentially the same.

```
instance Monad m => MonadFail (MaybeT m) where
  fail :: String -> MaybeT m a
  fail _ = mzero
```

It's time to try our sugary foo in the REPL.

```
> foo Nothing
Nothing
> foo (Just (3, 1))
Nothing
> foo (Just (7, 1))
Just 1
```

Good. We used Maybe as our MonadFail, but this works for other MonadFails too. For example, lists [], in which case an application of fail results in an empty list [], having the practical effect of skipping the list elements that can't be pattern-matched against.

```
> foo [(6, 1), (7, 2), (8, 3), (7, 4)]
[2, 4]
```

The foo application above becomes the equivalent of join [fail "...", pure 2, fail "...", pure 4], which reduces to join [[], [2], [], [4]]. That is, [2, 4]. Being able to filter lists using do notation can lead to code that's very easy to read.

There's no room for fail's informative String in lists, so we won't see it anywhere. One example of a MonadFail that uses said String is IO.

```
> foo (pure (4, 5) :: IO (Natural, Natural))
*** Exception: user error (Pattern match failure
                in do expression at <interactive>:1:2)
> foo (pure (7, 5) :: IO (Natural, Natural))
5
```

failing in IO does cause our program to crash, so it's not such a great idea to fail there unless we really intended for our program to crash or there are recovery mechanisms in place. But at least we get to see that informative message about why and where our program crashed. This kind of message, however, is mostly useful to us as software developers. End users won't be able to make much sense of it. Picture yourself trying to pay for something online and being told by that you can't because of a pattern match failure in a do expression. Not useful.

Three hundred eleven

Do we have a MonadEither? A MonadExcept? Yes. Kinda. Not exactly. It's complicated.

In mtl, the MonadError typeclass generalizes the short-circuiting behavior of Either e a, allowing one to either produce an a or abort the computation early with payload e, as if e was some sort of *error*. But MonadError goes a bit beyond that.

```
class Monad m => MonadError e m | m -> e where
    throwError :: e -> m a
    catchError :: m a -> (e -> m a) -> m a
```

MonadError e m says that it's possible to abort the computation early using throwError, yes. But, moreover, a catchError suggests that it is possible to *catch* those errors too. Let's have a look at the MonadError instance for Either to appreciate this in detail.

```
instance MonadError e (Either e) where
    throwError = Left
    catchError (Right a) _ = Right a
    catchError (Left e)  f = f e
```

Essentially, catchError allows us to *recover* from that error e by replacing the original *failing* functorial value with a different one, if the original does actually fail.

Now, we know that Either is a *general purpose* sum type where what we put on the Left side doesn't necessarily represent an error, but only the value with which to abort the computation early when Either e a is treated as a functorial value. However, when viewed through the lens of MonadError, nominally, Left does represent failure. Perhaps if MonadError was called MonadAbort or, you know, MonadEither, we wouldn't be having this conversation. It is what it is.

```
second :: [x] -> Either String x
second (_ : x : _) = Right x
second _           = Left "less than two elements"

half :: MonadError String m => Natural -> m Natural
half n | even n    = pure (div n 2)
       | otherwise = throwError "number is odd"
```

Suppose we want to compose the two functions above in order to calculate half of the second element of a list, by dividing it by 2. But if the list has less than two elements, we want to produce 0 instead.

```
halfOfSecond :: [Natural] -> Either String Natural
halfOfSecond nats = catchError
   (second nats >>= half)
   (\case "less than two elements" -> Right 0
          somethingElse -> Left somethingElse)
```

Perhaps not the best example, as usual. Are great examples just a myth? We'll never know. The first thing to notice about halfOfSecond is that while second mentions Either String explicitly, half is a bit more general and talks only about a MonadError String m. For no fundamental reason, mind you. This is just an example. But, despite these differences, we can still bind these functorial values just fine using >>= because we know that Either String is one such MonadError String m.

```
> halfOfSecond [8]
Right 0
> halfOfSecond [4, 5]
Left "number is odd"
```

When we apply halfOfSecond to [8], a list with less than two elements, the internal use of second [8] fails with that "less than two elements" error message, which of course causes the entire composition second [8] >>= half to fail in the same way. However, in the second argument to catchError, this particular error is intercepted and transformed to an alternative functorial value Right 0, which is why we see that as final result.

In the second example, the use of half fails. This time, however, its error is not intercepted by catchError. That's why we see the Left "number is odd" output. Finally, we have the happy scenario where no error is ever thrown so the Rightful result is preserved.

```
> halfOfSecond [1, 6]
Right 3
```

Either is not the only MonadError out there, of course. ExceptT, for example, is too.

```
instance Monad m => MonadError (ExceptT e m) where
  throwError = ExceptT . pure . Left
  catchError x f = ExceptT (do
    runExceptT x >>= \case
      Right a -> pure (Right a)
      Left  e -> runExceptT (f e))
```

And similarly, other monad transformers that simply wrap another
MonadError. For example, StateT.

```
instance MonadError m => MonadError (StateT s m) where
  throwError = lift . throwError
  catchError x f = StateT (\s0 ->
    catchError (runStateT x s0)
               (\e -> runState (f e) s0))
```

In the market, despite its name, MonadError is not used much for error
handling as intended, but for its early termination features. Usually, when
dealing with errors, the mechanisms offered by the raw Either and ExceptT
are sufficient whenever Either or ExceptT are part of our Monad. And in
most other scenarios, by errors we probably mean *exceptions*. Exceptions
such as the ones we've often encountered when using partial functions, or
the ones we'd encounter if, say, we want to send an email from Haskell
but the computer is offline at the moment, so the operating system *throws*
an exception making a fuss about the situation, making our program crash
unless we *catch* that exception and do something about it. But such a
scenario implies that we are dealing with IO interactions, and in those cases
we prefer to use mechanisms similar to MonadError which don't really limit
the type of errors our Monad can catch. Remember the m -> e functional
dependency on MonadError e m? That's limiting the type of errors our m can
catch. However, when dealing with IO, *anything* can go wrong, so we
better be ready to catch all possible errors, not just those represented by
one blessed type e.

In this book we won't talk about exceptions beyond this. It's a tricky topic
that requires approximately a bookful of words and examples to get right.
But don't worry, we do have good tools to deal with exceptions in Haskell,
so you'll be fine. Let's put MonadError and exceptions aside for now.

Three hundred twelve

Functions are Functors, we know this.

```
instance Functor ((->) a) where
  fmap :: (b -> c) -> (a -> b) -> (a -> c)
  fmap = (.)
```

We also know that functions are Applicatives. Or maybe we don't, because
when we learned about this, we learned it by a different name. We talked
about functions being monoidal functors rather than Applicatives. But

eventually we also learned that monoidal functors and Applicatives are
essentially the same thing. So, yes, functions are Applicatives too.

```
instance Applicative ((->) a) where
  pure :: b -> (a -> b)
  pure = const
  liftA2 :: (x -> y -> z)
          -> (a -> x)
          -> (a -> y)
          -> (a -> z)
  liftA2 g fx fy = \a -> g (fx a) (fy a)
```

For any Applicative functor f, liftA2 lifts a function x -> y -> z into a
function f x -> f y -> f z. Our f happens to be (->) a here, that's why
we end up with so many arrows in the type of liftA2.

```
> liftA2 (,) length reverse [1, 2, 3, 4, 5]
(5, [5, 4, 3, 2, 1])
```

What liftA2 does in this example is provide the same input list to both
functorial values length and reverse, and then combine their respective
functorial outputs using (,). Remember, length and reverse are functions,
but also they are functorial values in the (->) a functor. The important bit
to notice is that exactly the same input is being provided to *both* functorial
values. That is, to both functions. The essence of what it means for
functions to be Applicative functors is that their input will be fanned out
to other functions. That is, in f = liftA2 g fx fy, when f is applied to
something, that something will be supplied to fg and to fy too.

But the story doesn't end there. It turns out that functions are Monads too.

```
instance Monad ((->) a) where
  (>>=) :: ((->) a x) -> (x -> (->) a y) -> (->) a y
```

Man can't read that. Let's put the arrows in infix form and drop some
redundant parentheses.

```
instance Monad ((->) a) where
  (>>=) :: (a -> x) -> (x -> a -> y) -> (a -> y)
  f >>= g = \a -> g (f a) a
```

Bind >>= says that when we finally apply a -> y to a value of type a, this a
will be provided to a -> x first, and then both this same a and the newly
obtained x will be provided to x -> a -> y. That is, the initial a is still being
fanned out to all the functorial values involved, but this time, x -> a -> y
can also depend on the functorial output from a -> x. Future functorial
values depending on past functorial outputs, that's what Monads are all
about.

```
> (length >>= lookup) [(2, "two"), (7, "seven")]
Just "two"
```

Here, >>= first applied length to the list, which produced 2 as output, and subsequently applied `lookup` to this 2 and the input list once again, resulting in `Just "two"`. In other words, `length`'s output is being provided to `lookup`, as it would have been the case in any other `Monad`, but the original input list is being implicitly provided to both `length` and `lookup` as well. This example fully exploits the `Monadic` features of functions by having `lookup` depend not only on the input list, but also on the output from `length`. We can't implement this code relying solely on the `Applicative` features of functions, we need bind `>>=`.

Oh, and in case it wasn't obvious, `lookup` does what its name suggests. Given a key and a list pairing keys to values, it returns the leftmost value associated to that key, if any. It *looks up* the key in the list, see.

```
lookup :: Eq k => k -> [(k, v)] -> Maybe v
lookup x ((k, v) : kvs)
  | x == k = Just v
  | otherwise = lookup x kvs
lookup _ _ = Nothing
```

Traditionally, we call these lists pairing keys to values *association lists*. They are a rudimentary structure for organizing values identified by a particular key.

Going back to our main topic, there's something important we must highlight. Functions are `Monads`, true, but we can't emphasize enough how *weird* code that exploits this feature looks like, so it's unlikely that we'll encounter much of it out there in the wild. The `length >>= lookup` example above was already a bit tricky to figure out, but things get even funnier if, for example, we start using do notation for functions. Just look at this. It's exactly the same example as above, but written using do notation.

```
foo :: [(Natural, String)] -> Maybe String
foo = do e <- length
         lookup e
```

These uses of the normal functions `length` and `lookup` seem to be partially applied, but they aren't. Behind the scenes, bind `>>=` is applying the `[(Natural, String)]` input to every function, every effect, every functorial value in this computation. Moreover, seeing as the whole function returns a `Maybe String`, by inertia we assume that this do notation is constructing a functorial value in `Maybe`. Alas, it isn't. It is constructing a functorial value in `(->) [(Natural, String)]`. Confusing, yes. Perhaps not so much here in `foo` because it's such a small example. But imagine dealing with a bigger computation where we need access the same input from multiple places. On the one hand, it would be a shame to have to pass the input around explicitly everywhere, so we appreciate being able to rely on a function's `Monadic` features to avoid that. But on the other hand, we won't be able to read much of the code we write using said technique, so maybe let's just avoid it. Luckily for us, we won't see a lot of code exploiting functions as

Monads out there. Understandably, folks don't seem to enjoy reading nor writing code this way. It's just weird.

Three hundred thirteen

Functions are Monads, but we don't like using them directly as such because it's confusing. However, we do like this idea of an input being readily available, implicitly, to every functorial value in a computation. To generalize this idea beyond functions, mtl proposes a MonadReader typeclass.

```
class Monad m => MonadReader e m | m -> e where
  ask :: m e
  local :: (e -> e) -> m a -> m a
```

The input in question is most often described as the *environment* of m, that's why we chose to name our type variable e. This name evokes the idea that a computation in r takes place while having access to an environment e. What is an environment? Whatever we need it to be. Maybe it's a Natural number, maybe it's the color of this room, maybe the contents of a book. It's just an extra input value that will always be readily available to us, implicitly.

Notice the functional dependency there, similar to the one MonadState had. It says that e depends on m. That is, knowing the type of m is enough information for the type-checker to uniquely determine what e is. This suggests that, in practice, we can expect there will be one MonadReader per Monad at most as it was the case in MonadState, assuming the authors of the MonadReader instances keep that m as polymorphic as it can be.

ask is a functorial value producing that input environment, e. Wanna know what e is? All you have to do is ask. This method is very similar to MonadState's get, which allows us to take a look at what the current state is without transforming it in any way. Except MonadReader has nothing to do with state transformations, but only with taking a look at what e is. It's a lot more boring. We can think of MonadReader as the read-only version of MonadState. *Read-only*, that's where the *reader* name comes from. Probably. We can read but we can't write. We can look but we can't touch. Or, you know, we can touch but we can't change. You get the idea.

And then there's local, which can be used to temporarily modify this e within a limited scope. It's hard to see this in the type of local because, due to an unfortunate lack of parametric polymorphism, we can't understand local's intentions just from its type. Don't worry, we'll look into it soon.

```
bar :: MonadReader [(Natural, String)] m
    => m (Maybe String)
bar = do e <- ask
         pure (lookup (length e) e)
```

Here we have a small `MonadReader` example. `bar` achieves the same as the `foo` example from last chapter, except `bar` is now polymorphic in `m`, and it can be read with fewer surprises. An environment of type `[(Natural, String)]` is available everywhere, implicitly. This is made clear by the `MonadReader` constraint. But if we want to *access* the environment, we must explicitly ask for it. Before, every single functorial value in this computation would have been applied to the environment automatically. Crazy.

Anyway, seeing as `MonadReader` is supposed to generalize the `Monadic` capabilities of functions, we'll find that there is a `MonadReader` instance for functions.

```
instance MonadReader e ((->) e) where
  ask :: e -> e
  ask = id
  local = ...more about this later...
```

With this instance in place, we can run the recent `bar` example just like we ran last chapter's `foo`, getting exactly the same results.

```
> foo [(2, "two"), (7, "seven")]
Just "two"
> bar [(2, "two"), (7, "seven")]
Just "two"
```

Here, `bar` takes the same type as `foo` when we attempt to perform a normal function application with it. That is, the type of `bar` becomes `[(Natural, String)] -> Maybe String`.

Three hundred fourteen

What about `local`? It's easier to understand what it's supposed to do by just looking at an example.

```
tak :: MonadReader Natural m => m (Natural, Natural, Natural)
tak = do a <- ask
         b <- local (+1) ask
         c <- ask
         pure (a, b, c)
```

A quick interaction in the REPL shows what `local` is all about.

```
> tak 5
(5, 6, 5)
```

That is, `local` modifies the environment *within a limited scope*. When we say `local (+1) ask`, we are saying that the environment made available to the second parameter to `local`, here `ask`, shouldn't be the original one, but rather it should be modified by `(+1)`. After `local` produces its output `b`, the environment goes back to being what it was before. This is unlike the `put` action in `MonadState`, which keeps the modified state. Of course, we could implement something that behaves as `local` using state transformation

vocabulary. In fact, we did something like that when we implemented the
ParserT's rollback behavior. But that's a story for another day. Let's look at
local's type.

```
local :: MonadReader e m => (e -> e) -> m a -> m a
```

As we said before, due to the lack of parametric polymorphism in this type
signature, the type of local doesn't fully explain what's going on. Not
even if we pick functions as our m.

```
local :: (e -> e) -> (e -> a) -> (e -> a)
```

But look at what an hypothetical fully parametrically polymorphic local
tells us.

```
local :: (e -> x) -> (x -> a) -> (e -> a)
```

It says there, clearly, that if we have a function x -> a taking a modified
environment x as input, and we have a function e -> x allowing us to
obtain said modified environment x from e, then we can use local to
transform x -> a into an e -> a computation.

Why doesn't MonadReader provide a fully parametrically polymorphic type
for local? Consider this attempt at making its type just a bit more
polymorphic.

```
class Monad m => MonadReader e m | m -> e where
    ask :: m e
    local :: (e -> x) -> m a -> m a
```

Due to the functional dependency in this typeclass, as soon as we pick a
specific m as our Monad, the type of the input environment must be a
specific e uniquely determined by m. So, even though in local we would
have access to a function of type e -> x that could be used to modify the
original e environment, we wouldn't be able to use a value of type x as the
new environment for m, simply because x is not e. For this reason, the real
type of the local method in MonadReader requires the environment
transformation function to maintain the environment type. We can
change the environment value, but not the environment type. That's what
the e -> e type says. If the functional dependency wasn't there, we
wouldn't have this limitation, but we'd suffer from very poor type-
inference instead, a scenario we considered when discussing MonadState.
Pick your poison.

Having this new understanding about local, we can finally give a full
implementation to the MonadReader instance for functions.

```
instance MonadReader e ((->) e) where
    ask :: e -> e
    ask = id
    local :: (e -> e) -> (e -> a) -> (e -> a)
    local = flip (.)
```

The implementation of ask and local couldn't be simpler. And the idea of MonadReader is very simple too. Read-only access to some input, that's it.

Three hundred fifteen

By relying on MonadReader, we can avoid writing ugly code that uses the Monadic features of functions, yet we can still reap the benefits of what doing so would allow. But that's only part of the picture. The whole motivation for these MonadWhatever beasts, remember, was using them in combination with other such beasts, composing their effects without explicitly mentioning lift nor the actual types implementing said instances. We could, for example, have some code making use of MonadReader and MonadParser at the same time.

```
abc :: (MonadReader Char m, MonadParser m) => m Natural
abc = do x <- ask
         expect1 x
         a <- parseNatural
         expect1 x
         pure a
```

This abc example parses a Natural number surrounded by a Char that's supposed to be available as m's environment.

In order to try abc, we could for example pick an m where ParserT is on top of a function providing that Char, and tear down that monad tower by first using runParserT and then applying the result to the environment Char. In this case, abc would have the type ParserT ((->) Char) Natural.

```
> runParserT abc "x123x" 'x'
Right (5, "", 123)
```

Here, we are applying all of runParserT abc "x123x" to 'x'. If you can't see it, imagine redundant parentheses. Or perhaps imagine using a function called runFunction to run the function layer, instead of the merely juxtapositioning the function and 'x'.

```
runFunction :: (a -> b) -> a -> b
runFunction = id
```

With runFunction, it should be easier to see how m is run.

```
> runFunction (runParserT abc "x123x") 'x'
Right (5, "", 123)
```

Everything works perfectly. However, we are limited to using our Monad layers in exactly this order. We can have ParserT on top of a function, but we can't have a function on top of ParserT because functions have no room for an underlying Monad in their type. Functions are not MonadTransformers. If we look at StateT s m, ParserT m or MaybeT m, all of them have room for an m there, but no such luck in (->) e. So, if we are

planning to ride an m with a function, we better come up with something
else.

```
data ReaderT e m a = ReaderT (e -> m a)
```

In Haskell we have ReaderT, a MonadTransformer that can extend m with
access to some environment of type e. That's all. Very boring transformer.
Second only to IdentityT in terms of how boring it is.

```
instance Monad m => MonadTrans (ReaderT e m) where
  lift :: m a -> ReaderT e m a
  lift = ReaderT . const
```

And being a MonadTransformer implies that ReaderT e m is a Monad too.

```
instance Monad m => Monad (ReaderT e m) where
  (>>=) :: ReaderT e m a -> (a -> ReaderT e m b) -> ReaderT e m b
  m >>= g = ReaderT (\e -> runReaderT m e >>= \a ->
                             runReaderT (g a) e)
```

runReaderT does the expected thing.

```
runReaderT :: ReaderT e m a -> e -> m a
runReaderT (ReaderT x) = x
```

It's an Applicative, it's a Functor, and most importantly for our current
concerns, it's a MonadReader too.

```
instance Monad m => MonadReader e (ReaderT e m) where
  ask :: ReaderT e m e
  ask = ReaderT pure
  local :: (e -> e) -> ReaderT e m a -> ReaderT e m a
  local f m = ReaderT (runReaderT m . f)
```

Other than having access to an underlying Monad, nothing in ReaderT is
fundamentally different from how things were with functions.

And as probably expected by now, there's a runReader too, accompanied by
a non-transformer type synonym of ReaderT called Reader, hardcoding
Identity as the underlying m for those times when we don't care about the
MonadTransforming features of ReaderT.

```
type Reader e a = ReaderT e Identity a

runReader :: Reader e a -> e -> a
runReader = fmap runIdentity . runReaderT
```

This Reader is isomorphic to a function, as proved by runReader and its
inverse reader, a handy function for constructing a MonadReader out of a
pure function.

```
reader :: MonadReader e m => (e -> a) -> m a
reader = ReaderT . fmap pure
```

The type of reader is a bit more general than what an inverse of runReader
needs, but nonetheless, their composition proves that e -> a and Reader e

a are indeed isomorphic.

```
id == reader . runReader

id == runReader . reader
```

By the way, unrelated, when defining that Reader type synonym, there's no need to fully apply the ReaderT type constructor. The following Reader definition works just as fine.

```
type Reader e = ReaderT e Identity
```

In both cases, the kind of Reader is Type -> Type -> Type, whether we explicitly mention that last type parameter a or not. See, type Reader e a = ReaderT e Identity a is eta-expanded in the same way a similarly shaped function definition foo e a = baz e m a is. And we know that we can eta-reduce that function definition by removing the trailing as, keeping just foo e = bar e m. That's what we did in Reader. No magic.

Anyway, having ReaderT and friends, we can finally achieve our goal of having a function on top of a Parser. Kinda.

```
> runParser (runReaderT abc 'x') "x123x"
Right (5, "", 123)
```

Here, abc took the type ReaderT Char Parser Natural. We have a function-like thing, ReaderT, on top of a Parser.

Three hundred sixteen

Reader e b is just a function in disguise, but ReaderT e m b is no stranger either. Look at these two side by side.

```
data ReaderT e m b = ReaderT (e -> m b)

data Kleisli m e b = Kleisli (e -> m b)
```

That's right. ReaderT is just a Kleisli arrow with its type parameters in a different order. This makes their kinds different, which means one may fit in places where the other doesn't. Only ReaderT can have a MonadTrans instance, say, and only Kleisli can implement Category or Profunctor instances. Other than that, there's no fundamental difference between Kleisli and ReaderT.

Three hundred seventeen

Oh, yeah, Kleisli arrows are Profunctors too. We haven't talked about that.

```
class Profunctor p where
  dimap :: (a -> c) -> (b -> d) -> p c b -> p a d
```

Profunctor, remember, generalizes the idea of something being able to consume inputs and produce outputs. A Profunctor is a functor that's contravariant with its first type parameter and covariant with its second one. If p is a Profunctor, then p i o says that p can consume input of type i and produce output of type o.

This is not the most precise definition, but the intuition serves us well in Haskell so we take it. In Category Theory, the branch of mathematics exploring categories and friends, profunctors are a bit more abstract than what we see here in our Profunctor typeclass. And the same is true for Functors.

What we call a Functor in Haskell is not really the maximum expression of what a functor can be. In Category Theory, we would call Functor an *endofunctor*. The Greek word *endo*, as you might remember from its appearance in the name of the Endo monoid, meant something along the lines of *within*. What Category Theory says about endofunctors is that they are functors that take objects from one category to that same category. They stay *within* the category, they never leave. When viewed through our restrictive Haskelly eyes, this means that in a Functorial type such as Maybe x, both the x and the fully applied type Maybe x are objects in the same category *Hask*. We can see this very easily in Maybe's kind. Type -> Type says that when applied to a Type, that is, to an object in the *Hask* category, we obtain another Type, another object in the *Hask* category. Both x and Maybe x are within the same category. Thus, Maybe is an *endo*Functor.

All of this implies that in Category Theory there are *functor* things that are not *endo*functors. In this book, however, as well as in day-to-day Haskell, we don't care too much about the more categorical definitions. We are stuck in *Hask*, so mostly we limit ourselves to what we can do and say there. Mostly. But, for the record, the real functors take objects and morphisms in one category into objects and morphisms in a potentially different category, respecting the identity and composition laws we know and love from the Functor class. For example, there could be a functor taking objects and morphisms in *Hask* into texts and translations in that funny multilingual category from before. It would be a *functor*, as long as the functor laws are respected, but not an *endo*functor.

Anyway, functions are Profunctors. We knew that. And their Monadic counterpart, Kleisli arrows, are too.

```
instance Monad m => Profunctor (Kleisli m) where
    dimap :: (a -> c) -> (b -> d) -> Kleisli m c b -> Kleisli m a d
    dimap f h (Kleisli g) = Kleisli (fmap h . g . f)
```

What does Kleisli arrows being Profunctors mean? For our non-categorical Haskell purposes, it means they represent programs of some sort, consuming inputs and producing outputs. That's all. The difference between Kleisli arrows and functions is that only one of them gets to play in m while doing so. As for us, let's go back to playing in *Hask*.

Three hundred eighteen

What's wrong with head?

```
head :: [x] -> Maybe x
head (x : _) = Just x
head []      = Nothing
```

Nothing. head's great. However, it might surprise you to know that, for the longest time, in Haskell, head has been the name given to this other creature.

```
head :: [x] -> x
head (x : _) = x
head _       = undefined
```

A partial function. Crazy. Convenient, sure, but crazy. The fake head that returns a Maybe solves head's partiality by extending its codomain so that Nothing conveys the idea that a first element can't be obtained because the list is empty. However, this problem could be solved at the other end, too. Rather than extending head's codomain, we could *restrict its domain*, its input. That is, rather taking a list as input, it could take a *non-empty* list.

```
data NonEmptyList x = NonEmptyList x [x]
```

This NonEmptyList constructor carries as payload the first element, followed by the rest of the elements of the list. Clearly, in a value of type NonEmptyList x, there is always at least one x. NonEmptyList is a straightforward datatype, there's nothing fancy about it. Yet, it brings a lot of clarity to our programming life.

```
head :: NonEmptyList x -> x
head (NonEmptyList x _) = x
```

Both this new version and the Maybe version of head are *safe*, but the one embracing NonEmptyList is just better. Users of the Maybe version must analyze the resulting Maybe in order to determine if the scrutinized list *was* empty or not, potentially more than once if used in multiple places. Users of the head that takes a NonEmptyList, on the other hand, need not worry about anything. And, as usual, it's not so much *things* that are annoying, but the *relationship* between things.

```
multiplyHeads :: [x] -> [x] -> Maybe x
multiplyHeads as bs = liftA2 (*) (head as) (head bs)
```

In multiplyHeads we are assuming the Maybe version of head. As the name suggests, this function multiplies the heads of the given lists together. But look at how much information is lost. Was it the first list that was empty or was it the second one? Sure, this function could return something richer than Maybe to communicate which, but do we even care about that? These are questions better not posed. multiplyHeads should just do *its* thing correctly, and let somebody else worry about the emptiness of lists. Somebody like the head that takes NonEmptyLists as input.

```
multiplyHeads :: NonEmptyList x -> NonEmptyList x -> x
multiplyHeads as bs = head as * head bs
```

At some point before making use of the better head or the better multiplyHeads, somebody had to construct said NonEmptyLists. And sure, at that point there might have been a check regarding the contents of the list, possibly through a smart constructor like fromList here:

```
fromList :: [x] -> Maybe (NonEmptyList x)
fromList (x : xs) = Just (NonEmptyList x xs)
fromList _        = Nothing
```

However, whether through fromList or through a direct use of the NonEmptyList constructor, once we get our hands on a NonEmptyList, we won't ever have to check *again* whether the underlying list was empty or not. That term-level knowledge has been promoted to the type-level now. This is why we say the NonEmptyList version of head is better. It's not so much about head itself, but about the discipline it encourages and the meaning NonEmptyList conveys.

Three hundred nineteen

There is another kind of scenario where we would benefit from knowing that we are dealing with a non-empty list, but in a somewhat different way.

```
some :: Alternative f => f x -> f [x]

many :: Alternative f => f x -> f [x]
```

We met some and many a while ago, as part of our Alternative studies. some will collect the functorial outputs of *one or more* successful executions of the given f x, and many will collect *zero or more* of them. Or was it the other way around? Unclear. The functorial output [x] doesn't help us see which is which. Wouldn't it be better to use NonEmptyList instead?

```
some :: Alternative f => f x -> f (NonEmptyList x)

many :: Alternative f => f x -> f [x]
```

See? Much better now. Before, we improved head by restricting its domain, what it consumes. Here, we improve some by restricting its codomain, what it produces.

Having head take a NonEmptyList as input makes it safer for its users, for they never have to worry about whether they are applying it to an empty list. Attempting to do so just won't type-check. Having some produce NonEmptyList, on the other hand, is mostly about forcing the *authors* of some to produce a non-empty list as promised, a fact that gets passed on to the users of some, who can then benefit from having a NonEmptyList rather than a potentially empty list insofar as its type is concerned.

NonEmptyList comes out of the box with Haskell, but it's called NonEmpty instead, as if only lists could ever be non-empty. Go figure. It's constructor is called :| rather than whatever we called it.

```
data NonEmpty x = x :| [x]
```

But these naming differences are superficial. Everything we said about our NonEmptyList remains true. Additionally, NonEmpty comes with Monad, Applicative and Functor instances behaving just as their standard list counterparts, so we get to use fmap, do notation, and all those nice things.

```
> fmap (+1) (pure 8) :: NonEmpty Natural
9 :| []
> pure (+1) <*> pure 8 :: NonEmpty Natural
9 :| []
> do x <- pure 8
     pure (x + 1) :: NonEmpty Natural
9 :| []
```

Three hundred twenty

NonEmpty lists and normal lists have a lot in common, but being a Monoid is not one of them.

```
instance Monoid (NonEmpty x) where
  mappend :: NonEmpty x -> NonEmpty x -> NonEmpty x
  mappend (a :| as) (b :| bs) = a :| (as ++ (b : bs))
  mempty :: NonEmpty x
  mempty = ⁇
```

mappend is easy, but what about mempty? How would we ever come up with a NonEmpty x when we don't know what x is, yet we are supposed to put at least one of those inside our NonEmpty x? Impossible. A NonEmpty can't possibly be a Monoid. And that's sad, because Monoids are beautiful. They make it really easy to reason about problems and solutions, about composition. Sad, so sad. We'll miss this. But, not everything is lost.

It turns out that the Monoid typeclass we know and love was fake too. The real thing is split into two halves. First, there is the mappend half, called Semigroup.

```
class Semigroup a where
  (<>) :: a -> a -> a
```

A while ago we said that <> behaved the same as mappend. Remember? That is true. What we didn't say, though, is that <> lived in the Semigroup class. Semigroups are types with an associative binary operation. That's all. The Semigroup class comes with just one law, the *associativity* law, exactly as it was for mappend.

```
x <> (y <> z)  ==  (x <> y) <> z
```

The second half of our beloved old Monoid class is still called Monoid. But this time, rather than mentioning both mappend and mempty all by itself, it mentions only mempty, and then requires that there be a Semigroup instance for which mempty can act as identity element.

```
class Semigroup a => Monoid a where
  mempty :: a
```

All Monoids are Semigroups, as the superclass mandates, but not all Semigroups are Monoids. For example, a normal list [x] is both a Semigroup and a Monoid, we know this.

```
instance Semigroup [x] where
  (<>) = (++)

instance Semigroup [x] => Monoid [x] where
  mempty = []
```

But a NonEmpty list can only be a Semigroup, not a Monoid. We known this too.

```
instance Semigroup (NonEmpty x) where
  (a :| as) <> (b :| bs) = a :| (as ++ (b : bs))
```

Great. We can use <> to concatenate NonEmpty lists, even if they are not Monoids. And, because we know that <> must always be an *associative* operation, the result will always be the same no matter where we put our parentheses.

```
> (1 :| [2, 3]) <> (4 :| [5, 6])
1 :| [2, 3, 4, 5, 6]
> (1 :| [2]) <> ((3 :| [4]) <> (5 :| [6]))
1 :| [2, 3, 4, 5, 6]
> ((1 :| [2]) <> (3 :| [4])) <> (5 :| [6])
1 :| [2, 3, 4, 5, 6]
```

So, why this division between Semigroup and Monoid? And, perhaps more importantly, why didn't we make this division clear from the beginning? Mostly because talking about an associative binary operation without relating it to an identity element is not very interesting. We've seen how some of the most fascinating creatures we've encountered during our journey are Monoids, but only now we've come to encounter NonEmpty lists, our first Monoid-like thing that's not really a Monoid.

The split of the Monoid is a relatively recent event in Haskell history.

Previously there was only a `Monoid` class, just as we learned before, containing both `mempty` and `mappend`. So, related literature and names might be outdated at times. Sometimes by accident, and sometimes deliberately like here.

Three hundred twenty-one

What other things are a `Semigroup` but not a `Monoid`? Quite a few. For example, consider `First`.

```
data First a = First a
```

`First` merely keeps the first, that is, the leftmost `First` that is composed with `<>`, and discards the rest.

```
instance Semigroup (First a) where
  x <> _ = x
```

Playing with `First` in GHCi, we can see it does exactly what we expect. And notice how it doesn't matter where we put our parentheses because `<>`, by law, must be associative.

```
> First 3 <> First 1 <> First 2
First 3
> (First 3 <> First 1) <> First 2
First 3
> First 3 <> (First 1 <> First 2)
First 3
```

As a bonus, look how `First` is lazy on its second parameter.

```
> First 3 <> undefined
First 3
```

Interesting. Irrelevant, but interesting. Anyway, `First` can't possibly be a `Monoid` because `mempty` would have nothing to use as payload in the `First` constructor. Nevertheless, it can play the `Semigroup` role just fine.

Similarly, we have a `Semigroup` named `Last` doing exactly the opposite of `First`. That is, it keeps the rightmost value, the *last*, rather than the leftmost.

```
> Last 3 <> undefined <> Last 2
Last 2
```

Three hundred twenty-two

And we have `Min` and `Max` too, doing what their names suggest, as GHCi readily confirms.

```
> Min 3 <> Min 1 <> Min 2
1
> Max 3 <> Max 1 <> Max 2
3
```

The Semigroup instances for Min and Max are a bit more complex than First and Last. But only just a bit.

```
data Max a = Max a

instance Ord a => Semigroup (Max a) where
  Max x <> Max y = Max (max x y)
```

Notice the Ord constraint. It comes from the use of max, a function that returns the *maximum* of its two inputs. There are Ord instances for numeric types like Natural and Integer, of course. Otherwise, our example wouldn't have worked.

```
max :: Ord a => a -> a -> a
max x y = if x > y then x else y
```

In case you missed elementary school, x > y means x *is more than y*. So, if x is more than y, then max x y returns x, otherwise y. Similarly, we have min x y keeping the minimum of x and y.

```
class Eq a => Ord a where
  (<=) :: a -> a -> Bool
```

The Ord typeclass is a bit more complicated than we see here, but having a definition for <=, which stands for *less than or equal to*, suffices. The Haskell base library defines the other comparison operators in terms of <= by default, anyway.

```
(<), (>=), (>) :: a -> a -> Bool
a <  b = not (b <= a)
a >= b =      b <= a
a >  b = not (a <= b)
```

Three hundred twenty-three

An additional tool we get with Semigroup is sconcat.

```
sconcat :: Semigroup x => NonEmpty x -> x
sconcat (x :| (y : ys)) = x <> sconcat (y :| ys)
sconcat (x :| []       ) = x
```

sconcat is like mconcat, which we already met before, but for Semigroups instead of Monoids. Naturally, sconcat takes a NonEmpty x list as input in order to convey the fact that at least one such x must be provided. mconcat doesn't impose that restriction, but it requires that x be a Monoid instead.

```
mconcat :: Monoid x => [x] -> x
mconcat (x : xs) = x <> mconcat xs
mconcat []       = mempty
```

Great. But we've learned a enough about `Semigroups` for now, let's learn about financial markets instead. In financial markets, where one can buy or sell financial assets such as currency or company shares, the prices of those assets are often visualized through *candlesticks*, which, for a given window of time, track the prices at the beginning and end of that window, as well as the minimum and maximum price during that time. Well, surprise, candlesticks are `Semigroups` too.

```
data Candlestick x = Candlestick (First x) (Max x) (Min x) (Last x)
```

These extreme price points are called the *open, high, low* and *close* of a candlestick. Their types have other names here, but their meaning is the same. Now, a `Candlestick` cannot be a `Monoid`, for we wouldn't have an initial price to provide to any of `First`, `Max`, etc. But as a product type fully made out of `Semigroups`, it can surely be a `Semigroup`.

```
instance Ord x => Semigroup (Candlestick x) where
  Candlestick a b c d <> Candlestick a' b' c' d' =
    Candlestick (a <> a') (b <> b') (c <> c') (d <> d')
```

If we did have an initial price, though, we could definitely make a `Candlestick` out of it, where said price was all of the `First`, `Max`, `Min` and `Last` price, all happening at the same point in time. Sure, why not.

```
candlestick :: x -> Candlestick x
candlestick x = Candlestick (First x) (Max x) (Min x) (Last x)
```

Now, imagine we have a `NonEmpty` list containing all the prices at which a particular financial asset was traded during a single day, ordered from oldest to newest, and we wanted to make a single `Candlestick` out of them, a summary of the price extremes for that day.

```
prices :: NonEmpty Natural
prices = 50 :| [52, 49, 51, 53, 55, 57, 52]
```

We have all the right tools, so let's just go ahead and do it. Let's make our big `Candlestick` with the help of sconcat.

```
> sconcat (fmap candlestick prices)
Candlestick (First 50) (Max 57) (Min 49) (Last 52)
```

Isn't it nice? We've said it time and time again, *correctness composes*. All of our `Semigroups` and their related vocabulary are correct, so a `Candlestick`, which is significantly larger than its constituent pieces but still made out of them, is correct too. Imagine how terrible, how sad programming must be without composable ideas.

Three hundred twenty-four

One last thing about our `Candlestick`.

```
data Candlestick x = Candlestick (First x) (Max x) (Min x) (Last x)
```

When we sconcat multiple Candlesticks together, when folding into one, we are treating the ever growing Candlestick as an accumulator. And, we know, it's pretty much always a good idea to make accumulators strict, so that we don't create an unnecessary amount of unevaluated thunks as we go about our accumulation. We know how to do this, we have different tools such as seq or using ! when binding a formal function parameter to a name. But another alternative, perhaps a bit easier to work with, is to make the *fields* of the datatype themselves strict.

```
data Candlestick x = Candlestick
                   !(First x) !(Max x) !(Min x) !(Last x)
```

By prepending a field type with a bang ! at its definition site, we are saying that when the Candlestick is evaluated to weak head normal form, then the field will be evaluated to weak head normal form too. Compare the following two datatypes.

```
data Foo = Foo Bool
```

```
data Bar = Bar !Bool
```

If we use seq to force the evaluation of Foo undefined to weak head normal form, the undefined goes unnoticed, for it's never evaluated.

```
> seq (Foo undefined) ()
()
```

On the other hand, if we use Bar, we see that Haskell tries to evaluate undefined, leading to an exception.

```
> seq (Bar undefined) ()
*** Exception: Prelude.undefined
```

None of this is magic, though. It's just some syntactic sugar to make dealing with laziness in Haskell easier. Making a field strict is comparable to using seq on the value of the field *before* applying the datatype constructor to it. That is, if we wanted, we could manually achieve the same strictness semantics of Bar on Foo, just without the nice syntax.

```
> let x = undefined
  in seq (seq x (Foo x)) ()
*** Exception: Prelude.undefined
```

With this small change to Candlestick, we can be sure that no unnecessary thunks will be accumulated as we go about populating its fields, either through <>, sconcat or through any other means.

It's generally a good idea to make the fields of our datatypes strict, except perhaps in situations where we deal with recursive datatypes. So, bang ! more of them. There is a way to tell GHC to make all datatype fields strict by default, which is generally a good idea, but we won't talk about that here.

Three hundred twenty-five

We still have thousands of programming things we'd like to talk about. Programming is a never ending topic, you see. Unfortunately, we are getting close to the end of our travels, so we'll have to make some sacrifices. But let's take care of some of the most important Haskelly things at least, like sequence, which we met many chapters ago.

sequence traversed a list of Maybes and, if all the elements in it were Just, then returned Just a list of the payloads, in the same order.

```
sequence :: [Maybe a] -> Maybe [a]
sequence (Just a : ays) = fmap (a :) (sequence ays)
sequence (Nothing : _)  = Nothing
sequence []             = Just []
```

And then we learned that sequence wasn't really about Maybe, but about Applicative functors. That is, sequence traverses a list of functorial values, Applicatively composing them as it goes, collecting their respective functorial outputs in a list that preserves the original order.

```
sequence :: Applicative f => [f a] -> f [a]
sequence = foldr (liftA2 (:)) (pure [])
```

In the Maybe case, composing the Applicative functorial values meant that the output would be Just only if all the list elements were Just. But other Applicative functors compose differently, and sequence supports them too. For example, in the past we used sequence to exploit the Applicative fan-out nature of functions, and it was fine.

```
> sequence [(+ 1), (* 10), \x -> x * x] 3
[4, 30, 9]
```

But in truth, sequence is even more general than this. sequence is not really about lists, but about Traversable structures.

```
sequence :: (Applicative f, Traversable t)
         => t (f a) -> f (t a)
```

A Traversable structure is one where each element in it can be transformed through an Applicative functor while the overall the shape of the structure is preserved. Perhaps this transformation is not very obvious from looking at the type of sequence, but if we consider its related function, traverse, it becomes more apparent.

```
traverse :: (Applicative f, Traversable t)
         => (a -> f b) -> t a -> f (t b)
traverse g = sequence . fmap g
```

traverse, easily defined in terms of sequence and fmap, makes it a bit more clear that each a inside t is being transformed into a b by means of a -> f b, and all of those individual functorial values f b are being Applicatively composed into one big functorial value f (t b), producing the same structure t, but containing the new bs rather than the original as.

Alternatively, we can look at traverse as a different version of fmap that allows us to perform functorial effects while transforming each element. And like fmap, it preserves the shape of the structure being transformed. This is easy to see if we compare the types of fmap and traverse side by side, leaving out their respective constraints.

```
fmap     :: (a ->   b) -> t a ->    t b
traverse :: (a -> f b) -> t a -> f (t b)
```

Very similar. In fact, if we are smart about our choice of f, we can implement fmap by relying on traverse only.

```
fmapDefault :: Traversable t => (a -> b) -> t a -> t b
fmapDefault g = runIdentity . traverse (Identity . g)
```

Great. That definitely looks like fmap. And we known that Identity is the Functor that doesn't add anything interesting, so we can just ignore its use here. We see an unexpected Traversable constraint on fmapDefault, however. What is that? As the name suggests, traverse lives inside the Traversable typeclass, so whenever we use traverse in our code, the Traversable constraint will need to be satisfied.

```
class (Functor t, Foldable t) => Traversable t where
    traverse :: Applicative f => (a -> f b) -> t a -> f (t b)
```

The superclass constraints mandate that t be a Functor and a Foldable. The Functor part is easily justified by the experiment we just did, where we showed how fmapDefault implemented a correct fmap in terms of traverse. Knowing that every Traversable can be seen as a Functor through fmapDefault, Traversable just goes ahead and mentions Functor as a superclass so that fmap can be used directly on Traversable structures, instead of going through fmapDefault.

The justification for the Foldable superclass is similar, but we'll leave it for later. For now, the intuition that Traversable structures are *containers* of some sort should suffice. Remember, Foldable types are containers, as readily witnessed by Foldable s toList method which says that we are always able to extract the elements of the container into a list. But not all Functors contain elements. Some, like functions, merely *produce* elements, and those just can't be Traversable. The Foldable superclass here makes this requirement clear.

The real Traversable class is a bit more contrived than this. Both traverse and sequence live in the Traversable class, defined in terms of each other, with a MINIMAL pragma requiring us to define at least one of them by hand so that using them doesn't lead to an infinite loop. sequence is just traverse id, in case you hadn't figured that out yet. Additionally, for historical reasons, said sequence works with Monads rather than the simpler

Applicatives. The Applicative version exists in the Traversable class, yes, but it is called sequenceA. However, seeing as all of this is just noise, we'll pretend that Traversable is as cute as we say it is and move on.

Another interesting curiosity about the relationship between fmap and traverse is how they compare when they both are made to take a same function a -> f b as input:

```
fmap     :: (a -> f b) -> t a -> t (f b)
traverse :: (a -> f b) -> t a -> f (t b)
```

fmap doesn't make use of the fact that f is an Applicative functor, it just leaves the f b functorial values in there, contained in t. traverse, on the other hand, Applicatively composes all those functorial effects into single one producing the same t structure with the plain b elements. Fascinating. Anyway, let's make lists Traversable.

```
data Traversable [] where
  traverse :: Applicative f => (a -> f b) -> [a] -> f [b]
  traverse g = foldr (liftA2 (:) . g) (pure [])
```

Nothing of essence is new here. We defined sequence for lists before, and here we define the very similar traverse instead. Notice how we preserve the structure of the original list as Traversable requires. That is, if and when a list is finally produced, then its length and the position of its elements are as they were before. The given function g, in turn, knows nothing about the list. Only about a, b and some chosen Aplicative functor f.

```
> let g :: Natural -> Maybe Natural
      g = \x -> if odd x then Just (x + 1) else Nothing
> traverse g [1, 2]
Nothing
> traverse g [1, 3]
Just [2, 4]
```

Depending on the chosen Applicative functor, it can be a bit tricky to see the structure being preserved. For example, if we pick [] as both the Traversable structure and the Applicative functor, it gets weird.

```
> let g :: Natural -> [Natural]
      g = \x -> if odd x then enumFromTo 0 x else []
> traverse g [1, 2]
[]
> traverse g [1, 3]
[[0,0], [0,1], [0,2], [0,3], [1,0], [1,1], [1,2], [1,3]]
```

However, that alleged weirdness is just the normal behavior of the Applicative composition of lists, resulting in their cartesian product. If we ignore the outer list, which corresponds to the Applicative functor used by traverse, and instead pay close attention to the inner lists, which correspond to the Traversable structure, we'll see that the two-element list shape has indeed been preserved.

Three hundred twenty-six

What other Traversable structures are there? Maybe perhaps? We know Maybe is Functor and Foldable already, so it's likely Traversable too.

```
data Traversable Maybe where
   traverse g = maybe (pure Nothing) (fmap Just . g)
```

Great. And what about tuples?

```
data Traversable ((,) w) where
   traverse g (w, a) = fmap (\b -> (w, b)) (g a)
```

Excellent. So, what does this all mean? What have we achieved? What can we do with traverse and sequence? Well, exactly what we've been saying. Without really knowing which structure we are working with, we can functorially transform the individual elements of a Traversable structure while preserving its shape. That's all. traverse takes care of iterating over the Traversable structure, functorially transforming its elements as it goes, much like fmap does it nonfunctorially.

```
> siblings = ["Bart", "Lisa", "Maggie"]
> :t traverse askAge siblings
traverse askAge siblings :: IO [Natural]
> ages <- traverse askAge siblings
How old is Bart?
10
How old is Lisa?
8
How old is Maggie?
1
> ages
[10, 8, 1]
```

In the example above we are transforming the siblings list, a Traversable structure of type [String], into a list of type [Natural] by means of traverse and askAge, a made up function of type String -> IO Natural which does what its name suggests.

```
askAge :: String -> IO Natural
```

It doesn't really matter how askAge is implemented, what matters is that askAge is completely unaware of the Traversable structure. askAge only cares about transforming a String value into a Natural value while preforming some effect in its chosen Applicative functor, IO. We can traverse the same askAge over other structures as well, and it would work just as fine.

```
> traverse askAge Nothing
Nothing
> traverse askAge (Just "Marge")
How old is Marge?
34
Just 34
> traverse askAge ("Patty", "Selma")
How old is Selma?
35
("Patty", 35)
```

These people don't age well, so the data may be wrong, but you get the idea. By the way, unrelated, notice how we used <- in the siblings example without explicitly mentioning do. This is fine. GHCi allows us to do this. Remember that GHCi, the Grand Haskell Computer Interactivator, is intended to be used as a REPL, and as such it has some handy REPL-only features like allowing us to skip writing do at times, features that are not necessarily available when writing Haskell code outside of GHCi.

traverse gets much more interesting when operating on elements contained in nested structures. By composing traverse with itself, we can target a deeper Traversable structure.

```
> runReader ( (traverse . traverse . traverse)
                (\x -> fmap (x *) ask)
                (False, Right [2, 3])
            ) 2
(False, Right [4, 6])
```

Here, for example, we have Naturals inside of a list, inside of an Either, inside of a tuple. And by composing traverse the right number of times, we are able to functorially double the deepest Naturals through our chosen Applicative functor, Reader. The whole structure remains the same, and our transforming function \x -> ... remains completely oblivious about the structure we are dealing with.

traverse seems simple and redundant, but you'll be amazed at how often traverse is exactly what we are looking for, no matter which problem we are trying to solve.

Three hundred twenty-seven

Let's meet another Functor. One of the funny ones this time.

```
data Const a b = Const a
```

Const a b carries an a as its payload and completely ignores the b that its type mentions, its functorial output. Very funny.

```
instance Functor (Const a) where
    fmap _ (Const a) = Const a
```

The type `Const a b` should remind us of the expression `const a b`. Both of them preserving the `a` and ignoring the `b`. Strange `Functor`, isn't it? And in an even stranger way, it's an `Applicative` too.

```
instance Monoid m => Applicative (Const a) where
  pure _ = Const mempty
  Const x <*> Const y = Const (x <> y)
```

The `Applicative` instance does nothing with the `b` in `Const a b`, for there isn't one. It just focuses on the `a` by treating it as a `Monoid`, with `pure` corresponding to `mempty` and `<*>` behaving as `<>`.

```
> pure 8 :: Const String Natural
Const ""
> Const "has" <*> Const "kell" :: Const String Natural
Const "haskell"
```

Is it a good idea to just add `Monoid` constraint on `a` so that we have a way of implementing `pure` and `<*>`? Couldn't we have chosen something other than `Monoid`? Maybe. But long as we are not violating the `Applicative` laws, everything is fine. And, seeing as we are not going to find anything smaller than a `Monoid` to satisfy the needs of `pure` and `<*>`, we can say this solution is perfect.

What is all of this `Const` nonsense good for, anyway? We'll see, but first let's take a look at something completely unrelated. Remember how we said before that we could always use `liftA` as a free implementation of `fmap` relying only on `Applicative` vocabulary? Well, it's not so free this time, is it?

```
instance Monoid a => Functor (Const a) where
  fmap = liftA
```

It is true that we can use `liftA` as the implementation of `fmap`. However, since the `Applicative` instance for `Const a` requires `a` to be a `Monoid`, this `Functor` instance relying on it will require so too. But we don't like having unnecessary constraints anywhere, so we keep the manually written version of `fmap` instead, as the standard Haskell version of this `Functor` instance does.

Anyway, back to doing something useful with `Const`. Do you remember how `fmapDefault` was an implementation of `fmap` relying solely on `Traversable` and `Identity`?

```
fmapDefault :: Traversable t => (a -> b) -> t a -> t b
fmapDefault g = runIdentity . traverse (Identity . g)
```

Nice little function. Now, see what happens if we use `Const` instead of `Identity`.

```
foldMapDefault
   :: (Traversable t, Monoid b) => (a -> b) -> t a -> b
foldMapDefault g = runConst . traverse (Const . g)
```

That's right, using Traversable and Const together in this way gives any Traversable a suitable implementation of foldMap, of Foldable fame. How does it work? Each a is converted into a b that gets wrapped in a Const b whatever by traverse, which Applicatively composes these Consts, causing the bs to be mappended because that's what Applicatively composing Consts does. Then, runConst removes the Const wrapper to reveal the b, et voilà.

```
runConst :: Const a b -> a
runConst (Const a) = a
```

Actually, runConst is not called runConst in Haskell's standard library, but getConst. Naming can be a bit inconsistent at times. Is getConst a better name than runConst? Who knows, who cares, it's not a big deal within these pages.

So, foldMapDefault explains why Traversable has Foldable as its superclass. Strictly speaking, Traversable doesn't *need* the Foldable superclass just as it doesn't need the Functor superclass, but seeing as Foldable instances can be defined for any Traversable by means of foldMapDefault, having Traversable imply Foldable is a nice addition. That way, functions that make use of both Foldable and Traversable features can just mention Traversable in its constraints, instead of mentioning both of them. And the same is true for the Functor superclass, but we already knew that.

Three hundred twenty-eight

Const is a funny Functor and a funny Applicative. But funnily, it ain't no Monad, no. Our attempt at implementing an instance for it satisfies the type-checker, but that's not enough.

```
instance Monoid a => Monad (Const a) where
   (>>=) :: Const a b -> (b -> Const a c) -> Const a c
   Const a >>= f = Const a
```

First, unrelated, notice how we see the Monoid constraint on a once again. Even though >>= doesn't make use of any Monoidal features, we know that for Const a to be a Monad, it must be an Applicative too. And we know that for Const a to be an Applicative, the chosen a must be a Monoid. This is why we see the Monoid constraint on a here. We can't avoid it, and we shouldn't want to avoid it anyway.

Second, we can see that >>= is not using f. This, on its own, is not necessarily a problem. After all, the implementation of >>= for Maybe, say, did something similar.

```
(>>=) :: Maybe a -> (a -> Maybe b) -> Maybe b
Just a  >>= f = f a
Nothing >>= f = Nothing
```

Is it problematic that we are not using f? Is it not? Let's see what the Monad laws have to say about it. If our implementation of >>= abides by them, we can confidently claim that Const a is a Monad. Otherwise, *afuera!*

First, we have an *identity law* stating that pure a >>= f must equal f a. But we know that our implementation of >>= for Const doesn't use of that f at all, so how could they possibly be equal? This alone already proves that Const can't ever be a Monad. *Afuera!* In the Maybe case, the fact that f is not being used in the Nothing scenario is irrelevant to the identity law because pure a >>= f translates to Just a >>= f, and the Just branch in >>= does indeed use f. The law is just not interested in the Nothing branch. Anyway, no Monad for Const. Look at this, though:

```
instance Monad (Const ()) where
  Const () >>= _ = Const ()
```

While Const a can't be a Monad *for all choices of* a, it can be one for *some* choices of a. Namely, types whose cardinality is one, like (). Why? Well, because even without using f, we know that it will return the value Const (), seeing as that's all its type allows. So, the Monad identity law is respected in this case. However, we don't pay too much attention to this scenario because it is super boring. Generally speaking, we just say Const is not a Monad.

This is the first time we meet an Applicative functor that's not a Monad. Hopefully we can better appreciate now why it makes sense for these related but different concepts to exists in different typeclasses. In foldMapDefault we proved the value of the Applicative features of Const without even considering the possibility of its being a Monad. If Applicative and Monad weren't separate things, as they weren't before man discovered Applicatives, maybe foldMapDefault wouldn't even exist.

Three hundred twenty-nine

It might not seem like it, but even after all we've been through, we've barely scratched the surface of what the Haskell type-system is capable of. We are running out of book, though, so that type of programming will have to wait. Still, let's take a quick peek just to know what we are missing.

```
data List (a :: Type) = Nil | Cons a (List a)
```

Here is our well-known List datatype once again. We know that in Haskell, List, Nil and Cons go by the names [] and :, but let's ignore that for now. We are explicitly mentioning that a is of kind Type just for the sake of being explicit, but we could skip mentioning it and Haskell would infer it as necessary. This is how we've usually defined the List datatype in Haskell. However, it's not the only syntax available for defining a datatype like this one. We could write this other thing instead, and achieve exactly the same:

```
data List (a :: Type) where
   Nil :: List a
   Cons :: a -> List a -> List a
```

Nothing has changed in terms of what List means or how it will be used. It's just that this syntax makes the types of the Nil and Cons constructors a bit more explicit. It says that Nil is a constructor for List a, for all choices of a, carrying no payload. And it also says that Cons is a constructor for List a that carries two payloads a and List a. In all of these cases, a is expected to be a type of kind Type, as constrained on the first line. The syntax for these constructors looks exactly like their types. We can ask GHCi to infer the types of Nil and Cons, and we'll see exactly what we've written down.

```
> :t Nil
Nil :: List a
> :t Cons
Cons :: a -> List a -> List a
```

So, why do we have two ways of defining the same datatype? The truth is that we don't. Not really. Well, maybe just a bit. OK, fine, yes, both the old and the new syntax can do this, but the fancy one can also do *so much more*. What matters is not how these things are similar, but how they are different, so let's focus on that. Perhaps you are right, though, and maybe shouldn't be bothering with the lesser syntax at all. Oh well, maybe in a different lifetime.

We call datatypes like List, Bool, Maybe and everything else we've defined so far with the typical data ... = ... syntax, *algebraic data types*, or *ADTs*. We know this. We learned about it a long time ago. And while this new data ... where ... syntax allows us to readily define ADTs as we just did, it also allows us to define *GADTs*, *generalized algebraic data types*.

What's *general* about these *generalized* ADTs is that the constructors can decide, to a good extent, which type they want to construct. In our recent List example, for a chosen a, both Nil and Cons constructed value of type List a. But look at this new List:

```
data List (n :: Natural) (a :: Type) where
   Nil  :: List 0 a
   Cons :: a -> List n a -> List (n + 1) a
```

There are quite a few new things here. Let's explore them them line by line.

```
data List (n :: Natural) (a :: Type) where
```

First line. By saying `data ... where ...`, Haskell knows that we are going to be using the GADT syntax to define whatever it is that we are going to define. In this case, it will be a datatype named `List`. Or, more precisely, `List` is the type constructor of kind `Natural -> Type -> Type` that we are defining. We will be using the type `List n a` to represent lists containing n elements of type a.

Seeing as n will be used to count things, it makes sense for it to be some *kind* of number. `Natural` is a decent choice. It's the first time we encounter the `Natural` *kind*, but don't worry, it's more or less what I'm sure you expect it to be. We'll see more about `Natural` later on. Then, after n, we have the type parameter a describing the type of each element contained in the list. The kind of a is `Type` because that's the kind of types for values that can exist at the term-level, and we do want the elements of our list to exist at the term-level, so `Type` it is. At least for now. In the next line, things get more interesting.

```
Nil :: List 0 a
```

`Nil` is a nullary constructor that constructs a value of type `List 0 a`, conveying the idea of an empty list of as. That is, a list with 0 elements in it. And by the way, in case it wasn't obvious from its name, a *nullary* constructor is one that takes no payloads.

`Nil` *chooses* to return this type, `List 0 a`, for all choices of a. It could have chosen something else like `List 32 Bool` or even the fully polymorphic `List n a`, but it didn't. GADT constructors can choose to return any type they want as long as said type is a fully applied version of the type constructor being introduced as a GADT. That's what makes GADTs different from ADTs. In our example, the `List` constructors are expected to return a fully applied `List`, of kind `Natural -> Type -> Type`. `Nil`'s choice, `List 0 a`, for all choices of a, fits. Now, see what happens if we try to lie to Haskell, saying `Nil` is a list of a different length.

```
> Nil :: List 3 a
<interactive>:7:1: error:
  • Couldn't match type '0' with '3'
    Expected type: List 3 a
      Actual type: List 0 a
```

Lies don't type-check. Isn't that nice? Let's take a look at the other constructor now, `Cons`. That's where things get really interesting.

```
Cons :: a -> List n a -> List (n + 1) a
```

`Cons` allows us to prepend an element of type a to a list of type `List n a`, resulting in a `List` one element longer than the original. What's interesting is that this new length is mentioned in the type `Cons` chooses to return, `List (n + 1) a`, where n is the length of the list given to `Cons` as second argument.

```
> :t Cons 'a' Nil
Cons 'a' Nil :: List 1 Char
```

If we apply Cons to a List 0 something, then we get a List 1 something in return. If we apply it to a List 4 whatever, we get a List 5 whatever in return, and so on.

```
> :t Cons 'b' (Cons 'a' Nil)
Cons 'b' (Cons 'a' Nil) :: List 2 Char
```

In other words, contrary to Nil, the type Cons chooses to return is directly tied to the type of one of its payloads. And, again, lying doesn't type-check.

```
> Cons 'a' Nil :: List 9 Char
<interactive>:1:1: error:
    • Couldn't match type '1' with '9'
      Expected: List 9 Char
        Actual: List (0 + 1) Char
```

But what is the Natural kind, exactly? The Natural kind is to the type 3 as the Natural type is to the value 3. And likewise for any other natural number. We've seen similar relationships between terms, types and kinds in the past. For example, we learned that the type called Bool, with value constructors named True and False, could be promoted to a kind also named Bool with type constructors also named 'True and 'False. Well, Natural is not so different, except for the syntactic convenience of not needing to prepend that tick ' to promote the numbers. Haskell makes the syntax for type-level numbers special, just as it makes it for term-level numbers. But that's about as special as Natural gets.

Three hundred thirty

Let's use these brand new GADT Lists for something. For example, we recently learned how the NonEmpty list type improved head by keeping it from ever having to deal with empty lists. Let's try something similar, but relying our new List instead.

```
head :: (1 <= n) => List n a -> a
head (Cons a _) = a
```

Behold, our newest safe head. Notice the constraint mandating that n be at least 1. This constraint makes it impossible for head to be used on empty lists, on Nil.

```
> head Nil
<interactive>:1:1: error:
    • Cannot satisfy: 1 <= 0
    • In the expression: head Nil
```

head `Nil` doesn't compile, and we get a descriptive type-checking error instead. Notice how we didn't write an equation for `Nil` when implementing head. Had we been dealing with a normal list, Haskell would have warned us that we didn't deal with `Nil`. This time, however, Haskell knows that `Nil` is an impossible situation that doesn't need to be dealt with. Haskell is satisfied, Haskell doesn't whine about it. In fact, see what happens if we attempt to add an equation for `Nil`.

```
head :: (1 <= n) => List n a -> a
head (Cons a _) = a
head Nil        = undefined
```

undefined is there so as to please syntax rules. We know, however, that this code will never be executed, so there's nothing to worry about. But more importantly, the type-checker knows. We can hear Haskell's lament about it when we compile this.

```
<interactive>:3:1: warning: [-Woverlapping-patterns]
    Pattern match is redundant
    In an equation for 'head': head Nil = ...
```

Warning, pattern match is redundant in `Nil`. How much more specific can this get? Not only is the type of head keeping users from accidentally applying it to an empty list, it's also keeping us authors from even worrying about it. Beautiful. Here is another example.

```
zip :: [a] -> [b] -> [(a, b)]
```

We met zip a long time ago. zip x y pairs the first element in x with the first element in y, the second element in x with the second element in y, the third with the third, and so on. But there is a funny situation with zip. If one list is shorter than the other, then the extra elements in the longer list are discarded. This is arguably a reasonable behavior. However, what if it was by accident that we applied zip to lists of different lengths? The truncation will go silently unnoticed. How can we avoid this? Successfully executed wrong programs are the worst kind of programs.

```
zip :: List n a -> List n b -> List n (a, b)
```

Problem solved. The type of this new zip requires all the lists involved to have the same length. We don't need to go into the boring details of how to implement this to be able to appreciate the idea. The more type-level information we have about term-level facts, the better. How much is enough? We'll never know.

Three hundred thirty-one

Working with `Natural` numbers comes so naturally to us that we didn't stop to question why some of our length-aware `List`s expressions type-checked at all.

```
> :t Cons 'a' Nil
Cons 'a' Nil :: List 1 Char
```

Nothing seems out of place there. But look again, very carefully, at the type of Cons.

```
Cons :: a -> List n a -> List (n + 1) a
```

It says there that Cons constructs a value of type List (n + 1) a. In our example, we are applying Cons to a value of type Char and another one of type List 0 Char. So, shouldn't this expression have type List (0 + 1) Char rather than List 1 Char? That is, 1 and 0 + 1 look like different types.

It turns out that 0 + 1 is not a type. Or rather, it is, but it's not the infix application of some type constructor named + to the Natural-kinded types 0 and 1 as we would expect. Instead, + is what we call a *type family*, which is mostly a funny way of saying *type-level function*. Yes, that's right, type-level functions. In Haskell we can have functions at the type level. Functions transforming types, rather than values. As an example, consider the normal term-level function not.

```
not :: Bool -> Bool
not True  = False
not False = True
```

Very straightforward. not operates on *values* of type Bool. Nothing new for us here. Now, let's implement its analogous *type family*, let's call it Not, operating on *types* of kind Bool.

```
type family Not (x :: Bool) :: Bool where
  Not 'True  = 'False
  Not 'False = 'True
```

The Not type family achieves at the type level the same thing that the normal not function achieves at the term level. On the first line we are saying that Not is a type family that, when applied to a *type* of kind Bool, returns yet another type of kind Bool. We gave the name x to the input Bool in order to satisfy Haskell's syntax requirements, but naming is irrelevant in this example. On the next lines, we define equations for Not just as we did in the not function. GHCi's :kind! confirms that everything behaves as expected.

```
> :kind! Not
Not :: Bool -> Bool
= Not
> :kind! Not 'True
Not 'True :: Bool
= 'False
> :kind! Not 'False
Not 'False :: Bool
= 'True
```

The way the + type family is defined is a bit different because of Haskell's special literal syntax for writing numbers, but the fundamental idea is the same. The + type family of kind `Natural -> Natural -> Natural` simply transforms two given `Natural`s into a third one.

There's no requirement that the inputs and outputs of a type family be of the same kind as we see in + and `Not`, that was just a coincidence. For example, we could have a type family that gave us the first element of a type-level list.

```
type family Head (xs :: [k]) :: k where
  Head (a : _) = a
```

Wait, what? Type-level lists? Yes, Haskell can do that.

```
> :kind! Head [1, 2, 3]
Head [1, 2] :: Natural
= 1
> :kind! Head ['a', 'b', 'c']
Head ['a', 'b', 'c'] :: Char
= 'a'
> :kind! Head [Integer, String, [Natural]]
Head [Integer, String, [Natural]] :: Type
= Integer
```

Why would we want type-level lists? Well, why not. We have `Type`s at the type level, we have type-level `Natural`s, `Bool`s and `Char`s, we have type families, we have so many type-level things now that putting some of them together in a type-level list is starting to look like the most obvious next thing to do.

```
> "this is a normal string" :: Head [String, Natural]
this is a normal string
```

Notice that in our implementation of `Head` we didn't explicitly handle the empty list case. This is fine, because trying to use `Head` on an empty type-level list —written `'[]`, with that mandatory leading tick— just won't type-check anyway.

```
> "this does not work" :: Head '[]
<interactive>:1:1: error:
  • Couldn't match expected type 'Head '[]'
    with actual type 'String'
```

However, wouldn't it be nice if the error message from the type-checker was a bit nicer? Well, what do you know, in Haskell we can create our own type-checking error messages too.

```
type family Head (xs :: [k]) :: k where
  Head (a : _) = a
  Head _ = TypeError (Text "Headless list!")
```

`TypeError` is itself a type family of kind `∀ x. ErrorMessage -> x`. So here we have an example of a type family, `Head`, making use of another type family,

TypeError. What is x? Whatever we want it to be. That's what ∀, *for all*, is saying. TypeError is a bit like a type-level version of undefined, which also promises to be a value of *any* type. Except it's a bluff. We know that a program will diverge as soon as it tries to evaluate an undefined value. TypeError is similar. We can make use of TypeError anywhere a type is expected, but it will be a bluff too. As soon as the type-checker attempts to type-check anything tainted with a TypeError, it fails with our desired message.

```
> "whatever" :: Head '[]
<interactive>:66:1: error:
    • Headless list!
```

It's not necessary for TypeError to be used inside another type family like Head, it can stand alone anywhere a type is expected.

```
> 123 :: TypeError (Text "Superproblem.")
<interactive>:68:30: error:
    • Superproblem.
```

Isn't that nice? Never mind Text, it's just the type constructor we use to transform a type-level String into the ErrorMessage expected by TypeError.

Type families are among the most sophisticated tools available in Haskell for making sure that our programs are safe. They allow us to push very far the idea that only correct programs should type-check. Here we've seen just a bit of what can be accomplished through them.

Three hundred thirty-two

The length, it is known. List n a says that a List contains exactly n elements of type a. This knowledge exists at the type level. But what if we need to use n as a term-level value, too? For example, let's say we want to show the length of the list somewhere. We'll surely benefit from having a function of type List n a -> Natural for cases like that one. So, knowing that our new List is structurally the same as Haskell's standard list [], we proceed to implement a new length just like the one for [].

```
length :: List n a -> Natural
length Nil          = 0
length (Cons _ rest) = 1 + length rest
```

Indeed, this works. However, this recursion is a waste of time and space. By the time GHC compiles an application of length to a List n a, the type system knows exactly what n is. So why not just reuse n at the term level, instead of recursing into a possibly very long list just to reach to the same conclusion?

```
length :: ∀ n a. KnownNatural n => List n a -> Natural
length _ = naturalVal (Proxy :: Proxy n)
```

There are many new things going on here, let's take them step by step. First, naturalVal.

```
naturalVal :: KnownNatural n => Proxy n -> Natural
```

naturalVal, given a Proxy n, where n is a type of kind Natural, returns a term-level representation of said n as a term-level value of type Natural. The fact that n must be of kind Natural is not mentioned explicitly here, but if we ask GHCi about the kind of the KnownNatural constraining n, we'll see that it is the case.

```
> :kind! KnownNatural
KnownNatural :: Natural -> Constraint
= KnownNatural
```

In principle, we expect all types of kind Natural to be representable as values of type Natural through naturalVal. In practice, the Natural number in question needs to satisfy the KnownNatural constraint. This constraint is automatically satisfied by every Natural-kinded number written literally such as 2 or 81. It gets a bit trickier when we refer to the number through a variable name or through a type family, rather than literally, but we won't worry about that in these remaining pages.

By the way, did you notice the resulting Constraint in the kind of KnownNatural? It's the first time we see this explicitly. As the name suggests, Constraint is the kind of *constraints*, the things that often appear to the left of the fat arrow => in places such as function types or superclasses, *constraining* the types that can substitute the type variables involved. For example, Eq Bool is of kind Constraint. And Ord Integer, and Show String, and so on. Constraint is the kind of fully applied typeclasses, which suggests that KnownNatural could be a typeclass too.

```
class KnownNatural (n :: Natural) where
    naturalVal :: Proxy n -> Natural
```

The actual KnownNatural typeclass is a bit more complex than this, a price paid for having access to some extra features not discussed here. However, for our purposes, this is sufficient. Moreover, for historical reasons, neither KnownNatural nor naturalVal are called that. They go by the names KnownNat and natVal, but we'll pretend they don't. We'll pretend our fake KnownNatural is the real thing. Great. So, how does naturalVal work? We need to understand Proxy first.

```
data Proxy a = Proxy
```

Proxy is a strange thing. Proxy a is a type whose only constructor carries no payload. The Proxy constructor alone is enough to construct values of such Proxy a type, for all choices of a.

```
> Proxy :: Proxy Bool
Proxy
> Proxy :: Proxy 4
Proxy
> Proxy :: Proxy 32
Proxy
```

Notice how we must always explicitly mention the type we want Proxy to have. Haskell can't possibly infer it from a term-level value of type a because there isn't one. It might aid our understanding to see Proxy a as isomorphic to Const () a. Or not. It depends on how comfortable we are with Const.

```
fromProxy :: Proxy a -> Const () a
fromProxy _ = Const ()

toProxy :: Const () a -> Proxy a
toProxy _ = Proxy
```

So, what's this Proxy contraption for? Recall how we got here. We wanted to efficiently measure the length of a list of type List n a. We knew that n, a type of kind Natural, already described the length of the list. But n was a type, not a term-level value, so we needed a way to obtain a term-level representation of that type-level Natural. That is, we needed to use naturalVal, a function that given a type-level Natural returns its term-level representation. Well, how do we tell naturalVal the type-level Natural we are interested in? How do we let naturalVal know about n? How do we apply naturalVal to n when n is not a term-level value but a type? This is exactly what Proxy is for. Proxy is just a vehicle for passing types around when the type in question can't be inferred by other means.

```
> naturalVal (Proxy :: Proxy 2)
2
> naturalVal (Proxy :: Proxy (3 :: Natural)) :: Natural
3
```

Is KnownNatural magical? Kind of. Not really.

```
instance KnownNatural 0 where
  naturalVal _ = 0

instance ∀ n. KnownNatural (n - 1) => KnownNatural n where
  naturalVal _ = 1 + naturalVal (Proxy :: Proxy (n - 1))
```

Easy, but very inefficient. In truth, because of the special literal syntax and non-inductive nature of Haskell's Naturals, GHC implements KnownNatural by other magical means in order to avoid the inefficient inductive recursion. We are thankful for that, but it doesn't change the meaning of what we just discussed. Let's look back at length now.

```
length :: ∀ n a. KnownNatural n => List n a -> Natural
length _ = naturalVal (Proxy :: Proxy n)
```

By deferring all work to the efficient `naturalVal`, `length` achieves the goal of not having to iterate over the entire `List` in order to determine its length. In fact, the list is not scrutinized at all.

```
> length (undefined :: List 9 Bool)
8
```

Three hundred thirty-three

With tools like `naturalVal` we can obtain term-level representations for some types. Great. However, this shouldn't surprise us too much, seeing how it's only natural to expect to be able to keep track of early compile-time knowledge for later use at runtime. It's simply a matter of time, of *when* things happen, of not forgetting. What may surprise us, on the other hand, is that we can also go in the other direction. In Haskell it's possible to introduce new types at runtime from their corresponding term-level representation. How strange is that? Here's a first attempt.

```
wrong :: Natural -> Proxy (n :: Natural)
```

Wrong is `wrong`. We don't even need to attempt to implement it in order to understand that this function couldn't possibly help us obtain the type-level `Natural` representation of a given term-level `Natural`. However, to the untrained eye, it might not be obvious why `wrong` is wrong. So, let's train the eye. The first thing to notice is that n is a type universally quantified in the usual way. We can make this a bit more explicit in case it's not obvious.

```
wrong :: ∀ (n :: Natural). Natural -> Proxy n
```

We didn't need to move n's explicit `Natural` kind annotation to where we did, but we like it more this way.

Throughout our journey, we haven't paid much attention to the topic of type variable quantification because we've always allowed types to prescribe terms, even if that was sometimes hard to see in light of type inference. However, we are doing things backwards now. We want the opposite. We want terms to prescribe types. We want the `Natural`-kinded type n to be fully determined by a term-level `Natural` value.

A universally quantified n like the one above allows the users of `wrong` to choose what n will be. Or more precisely, it *requires* its user to choose an n. We know this already, we learned about it a long time ago. But, seeing as this can be tricky to see whenever the type in question appears in positive position, here is an example of yore to remind us.

```
fromInteger :: ∀ (x :: Type). Num x => Integer -> x
```

He who dares call `fromInteger` must choose x, for `fromInteger` won't itself choose x. `fromInteger` can't. Had `fromInteger` chosen, as polymorphic as it is, x would not anymore be.

Three hundred thirty-four

For our purposes, it is a problem that the caller of wrong can choose n during the type-checking phase. We want wrong itself, not its caller, to fully determine what n shall be after having scrutinized the given term-level Natural number at runtime. We need to do things differently.

```
better :: ∀ (z :: Type)
         . Natural
        -> (∀ (n :: Natural). Proxy n -> z)
        -> z
```

Better is better. Awkward, yes, but better. Notice how there are two separate quantifications taking place now. First, we see the typical universal quantification of z. There's nothing strange about z. What's strange is the quantification of n, which rather than taking place alongside that of z, takes place inside those parentheses surrounding the function that's given as input to better. This is new to us. What does it mean?

We know that better will be making use of the given function because, eventually, better returns a value of type z, and the only way to obtain one such z is through said function. Next, we see that said function takes a Proxy n as input. We know what Proxy is, but what is n? Well, that's the thing. *We don't know.* We can't know until later, until the term-level Natural number is scrutinized at runtime. *We don't want to know.* It's not for us, as users of better, to decide during type-checking what n shall be. It's better itself, the caller of said function, not us, who makes a choice about n *during runtime.* Better is better.

The n type is still universally quantified within those parentheses, but only from the point of view of the function to which we are applying better. This implies two things. First, that *any* type of kind Natural could take the place of n, in turn requiring that the given function be able to deal with *all* possible choices of n. *Any, all, universal.* Remember, we read ∀ as *for all.* Second, and new to us, n can't exist outside these parentheses. See what happens if we try to let n escape this scope, for example, by simply having the given Proxy n become our z by means of id.

```
> x = f 123 id
<interactive>:31:11: error:
  • Couldn't match type 'z' with 'Proxy n'
    Expected: Proxy n -> z
      Actual: Proxy n -> Proxy n
    'z' is a rigid type variable bound by
      the inferred type of x :: z
  • In the second argument of 'f', namely 'id'
```

The error message is not as straightforward as it could be, but it says there that it's not possible to unify the types Proxy n with z. Why? Well, we know why. It's because n and z exist in different places. n exists only within those parentheses, where it's been quantified, and we are trying to let it out. z exists within those parentheses too, but more generally, it exists everywhere else in the type of better. By trying to unify Proxy n with z we are saying that both n and z must exist together everywhere in better, but that's just not how things are

We say that z and n, as they appear in the type of better, are types of different *ranks*. z and every other similarly quantified type in this book so far are *first rank types*. On the other hand, n, being one level of quantification away from that of z, is a *second rank type*. And we can keep going. Third rank, sixth rank, whatever. Those ∀ can be nested as necessary by simply adding more parentheses.

```
uncommon :: ∀ uno. (∀ dos. (∀ tres. ...
```

That's a pretty uncommon thing to do, though. In any case, generally speaking, we refer to type variables of ranks higher than one as *higher rank types*. That's what this chapter is all about.

Three hundred thirty-five

So, how do we solve our problem with better? How do we convert our term-level Naturals to type-level Naturals? Easy, we just need better to choose the correct n when applying the passed in function. Remember, we rearranged things so that now better gets to choose n. That's the whole point of this awkward type of programming.

```
better :: ∀ (z :: Type)
       . Natural
       -> (∀ (n :: Natural). Proxy n -> z)
       -> z
better 0   f = f (Proxy :: Proxy 0)
better 1   f = f (Proxy :: Proxy 1)
better ... f = f ...
```

Of course, better is not implemented like that exactly. First, because it would be impossible to list all of the Natural numbers. But even if we could work around that through induction somehow, it still wouldn't work. While better is better, it's also insufficient. See, the function that returns z is being applied to a Proxy n, where all that's known about n is that it is a type of kind Natural. That's *all* that is known about n, and we can't do anything useful with that information while trapped within the confines of n's quantification. If at least we knew that n implemented some typeclass, then we could try to do something through it. Alas, we know nothing. Except our answer. We know our answer now. We just uncovered it ourselves.

```
withSomeNatural :: ∀ z
                 . Natural
                -> (∀ n. KnownNatural n => Proxy n -> z)
                -> z
```

Universally quantifying a Natural-kinded type n within a limited scope and sticking it on a Proxy is just not enough. In order for n to be useful, we must say something else about it. So, in withSomeNatural we say that n is a KnownNatural. This knowledge allows the passed in function to make use of n through any of the KnownNatural features without knowing what type n is. For example, we know of at least one thing KnownNatural allows us to do. Namely, turning this type-level Natural number n back to is term-level Natural representation by means of naturalVal.

```
id :: Natural -> Natural
id x = withSomeNatural x naturalVal
```

Look at that, yet another contrived way of writing the identity function for Naturals. Boring, yes, but consider what's going on. Each time this funny id function is used, withSomeNatural creates a new type of kind Natural known to satisfy a KnownNatural constraint, and then turns that type back into its term-level representation by means of naturalVal. Isn't that something?

In the implementation of withSomeNatural, everything is as it was in better, except for that extra KnownNatural constraint.

```
withSomeNatural :: ∀ z
                 . Natural
                -> (∀ n. KnownNatural n => Proxy n -> z)
                -> z
withSomeNatural 0   f = f (Proxy :: Proxy 0)
withSomeNatural 1   f = f (Proxy :: Proxy 1)
withSomeNatural ... f = f ...
```

The KnownNatural constraint on f requires withSomeNatural to ensure that a KnownNatural instance for the chosen n be available by the time f is applied. Otherwise, the code just doesn't type-check. Luckily for us, as we learned before, these instances are readily satisfied by every Natural type written down literally, so we don't have to worry too much about it.

And yes, once again, we know that withSomeNatural can't possibly be implemented *exactly* like this because of how infinite and special the syntax for Natural numbers is. The actual implementation is a bit more magical, but it's alright, we can nonetheless pretend that this implementation is the one. Semantically, it is. And in any case, withSomeNatural doesn't even exist with this exact type nor name in Haskell's base library, so what are we even rambling about? Similarly to how it was for naturalVal, the version of withSomeNatural available in base is a bit more general. Its idea, however, remains.

Three hundred thirty-six

It's ugly, though, isn't it?

```
withSomeNatural :: ∀ z
                 . Natural
                -> (∀ n. KnownNatural n => Proxy n -> z)
                -> z
```

Compare withSomeNatural with our first and wrong attempt. It was wrong, yes, but at least it wasn't ugly.

```
wrong :: ∀ (n :: Natural). Natural -> Proxy n
```

Can we improve withSomeNatural a bit? Make it less ugly? Make it more like wrong somehow? Let's see. The problem with wrong was that, contrary to what we wanted, its users were responsible for choosing n. We wanted wrong itself to choose n. So we moved things around, and using higher rank types we made things better. But what if we could just hide n altogether instead? Wouldn't that solve the issue too? Interesting. Maybe.

```
someNatural :: Natural -> SomeNatural
```

We expect someNatural to achieve fundamentally the same as withSomeNatural, except without its confusing type signature. The idea is that, as its name suggests, *some natural* will exist within SomeNatural as a type of kind Natural, but its existence will be a secret. Hush hush.

```
data SomeNatural where
   SomeNatural :: Proxy (n :: Natural) -> SomeNatural
```

Using GADT syntax, we are defining a new datatype called SomeNatural having a single constructor also called SomeNatural. What's novel about this whole definition is that the SomeNatural constructor mentions a type n as part of its payload, yet n doesn't appear anywhere in the type of SomeNatural. Where is n coming from? Does this type-check? Excellent questions.

n, as it appears within the SomeNatural constructor, is what we call an *existentially quantified* type. n exists, sure, but we don't really know what n is. After putting n inside the SomeNatural constructor, the outside world can forget everything about it. Consequently, n doesn't need to be mentioned in the SomeNatural type.

Useless, isn't it? A type *exists* yet we know nothing about it. What could we ever do with it? Well, isn't dealing with the unknown what we have been doing for the last thousand chapters? I'm sure we'll figure something out. We did it in withSomeNatural before. We knew nothing about its n, yet by sprinkling some liberating KnownNatural restrictions we managed to make something useful out of it. Could we do something like that here?

```
data SomeNatural where
    SomeNatural :: KnownNatural n => Proxy n -> SomeNatural
```

Why, look at that. We are specifying a constraint within a value constructor. What does it mean? Can we do it? As we go about our Haskell journey, we'll find that the boundaries between supposedly different concepts tend to become mere suggestions. Here, in the SomeNatural value constructor, alongside a constraint, we carry a type of an unrelated kind through an empty value vessel. Liberating.

By the way, we don't need to use GADT syntax for any this. We can use the normal ADT syntax. However, it is uglier, so we don't. See for yourself.

```
data SomeNatural
  = ∀ n. KnownNatural n => SomeNatural (Proxy n)
```

This SomeNatural means and behaves the same as the other one. It's just a different syntax. A confusing one, seeing how the universal quantifier there, ∀, should probably be replaced with the *existential* quantifier, ∃. Alas, Haskell's syntax doesn't support that.

Anyway, using SomeNatural to construct a value of type SomeNatural functions exactly the same as applying f did in our withSomeNatural example.

```
withSomeNatural :: ∀ z
                .  Natural
                -> (∀ n. KnownNatural n => Proxy n -> z)
                -> z
withSomeNatural 0   f = f (Proxy :: Proxy 0)
withSomeNatural 1   f = f (Proxy :: Proxy 1)
withSomeNatural ... f = f ...
```

That is, in order to call f, not only did withSomeNatural need to supply the expected Proxy n input, but also it needed to ensure that a KnownNatural instance for n existed at that time. This was required by f then, and it is required by the SomeNatural constructor now. It says so clearly on its type, which is essentially the same as f's but without the polymorphic return. This suggests that a partially applied withSomeNatural, with the SomeNatural constructor substituting f, could be our longed someNatural function, the less ugly one.

```
someNatural :: Natural -> SomeNatural
someNatural x = withSomeNatural x SomeNatural
```

Beautiful. We've trapped all the noise inside the SomeNatural constructor. Now what? What do we do with a value of this SomeNatural type once we have one? How do we reveal its secrets? Easy. Like every other constructor, we pattern match on it.

```
fromSomeNatural :: SomeNatural -> Natural
fromSomeNatural (SomeNatural p) = naturalVal p
```

Remember, naturalVal takes a Proxy n as input and expects a KnownNatural n constraint to be satisfied. So, if the mystery type and its constraint weren't there, fromSomeNatural just wouldn't type-check. We expect the composition of these two functions to behave as the identity function, of course.

```
> fromSomeNatural (someNatural 2)
2
> fromSomeNatural (someNatural 47)
47
```

Existential quantification and *higher rank types* are closely related ideas. We'll rarely encounter one without the other. In this book we don't go into details about why, when and how to use these things beyond these small examples. We just wanted to show that, despite our love for types, sometimes, we may need to forget them.

Three hundred thirty-seven

We have the SomeNatural type now, which hides an existential type inside. And we have the usual Natural type too, which for all we know may do so too. How could we know? It's up to the Haskell chiefs to decide how to implement the Natural type, and maybe they do just that. Existentialize. So, what have we gained with SomeNatural? Nothing too practical, of course. This is not *that* kind of book. But we know more things now. And this new knowledge need not be restricted by a Natural selection.

Say we don't want to write programs anymore. We'll just go to the countryside instead. Grow some vegetables, fish, hunt. Build stuff, fix stuff, shoot some cans. Care for our family, neighbours and friends. Cook, exercise, celebrate, love. You know, the good stuff, what we are here for. But we won't leave without taking with us some Souvenirs, keepsakes from our adventures to remember things by. We are, after all, fond of these times.

```
data Souvenir where
  Souvenir :: Show x => x -> Souvenir
```

Unfortunately, our mementos have different types. Bools, Seasons, you name it. So, we can't just group them all together in a list [] for safekeeping. However, once we acknowledge that we don't need to take the actual values, but just something to Show for them, we can wrap them in the Souvenir constructor so as to existentialize their types, keeping only what matters to us. A visual description of the moment, a Souvenir.

```
souvenirs :: [Souvenir]
souvenirs = [ Souvenir True
            , Souvenir ()
            , Souvenir Summer
            , Souvenir "thing" ]
```

The SomeNatural and Souvenir constructors are a bit different. Whereas in
SomeNatural we were interested in the existentialized type itself, in Souvenir
we are interested in a *value* of that existentialized type. But that's as far as
the differences go. Really, it's the existential quantification what matters.
There *exists* a type within these constructors, we just don't know which.
It's forgotten. For our needs, however, this could be enough.

```
> traverse (\(Souvenir x) -> putStrLn (show x)) souvenirs
True
()
Summer
"thing"
```

Just look at them, so beautiful. As in the SomeNatural case, we must
pattern-match on the Souvenir constructor in order to reveal to the type-
checker the Show constraint trapped inside. Only then, and after having
given a name to the payload value, can we show it. And that's all we can do
with the value, for we know nothing about its type beyond what Show tells
us. Is this lack of knowledge a problem for our Souvenirs? Maybe. History
matters, you see. When we don't know all the facts, the best we can do is
trust in what we are shown, which may or may not be true.

```
data Fact = Fake | Real

instance Show Fact where
  show Fake = "Real"
  show Real = "Fake"
```

How sad, our Souvenirs may not mean anything after all. Luckily, in our
souvenirs example, we control all the Show instances involved. We chose
them explicitly when constructing each Souvenir, and we even wrote some
of them ourselves throughout this book. So, *we* get to decide the truth that
will be remembered through our Souvenirs. See? Everything is fine.

Alternatively, we could remember the forgotten types by means other
than the type-system. For example, by also carrying in the Souvenir the
name of the soon to be forgotten type. Then, provided we have some
machinery capable of telling us the type to which said name corresponds,
we could recreate that type somehow. We could, yes we could, but we
won't. That would be another type of programming, and there's no room
for that in this book.

Now, about that country life...